# SPEECH AND AUDIO SIGNAL PROCESSING

## Processing and Perception
## of Speech and Music

# SPEECH AND AUDIO SIGNAL PROCESSING
## Processing and Perception of Speech and Music

**BEN GOLD**

*Massachusetts Institute of Technology*
*Lincoln Laboratory*

**NELSON MORGAN**

*University of California at Berkeley*
*International Computer Science Institute*

*with contributions from*
Hervé Bourlard
Eric Fosler-Lussier
Jeff Gilbert

**JOHN WILEY & SONS, INC.**
*New York / Chichester / Weinheim / Brisbane / Singapore / Toronto*

Acquisitions Editor     *Bill Zobrist*
Marketing Manager     *Katherine Hepburn*
Senior Production Editor     *Robin Factor*
Senior Designer     *Laura Boucher*
Illustration Editor     *Gene Aiello*
Electronic Illustrations     *Radiant*

This book was set in 10/12 *Times Roman* by *TechBooks*, and printed and bound by *Quebecor/Fairfield*. The cover was printed by *Lehigh Press*.

The book is printed on acid-free paper. ∞

*Library of Congress Cataloging in Publication Data:*

Gold, Ben, 1923–
   Speech and audio signal processing : processing and perception of speech, and music / Ben Gold, Nelson Morgan ; with contributions from Hervé Bourlard, Eric Fosler-Lussier, and Jeff Gilbert,
      p.     cm.
   Includes Index.
   ISBN 0-471-35154-7 (alk. paper)
   1. Speech processing systems.  2. Signal processing—Digital techniques.  3. Electronic music.  I. Morgan. Nelson.
TK7882.S65G65   1999
621.382′2—dc21                                          99-16025
                                                           CIP

ISBN 0-471-35154-7

Printed in the United States of America

10 9 8 7 6 5 4 3 2

*This book is dedicated to*
*our families*
*and our students*

# CONTENTS

**CHAPTER 5**     *SPEECH-RECOGNITION OVERVIEW*   **56**

**PART II**

# MATHEMATICAL BACKGROUND

**CHAPTER 6**     *DIGITAL SIGNAL PROCESSING*   **69**

**PART IV**

## AUDITORY PERCEPTION

**CHAPTER 14**    *EAR PHYSIOLOGY* **189**

**CHAPTER 15**    *PSYCHOACOUSTICS* **205**

**CHAPTER 16**    *MODELS OF PITCH PERCEPTION* **214**

**CHAPTER 17**    *SPEECH PERCEPTION* **228**

**CHAPTER 30**        *PITCH DETECTION*    **415**

**CHAPTER 31**        *VOCODERS*    **431**

**CHAPTER 32**        *LOW-RATE VOCODERS*    **451**

# CHAPTER 1

# INTRODUCTION

We are confronted with insurmountable opportunities.

—Walt Kelly

## 1.1 WHY WE WROTE THIS BOOK

Speech and music are the most basic means of adult human communication. As technology advances and increasingly sophisticated tools become available to use with speech and music signals, scientists can study these sounds more effectively and invent new ways of applying them for the benefit of humankind. Such research has led to the development of speech and music synthesizers, speech transmission systems, and automatic speech recognition (ASR) systems. Hand in hand with this progress has come an enhanced understanding of how people produce and perceive speech and music. In fact, the processing of speech and music by devices and the perception of these sounds by humans are areas that inherently interact with and enhance each other.

Despite significant progress in this field, there is still much that is not well understood. Speech and music technology could be greatly improved. For instance, in the presence of unexpected acoustic variability, ASR systems often perform much worse than human listeners. Speech that is synthesized from arbitrary text still sounds artificial. Speech-coding techniques remain far from optimal, and the goal of transparent transmission of speech and music with minimal bandwidth is still distant. All fields associated with the processing and perception of speech and music stand to benefit greatly from continued research efforts. Finally, the growing availability of computer applications incorporating audio (particularly over the Internet and in portable devices) has increased the need for an ever-wider group of engineers and computer scientists to understand audio signal processing. For all of these reasons, as well as our own need to standardize a text for our graduate course at UC Berkeley, we have written this book.

The notes on which this book is based have proved beneficial to graduate students for close to a decade; during this time, of course, the material has evolved to its present form, including a problem set for each chapter. The material includes coverage of the physiology and psychoacoustics of hearing as well as the results from research on pitch and speech perception, vocoding methods, and information on many aspects of ASR. To this end, the authors have made use of their own research in these fields, as well as the methods and results of many other contributors.

In many chapters, the material is written in a historical framework. In some cases, this is done for motivation's sake; the material is part of the historical record, and we hope that the reader will be interested. In other cases, the historical methods provide a convenient introduction to a topic, since they often are simpler versions of more current approaches. Overall, we have tried to take a long-term perspective on technology developments, which in our view requires incorporating a historical context. The fact that otherwise excellent books on this topic have typically avoided this perspective was one of our major motivations for writing this book.

## 1.2  HOW TO USE THIS BOOK

This text covers a large number of topics in speech and audio signal processing. While we felt that such a wide range was necessary, we also needed to present a level of detail that is appropriate for a graduate text. Therefore, we have elected to focus on basic material with advanced discussion in selected subtopics. We have assumed that readers have prior experience with core mathematical concepts such as difference equations or probability density functions, but we do not assume that the reader is an expert in their use. Consequently, we will often provide a brief and selected introduction to these concepts to refresh the memories of students who have studied the background material at one time but who have not used it recently. The background topics are selected with a particular focus, namely, to be useful to both students and working professionals in the fields of ASR and speaker recognition, speech bandwidth compression, speech analysis and synthesis, and music analysis and synthesis. Topics from the areas of digital signal processing, pattern recognition, and ear physiology and psychoacoustics are chosen so as to be helpful in understanding the basic approaches for speech and audio applications.

The remainder of this book comprises 35 chapters, grouped into eight sections. Each section or part consists of three to seven chapters that are conceptually linked. Each part begins with a short description of its contents and purpose. These parts are as follows:

**I. Historical Background.** In Chapters 2 through 5 we lay the groundwork for key concepts to be explored later in the book, providing a top-level summary of speech and music processing from a historical perspective. Topics include speech and music analysis, synthesis, and speech recognition.

**II. Mathematical Background.** The basic elements of digital signal processing (Chapters 6 and 7) and pattern recognition (Chapters 8 and 9) comprise the core engineering mathematics needed to understand the application areas described in this book.

**III. Acoustics.** The topics in this section (Chapters 10–13) range from acoustic wave theory to simple models for acoustics in human vocal tracts, tubes, strings, and rooms. All of these aspects of acoustics are significant for an understanding of speech and audio signal processing.

**IV. Auditory Perception.** This section (Chapters 14–18) begins with descriptions of how the outer ear, middle ear, and inner ear work; most of the available information comes from experiments on small mammals, such as cats. Insights into human hearing are derived from experimental psychoacoustics. These fundamentals then lead to the study of human pitch perception as applied to speech and music, as well as to studies of human speech perception and recognition.

V. **Speech Features.** Systems for ASR and vocoding have nearly always incorporated filter banks, cepstral analysis, linear predictive coding, or some combination of these basic methods. Each of these approaches has been given a full chapter (19–21).

VI. **Automatic Speech Recognition.** Seven chapters (22–28) are devoted to this study of ASR. Topics range from feature extraction to statistical and deterministic sequence analysis, with coverage of both standard and discriminant training of hidden Markov models (including neural network approaches). Part VI concludes with an overview of a complete ASR system.

VII. **Synthesis and Coding.** Speech synthesis (culminating in text-to-speech systems) is first presented in Chapter 29. Chapter 30 is devoted to pitch detection, which applies to both speech and music devices. Many aspects of vocoding systems are then described in Chapters 31–33, ranging from very-high-quality systems working at relatively high bit rates to extremely low-rate systems.

VIII. **Other Applications.** In Chapters 34–36 we present several application areas that were not covered in the bulk of the book. Modifications of the time scale, pitch, and spectral envelope can transform speech and music in ways that are increasingly finding common applications (Chapter 34). Chapter 35 is a review of major issues in music synthesis. The final chapter (36) is an overview of speaker recognition, with an emphasis on speaker verification. With increasing access to electronic information and expansion of electronic commerce, verification of the identity of a system user is becoming increasingly important.

Readers with sufficient background may choose to focus on the application areas described in Parts V–VIII, as the first four parts primarily give preparatory material. However, in our experience, readers at a graduate or senior undergraduate level in electrical engineering or computer science will benefit from the earlier parts as well. In teaching this course, we have also found the problem sets to be helpful in clarifying understanding, and we suspect that they would have similar value for industrial researchers. Another useful study aid is provided by a collection of audio examples that we have used in our course. These examples will be made freely available via the book's World-Wide Web site which can be found at http://catalog.wiley.com/. This Web site will also be augmented over time to include links to errata and addenda for the book.

Other books on a similar topic but with a different emphasis can also be used to complement the material here; in particular, we recommend [6] or [8]. Additionally, a more complete exposition on the background material introduced in Parts II–IV can be found in such texts as the following:

- [5] for digital signal processing
- [1] or [2] for pattern recognition (note that [2] is expected to be re-released in a new edition soon)
- [3] for acoustics

- [7] for auditory physiology
- [4] for psychoacoustics

## 1.3   A CONFESSION

The authors have chosen to spend much of their lives studying speech and audio signals and systems. Although we would like to say that we have done this to benefit society, much of the reason for our vocational path is a combination of happenstance and hedonism; in other words, dumb luck and a desire to have fun. We have enjoyed ourselves in this work, and we continue to do so. Speech and audio processing has become a fulfilling obsession for us, and we hope that some of our readers will adopt and enjoy this obsession too.

## 1.4   ACKNOWLEDGMENTS

Many other people contributed to this book. Students in our graduate class at Berkeley contributed greatly over the years, both by their need for a text and by their original scribe notes that inspired us to write the book. Two students in particular, Eric Fosler and Jeff Gilbert, ultimately wrote material that was the basis of Chapters 22 and 33, respectively. Hervé Bourlard came through very quickly with the core of Chapter 36 when we realized that we had no material on speaker identification or verification. Su-Lin Wu was an extremely helpful critic, both on the speech recognition sections and on some of the psychoacoustics material. Anita Bounds-Morgan provided very useful editorial assistance. The anonymous reviewers that our publisher used were also quite helpful.

We certainly appreciate the indulgence of the International Computer Science Institute and Lincoln Laboratory for permitting us to develop this manuscript on the equipment of these two labs. Devra Polack and Elizabeth Weinstein of ICSI also provided a range of secretarial and artistic support that was very important to the success of this project. We also thank Bill Zobrist of Wiley for his interest in the project.

Finally, we extend our special thanks to our wives, Sylvia and Anita, for putting up with our intrusion on family time as a result of all the days and evenings that were spent on this book.

## BIBLIOGRAPHY

1. Bishop, C., Neural Networks for Pattern Recognition, Oxford Univ. Press, London/New York, 1996.
2. Duda, D., and Hart, P., Pattern Classification and Scene Analysis, Wiley–Interscience, New York, 1973.
3. Kinsler, L., and Frey, A., Fundamentals of Acoustics, Wiley, New York, 1962.

4. Moore, B., An Introduction to the Psychology of Hearing, 3rd ed. Academic Press, New York, 1989.

5. Oppenheim, A., and Schafer, R., Discrete-Time Signal Processing, Prentice–Hall, Englewood Cliffs, N.J., 1989.

6. O'Shaughnessy, D., Speech Communication, Addison–Wesley, Reading, Mass., 1987.

7. Pickles, J., An Introduction to the Physiology of Hearing, Academic Press, New York, 1982.

8. Rabiner, L., and Juang, B.-H., Fundamentals of Speech Recognition, Prentice–Hall, Englewood Cliffs, N.J., 1993.

# HISTORICAL
# BACKGROUND

The future is a lot like the present, only it's longer.
—Dan Quisenberry

**A**N **UNDERSTANDING** of the goals and methods of the past can help us to envision the advances of the future. Ideas are often rediscovered or reinvented many times. Often the new form has much greater impact, though sometimes for fairly mundane reasons, such as greater accessibility to larger computational capabilities. It would be obvious to most that a study of the social sciences would be incomplete without inclusion of the methods and beliefs of the past. In our view, a study of current methods in speech and audio engineering without any historical context would be similarly inadequate.

For these reasons, we introduce the basic concepts of speech analysis, synthesis, and recognition in Part I, using a historical frame of reference. Later parts will provide greater technical detail in each of these areas. We begin in Chapter 2 with a brief history of synthetic audio, starting with 18th Century mechanical devices and proceeding through speech and music machines from the first half of the 20th Century. The discussion continues in Chapter 3 with a discussion of systems for analysis and synthesis, including a brief introduction to the concept of source–filter separation. Speech recognition is a 20th Century invention, and the Chapter 4 discussion of the history of research in this area is largely confined to the past 50 years. Finally, Chapter 5 introduces speech recognition technology, discussing such topics as the major components of a recognizer and the sources of difficulty in this problem. Overall, Part I is intended to provide a light overview that will give the reader motivation for the more detailed material that follows.

# CHAPTER 2

# SYNTHETIC AUDIO: A BRIEF HISTORY

## 2.1 VON KEMPELEN

Many years ago, von Kempelen demonstrated that the speech-production system of the human being could be modeled. He showed this by building a mechanical contrivance that "talked." The paper by Dudley and Tarnoczy [2] relates the history of von Kempelen's speaking machine. This device was built about 1780, at a time when the notion of building automata was quite popular. Von Kempelen also wrote a book [7] that dealt with the origin of speech, the human speech-production system, and his speaking machine. Thus, for over a century, an existence proof was established that one could indeed build a machine that spoke. (Von Kempelen's work brings to mind that of another great innovator, Babbage, who also labored for many years with mechanical contrivances to try to build a computing machine.)

Figure 2.1 shows the speaking machine built by Wheatstone that was based on von Kempelen's work. The resonator of leather was manipulated by the operator to try to copy the acoustic configuration of the vocal tract during the sonorant sounds (vowels, semivowels, glides, and nasals); the bellows provided the air stream; the vibrating reed produced the periodic pressure wave; and the various small whistles and levers shown controlled most of the consonants. (Much later, Riesz [6] built a mechanical speaking machine that was more precisely modeled after the human speech-producing mechanism. This is depicted in Fig. 2.2, shown here for comparison to the von Kempelen–Wheatstone model of Fig. 2.1).

## 2.2 THE VODER

Modern methods of speech processing really began in the U.S. with the development of two devices. Homer Dudley pioneered the development of the channel vocoder (voice coder) and the Voder (voice-operated demonstrator) [1]. We know from numerous newspaper articles that the appearance of the Voder at the 1939 World's Fair in San Francisco and New York City was an item of intense curiosity. Figure 2.3 is a collage of some clippings from that period and reflects some of the wonder of people at the robot that spoke.

It is important to realize that the Voder did not speak without a great deal of help from a human being. The operator controls the Voder through a console, which can be compared

9

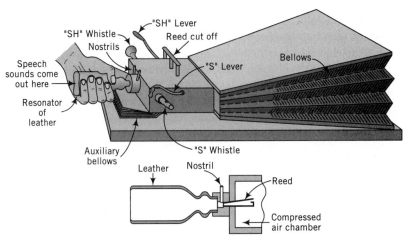

**FIGURE 2.1**   Wheatstone's speaking machine. From [2].

to a piano keyboard. In the background is the electronic device that does the speaking. Operator training proved to be a major problem. Many candidates for this job were unable to learn it, and the successful ones required training for periods of 6 months to 1 year. Figure 2.4 shows an original sketch by S. W. Watkins of the Voder console.

The keys were used to produce the various sounds; the wrist bar was a switch that determined whether the excitation function would be voiced or unvoiced, and the pitch pedal supplied intonation information. Figure 2.5 is a close-up of the controls in the console and shows how these relate to the articulators of a human vocal tract.

The keys marked 1 through 10 control the connection of the corresponding bandpass filters into the system. If two or three of the keys were depressed and the wrist bar was set

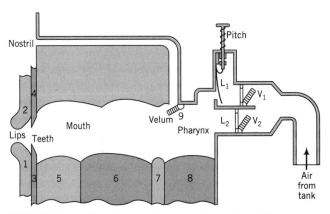

**FIGURE 2.2**   Riesz's speaking machine. From [3].

**FIGURE 2.3** News clippings on the Voder.

to the buzz (voicing) condition, vowels and nasals were produced. If the wrist bar were set to hiss (voiceless), sounds such as the voiceless fricatives (e.g., f) were generated. Special keys were used to produce the plosive sounds (such as p or d) and the affricate sounds (ch as in cheese; j as in jaw).

## 2.3 TEACHING THE OPERATOR TO MAKE THE VODER "TALK"

The Voder was marvelous, not only because it "talked" but also because a person could be trained to "play" it. Speech synthesis today is done by real-time computer programs or specialized hardware, and the emphasis is either on voice answer-back systems, in which the synthesizer derives information from a stored vocabulary, or on text-to-speech systems, in which text that is either typed or electronically scanned is used to control the synthesizer

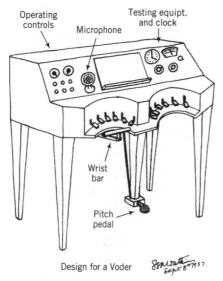

Design for a Voder

**FIGURE 2.4**   Sketch of the Voder.

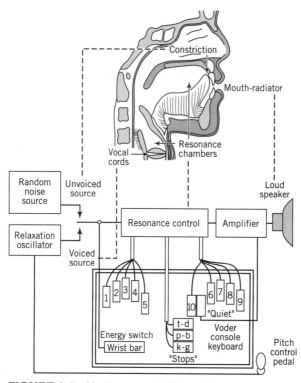

**FIGURE 2.5**   Voder controls. From [2].

| Sounds | Wrist Bar | Keys |
|--------|-----------|------|
| S | up | 9. |
| Sh | up | 7.8. Light & Smooth |
| M | down | 1. |
| ē (seen) | down | 1.8. |
| aw (dawn) | down | 3. |

She saw me.
See me seesaw.

WORDS   Think of how they are <u>pronounced</u>, not how they are
written.

| Cease | Sauce | See | She |
|-------|-------|-----|-----|
| Me | Saw | Seesaw | |

**FIGURE 2.6**:   Lesson 1 of the Voder instructions.


parameters. It is a pity that further work on real-time control by a human operator has not been seriously pursued.

Figures 2.6, 2.7, and 2.8 describe Lessons 1, 9, and 37 of the Voder Instruction Manual.

Relatively few of the candidate operators were successful, but one young woman (Mrs. Helen Harper) was very proficient. She performed at the 1939 New York World's


| Sounds | Wrist Bar | Keys |
|--------|-----------|------|
| Initial R | down | 2.5. before ō. |
| Initial R | down | 2.5.6. before every-<br>thing else |
| a (take) | down | 3.7.--2.8. This is<br>a dipthong (e–i) |

Make it safer.
She's sorry for me.

WORDS   Think of how they are <u>pronounced</u>, not how they are
spelt.

| Chase | Nearest | Rich | Shakes |
|-------|---------|------|--------|
| Chases | Raise | Richer | Sorrier |
| Error | Raises | Riches | Sorriest |
| Errors | Rate | Richest | Sorry |
| Face | Rates | Rose | Wrote |
| Faces | Raw | Safe | |
| Make | Ray | Safer | |
| Makes | Reach | Safest | |
| May | Reaches | Sake | |
| Nature | Rest | Say | |
| Nearer | Rests | Shake | |

PRACTICE SENTENCES

Your chef makes rich sauces.
Shake it off.
Is nature fair?
She wrote to Rose.

**FIGURE 2.7**:   Lesson 9.

| Sounds | Wrist Bar | Keys |
|---|---|---|
| Initial Bl | | |
| Final -sl | up--down | 9--2 |
| Final -tl | up--down | 6.8.--2. Very light tap on 6.8. |

WORDS    Think of how they are <u>pronounced</u>, not how they are spelt.

| Ably | Blow | Mental | Whistle |
|---|---|---|---|
| Black | Blue | Metal | |
| Blade | Establish | Pestle | |
| Blood | Little | Total | |

<u>PRACTICE SENTENCES</u>

The wind blows cold tonight.
Her hair is quite black.
Whistle and I'll come to you.
Make a mental note of it.
What a little dog it is.
How much is the total amount?

**FIGURE 2.8:**   Lesson 37.

Fair. Many years later (in the 1960's) a highlight of Dudley's retirement party was the Voder's speaking to Mr. Dudley, with the help of Mrs. Harper.

## 2.4  SPEECH SYNTHESIS AFTER THE VODER

Many speech-synthesis devices were built in the decades following the invention of the Voder, but the underlying principle, as captured in Fig. 2.5, has remained quite fixed. For many cases, there is a separation of source and filter followed by the parameterization of each. As we shall see in the following sections, the same underlying principles control the design of most music synthesizers. In later chapters, the field of speech synthesis from the past to the present is explored in some detail, including advanced systems that transform printed text into reasonable-sounding speech.

## 2.5  MUSIC MACHINES

Figure 2.9 shows a 17th Century drawing of a water-powered barrel organ. Spring-powered barrel organs may have existed as long ago as the 12th Century. Barrel organs work on the same concepts as present-day music boxes; that is, once the positions of the pins are chosen, the same music will be played for each complete rotation. Keys can be depressed or strings can be plucked, depending on the overall design of the automatic instrument.

The barrel organ is a form of read-only memory, and not a very compact form at that. Furthermore, barrel organs could not record music played by a performer. In the late 18th Century, both of these problems were overcome by melography, which allowed music

**FIGURE 2.9** 17th Century drawing of a water-powered barrel organ.

to be both recorded and played back, using the medium of punched paper tape or cards. The idea originated for the automation of weaving and was developed fully by Joseph Marie Jacquard, who designed a device that could advance and register cards. (Punched cards were used by Babbage in the design of his computing machine and, in our time, were used by many computer manufacturers such as IBM.) Card-driven street organs made use of this technology. Card stacks were easy to duplicate; also, different stacks contained different music, so that music machines became very marketable. By the beginning of this century, the concept had been applied to the player piano. A roll of paper tape could be made and the holes punched automatically while a master pianist (such as Rachmaninoff or Gershwin) played. This paper roll could then actuate the playback mechanism to produce the recorded version. Since the piano keys were air driven, extra perforations in the paper roll allowed variable amounts of air into the system, thus changing volume and attack in a way comparable to that of the human performer. Until the development of the high-fidelity microphone, player pianos offered greater reproduction fidelity than the gramophone— but of course they could only record the piano, whereas the gramophone recorded all sounds.

A modern example of a player piano is the solenoid-controlled Bosendorfer at the MIT Media Laboratory. Using this system, Fu [4] synthesized a Bosendorfer version from an old piano roll by Rachmaninoff.

At the beginning of the century, a mighty device called the telharmonium was constructed by Thaddeus Cahill. Remember that this was built *before* the development of electronics; nevertheless, Cahill had the ingenuity to realize that any sound could be synthesized by the summation of suitably weighted sinusoids. He implemented each sinusoid by actuating a generator. To create interesting music, many such generators (plus much additional equipment) were needed, so the result was a monster, weighing many tons. Cahill's concept of additive synthesis is still an important feature of much of the work in electronic music synthesis. This is in contrast to many later music synthesizers that employ subtractive synthesis, in which adaptive filtering of a wideband excitation function generates the sound. (The additive synthesis concept was used by McCaulay and Quatieri [5] to design and build a speech-analysis-synthesis system; we discuss this device in later chapters.)

The player piano is only partially a music machine, since it requires a real piano to be part of the system. The telharmonium, by contrast, is a complete synthesizer, since music is made from an abstract model, that is, sine generators. Another, although totally different, complete synthesizer is the theremin, named after its inventor, the Russian Lev Termin. In this system, an antenna is a component of an electronic oscillator circuit; moving one's arm near the antenna changes the oscillator frequency by changing the capacitance of the circuit, and this variable frequency is mixed with a fixed-frequency oscillator to produce an audio tone whose frequency can be varied by arm motion. Thus the theremin generates a nearly sinusoidal sound but with a variable frequency that can produce pitch perceptions that don't exist in any standard musical scale. In the hands of a trained performer, the theremin produces rather unearthly sounds that are nevertheless identifiable as some sort of (strange) music. A trained performer could play recognizable music (e.g., Schubert's Ave Maria). Figure 2.10 shows Clara Rockmore at a theremin. Her right hand controls the frequency of the straight antenna while her left hand controls the amplitude by changing the capacitance of a different circuit.

The theremin is again in the news. In 1994 a film called "Theremin:   An Electronic Odyssey" was released. In 1995, more than one thousand instruments were sold. The New York Times of July 7, 1996 had a feature article on the theremin in the Arts and Entertainment Section.

**FIGURE 2.10**  Clara Rockmore at the theremin.

## 2.6 EXERCISES

**2.1** The Voder was which of the following:

(a) a physical model of the human vocal apparatus,

(b) an early example of subtractive synthesis,

(c) an early example of additive synthesis, or

(d) a member of the electorate with a head cold.

**2.2** Shown in Fig. 2.11 is Dudley's speech-sound classification for use with Voder training. Find the Voder sequence for any of the practice sentences of Fig. 2.8 (Lesson 37). Break the sentence into a phoneme sequence, using the notation of Fig. 2.11. Note that the BK1, BK2, and BK3 keys in Fig. 2.11 are the k-g, p-b, and t-d keys of Fig. 2.5. A sample is shown below for the sentence "The Voder can speak well."

| | | |
|---|---|---|
| /th/ | Voiced | 10Q |
| /ŭ/ | Voiced | 3458 |
| /v/ | Voiced | 67Q |
| /o/ | Voiced | 3–2 |
| /d/ | Voiced | BK3 |
| /r/ | Voiced | 36 |
| /k/ | Unvoiced | BK1 |
| /á/ | Voiced | 457 |
| /n/ | Voiced | 1 |
| /s/ | Unvoiced | 9 |
| /p/ | Unvoiced | BK2 |
| /e/ | Voiced | 18 |
| /k/ | Unvoiced | BK1 |
| /w/ | - - - - - | - - |
| /e/ | Unvoiced | 37 |
| /l/ | Voiced | 2 |

Voder example: "The Voder can speak well."

**2.3** Compare von Kempelen's speaking machine with Dudley's Voder.

(a) What are the chief differences?

(b) What are the chief similarities?

(c) How would you build a von Kempelen machine today?

**2.4** Figures 2.12 and 2.13 show spectrograms of the saying "greetings everybody" by the announcer and the Voder.

(a) What do you perceive to be the main difference between the natural and the synthetic utterances?

(b) Estimate the instants when the operator changes the Voder configuration.

**2.5** Synthesizers can be classified as articulatory based or auditory based. The former type works by generating sounds that are based on a model of how the sound is produced. The latter type relies on the properties of the ear to perceive sounds that are synthesized by different methods than the natural sounds that they imitate.

**Classification of the Speech Sounds**

## 1. Vowels

(2) (tool) ū    w    y    ē (tee) (18)
(2) (shorter) (book) u    h    i (it) (28)
(3··2) (tone) o    ou    ī    ā (ape) (37··28)
(3) (awl) ȯ    e (ten) (37)
(4) (not) ŏ    ȧ (at) (457)
(3458) (nut) ŭ
(458) a (far) (458)

## 2. Combinational and Transitional Sounds

w·y·ou·ī·h·ā (pay)·ō·(woe)

## 3. Semi Vowels

Initial l (268)    final l (2)
Initial r (256)    final r (36)

## 4. Stop Consonants

| Voiced | | Unvoiced | | Nasalized | Formation of Stop |
|---|---|---|---|---|---|
| (BK2) | b | (BK2) | p | m (1) | lip against lip |
| (BK3) | d | (BK3) | t | n (1) | tongue against teeth |
| (789) | j | (789) | ch | – | tongue against hard palate |
| (BK1) | g | (BK1) | k | ng(18·1) | tongu against soft palate |

## 5. Fricative Consonants

| Voiced | | Unvoiced | Formation of Air Outlet |
|---|---|---|---|
| (67Q) | v | f (67Q) | lip to teeth |
| (9) | z | s (9) | teeth to teeth |
| (10Q) | th (then) | th (thin) (10Q) | tongue to teeth |
| (789) | zh (azure) | sh (78) | tongue to hard palate |

(Numbers in parentheses indicate key designations on VODER console)

**FIGURE 2.11**   Classification of speech sounds for Voder use.

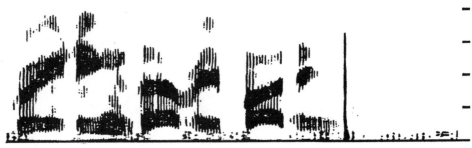

**FIGURE 2.12**   Spectrogram of "greetings everybody" by an announcer.

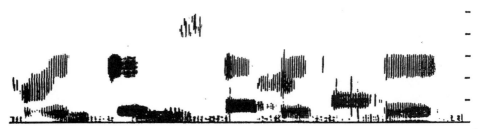

**FIGURE 2.13**    Spectrogram of "greetings everybody" by the Voder.

Categorize each of the following as an articulatory-based or auditory-based synthesizer:

**(a)** telharmonium,

**(b)** Wheatstone–von Kempelen speaking machine,

**(c)** Voder,

**(d)** theremin; and

**(e)** player piano.

## BIBLIOGRAPHY

1. Dudley, H., Riesz, R., and Watkins, S., "A synthetic speaker," *J. Franklin Inst.* **227**: 739, 1939.
2. Dudley, H., and Tarnoczy, T. H., "The speaking machine of Wolfgang von Kempelen," *J. Acoust. Soc. Am.* **22**: 151–166, 1950.
3. Flanagan, J. L., *Speech Analysis Synthesis and Perception,* 2nd ed., Springer-Verlag, New York/Berlin, 1972.
4. Fu, A. C., "Resynthesis of acoustic piano recordings," M.S. Thesis, Massachusetts Institute of Technology, 1996.
5. McAulay, R. J., and Quatieri, T. F., "Speech analysis/synthesis based on a sinusoidal representation," *IEEE Trans. Acoust. Speech Signal Proc.* **34**(4): 744–754, Aug. 1986.
6. Riesz, R. R., personal communication to J. L. Flanagan, 1937. (Details of this work are described by Flanagan in Ref. 3, pp. 207–208.)
7. Von Kempelen, W., *Le Mechanisme de la pavola, suivi de la Description d'une machine parlante.* Vienna: J.V. Degen, 1791.

# CHAPTER 3

# SPEECH ANALYSIS AND SYNTHESIS OVERVIEW

"If I could determine what there is in the very rapidly changing complex speech wave that corresponds to the simple motion of the lips and tongue, if I could then analyze speech for these quantities, I would have a set of speech defining signals that could be handled as low frequency telegraph currents with resulting advantages of secrecy, and more telephone channels in the same frequency space as well as a basic understanding of the carrier nature of speech by which the lip reader interprets speech from simple motions."

—Homer Dudley, 1935

## 3.1 BACKGROUND

If we think, for the moment, of speech as being a mode of transmitting word messages, and telegraphy as simply another mode of performing the same action, this immediately allows us to conclude that the intrinsic information rate of speech is exactly the same as that of a telegraph signal generating words at the same average rate. Speech, however, conveys emphasis, emotion, personality, etc., and we still don't know how much bandwidth is needed to transmit these kinds of information.

In the following sections, we begin with some further historical background on speech communication.

### 3.1.1 Transmission of Acoustic Signals

Perhaps the earliest network for speech communication at long distances was a system that we'll call the "stentorian network," which was used by the ancient Greeks. It consisted of towers and men with very loud voices. The following excerpts were found in Homer Dudley's archives:

> *Homer has written that the warrior Stentor, who was at the siege of Troy, had such a loud voice that it made more noise than fifty men all shouting at once. Alexander the Great (356–325 B.C.) seems to have had a method whereby a stentor's voice could be heard by the whole army. Did it consist of acoustical signals which were repeated from one soldier crier to another, organized as a transmitting group?*
>
> *We quote the following from Caesar's commentaries:   when extraordinary events happened, the Gauls relayed the information by shouting from one place to another:*

*for example, the massacre of the Romans which took place at Orleans at sunrise was known at nine o'clock the same evening at Auvergne, forty miles away.*

*Diodorus of Sicily, a Greek historian living in the age of Augustus, said that at the order of the King of Persia, sentinels, who shouted the news which they wished to transmit to distant places were stationed at intervals throughout the land. The transmission time was 48 hours from Athens to Susa, over 1500 miles apart.*

We also note that, in addition to speech, flare signals were used as a communications medium. The Appendix to this chapter illustrates this with an excerpt from the Greek play *Agamemnon* (by Aeschylus) that describes the transmission of information about the fall of Troy (also see Fig. 3.13 in the Appendix).

## 3.1.2 Acoustical Telegraphy before Morse Code

The Dudley archives provide some fascinating examples of pre-Morse code communications:

*Later a group of inventors, among whom we find Kircher(1601–1680), Scheventer (1636) and the two Bernoulli brothers, sought to transmit news long distances by means of musical instruments each note representing a letter. One of the Bernoullis devised an instrument, composed of five bells, which permitted the principal letters of the alphabet to be transmitted.*

*It is told that the King of England was able to hear news transmitted 1.5 English miles to him by means of a trumpet. He had this trumpet taken to Deal Castle, whose commander said that this instrument permitted a person to make himself understood over a distance of three nautical miles. It was invented by the "genial mechanic" of Hammersmith, Sir Samuel Morland (1626–1696). It's [sic] mouthpiece was designed so that no sound could escape from either end. Morland published a treatise on this instrument entitled "Tube Stentorophonica" and in 1666 he wrote a report on "a new cryptographic process."*

*In 1762 Benjamin Franklin experimented with transmitting sound under water. In 1785 Gauthoy and Biot transmitted words through pipes for a distance of 395 meters. But at a distance of 951 meters speech was no longer intelligible.*

*We can also regard the ringing of bells as acoustical telegraphy or telephony, if we consider that in certain Swiss villages the inhabitants recognize from their tone whether the person who has just died is a man or a woman, a member of a religious order, etc. Moreover, every Sunday the inhabitants of these villages follow the principal passages of the divine service with the aid of the pealing of the different bells. We have seen old people, prevented from attending the service because of their infirmities, with prayer book in hand, follow at a distance the priest's various movements.*

*Our story would be incomplete if we did not mention the African tom-tom, which some people consider a sort of acoustical telegraphy. The African explorer, Dr. A. R. Lindt, has written a short report on the tom-tom. We quote the following from his work: "There is no key to the acoustical telegraphy of the Africans. Since they have no written language, they are unable to divide their words into letters. The tom-tom therefore does not translate letter by letter or even word by word, but translates a series of well-defined thoughts into signals. There are different signals for all acts interesting to the tribe:   mobilization, death of the chief, and summons to a judicial convocation. However, the tom-tom also serves to transmit an order to a definite person. Thus, when a young man enters the warrior class, he receives a kind of call signal which introduces him and enables him to be recognized at a distance.*

*As yet, explorers have not been able to discover how intelligible the same signals are to different tribes. It is certain, however, that friendly tribes use the same signals. A settlement receiving a signal transmits it to the next village, so that in a few minutes a communication can be sent several hundred kilometers.*

*Acoustical telegraphy is still used today by certain enterprises such as railroads, boats, automobiles, fire fighting services and alarm services.*

This completes our quotations from the Dudley archives. We see that the concept of long-distance communication has a long history and that there is some evidence that speech communication at a distance was practiced by the ancients.

### 3.1.3   The Telephone

Proceeding more or less chronologically, we come to that most important development, the invention of the telephone by Alexander Graham Bell. There is no need to chronicle the well-known events leading to this invention and the enormous consequent effect on human communication; we restrict ourselves to several comments. It is interesting that Bell's primary profession was that of a speech scientist who had a keen understanding of how the human vocal apparatus worked, and, in fact, Flanagan [5] describes Bell's "harp telephone," which showed that Bell understood the rudiments of the speech spectral envelope. Nevertheless, telephone technology has been mostly concerned with transmission methods. Recently, however, with the growing use of cellular phones in which transmission rate is limited by nature, efficient methods of speech coding have become an increasingly important component of speech research at many laboratories.

### 3.1.4   The Channel Vocoder
### and Bandwidth Compression

In a *National Geographic* magazine article [2], Colton gives an engrossing account of the history of telephone transmission. Figure 3.1, taken from that article, shows the telephone wires on lower Broadway in New York City in the year 1887. It is clear that progress in telephony could easily have been brought to a halt if not for improvements, such as underground cables, multiplexing techniques, and fiber-optical transmission. Dudley pondered this traffic problem in a different way, that is, through coding to reduce the intrinsic bandwidth of the source, rather than increasing the capacity of the transmission medium.

Just as the Voder was the first electronic synthesizer, so the channel vocoder [3] was the first analysis-synthesis system. The vocoder analyzer derived slowly varying parameters for both the excitation function and the spectral envelope. To quote Dudley, this device could lead to the advantages of "secrecy, and more telephone channels in the same frequency

**FIGURE 3.1**   Lower Broadway in 1887.

space." Both of these predictions were correct, but the precise ways in which they came to pass (or are coming to pass) probably differ somewhat from how Dudley imagined them. In 1929, there was no digital communications. When digitization became feasible, it was realized that the least-vulnerable method of secrecy was by means of digitization. However, digitization also meant the need for wider-transmission bandwidths. For example, a 3-kHz path from a local telephone cannot transmit a pulse-coded modulation (PCM) speech signal coded to 64 kbits/s (the present telephone standard). The channel vocoder was thus quickly recognized as a means of reducing the speech bit rate to some number that could be handled through the average telephone channel, and this led eventually to a standard rate of 2.4 kbits/s.

With respect to the second prediction, given that the science of bandwidth compression is now approximately 50 years old, one might assume that "more telephone channels in the same frequency space" would by now be a completely realized concept within the public telephone system. Such, however, is not the case. Although it is our opinion that Dudley's second prediction will eventually come true, it is fair to ask why it is taking so long. With the recent boom in wireless telephony, the bandwidth is now an issue of even greater importance.

We conclude this section with a reference to an informative and entertaining paper by Bennett [1] This paper is a historical survey of the X-System of secret telephony that was used during World War II. Now totally declassified, the X-System turns out to be a quite sophisticated version of Dudley's channel vocoder! It included such features as PCM transmission, logarithmic encoding of the channel signal, and, of course, enciphered speech. Bennett has many interesting anecdotes concerning the use of the X-System during the war.

## 3.2  VOICE-CODING CONCEPTS

To understand why a device such as a vocoder reduces the information content of speech, we need to know enough about human speech production to be able to model it approximately. Then we must convince ourselves that the parameters of the model vary sufficiently slowly to permit efficient transmission. Finally, we must be able to separate the parameters so that each one is coded optimally. The implementation of these concepts is captured by the phrase "analysis-synthesis system." The analysis establishes the parameters of the model; these parameters are transmitted to the receiving end of the system and used to control a synthesizer with the goal of reproducing the original utterance as faithfully as possible.

A convenient way to understand vocoders is to begin with the synthesizer. A concise statement that helps define a model of speech is given by Fant [4]: "The speech wave is the response of the vocal tract to one or more excitation signals." This concept leads directly to engineering methods to separate the *source* (the excitation signal) from the *filter* (the time-varying vocal tract). The procedures (and there are many) for implementing this separation can be called deconvolution, thus implying that the speech wave is a linear convolution of source and filter.[1] In spectral terms, this means that the speech spectrum can be treated as the *product* of an excitation spectrum and a vocal tract spectrum. Figure 3.2 is a simplified illustration of the spectral cross section for sustained vowels. Numerous experiments have shown that such waveforms are quite periodic; this is represented in the figures by the lines. In (a) the lines are farther apart, representing a higher pitched sound; in (b) and (c) the fundamental frequency is lower.

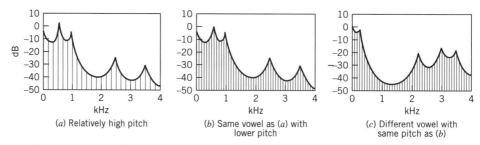

(a) Relatively high pitch     (b) Same vowel as (a) with lower pitch     (c) Different vowel with same pitch as (b)

**FIGURE 3.2**   Fine structure and spectral envelope of sustained vowels.

---

[1] Acoustic speech or music production often involves varying degrees of nonlinear behavior, usually at the interface between excitation and filter. New research is now being directed at this subject. In many cases we expect that the resulting effects will be minor, but there could be surprises.

The spectral envelope determines the relative magnitudes of the different harmonics, and it, in turn, is determined from the specific shape of the vocal tract during the phonation of that vowel. Deconvolution is the process of physically separating the spectral envelope from the spectral fine structure, and in later chapters we describe methods of implementing such a process. Once this separation is accomplished, we can hypothesize, with some confidence, that both the spectral envelope and spectral fine structure can be efficiently parameterized, with consequent bandwidth savings during transmission.

The parameters, if appropriately obtained, must vary relatively slowly because ultimately they depend on the articulator motions of the speech-producing mechanisms. Since these are human motions they obey the mechanical constraints imposed by the flesh-and-blood properties of the pertinent human organs, which move relatively slowly compared to typical speech bandwidths of 5 kHz.

The human vocal tract has been represented as a time-variable filter excited by one or more sources. The mechanism for this production varies according to the type of speech sound. Air pressure is supplied by the lungs. For vowel production, the cyclic opening and closing of the glottis creates a sequence of pressure pulses that excite resonant modes of the vocal tract and nasal tract:   the energy created is radiated from the mouth and nose to the listener.

For voiceless fricatives (e.g., s, sh, f, and th), the vocal cords are kept open and the air stream is forced through a narrow orifice in the vocal tract to produce a turbulent, noiselike excitation. For example, the constriction for "th" is between tongue and teeth; for "f" it is between lips and teeth.

For voiceless plosives (e.g., p, t, and k), there is a cross section of complete closure in the vocal tract, causing a pressure buildup. The sudden release creates a transient burst followed by a lengthier period of aspiration.

A more extensive categorization of speech sounds is given in Chapter 23, including some additional material about the articulator positions (tongue, lips, jaw, etc.) corresponding to these categories.

Several basic methods of source–filter separation and subsequent parameterization of each have been developed over the past half-century or so. We limit our discussion to four such methods:   (a) the channel vocoder, (b) linear prediction, (c) cepstral analysis, and (d) formant vocoding. Details of these methods will be examined in later chapters; for now we discuss the general problem of source–filter separation and the coding of the parameters.

One way to obtain an approximation of the spectral envelope is by means of a carefully chosen bank of bandpass filters. Looking at Fig. 3.2, we see that the complete spectrum envelope is not available; only the *samples* of this envelope at frequencies determined by the vertical lines are available. We assume that the fundamental frequency is not known so that we have no *a priori* knowledge of the sample positions. However, by passing the signal through a filter bank, where each filter straddles several harmonics, one can obtain a reasonable approximation to the spectral envelope. If the filter bandwidths are wide enough to encompass several harmonics, the resulting intensity measurements from all filters will *not* change appreciably as the fundamental frequency varies, as long as the envelope remains constant. This is the method employed for spectral analysis in Dudley's channel vocoder. The array of (slowly varying) intensities from the filter bank can now be coded and transmitted.

Linear prediction is a totally different way to approximate the spectral envelope. We hypothesize that a reasonable estimate of the $n$th sample of a sequence of speech samples is given by

$$\tilde{s}(n) = \sum_{k=1}^{p} a_k s(n - k). \tag{3.1}$$

In Eq. 3.1, the $a_k$'s must be computed so that the error signal

$$e(n) = s(n) - \tilde{s}(n) \tag{3.2}$$

is as small as possible. As we will show in the Chapter 21, Eq. 3.1 and the minimizing computational structure used lead to an all-pole digital synthesizer network with a spectrum that is a good approximation to the spectral envelope of speech.

Source–filter separation can also be implemented by cepstral analysis, as illustrated in Figure 3.3. Figure 3.3(a) shows a section of a speech signal, Fig. 3.3(b) shows the spectrum of that section, and Fig. 3.3(c) shows the logarithm of the spectrum. The logarithm transforms the multiplicative relation between the envelope and fine structure into an additive relation. By performing a Fourier transform on Fig. 3.3(c), one separates the slowly varying log spectral envelope from the more rapidly varying (in the frequency domain) spectral fine structure, as shown in Fig. 3.3(d). Source and filter may now be separately coded and transmitted.

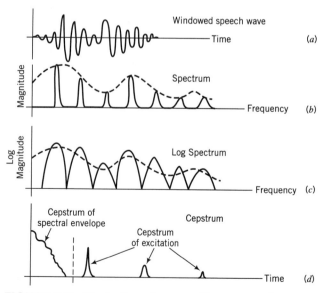

**FIGURE 3.3**   Illustration of source–filter separation by cepstral analysis.

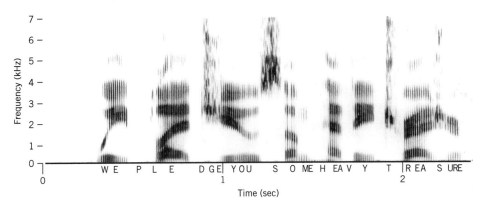

**FIGURE 3.4** Wideband spectrogram.

Finally, formant analysis can be used for source–filter separation. In Chapters 10 and 11 (Wave Basics and Speech Production), the theory of vocal-tract resonance modes is developed. However, we can to some extent anticipate the result by studying the speech spectrograms of Figs. 3.4 and 3.5. These figures are three-dimensional representations of time (abscissa), frequency (ordinate), and intensity (darkness). Much can be said about the interpretation of spectrograms; here we restrict our discussion to the highly visible resonances or *formants* and to the difference between Fig. 3.4 (wideband spectrogram) and Fig. 3.5 (narrow-band spectrogram).

We see from Fig. 3.4 that during the vowel sounds, most of the energy is concentrated in three or four formants. Thus, for vowel sounds, an analysis could entail tracking of the frequency regions of these formants as they change with time. Many devices have been invented to perform this operation and also to parameterize the speech for other sounds, such as fricatives (s, th, sh, f) or plosives (p, k, t); again we defer detailed descriptions for later.

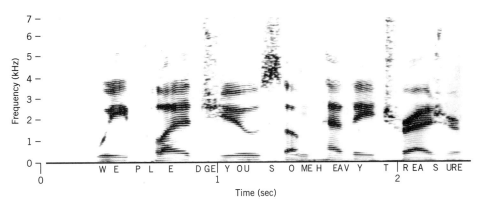

**FIGURE 3.5** Narrow-band spectrogram.

Formant tracks are also visible in Fig. 3.5, but there is a significant difference between the two figures. Whereas Fig. 3.4 displays the periodicity of the signal during vowels as vertical striations, Fig. 3.5 displays the periodicity horizontally. An explanation of this difference is left as an exercise.

## 3.3  HOMER DUDLEY (1898–1981)

Homer Dudley's inventions of the channel vocoder and Voder triggered a scientific and engineering effort that is still in progress. On my first visit to the Bell Laboratories in 1961[2] I was hosted by Dr. Ed David, who then managed the speech-processing group. As we passed an office, David whispered to me, "that's Homer Dudley." I was not introduced to Mr. Dudley and on subsequent visits did not see him. At that time he was near retirement age and, I suppose, not in the mainstream of Bell Laboratories' work. Quite a few years later (the late 1960's), Lincoln Laboratory was privileged to have the then-retired inventor as a consultant. We mention several items of interest from his brief stay there.

Dudley had a strong feeling that we should study speech waveforms as much, and perhaps more, than speech spectrograms. He felt that with practice, one could learn to read these waveforms. Dudley's speculation remains unproven. However, in an effort to augment his claim, Dudley, with the help of Everett Aho, produced photographs that are very informative and aesthetically pleasing. They are reproduced here as Figs. 3.6–3.11. Observing these waveforms, one develops a good feeling for the relative duration and amplitude of the vowels versus the consonants. In addition, we see the precise timing of the burst and voice-onset time of the voiced plosive sounds, b, d, and g. An inspection of the vowel sound I as in "thin" or "fish" illustrates the high frequency of the second resonance and the low frequency of the first resonance. We also note that the energy of the sh sound in "fish" is much stronger than the "f" sound in fish. Many other relationships among the acoustic properties of the phonemes can be found by careful observation of good-quality speech waveforms.

In 1967, a vocoder conference was organized under the auspices of the U.S. Air Force Cambridge Research Laboratory (AFCRL). Dudley was honored at this conference. Figure 3.12 shows Dudley displaying the plaque to the audience.

In 1969, when Dudley discontinued his consultancy at Lincoln Laboratory, he entrusted one of the authors (Gold) with two boxes filled with various technical information, plus a large number of newspaper clippings on his inventions. These have been used freely in this chapter, and they have been donated to the archives of the Massachusetts Institute of Technology.

In 1981, we received the news of his death at age 83. Tributes to him were written by Manfred Schroeder and James L. Flanagan, who worked with Dudley at Bell Laboratories and appreciated his monumental contributions.

---

[2]Of course this is Gold speaking here. Morgan was 12 at the time.

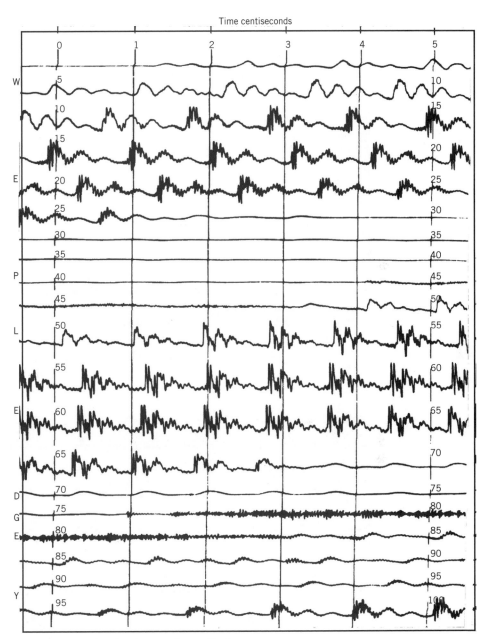

**FIGURE 3.6** Dudley's waveform display.

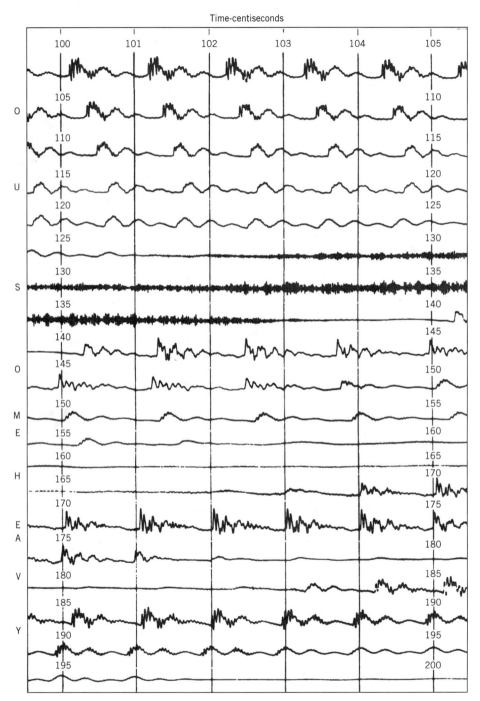

**FIGURE 3.7** Continuation of Dudley's waveform display.

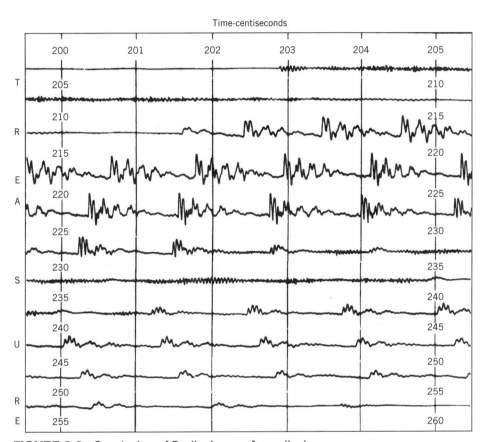

**FIGURE 3.8** Conclusion of Dudley's waveform display.

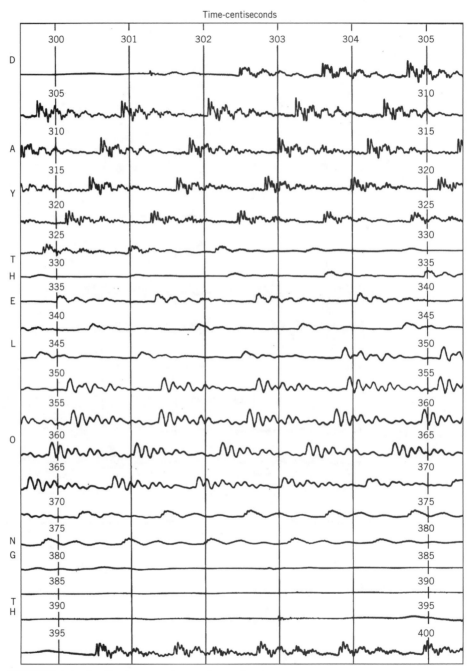

**FIGURE 3.9**  Dudley's second waveform display.

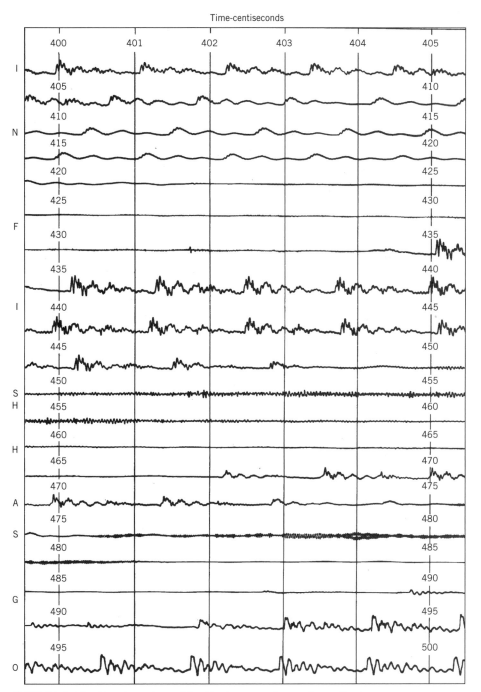

**FIGURE 3.10**  Continuation of Dudley's second waveform display.

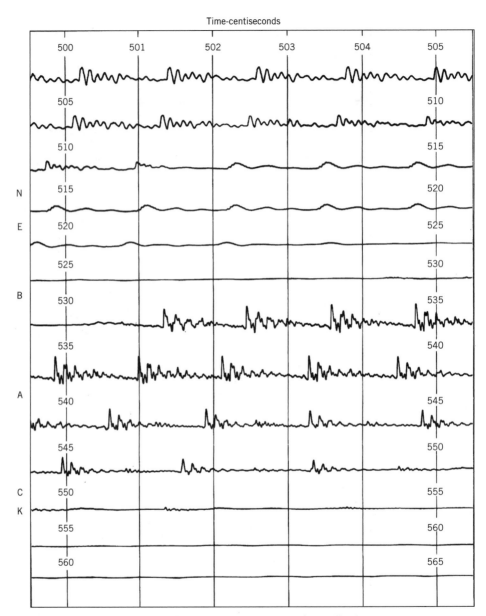

**FIGURE 3.11** Conclusion of Dudley's second waveform display.

**FIGURE 3.12** Dudley receiving an award.

# 3.4 EXERCISES

**3.1** Explain why wideband spectrograms show periodicity in time whereas narrow-band spectrograms show periodicity in frequency.

**3.2** Invent a display that shows periodicity in both time and frequency.

**3.3** Can you think of a reason why spectrograms are preferable visual displays to direct oscillographic waveforms?

**3.4** Which sounds are more likely to be better understood from waveforms? From spectrograms?

**3.5** Construct a table for the phonemes of the phrase "we pledge you some heavy treasure." The leftmost column should list the phonemes alternating with the transition regions; the next column should list your best estimate of the beginning; and the third column should list the end of the speech section. Base your estimates on Figs. 3.4 and 3.5.

**3.6** Construct a syllable table in the same manner as in the previous exercise.

**3.7** During World War II, Roosevelt and Churchill conversed by telephone between London and

Washington, using a channel vocoder. Explain why the vocoder was an important component of the communications link.

**3.8**  The phrase "carrier nature of speech" was proposed by Dudley as a way of explaining how a vocoder could represent speech with fewer bits (or less bandwidth). Explain how channel vocoders, linear predictive vocoders, and cepstral vocoders implement this concept and, as a result, represent the speech signal more efficiently than a standard telephone or PCM system.

## 3.5  APPENDIX: HEARING OF THE FALL OF TROY

LEADER OF CHORUS:
I come to do you reverence, Clytemnestra.
For it is right to give the king's wife honor,
A woman on a throne a man left empty
But if you know of good or only hope
to hear of good and so do sacrifice,
I pray you speak. Yet if you will, keep silence.
CLYTEMNESTRA:
With glad good tidings, so the proverb runs,
may dawn arise from the kind mother night.
For you shall learn a joy beyond all hope:
the Trojan town has fallen to the Greeks.
LEADER:
You say? I cannot hear–I cannot trust–
CLYTEMNESTRA:
I say the Greeks hold Troy. Do I speak clear?
LEADER:
Joy that is close to tears steals over me.
CLYTEMNESTRA:
Quite right. Such tears give proof of loyalty.
LEADER:
What warrant for these words? Some surety have you?
CLYTEMNESTRA:
I have. How not–unless the gods play tricks.
LEADER:
A fair persuasive dream has won your credence?
CLYTEMNESTRA:
I am not one to trust a mind asleep.
LEADER:
A wingless rumor then has fed your fancy?
CLYTEMNESTRA:
Am I some little child that you would mock at?

LEADER:
But when, *when*, tell us, was the city sacked?
CLYTEMNESTRA:
This night, I say, that now gives birth to dawn.
LEADER:
And what the messenger that came so swift?
CLYTEMNESTRA:
A god! The fire-god flashing from Mount Ida.
Beacon sped beacon on, couriers of flame.
First, Ida signaled to the island peak
of Lemnos, Hermes' rock, and swift from there
Athos, God's mountain, fired the great torch.
It leaped, it skimmed the sea, a might of moving light.
joy-bringing, golden shining, like a sun,
and sent the fiery message to Macistus.
Whose towers, then, in haste, not heedlessly
or like some drowsy watchman caught by sleep,
sped on the herald's task and flashed the beacon
afar, beyond the waters of Euripus
to sentinels high on Messapius' hillside,
who fired in turn and sent the tidings onward,
touching with flame a heap of withered heather.
So, never dimmed but gathering strength, the splendor
over the levels of Asopus sprang,
lighting Cithaeron like the shining moon,
rousing a relay there of travelling flame.
Brighter beyond their orders given, the guards
kindled a blaze and flung afar the light.
It shot across the mere of Gorgopis.
It shone on Aegiplanctus' mountain height,
swift speeding on the ordinance of fire,
where watchers, heaping high the tinder wood,
sent darting onward a great beard of flame
that passed the steeps of the Saronic Gulf
and blazing leaped aloft to Arachnaeus,
the point of lookout neighbor to our town.
Whence it was flashed here to the palace roof,
a fire fathered by the flame on Ida.
Thus did the they hand the torch on, one to other,
in swift succession finishing the course.
And he who ran both first and last is victor.
Such is my warrant and my proof to you:
my lord himself has sent me word from Troy.

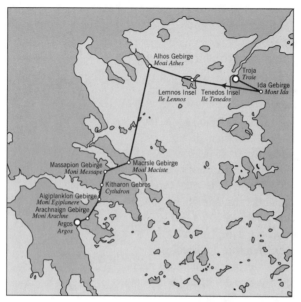

**FIGURE 3.13**  Map, showing the communications path described in *Agamemnon*.

## BIBLIOGRAPHY

1. Bennett, W. R., "Secret telephony as a historical example of spread-spectrum communication," *IEEE Trans. Commun.* **COM-31**: 98–104, 1983.
2. Colton, F. B., "The miracle of talking by telephone," *National Geographic* **70**(4): 395–433, 1937.
3. Dudley, H., "The vocoder," *Bell Labs Record* **17**: 122–126, 1939.
4. Fant, G., *Acoustic Theory of Speech Production*, Morton, S-Gravenhage, 1960.
5. Flanagan, J. L., *Speech Analysis Synthesis and Perception,* 2nd ed., Springer-Verlag, New York/Berlin, 1972.

# CHAPTER 4

# BRIEF HISTORY OF AUTOMATIC SPEECH RECOGNITION

**C**ONCEPTUALLY, the development of speech recognition is closely tied with other developments in speech science and engineering, and as such can be viewed as having roots in studies going back to the Greeks (as with synthesis). However, the history of speech recognition[1] per se in the 20th Century began with the invention of a small toy, Radio Rex.

## 4.1 RADIO REX

The first machine to recognize speech to any significant degree may have been a commercial toy named Radio Rex, which was manufactured in the 1920s. Here is a description from a 1962 review paper [14]:

> It consisted of a celluloid dog with an iron base held within its house by an electromagnet against the force of a spring. Current energizing the magnet flowed through a metal bar which was arranged to form a bridge with 2 supporting members. This bridge was sensitive to 500 cps acoustic energy which vibrated it, interrupting the current and releasing the dog. The energy around 500 cps contained in the vowel of the word Rex was sufficient to trigger the device when the dog's name was called.

It is likely that the toy responded to many words other than "Rex," or even to many nonspeech sounds that had sufficient 500-Hz energy. However, this inability to reject out-of-vocabulary sounds is a weakness shared by most recognizers that followed it. Furthermore, the toy was in some sense useful, since it fulfilled a practical purpose (amusing a child or playful adult), which was not often accomplished by many of the laboratory systems that followed. Although quite simple, it embodied a fundamental principle of speech recognizers for many years:    store some representation of a distinguishing characteristic of the desired sound and implement a mechanism to match this characteristic to incoming speech.

---

[1] As with any such brief historical review, we have been limited to discussing a small fraction of the many contributions and contributors to this extremely active field.

Radio Rex was later referred to in a famous letter to the Acoustical Society by John Pierce of Bell Labs [46], in which he strongly criticized the speech recognition research of that time (1969):

*What about the possibility of directing a machine by spoken instructions? In any practical way, this art seems to have gone downhill ever since the limited commercial success of Radio Rex.*

Although much of the work in vocoding and related speech analysis in the 1930s and 1940s was relevant to speech recognition, the next complete system of any significance was developed at Bell Labs in the early 1950s.

## 4.2 DIGIT RECOGNITION

A system built at Bell Labs and described in [15] may have been the first true word recognizer, as it could be trained to recognize digits from a single speaker. It measured a simple function of the spectral energy over time in two wide bands, roughly approximating the first two resonances of the vocal tract (i.e., formants). Although the system's analysis was crude, its estimate of a word-long spectral quantity may well have been more robust to speech variability than some of the later common approaches to estimating the time-varying speech spectrum. It tracked a rough estimate of formant positions instead of the spectrum itself. This is potentially resistant to irrelevant modifications of the overall speech spectrum. For instance, a simple turn of the talker's head away from a direct path to the listener often produces marked changes in the spectrum of the received speech (in particular, a relative reduction in the amplitude of the higher spectral components). The Bell Labs system's spectral estimation technique was, however, quite crude, histogramming low- and high-frequency spectral moments over an entire utterance, and thus timing information was lost. Although the idea was good, there was insufficient technology to develop it very far by modern standards; it used analog electrical components and must have been difficult to modify. Still, the inventors claimed that it worked very well, achieving a 2% error for a single speaker uttering digits that were isolated by pauses [15].

The system (see Fig. 4.1) worked generally as follows:  incoming speech was filtered into low- and high-frequency components and each component strongly saturated so that its amplitude was roughly independent of signal strength. The cutoff frequency in each case was roughly 900 Hz, which is a reasonable boundary between first and second formants for adult males.[2] Zero crossings were counted for each of the two bands, and the system used this value to estimate a central frequency for each band. The low-frequency number was quantized to one of six 100-Hz subbands (between 200 and 800 Hz), and the high-frequency number was quantized to be one of five 500-Hz subbands, beginning at 500 Hz. Together, these two quantized values correspond to one of 30 possible frequency pairs (in

---

[2]Children and adult women often have first formants above this frequency, and speakers of either gender can have second formants that are below 900 Hz for some sounds. Still, 900 Hz is a reasonable dividing point between major energy components in speech.

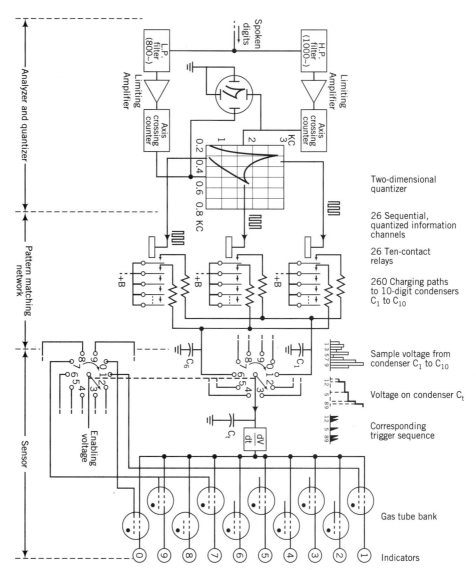

**FIGURE 4.1** Schematic for 1952 Bell Labs digit recognizer [15].

practice, only 28 were used, as the other two were rarely applicable). During a training period, capacitors were used to store charges associated with the time that the signal was mapped to a particular pair of frequencies. This distribution was learned for each digit. The resulting distributions were then used to choose conductances for RC circuits that would be used during recognition. When a new digit was uttered, a new distribution was determined in a similar way and compared to the stored distributions by switching between RC circuits corresponding to all possible digits (where the conductance corresponded to the template, the capacitances and charge time were all equal, and where the charging voltage for each frequency pair was determined by the new utterance). This procedure essentially implemented correlations between each stored distribution and the new distribution. The digits had distinguishable frequency-pair distributions and so could usually be discriminated from one another (See [15], Fig. 2, p. 639).

Note that even in 1952, researchers were reporting a speech recognizer that was 98% accurate! An examination of modern press releases suggests that this figure may be a constant for speech-recognition systems (those that are reported, anyway).

## 4.3   SPEECH RECOGNITION IN THE 1950s

In 1958, Dudley made a classifier that continuously evaluated spectra, rather than approximations to formants. This new paradigm was commonly used afterward; in fact, broadly speaking, the current dominant paradigm for speech recognition uses some function of a local spectral estimate varying over time as the representation of the incoming speech.

In 1959, Denes, from the College of London, added grammar probabilities in addition to acoustic information. In other words, he pointed out that the probability of a particular linguistic unit being uttered can also be dependent on the previous linguistic unit, so that the probability of a word need not be solely dependent on the acoustic input.

In 1962, David and Selfridge put together Table 4.1, which compared a number of speech-recognition experiments in the preceding decade [14] including the two recognizers mentioned above. In general, researchers performed spectral tracking, detected a few words and sounds, and performed tests on a small number of people.

## 4.4   THE 1960s

Throughout much of the 1960s, automatic speech-recognition research continued along similar lines. Martin deployed neural networks for phoneme recognition in 1964. Digit recognizers became better in the 1960s, achieving good accuracy for multiple speakers. Widrow trained neural networks to recognize digits in 1963 [64]. Phonetic features were also explored by a number of researchers. However, as noted earlier, in 1969 John Pierce wrote a caustic letter entitled "Whither Speech Recognition?" In it he argued that scientists were wasting time with simple signal-processing experiments because people did not do speech recognition, but rather speech understanding. He also pointed out the lack of scientific rigor in the experimentation at that time and he suggested that arbitrary manipulation of recognizer parameters to find the best performance was like the work of a "mad scientist," rather than that of a serious researcher. At the time, Pierce headed the Communications Sciences Division at Bell Labs, and his remarks were quite influential.

Although there may have been much that was correct about Pierce's criticism, there were a number of major breakthroughs in the 1960s that became important for speech-recognition research in the 1970s. First, as noted previously, prior to this period the primary approach to estimating the short-term spectrum was a filter bank. In the 1960s, three spectral

**TABLE 4.1  Pre-1962 Speech-Recognition Systems[a]**

| Investigators | Speech representation | Vocabulary | No. of talkers tested | Appox. error rate (%) | Additional facts and comments |
|---|---|---|---|---|---|
| Kersta | Selected entries from $\Delta f$–$\Delta t$ matrix (200 cps × 67 ms) | 10 digits | 9 men, 5 women | 0.2 | Spectrograms quantized into 2 levels |
| Davis, Biddulph, and Balashek | Formants 1 and 2 as a function of time | 10 digits | 1 talker | 2.0 | Correlation metric |
| Fry and Denes | Selected entries from $\Delta f$–$\Delta t$ matrix | 14 speech sounds in 139 words | 1 talker | 28.0 (sounds) 56.0 (words) | Phoneme diagram frequencies used to supplement primitive recognition from acoustic data |
| Olson and Belar | $\Delta f$–$\Delta t$ matrix | 10 words or syllables | 1 talker | 2.0 | |
| Dudley and Balashek | $\Delta f$–$\Delta t$ matrix | 10 digits | 2 men | 5 | Temporal sequence disregarded |
| Mathews and Denes | $\Delta f$–$\Delta t$ matrix | 10 digits | 6 men | 6 | Spectral pattern time and amplitude normalized |
| Hughes | Spectral features | 11 sound categories in 100 words | 4 men, 3 women | 30 | Feature-selection-based linguistic analysis |
| Shultz | Spectral features | 10 digits | 25 men, 25 women | 3 | |
| Petrick and Willett | $\Delta f$–$\Delta t$ matrix | 10 digits | 1 talker | <1.0 | Spectral petterns time normalized |
| Forgie and Forgie | Spectral features | 10 vowels | 11 men, 10 women | 7 | |
| Keith-Smith and Klem | $\Delta f$–$\Delta t$ matrix | 10 vowels | 11 men, 10 women | 6 | Statistical decision procedure used to select relevant spectral features |
| Sebestyen | $\Delta f$–$\Delta t$ matrix | 10 digits | 10 speakers | <1 | |
| Suzuki and Nakata | Formants 1 and 2 | 5 vowels in consonant contexts | 5 speakers | ≈20 | Additional experiments on vowels in bisyllable words and short sentences yield higher error rates |

[a]From [14].

estimation techniques were developed that were later of great significance for recognition, although their early applications to speech were for vocoding:   the Fast Fourier transform (FFT), cepstral (or homomorphic) analysis, and linear predictive coding (LPC). Additionally, new methods for the pattern matching of sequences were developed:   a deterministic approach called dynamic time warp (DTW), and a statistical one called the hidden Markov model (HMM).

### 4.4.1   Short-Term Spectral Analysis

As discussed in Chapter 7, Cooley and Tukey introduced the FFT [13]. This is a computationally efficient form of the discrete Fourier transform (DFT), which in turn can be interpreted as a filter bank. However, its efficiency was important for speech-recognition research, as it was for many other disciplines.

An alternative to filter banks and their equivalent FFT implementation was cepstral processing, which was originally developed by Bogert for seismic analysis [7] and applied later to speech and audio signals by Oppenheim, Schafer, and Stockham [42]. Cepstral processing will be discussed later (primarily in Chapter 20), but its significance for speech recognition is primarily as an approach to estimating a smooth spectral envelope. It ultimately became widely used for recognition, particularly in combination with other analysis techniques (see Chapter 22).

*Cepstrum*
*Chapt 20*

LPC is a mathematical approach to speech modeling that has a strong relation to the acoustic tube model for the vocal tract. Fundamentally, it refers to the use of an autoregressive (pole only) model to represent the generation of speech; each time point in sampled speech is predicted by a weighted linear sum of a fixed number of previous samples. In Chapter 21 we will provide a more rigorous definition, but for now the significance of LPC is that it provides an efficient way of finding a short-term spectral envelope estimate that has many desirable properties for the representation of speech, in particular the emphasis on the peak spectral values that characterize voiced sounds. Some of the early writings on this topic include [24], [1], and [35]. An excellent tutorial on the topic was written by Makhoul [33].

*LPC*
*Chapt 21*

### 4.4.2   Pattern Matching

Dynamic programming is a sequential optimization scheme that has been applied to many problems [6]. In the case of speech analysis for recognition, it was proposed as a method of time normalization – different utterances of the same word or sentence will have differing durations for the sounds, and this will lead to a potential mismatch with the stored representations that are developed from training materials. DTW applies dynamic programming to this problem. It was proposed by Sakoe around 1970 (but published in an English-language journal in 1978 [53]). Vintsyuk was among the first to develop the theory, and he also applied it to continuous speech [59]. DTW for connected word recognition was described by Bridle [10] and Ney [40]. Excellent review articles on the subject were written by White [63] and by Rabiner and Levinson [49].

DTW is a deterministic approach to the matching of the time sequence of short-term spectral estimates to stored patterns that are representative of the words that are being modeled [39]. Alternatively, one could imagine a statistical approach, in which the incoming time sequence is used to assess the likelihood of probabilistic models rather than speech examples or prototypes. The mathematic foundations for such an approach were developed in the 1960s, and they were built on the statistical characterization of the noisy communications channel as described in 1948 by Shannon [56]. Most notably, the work of Baum and colleagues at the Institute for Defense Analysis established many of the basic concepts, such as the forward–backward algorithm to compute the model parameters iteratively [5] (see Chapter 26). Briefly, hidden Markov modeling is a statistical approach that models an observed sequence as being generated by an unknown sequence of variables.

*HMM*

Towards the end of the 1960s, a number of researchers became interested in developing these ideas further for the case of a naturally occurring sequence, and in particular for speech recognition. Many of these ultimately joined a research group at IBM, which pioneered many aspects of HMM-based speech recognition in the 1970s. An early IBM report that influenced this work was [58], and a range of other publications followed through the early to mid-1970s, for example, [4], [2], [26], and [25]. The group developed an early HMM-based automatic speech-recognition system that was used for a continuous speech-recognition task referred to as New Raleigh Language. Baker independently developed an HMM-based system called Dragon while still a graduate student at Carnegie Mellon University (CMU) [3]. Many other researchers were working with this class of approaches by the mid-1980s (e.g., [54]).

## 4.5   1971–1976 ARPA PROJECT

As noted earlier, one of Pierce's criticisms of earlier efforts was that there was insufficient attention given to the study of speech *understanding*, as opposed to recognition. In the 1970s the Advanced Research Projects Agency (ARPA)[3] funded a large speech-understanding project. The main work was done at three sites:   System Development Corporation, CMU, and Bolt, Beranek & Newman (BBN). Other work was done at Lincoln, SRI International, and University of California at Berkeley. The goal was to perform 1000-word automatic speech recognition by using a few speakers, connected speech, and constrained grammar with less than a 10% semantic error. The funding was reported to be $15 million. According to Klatt, who wrote an interesting critique of this program [27], only a system called Harpy, built by a CMU graduate student (Bruce Lowerre), fulfilled the goals. He used LPC segments, incorporated high-level knowledge, and modified techniques from Baker's Dragon system, as well as from another CMU system, Hearsay.

## 4.6   ACHIEVED BY 1976

By 1976, researchers were using spectral feature vectors, LPC, and phonetic features in their recognizers. They were incorporating syntax and semantic information. Approaches incorporating neural networks, DTW, and HMMs were developed. A number of systems

---

[3]This U.S. government agency was originally known as ARPA but later became known as DARPA (the D standing for Defense), but after a few years it reverted back to ARPA; as of this writing it is DARPA again.

were built. Efforts on reducing search cost were explored. Techniques from artificial intelligence were often used, particularly for the ARPA program. HMM theory had been applied to automatic speech recognition, and HMM-based systems had been built. In short, many of the fundamentals were in place for the systems that followed.

## 4.7   THE 1980s IN AUTOMATIC SPEECH RECOGNITION

In the 1980s, most efforts were concentrated in scaling existing techniques (e.g., LPC and HMMs) to more difficult problems. New front-end processing techniques were also developed in this time period. For the most part, however, the structure of speech-recognition systems did not change; they were trained on a larger quantity of data and extended to more difficult tasks. This extension did require extensive engineering developments, which were made possible by a concerted effort in the community. In particular, there was a major effort to develop standard research corpora.

### 4.7.1   Large Corpora Collection

Prior to 1986 or so, the speech-recognition community did not have any widely accepted common data bases for training recognition systems. This made comparisons between labs difficult, since few researchers trained or tested on the same acoustic data. Many speech researchers were concerned with this problem. Industrial scientists (e.g., those with Texas Instruments and Dragon Systems) worked with NIST (National Institute of Standards and Technology)[4] and compiled large standard corpora.

In 1986, collection began on the TIMIT[5] corpus [41], which was to become the first widely used standard corpus. A 61-phone alphabet was chosen to represent phonetic distinctions. The sentences in TIMIT were chosen to be phonetically balanced, meaning that a good representation of each phone was available within the training set. There were 630 speakers that each said 10 sentences, including two that were the same for each speaker. The data were recorded at Texas Instruments and phonetically segmented at MIT, first by use of an automatic segmenter [31], followed by manual inspection and repair of the alignments by graduate students. This resulted in a data base in which the time boundaries of the phone in the speech signal are marked for every phone uttered by a speaker. Even though errors still undoubtedly exist in the TIMIT data base, it remains one of the largest and most widely used hand-labeled phonetic corpora.

With the advent of the second major ARPA speech program in the mid-1980s, a new task called Resource Management (RM) was defined, with a new data base [47] of speech. RM had much in common with the task from the first ARPA program in the 1970s. The major differences were that the grammar had a greater perplexity,[6] and the recordings were

---

[4]Formerly called the National Bureau of Standards (NBS).

[5]So called because the data were collected at Texas Instruments (TI) and annotated at MIT.

[6]Roughly speaking, perplexity is a measure of the uncertainty about the next word given a word history; a more precise definition will be given in Chapter 5.

made of read speech. Sentences were constructed from a 1000-word language model, so that no out-of-vocabulary words were encountered during testing. The corpus contained 21,000 utterances from 160 speakers. One important characteristic of the RM task was that it included speaker-independent recognition; that is, some systems were trained on many speakers, and they were tested on speakers not in the training set.

Later on in the program, the focus shifted to the Wall Street Journal Task – recognizing read speech from the Wall Street Journal.[7] The first test was constrained to be a 5000-word vocabulary test with no out-of-vocabulary words; later, a 20,000-word task with out-of-vocabulary words was developed. More recent tests used an essentially unlimited vocabulary, and researchers often used 60,000-word decoders for system evaluations.

Another task that was developed in parallel with the read speech program was Air Travel Information System (ATIS), which was based on spontaneous query in the airline-reservation domain. ATIS is a speech-understanding task (as opposed to a speech-recognition task). Systems not only had to produce word strings, but they also had to attempt to derive some semantic meaning from these word strings and perform an appropriate function. For instance, if the user said "show me the flights from Boston to San Francisco," the system should respond by showing a list of flights. Interaction continued with the system in order to reach some goal; in this case, ordering airline tickets. This domain was more practical than the Wall Street Journal task, but the vocabulary size was smaller. Systems today are now quite good at this task.

DARPA funded the collections of these corpora, and the collection processes were managed by NIST. NIST subcontracted much of the collection work to sites such as SRI and Texas Instruments. These and other corpora are now distributed through the Linguistic Data Consortium, which is based at the University of Pennsylvania in Philadelphia.

### 4.7.2  Front Ends

A number of new front ends, that is, subsystems that extract features from the speech signal, were developed in the 1980s. Of particular note are mel cepstrum [16], perceptual linear prediction [22], delta cepstral coefficients [17], and other work in auditory-inspired signal-processing techniques, for example, [55] and [21]. (See Chapter 22 for a discussion of many of these approaches.)

*Chapt 22*

### 4.7.3  Hidden Markov Models

As noted previously, the fundamentals of HMM methodology were developed in the late 1960s, with applications to speech recognition in the 1970s. In the 1980s, interest in these approaches spread to the larger community. Research and development in this area led to system enhancements from researchers in many laboratories, for example, BBN [54] and Philips [9]. By the mid-late 1980s, HMMs became the dominant recognition paradigm,

---

[7]The task was later called CSRNAB (Continuous Speech Recognition of North American Business News), which included data from other news sources.

with, for example, systems at SRI [37], MIT–Lincoln [44], and CMU. The CMU system was quite representative of the others developed at this time, and [29] provides an extended description.

Much of this activity focused on tasks defined in a new ARPA program. As in the 1970s, IBM researchers primarily worked with their own internal tasks, although ultimately they too participated in DARPA evaluations. See [48] for descriptions of the wide range of work done at Bell on HMMs for telephone speech, as well as on many other aspects of automatic speech recognition.

### 4.7.4  The Second (D)ARPA Speech-Recognition Program

In 1984, ARPA began funding a second program. The first major speech-recognition task in this program was the Resource Management task mentioned earlier. This task involved reading sentences derived from a 1000-word vocabulary. The sentences were questions and commands designed to manipulate a naval information data base, although the systems did not actually have to interface with any data base; ratings were based on word recognition. Sample sentences from the corpus [47] include the following:

- Is Dixon's length greater than that of Ranger?
- What is the date and hour of arrival in port for Gitaro?
- Find Independence's alerts.
- Never mind.

Evaluations of participating systems were held one to two times per year. Sites would receive a CD-ROM with test data, and send NIST the sentences produced by their recognizer, where the results would be officially evaluated.

The competition tended to make systems converge on good, similar systems, with each lab attempting to incorporate improvements that had been noted by the others. Although this led to a rapid set of improvements, this also led to a convergence of approaches for many systems.

The ARPA project fueled many engineering advances. As of 1998, many research systems can recognize read speech from new speakers (without speaker-specific training) with a 60,000-word vocabulary in real time, with less than a 10% word error.[8] The competition also inspired other sites that were not funded by the project, including laboratories in Europe. For example, Cambridge University in England participated in the evaluations, and developed HTK or HMM ToolKit, which has been widely distributed [65]. It is now possible to use HTK to get large vocabulary-recognition results close to those achieved by the major sites.

---

[8]The reader should keep in mind that this impressive performance is for read speech (that is, read from a page) in a limited domain with extensive language materials for training and relatively well-behaved acoustic input. The 1998 performance on tasks that are less constrained can be much worse.

It could be argued that the fundamentals of speech-recognition science have not greatly changed in the past decade; at least it is not clear that any major mechanisms (of the significance of dynamic programming, HMMs, or LPC) were developed during this period. However, there have been many developments that may ultimately prove to have been important – examples include front-end developments (mel or bark-scaled cepstral estimates, delta features, channel normalization schemes, and vocal tract normalization) and probabilistic estimation (e.g., maximum likelihood linear regression to adapt to new speakers or acoustics, schemes to improve discrimination with neural networks, or training paradigms to maximize the mutual information between the data and the models). Still, it is fair to say that the field has matured to the point that the efforts of many workers in the field are more oriented toward improving the engineering effectiveness of existing ideas rather than generating radically different ones. It is a matter of current controversy as to whether such an engineering orientation is sufficient to make major progress in the future, or whether radically different approaches will actually be required [8].

### 4.7.5  The Return of Neural Nets

The field of neural networks suffered a large blow when Minsky and Papert wrote their 1969 book *Perceptrons*, proving that the perceptron, which was one of the popular net architectures of the time,[9] could not even represent the simple exclusive or (XOR) function.[10] With the advent of backpropagation, a training technique for multilayer perceptrons (MLPs), in the early 1980s, the neural network field experienced a resurgence.

One application of neural networks to speech classification in the early 1980s was the use of a committee machine to judge whether a section of speech was voiced or unvoiced [20]. In 1983 Makino reported using a simple time-delayed neural network (a close cousin to a MLP in which the input layer includes a delayed version of itself in order to provide a simple context-delay mechanism) to perform consonant recognition [34]. This technique was later expanded by other researchers to add these delayed versions at multiple layers in the net [61]. Other researchers in the mid-1980s used Hopfield nets to classify both vowels and consonants [32].

By the late 1980s, many labs were experimenting with neural networks, both in isolated and continuous contexts. Only a few labs attacked large problems in automatic speech recognition with neural networks during this period; discrete probability estimators and mixtures of Gaussians were used in HMM recognizers for the majority of systems. Some sites have been using hybrid HMM–artificial neural network techniques, in which the neural network is used as a phonetic probability estimator, and the HMM is used to search through the possible space of word strings comprising the phones from the artificial neural network [36], [51].

---

[9]Although other network architectures were (and still are) available, including the perceptron's cousin, the MLP, the perceptron had properties that made it relatively easy to train.

[10]The XOR is a two-input logic function that returns true for inputs that are different (only one or the other is true) and false if the inputs are the same (either both true or both false).

### 4.7.6 Knowledge-Based Approaches

As noted previously, much of the work in the first ARPA speech project was strongly influenced by an artificial intelligence perspective. In the late 1970s and early 1980s, approaches based on the codification of human knowledge, typically in the form of rules, became widely used in a number of disciplines. Some speech researchers developed recognition systems that used acoustic–phonetic knowledge to develop classification rules for speech sounds; for instance, in [62], the consonants "k" and "g" following a vowel were discriminated on the basis of the proximity of the second and third resonances at the end of the vowel. This style of recognition was explained very well in [66].   One of the potential advantages of such an approach was that the speech characteristics used for discrimination were not limited to the acoustics of a single frame. Some of these points were explained in [12]. This reference, which is reprinted in [60], is also interesting because it includes a commentary from two BBN researchers (Makhoul and Schwartz), who took issue with the idea of focusing on the weak knowledge that we have about the utility of features chosen by experts. In this commentary, they suggested that systems should instead be focused on representing the ignorance that we have. In this case, they were really pointing to HMM-based approaches.[11] This dialog, and the personal interactions surrounding it at various meetings around this time, were extremely influential. By 1988 nearly every research site had turned to statistical methods. In the long term, however, the dichotomy might be viewed as elusive, since all of the researchers employing statistical methods continued to search for ways to include different knowledge sources, and the systems that attempted to use knowledge-based approaches also used statistical models.

## 4.8 RECENT WORK

It is too soon to judge the significance of developments from the past decade. However, we should comment very briefly on a few points.

**1.** The DARPA program continued, and moved on to a task referred to as Broadcast News. This is a significantly more realistic task than the Wall Street Journal transcription, since it includes a range of speaking styles (from read to spontaneous) and acoustic conditions (e.g., quiet studio to noisy street). It also is a *real* task, in the sense that the automatic transcription of broadcast data is closely related to several potential commercial applications.

**2.** The U.S. Defense Department also funded an effort to transcribe conversational speech. Two data bases collected for this work were Switchboard and Call Home; in the first case, talkers were asked to converse on the telephone on a selected topic (e.g., credit cards). In the second, callers were asked to telephone family members and discuss anything they wanted. These were, and are, extremely difficult tasks, and as of 1999 recognition performance is still very poor for the best systems.

---

[11] An earlier paper that made similar philosophical points, but that was not specifically concerned with speech recognition or HMMs, was [20].

**3.** Beginning in 1993, there has been an annual 6-week summer workshop that is focused on recognizing conversational speech. It was held for 2 years at Rutgers, and each summer since at Johns Hopkins.

**4.** Many of the first speech recognizers were segment based; that is, the recognizer hypothesized the boundaries of phone segments in the speech signal and then tried to do recognition based on this segmented speech. By the 1970s, most researchers turned to a more frame-based system, in which the base acoustic analysis regions were small, constant-duration sections, or frames, of speech. However, some researchers continue to work with segment-based systems, e.g., the MIT SUMMIT system [67], [45]. These systems developed ways of using statistical models, much as the frame-oriented systems had. Additionally, a number of researchers developed ways of extending HMM-based approaches to include segment statistics; see, for example, [43].

**5.** Through the 1980s, essentially every recognition system was extremely susceptible to a linear filtering operation (as one might experience from a telephone channel with a different frequency response than the one that was used to collect training data). In the past decade there was significant work to improve recognition robustness to different channels, as well as to variability in the microphone, and to acoustic noise [23], [57], [19], [28].

**6.** There has been an increased emphasis on issues of pronunciation [50], dialog modeling [11], long-distance dependencies within word sequences [52], and acoustic model adaptation [30], to mention just a few major topics.

**7.** There as a rapid expansion of research in other classification tasks related to automatic speech recognition. For instance, methods and systems were developed for speaker identification and verification ([18] and Chapter 36), as well as for language identification [38].

## 4.9  SOME LESSONS

Researchers often return to the same themes decade after decade – frame-based measures versus segment-based ones, statistical estimation of acoustic and language probabilities, incorporation of speech knowledge, and so on. With each return, the technology is more sophisticated. For instance, consumers can now purchase a dictation system that can recognize tens of thousands of words in continuous speech with a moderate error rate (after adaptation to the speaker), and the computers that can accomplish this are widely available.

However, the problems in speech recognition remain deep. Even five-word recognizers operate with significant errors under common natural conditions (e.g., moderate background noise and room reverberation, accent, and out-of-vocabulary words). In contrast, human performance is often far more stable under the same conditions, as discussed further in Chapter 18. We expect the general problem of the recognition and interpretation of spoken language to remain a challenging problem for some time to come.

## 4.10 EXERCISES

**4.1** How was the 1952 Bell Labs automatic-speech recognition system limited in comparison with a modern system? Is there any way in which it could potentially be better, while keeping the same basic structure?

**4.2** A new speech-recognition company is advertising their wonderful product. What percentage accuracy would you expect them to ascribe to their system? Describe some ways in which performance could be benchmarked in more realistic ways.

**4.3** Find a newspaper, magazine, or Web announcement about some speech-recognition system, either commercial or academic. Can you conclude anything about the structure and capabilities of these systems? If there is any content in the release information, try to associate your best guesses about the systems with any of the historical developments described in this chapter.

**4.4** In what way could Radio Rex be a better system than a recognizer trained to understand read versions of the Wall Street Journal?

## BIBLIOGRAPHY

1. Atal, B., and Hanauer, S., "Speech analysis and synthesis by prediction of the speech wave," *J. Acoust. Soc. Am.* **50**: 637–655, 1971.
2. Bahl, L., and Jelinek, F., "Decoding for channels with insertions, deletions, and substitutions with applications to speech recognition," *IEEE Trans. Inform. Theory* **IT-21**: 404–411, 1975.
3. Baker, J., "The DRAGON system – an overview," *IEEE Trans. Acoust. Speech, Signal Process.* **23**: 24–29, 1975.
4. Bakis, R., "Continuous-speech word spotting via centisecond acoustic states," IBM Res. Rep. RC 4788, Yorktown Heights, New York, 1974; abstract in *J. Acoust. Soc. Am.* **59** (Supp. 1): S 97, 1976.
5. Baum, L. E., and Petrie, T., "Statistical inference for probabilistic functions of finite state Markov chains," *Ann. Mathemat. Stat.* **37**: 1554–1563, 1966.
6. Bellman, R., "On the theory of dynamic programming," *Proc. Nat. Acad. Sci.* **38**: 716–719, 1952.
7. Bogert, B., Healy, M., and Tukey, J., "The quefrency analysis of time series for echos," in M. Rosenblatt, ed., *Proc. Symp. on Time Series Analysis*, Chap. 15, Wiley, New York, pp. 209–243, 1963.
8. Bourlard, H., Hermansky, H., and Morgan, N., "Towards increasing speech recognition error rates," *Speech Commun.* **18**: 205–231, 1996.
9. Bourlard, H., Kamp, Y., Ney, H., and Wellekens, C. J., "Speaker-dependent connected speech recognition via dynamic programming and statistical methods," in M. R. Schroeder, ed., *Speech and Speaker Recognition*, Karger, Basel, 1985.
10. Bridle, J., Chamberlain, R., and Brown, M., "An algorithm for connected word recognition," *Proc. IEEE Int. Conf. Acoust. Speech Signal Process.*, Paris, pp. 899–902, 1982.
11. Cohen, P., "Dialogue modeling," in R. Cole, J. Mariani, H. Uszkoreit, G. B. Varile, A. Zaenen, A. Zampoli, and V. Zue, eds. *Survey of the State of the Art in Human Language Technology*, Cambridge Univ. Press, London/New York, 1997.
12. Cole, R., Stern, R., and Lasry, M., "Performing fine phonetic distinctions:  templates versus

features," in J. S. Perkell and D. M. Klatt, eds., *Variability and Invariance in Speech Processes*, Erlbaum, Hillsdale, N.J., 1986.

13. Cooley, J. W., and Tukey, J. W., "An algorithm for the machine computation of complex Fourier series," *Math. Comput.* **19**: 297–301, 1965.

14. David, E., and Selfridge, O., "Eyes and ears for computers," *Proc. IRE* **50**: 1093–1101, 1962.

15. Davis, K., Biddulph, R., and Balashek, S., "Automatic recognition of spoken digits," *J. Acoust. Soc. Am.* **24**: 637–642, 1952.

16. Davis, S., and Mermelstein, P., "Comparison of parametric representations for monosyllabic word recognition in continuously spoken sentences," *IEEE Trans. Acoust. Speech Signal Process.* **28**: 357–366, 1980.

17. Furui, S., "Speaker independent isolated word recognizer using dynamic features of speech spectrum," *IEEE Trans. Acoust. Speech Signal Process.* **34**: 52–59, 1986.

18. Furui, S., "An overview of speaker recognition technology," in C. H. Lee, F. K. Soong, and K. K. Paliwal, eds., *Automatic Speech and Speaker Recognition*, Kluwer, Boston, Mass., 1996.

19. Gales, M., and Young, S., "Robust speech recognition in additive and convolutional noise using parallel model combination," *Comput. Speech Lang.* **9**: 289–307, 1995.

20. Gevins, A., and Morgan, N., "Ignorance-based systems," in *Proc. IEEE Int. Conf. Acoust. Speech Signal Process.*, San Diego, 39A.5.1–39A.5.4., 1984.

21. Ghitza, O., "Temporal non-place information in the auditory-nerve firing patterns as a front end for speech recognition in a noisy environment," *J. Phonet.* **16**: 109–124, 1988.

22. Hermansky, H., "Perceptual linear predictive (PLP) analysis of speech," *J. Acoust. Soc. Am.* **87**: 1738–52, 1990.

23. Hermansky, H., and Morgan, N., "RASTA processing of speech," *IEEE Trans. Speech Audio Process.* **2**: 578–589, 1994; special issue on robust speech recognition.

24. Itakura, F., and Saito, S., "Analysis-synthesis telephone based on the maximum-likelihood method," in Y. Konasi, ed., *Proc. 6th Int. Cong. Acoust.*, Tokyo, Japan, 1968.

25. Jelinek, F., "Continuous recognition by statistical methods," *Proc. IEEE* **64**: 532–555, 1976.

26. Jelinek, F., Bahl, L., and Mercer, R., "The design of a linguistic statistical decoder for the recognition of continuous speech," *IEEE Trans. Inform. Theory* **IT-21**: 250–256, 1975.

27. Klatt, D., "Review of the ARPA speech understanding project," *J. Acoust. Soc. Am.* **62**: 1345–1366, 1977.

28. Lee, C.-H., "On stochastic feature and model compensation approaches to robust speech recognition," *Speech Commun.* **25**: 29–48, 1998.

29. Lee, K.-F., *Automatic Speech Recognition – the Development of the Sphinx System*, Kluwer, Norwell, Mass., 1989.

30. Leggetter, C., and Woodland, P., "Maximum likelihood linear regression for speaker adaptation of continuous density hidden Markov models," *Comput. Speech Lang.* **9**: 171–185, 1995.

31. Leung, H., and Zue, V., "A procedure for automatic alignment of phonetic transcriptions with continuous speech," in *Proc. ICASSP'84*, San Diego, pp. 2.7.1–2.7.4, 1984.

32. Lippmann, R., and Gold, B., "Neural classifiers useful for speech recognition," in *Proc. IEEE First Int. Conf. Neural Net.*, San Diego, pp. 417–422, 1987.

33. Makhoul, J., "Linear prediction: a tutorial review," *Proc. IEEE* **63**: 561–580, 1975.

34. Makino, S., Kawabata, T., and Kido, K., "Recognition of consonants based on the perceptron model," *Proc. ICASSP'83*, Boston, Mass., pp. 738–741, 1983.

35. Markel, J., and Gray, A., *Linear Prediction of Speech*, Springer-Verlag, New York/Berlin, 1976.

36. Morgan, N., and Bourlard, H., "Continuous speech recognition: an introduction to the hybrid HMM/connectionist approach," *Signal Process. Mag.* **12**: 25–42, 1995.

37. Murveit, H., Cohen, M., Price, P., Baldwin, G., Weintraub, M., and Bernstein, J., "SRI's DECIPHER system," in *Proc. Speech Natural Lang. Workshop*, Philadelphia, pp. 238–242, 1989.

38. Muthusamy, Y. K., Barnard, E., and Cole, R. A., "Reviewing automatic language identification," *IEEE Signal Process. Mag.* **11**: 33–41, 1994.

39. Myers, C., Rabiner, L., and Rosenberg, L., "Performance tradeoffs in dynamic time warping algorithms for isolated word recognition," *IEEE Trans. Acoust. Speech Signal Process.* **28**: 623–635, 1980.

40. Ney, H., "The use of a one stage dynamic programming algorithm for connected word recognition," *IEEE Trans. Acoust. Speech Signal Process.* **32**: 263–271, 1984.

41. National Institute of Standards and Technology, *TIMIT Acoustic-Phonetic Continuous Speech Corpus*, Speech Disc 1-1.1, NIST Order No. PB91-505065, 1990.

42. Oppenheim, A. V., Schafer, R. W., and Stockham, T. G. Jr., "Nonlinear filtering of multiplied and convolved signals," *Proc. IEEE* **56**: 1264–1291, 1968.

43. Ostendorf, M., Bechwati, I., and Kimball, O., "Context modeling with the stochastic segment model," *Proc. IEEE Intl. Conf. Acoust. Speech Signal Process.*, San Francisco, pp. 389–392, 1992.

44. Paul, D., "The Lincoln continuous speech recognition system:  recent developments and results," in *Proc. Speech Natural Lang. Workshop*, Philadelphia, pp. 160–165, 1989.

45. Phillips, M., Glass, J., and Zue, V., "Automatic learning of lexical representations for sub-word unit based speech recognition systems," *Proc. Eurospeech*, Genova, Italy, pp. 577–580, 1991.

46. Pierce, J., "Whither speech recognition," *J. Acoust. Soc. Am.* **46**: 1049–1051, 1969.

47. Price, P., Fisher, W., Bernstein, J., and Pallett, D., "The DARPA 1000-word resource management database for continuous speech recognition," in *Proc. ICASSP'88*, New York, S.13.21, pp. 651–654, 1988.

48. Rabiner, L., and Juang, B.-H., *Fundamentals of Speech Recognition*, Prentice–Hall, Englewood Cliffs, N.J., 1993.

49. Rabiner, L., and Levinson, S., "Isolated and connected word recognition:  theory and selected applications," *IEEE Trans. Commun.* **29**: 621–659, 1981.

50. Riley, M., and Ljolje, A., "Automatic generation of detailed pronunciation lexicons," in C. H. Lee, F. K. Soong, and K. K. Paliwal, eds., *Automatic Speech and Speaker Recognition*, Kluwer, Boston, Mass., 1996.

51. Robinson, T., Hochberg, M., and Renals, S., "The use of recurrent neural networks in continuous speech recognition," in C. H. Lee, F. K. Soong, and K. K. Paliwal, eds., *Automatic Speech and Speaker Recognition*, Kluwer, Boston, Mass., 1996.

52. Rosenfeld, R., "A maximum entropy approach to adaptive statistical language modeling," *Comput. Speech Lang.* **10**: 187–228, 1996.

53. Sakoe, H., and Chiba, S., "Dynamic programming algorithm optimization for spoken word recognition," *IEEE Trans. Acoust. Speech Signal Process.* **26**: 43–49, 1978.

54. Schwartz, R., Chow, Y., Kimball, O., Roucos S., Krasner, M., and Makhoul, J., "Context-dependent modeling for acoustic-phonetic recognition of continuous speech," in *Proc. IEEE Int. Conf. Acoust. Speech Signal Process.*, Tampa, Fla., pp. 1205–1208, 1985.

55. Seneff, S., "A joint synchrony/mean-rate model of auditory speech processing," *J. Phonet.* **16**: 55–76, 1988.

56. Shannon, C., "A mathematical theory of communication," *Bell Sys. Tech. J.* **27**: 379–423, 623–656, 1948.

57. Stern, R., Acero, A., Liu, F.-H., and Ohshima, Y., "Signal processing for robust speech recognition," in C. H. Lee, F. K. Soong, and K. K. Paliwal, eds., *Automatic Speech and Speaker Recognition*, Kluwer, Boston, Mass., 1996.

58. Tappert, C., Dixon, N., Rabinowitz, A., and Chapman, W., "Automatic recognition of continuous speech utilizing dynamic segmentation, dual classification, sequential decoding, and error recovery," IBM Tech. Rep. RADC-TR-71-146, Yorktown Heights, NY, 1971.

59. Vintsyuk, T., "Element-wise recognition of continuous speech composed of words from a specified dictionary," *Kibernetika* **7**: 133–143, 1971.

60. Waibel, A., and Lee, K., eds., *Readings in Speech Recognition*, Morgan Kaufmann, San Mateo, Calif., 1990.

61. Waibel, A., Hanazawa, T., Hinton, G., Shikano, K., and Lang, K., "Phoneme recognition: neural networks vs. hidden Markov models," in *Proc. IEEE Int. Conf. Acoust. Speech Signal Process.*, New York, pp. 107–110, 1988.

62. Weinstein, C., McCandless, S., Mondshein, L., and Zue, V., "A system for acoustic-phonetic analysis of continuous speech," *IEEE Trans. Acoust. Speech Signal Process.* **23**: 54–67, 1975.

63. White, G., "Speech classification using linear time stretching or dynamic programming," *IEEE Trans. Acoust. Speech Signal Process.* **X**: 183–188, 1976.

64. Widrow, B., Personal Communication, Phoenix, Az., 1999.

65. Woodland, P., Odell, J., Valtchev, V., and Young, S., "Large vocabulary continuous speech recognition using HTK," in *Proc. IEEE Int. Conf. Acoust. Speech Signal Process.*, Adelaide, Australia, pp. II-125–128, 1994.

66. Zue, V., "The use of speech knowledge in automatic speech recognition," *Proc. IEEE* **73**: 1602–1615, 1985.

67. Zue, V., Glass, J., Phillips, M., and Seneff, S., "The MIT SUMMIT speech recognition system: a progress report," in *Proc. Speech Natural Lang. Workshop*, Philadelphia, pp. 179–189, 1989.

# CHAPTER 5

# SPEECH-RECOGNITION OVERVIEW

## 5.1  WHY STUDY AUTOMATIC SPEECH RECOGNITION?

Why do we study automatic speech recognition (ASR)? For one thing, there is a lot of money at stake: speech recognition is potentially a multi-billion-dollar industry in the near future. As of 1998, earnings (and savings) from simple telephone applications are reputed to be hundreds of millions of dollars per year.

There are many aspects of speech recognition that are already well understood. However, it is also clear that there is much that we still don't know. We don't have human-quality speech recognition; performance degrades rapidly when small changes are made to the speech signal, such as those that can be caused from switching microphones.

Speech recognition is potentially very useful. Sample applications include the following.

Telephone applications: For most current voice-mail systems, one has to follow a series of touch-tone button presses to navigate through a hierarchical menu. Speech recognition has the potential to cut through the menu hierarchy.

Hands-free operation: There are many situations in which hands are not available to issue commands to a device. Using a car phone and controlling the microscope position in an operating room are two examples for which some limited vocabulary systems already exist.

Applications for the physically handicapped: Speech recognition is a natural alternative interface to computers for people with limited mobility in their arms and hands, or for those with sight limitations.

For some aspects of computer applications, speech may be a more natural interface than a keyboard or mouse.

Dictation: General dictation is an advanced application, requiring a much larger vocabulary than, for instance, replacing a menu system. For instance, as of 1998, there are several dictation systems on the market that accept continuous speech input (e.g., from Dragon and IBM).

Translation: Another advanced application is translation from one language to another. The Verbmobil project in Germany is both a collaborative and competitive effort to provide language-to-language translation. The goal is to facilitate a conversation between native speakers of German and Japanese, using English as an intermediate language; the system is to act as an assistant to the German participant, translating words and phrases as needed from German into English (the speaker is assumed to be moderately competent in English).

## 5.2  WHY IS AUTOMATIC SPEECH RECOGNITION HARD?

There are many reasons why speech recognition is often quite difficult.

First, natural speech is continuous; it often doesn't have pauses between the words. This makes it difficult to determine where the word boundaries are, among other things. Also, natural speech contains disfluencies. Speakers change their mind in midsentence about what they want to say, will often accidentally switch phones (as in the phrase "teep a kape," which means "keep a tape"), and utter filled pauses (e.g., "uh" and "um") while they are thinking of their next message.

Second, natural speech can also change with differences in global or local rates of speech, pronunciations of words within and across speakers, and phonemes in different contexts. As a result, we can't just say that X is the spectral representation that corresponds to "uh." The spectrum will change, often quite dramatically, if any of these conditions are changed.

Third, large vocabularies are often confusable. A 20,000-word vocabulary is more likely to have more words that sound like each other than a 10-word vocabulary. There is also the issue of out-of-vocabulary words; for some tasks, no matter what words are in a vocabulary, recognition will always encounter words that have not been seen before. How to model these unknown words is an important unsolved problem.

Fourth, as noted previously, recorded speech is variable over room acoustics, channel characteristics, microphone characteristics, and background noise. In telephone speech, the channel used by the telephone company on any particular call (especially for analog segments) will have spectral and temporal effects on the transmitted speech signal. Background noise and acoustics in the environment that a telephone speaker is in will also have tangible effects on the signal. Different handsets, or in general different microphones, have different frequency responses; tilting a microphone at different angles will also change the frequency response. Nonlinear effects are particularly significant in carbon-granule microphones, but in general they can complicate the effects of using a particular handset. Some effects will be phone dependent; for instance, nasal sounds may be louder if the microphone is closer to the nose.

All of these factors can change the characteristics of the speech signal – a difference that humans can often compensate for, but that current recognition systems often cannot.

The algorithms for training recognition systems must be chosen carefully, for large training times are not practical for research purposes. Algorithms that take a year to run on available hardware may be of great theoretical interest, but since most programs have bugs, such a choice does not really permit the development of an experimental approach.

To replace other input modes with speech recognition, a high level of performance must be obtained. This does not necessarily mean near-perfect accuracy (although certainly too many errors can be very frustrating); perhaps it is just as important that recognition systems know when they are working well and when they are not, requiring some kind of confidence factor in designing a response to an input. This is difficult to do well when the recognition technology is known to be imperfect.

Another phenomenon that makes natural speech difficult to recognize is the effect of coarticulation. The physical realization of a phone can vary significantly depending on its phonetic context. For instance, consider the phone /t/ in the following words:

- take
- stake
- tray
- straight
- butter
- Kate

In the case of take, the /t/ is aspirated (i.e., there is a period of unvoiced sound after the release), whereas in stake it is not. The influence of r in tray and straight make /t/ come out more like a combination of t and ch. In butter, the /t/ is realized as a quick touch of the tongue against the alveolar ridge, also known as a flap. Finally, in Kate the /t/ is sometimes not released (especially in fast speech), so there is no large burst of noise at the end of the word (cf. take).

Besides differences in pronunciations of words within and between speakers, phonological variation often happens at the phrase level. For instance, the phrase "What are you doing?" often comes out sounding like "Whatcha dune?" In the same way, "Juwana eat?" is often the realization of "Do you want to eat?" In continuous speech, different strings of words can often sound like each other. Consider the following extreme examples:

- It's not easy to wreck a nice beach.
  It's not easy to recognize speech.
  It's not easy to wreck an ice beach.
- Moes beaches am big you us.
  Most speech is ambiguous.
- sly drool
  slide rule
- say s
  say yes

In addition, even if the word recognition is accurate, the semantic content may still not be clear. Consider these newspaper headlines [1]:

- Carter Plans Swell Deficit
- Farmer Bill Dies In House
- Nixon To Stand Pat On Watergate Tapes
- Stud Tires Out

We find most of these headlines funny because there is an alternate semantic interpretation one would not expect to see in the news (and sometimes it's the first interpretation we get).

---

[1] Courtesy of Ron Cole of the Oregon Graduate Institute.

## 5.3  AUTOMATIC SPEECH RECOGNITION DIMENSIONS

Given this general description of the difficulty of speech recognition, some dimensions of this difficulty can be defined. Claims of 98% accuracy in ASR are fairly meaningless without the specification of these task characteristics.

### 5.3.1  Task Parameters

One qualifier of an ASR task is whether it is speaker dependent (SD) or speaker independent (SI). A SD system is one that has been trained on one particular speaker and tested on the same speaker. A SI system is trained on many speakers and tested on a disjoint set of speakers. The Bell Labs digit recognizer discussed in [4], for instance, was trained and used in a SD manner, since it had to be adjusted for every speaker. The large ASR tasks that have been tackled under the U.S. ARPA program in the 1987–1995 period (Resource Management and Wall Street Journal dictation tasks) have both SD and SI components. Large vocabulary systems for use on personal computers have tended to be SD for greater accuracy. Although many systems have been ostensibly SI (at least they have been trained on many speakers and tested on others), many of them will perform quite poorly on speakers who are not native speakers of the target language.

Another descriptor is whether the task is to recognize isolated speech, recognize continuous speech, or spot keywords. The first type of task is to recognize words in isolation (demarcated by silence) and is in general less difficult than recognizing continuous speech, in which the word boundaries are not so apparent. A third type of task, which falls between the two earlier types, is keyword spotting. In this case the recognizer has a list of words that it tries to spot in the continuous speech input. The system must have a confidence factor about the match to prevent the false matching of words not in the keyword list. In general there is a trade-off between reducing the number of keyword occurrences that are missed and reducing the number of times that a non-keyword falsely triggers a keyword detection, and the cost of each of these kinds of errors must be considered in system optimization.

The lexicon (vocabulary) size also introduces another parameter. In general, a 20,000-word task is going to be much harder than a 10-word task. This is partly because there is a greater variability in the acoustics associated with each type of speech sound, but also because the larger task has many more words that are confusable with one another. However, even some small tasks have extremely confusable words. For instance, recognizing the E set of the alphabet (i.e., the letters b, c, d, e, g, p, t, v, z) may be harder than some tasks with a much larger vocabulary.

There is another reason why vocabulary size is not typically a reliable measure of task difficulty, which may also be strongly affected by constraints placed on the task. Systems that operate on the blocks world domain[2] will have a more constrained grammar than systems that attempt to understand radio news broadcasts. As noted previously, the perplexity of a grammar is a measure of how constrained it is. Perplexity is essentially the geometric mean

---

[2]The blocks world domain has been a favorite domain in artificial intelligence, in which objects are geometric solids such as blocks and are to be manipulated by a robot arm.

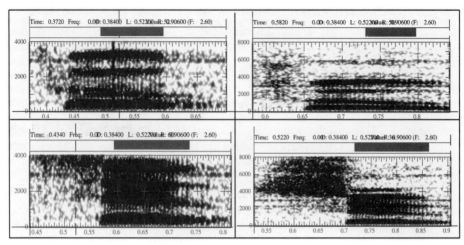

**FIGURE 5.1** Spectrograms of foo (top) and sue (bottom) sampled with 4-kHz and 8-kHz bandwidth restrictions. Courtesy of Eric Fosler.

*Perplexity*

of the branching factor (i.e., how many words can follow another word) of the grammar. More formally, it is $2^H$, where $H$ is the average entropy[3] associated with each word in the recognition grammar (i.e., the amount of uncertainty about the next word given the constraints and predictions of the grammar).

Speaking style has a strong effect on the difficulty of speech recognition. For instance, conversational speech is extremely difficult to transcribe. Speech that is carefully read from pre-existing text is comparatively easy. Fluent, goal-directed speech for a human–machine dialog is typically intermediate in difficulty; a motivated user will tend to speak more clearly, but the use of fluent speech will still be more difficult to recognize than read speech. Some of the characteristics of a more natural speaking style include a wider variability in speaking rate, an increase in disfluencies such as filled pauses or false starts, and a greater variability in vocal effort.

The recording conditions also play a part in determining the difficulty of an ASR task. The recording may range from wideband high-quality microphones to cellular phones in a moving car. The telephone channel typically has a bandwidth of less than 4 kHz, which means that it is more difficult to distinguish high-frequency consonants, such as /f/ and /s/. For example, Fig. 5.1 shows spectrograms of recordings of the words "foo" and "sue" at a 4-kHz (telephone) bandwidth and a 8-kHz bandwidth (high-quality microphone). Here /f/ and /s/ appear as the noise before the relatively straight lines, which represent the vowel. Note that it is more difficult to distinguish /f/ and /s/ in the 4-kHz spectral pictures than in the 8-kHz pictures. This is because most of the energy in /f/ and /s/ is above 4 kHz.

---

[3]The entropy of a random variable is just the negative log of its probability (conventionally in base 2 so that the result is in bits), and the average entropy is just the expectation of this quantity over the probability distribution.

Telephone speech also introduces other challenges. The range of speakers that have access to telephone speech have a greater variability than is typically observed in laboratory data bases (although this is also true for realistic data in many other cases as well). There is also a larger variability in background noise, and one must account for channel distortion from echo, cross talk, different spectral characteristics of the handset, and the communications channel in general. These sources of variability are particularly a problem for cellular and cordless telephones.

### 5.3.2  Sample Domain:   Letters of the Alphabet

As an example of a speech-recognition task, we present a classification of letters of the alphabet. This task has a vocabulary of 26 words, but many of them are confusable. For example, there are four major sets of letters that sound alike:

- E set:   B C D E P T G V Z
- A set:   J K
- EH set:   M N F S
- AH set:   I Y R

Recognition results from the Oregon Graduate Institute report an accuracy of 89% on letters of the alphabet and an accuracy of 87% when spelled names are dealt with from a telephone speech system trained on 800 speakers (12,500 letters) and tested on 400 speakers (4200 letters) [3]. A perceptual experiment was run in which 10 listeners identified 3200 letters from 100 alphabets and 100 spelled names. Human accuracy on the telephone speech data base was approximately 90–95%, with the average at approximately 93%. When the experiment was run with high-quality microphone speech, human accuracy jumps to approximately 99%. As noted previously, even small tasks can be relatively difficult, particularly if they involve typical users from the general public, and also particularly if the system is operating over the public-switched telephone network. Recognition can be even harder for cellular handsets and speakerphones.

## 5.4  COMPONENTS OF AUTOMATIC SPEECH RECOGNITION

Later in this book we will describe the characteristics and technological choices involved in systems for ASR. However, here we provide a preview, in which we describe very briefly the major components of these systems.

An ASR system may be described as consisting of five distinct subsystems (see Fig. 5.2): input acquisition, front end, local match, global decoder, and language model. This division is, of course, somewhat arbitrary. In particular, the first two subsystems are frequently described as one system that produces features for the classification stages. Here the functions are split out to emphasize their great significance (despite their relatively humble function). For instance, the microphone might seem to be a minor detail, one that is necessary but unworthy of discussion. However, as suggested earlier, some of the best

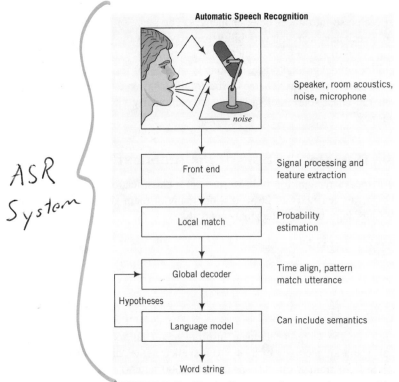

**FIGURE 5.2**  Block diagram of a speech-recognition system.

ASR systems have been brought to their knees, so to speak, by a change in microphone. It is important to understand, then, the dependence of an ASR system on the choice and position of the microphone (which can affect the overall spectral slope of the transduced speech, as well as the overall noise level and influence of room acoustics, with the latter being more pronounced for larger microphone–talker distances). Simple preprocessing may be used to partially offset such problems (e.g., adaptively filter to flatten the spectral slope, using a time constant much longer than a speech frame).

Feature extraction consists of computing representations of the speech signal that are robust to acoustic variation but sensitive to linguistic content. More simply, we wish to determine from speech some values that do not vary much when the same words are spoken many times (or by different talkers, for the speaker-independent case), but that change significantly when different things are said. One could argue that this is the entire problem, since finding separable speech representations would clearly mean that recognition could be accomplished. However, the correct speech representation can differ depending on the classification technique. Nonetheless, very simple data-examination techniques (e.g., scatter plots) can often be useful for screening features that are particularly bad or good for the discrimination of similar speech sounds.

Typically, speech analysis is done over a fixed length frame, or analysis window. For instance, suppose speech is sampled at 16 kHz after being low-pass filtered with a corner frequency lower than 8 kHz (e.g., 6.4 kHz) to prevent spectral aliasing. A window of length 32 ms (512 points) might be used as the input to a spectral analysis module, with one analysis performed every 10 ms (160 points). Since the analysis windows are not chosen to be synchronous with any acoustic landmark, the resultant features will be smeared over transition regions. However, methods that rely on presegmentation to establish analysis windows are difficult to do well (although in some instances they have worked well on specific tasks).

The local match module may either produce a label for a speech segment (e.g., word), or some measure of the similarity between a speech fragment and a reference speech fragment. This reference can be an explicit prototype of the same features that are extracted from speech during the recognition process (e.g., spectra). Alternatively, the input can be fit with statistical models, yielding probabilistic measures of the uncertainty of the fit.

Whatever the measure of similarity, one can imagine a matrix of distances between input features (the horizontal axis representing time), and the reference models or prototypes (the vertical axis representing a sequence of speech sounds within a reference utterance). Some form of temporal integration must be incorporated in order to find the utterance that in some sense is the minimum distance choice for what was said. This is the job of the global decoder.

It is generally insufficient to find a simple distance (e.g., Euclidean) between the reference model or prototype and the new speech features. One of the most obvious variations among multiple occurrences of the same utterance is the speed of the speech. A first-order solution for this is a linear normalization by utterance length. However, this does not compensate for the varying amount of time expansion and compression for different sounds. For instance, stop consonants such as "k" or "g" typically do not change their length much, whereas the length of sonorant sounds such as vowels tends to vary significantly with the speed of the speech.

These considerations led to the major algorithmic innovation (as noted in Chapter 4) known as dynamic time warp (DTW) [7]. For an isolated word example, for instance, given a local distance between each input frame and each reference frame, one would use the dynamic programming algorithm of Bellman [2] to determine the minimum cost match (defined as the match with the minimum sum of local distances plus any cost for permitted transitions). Pointers are retained at each step, so that the optimal path through the matrix can be backtraced once the best match is found. This path represents the best warping of the models to match the data.

The preceding section primarily addressed the case of deterministic distances between features such as spectra for reference sounds and those that are being recognized. However, a similar approach can be used for a statistical reference model. If one can estimate the probability of an observed spectrum for each hypothetical speech sound, as well as the probability of each permissible transition, the same procedure can be followed using a statistical distance measure (e.g., negative log probability). These distances are used in practical recognition systems based on hidden Markov models (HMMs). The use of these models was another fundamental advance in speech-recognition systems [1], [6], as noted in Chapter 4.

For isolated word recognition, a HMM for each vocabulary word can be used in place of the deterministic representation of a reference template for each word. For continuous speech recognition, phonemic HMMs can be concatenated to represent words, which in turn can be concatenated to represent complete utterances. Model variations can also be introduced to represent common effects of coarticulation between neighboring phonemes or words. Word transition costs can also be introduced to permit the use of a grammar. When dynamic programming is used to get the best match between the data and a statistical model as described earlier, the resulting best-path calculation is called a Viterbi decoding [8], [5]; this is the dominant approach used in continuous speech recognition.

For most speech-recognition tasks of interest, the acoustic information by itself is insufficient to uniquely determine what was said. In fact, in the human example, what was spoken is determined by a complex combination of mechanisms, including the incorporation of knowledge about the syntax and semantics of a language, as well as the pragmatic expectations from a situation. Likewise, our speech-recognition systems in general must make use of some information about the language that is available prior to the reception of the new acoustic information. In most current systems, however, only very simple linguistic information is incorporated, such as the frequency of pairs or triplets of words. In some systems, however, particularly those that operate in some limited application domain, deeper knowledge about probable paths in the task dialog can be used, sometimes incorporating structured models for natural language.

## 5.5    FINAL COMMENTS

Given the relatively long history of research into ASR, the variety of techniques alluded to in earlier sections, and the widely reported successes with fairly difficult tasks, one might wonder why ASR is considered a research topic at all. In fact, despite press reports to the contrary,[4] speech recognition by machine is still a difficult and largely unsolved problem, and there are a number of areas of active research that are being explored in the attempt to conquer the remaining serious problems. In particular, although ASR is good enough to be used for many practical tasks, recognizers as of 1999 are still often brittle, providing unreliable performance under conditions that are handled quite well by human listeners. A short parable (courtesy of John Ohala) may illustrate the current state of affairs.

Stanford[5] artificial intelligence researchers perfected a talking and listening handyman robot, which was then sent out to solicit research funds door to door. The robot rolled up to its first house, and rang the bell:

ROBOT:
I am Stanford's handyman robot. Tell me a task, and I will do it for $5 per hour. This money will be applied to further research in artificial intelligence.
HUMAN:
$5 an hour? Sounds great! Can you paint?

---

[4]It has often been suggested that the solution to machine speech-recognition is five years away; in fact, this has been suggested so repeatedly that it *must* be true!

[5]Remember, this is a parable.

ROBOT:

My painting is of the highest quality.

HUMAN:

OK. See that paint brush and bucket of paint? Take them out back and paint the porch.

ROBOT:

Your request will be fulfilled, courtesy of Stanford.

(The robot trundles off to do his job, and returns in an hour).

ROBOT:

The task is complete. Please deposit $5 to aid in further research.

HUMAN:

(Handing over the cash) This was a great deal! Come back again!

ROBOT:

(While leaving) Oh, by the way, it wasn't a Porsche. It was a BMW.

Given the human example, perhaps we can get some good ideas about how to build artificial systems that will not perform so poorly when handling situations that people find so straightforward. Although naive mimicry of the human systems is likely to be an insufficient tactic, we believe that there is much to be learned from human speech perception.

Before we proceed to further detail on either human or machine systems for audio signal processing, we must first provide some technical background. We will return at a later point to the aspects of speech-recognition technology that have been briefly alluded to in this chapter.

# 5.6 EXERCISES

**5.1** If the acoustic environment is noisy, how could you imagine each of the blocks of Fig. 5.2 being modified to help with speech recognition?

**5.2** You have a recognition system that can recognize strings of up to 16 digits, in which approximately 15% of the digits are incorrect. Describe some scenarios in which the system could be useful.

**5.3** Describe some situations in which a five-word recognizer can accomplish a more difficult task than a 1000-word recognizer.

**5.4** Suppose that any one of 16 distinct symbols could occur at any point in a sequence, with equal probability:

(a) What is the entropy associated with an occurrence of a symbol?

(b) What is the perplexity?

Now suppose that four of the symbols could occur with the same probability as before, but four were twice as probable and eight were half as probable:

(c) What is the entropy?

(d) What is the perplexity?

# BIBLIOGRAPHY

1. Baker, J. K., "The DRAGON system – an overview," *IEEE Trans. Acoust. Speech Signal Process.* **23**: 24–29, 1975.
2. Bellman, R., and Dreyfus, S., *Applied Dynamic Programming*, Princeton Univ. Press, Princeton, N.J., 1962.
3. Cole, R. A., Fanty, M., Gopalakrishnan, M., and Janssen, R. D. T., "Speaker-independent name retrieval from spellings using a database of 50,000 names," in *Proc. IEEE Int. Conf. Acoust. Speech Signal Process.* Toronto, Canada, pp. 325–328, 1991.
4. Davis, K., Biddulph, R., and Balashek, S., "Automatic recognition of spoken digits," *J. Acoust. Soc. Am.* **24**: 637–642, 1952.
5. Deller, J., Proakis, J., and Hansen, J., *Discrete-Time Processing of Speech Signals*, Macmillan Co., New York, 1993.
6. Jelinek, F., "Continuous recognition by statistical methods," *Proc. IEEE* **64**: 532–555, 1976.
7. Sakoe, H., and Chiba, S., "Dynamic programming algorithm optimization for spoken word recognition," *IEEE Trans. Acoust. Speech Signal Process.* **26**, 43–49, 1978.
8. Viterbi, A., "Error bounds for convolutional codes and an asymptotically optimal decoding algorithm," *IEEE Trans. Inf. Theory* **13**: 260–269, 1967.

# MATHEMATICAL
# BACKGROUND

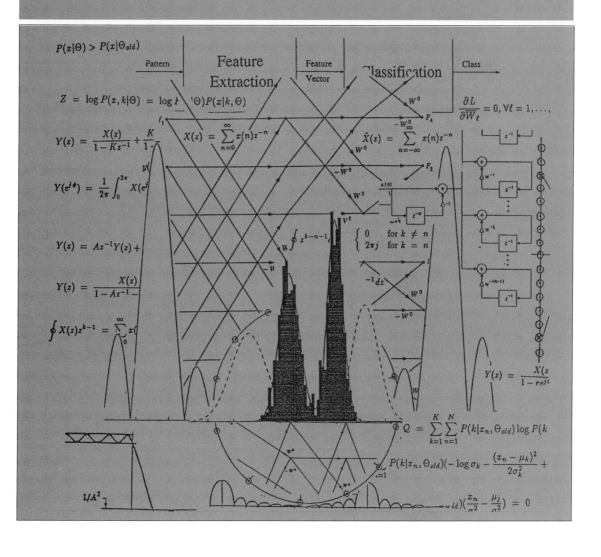

Baseball is 90% mental. The other half is physical.

—Yogi Berra

---

**I**T IS not possible to understand speech and audio signal processing in any depth without having a solid background in the mathematical underpinnings of signal processing and pattern recognition. There are many good books on these topics, and this text is not intended to replace them. However, in Part II we have gathered together some major points and concepts that we feel the reader must grasp. While distributing earlier drafts of this text, we found that the backgrounds of interested students and colleagues varied greatly. Therefore, even though we have limited the discussion to a few principal relevant points, we have included material at several levels of difficulty. A very brief introduction into the most relevant engineering mathematics for signal processing and pattern recognition is provided here. Readers who have studied this material previously may still wish to refresh their memories, or at least touch on familiar material lightly in order to get the feeling for our style of notation. Some advanced sections are also provided, such as the discussion of the Expectation maximization algorithm in Chapter 9.

Chapters 6 and 7 focus on digital signal processing, and Chapters 8 and 9 provide basic material for pattern recognition. We begin in Chapter 6 with a brief description of the basic mathematical descriptive device for discrete-time systems, the $z$ transform. We then apply this approach to convolution and ultimately to resonance, a form of filtering that is often useful in speech and music processing. Chapter 7 extends this discussion to more general types of digital filters, concluding with the discrete Fourier transform and its efficient implementation as the fast Fourier transform. Chapter 8 introduces some of the fundamentals for pattern-recognition systems, including approaches that range from minimum distance classifiers to multilayer perceptrons. Chapter 9 extends this discussion to statistical systems, starting with a brief reminder about probability densities and their properties, and proceeding through descriptions of several styles of probability estimators that are used in pattern-recognition problems such as speech recognition. The chapter concludes with a discussion of expectation maximization, which is a somewhat more advanced topic than the earlier material. However, it is included here because it has become a key method for modern statistical pattern recognition.

---

# CHAPTER 6

# DIGITAL SIGNAL PROCESSING

## 6.1 INTRODUCTION

Von Kempelen spent 20 years building his speech synthesizer. He used the most viable method of implementation for his time (~1780): mechanical devices. In the first half of the 20th century Fant and others built speech synthesizers from analog electronic components. When the digital computer arrived, speech researchers recognized its potentiality for speech-processing tasks, but it was not until recently that computational power became sufficiently great and cost became sufficiently low that even complex algorithms could be implemented cheaply and in real time. So, advances in speech processing owe much to advancing computer technology; but, in addition, this progress has been dependent on the mathematical discipline of digital signal processing – also called discrete-time signal processing.

The connection between speech and digital signal processing is straightforward. Speech depends greatly on filtering, both in production and perception. The vocal tract is a complicated arrangement of acoustic tubes; understanding the behavior of vocal tracts relies on physical models of these acoustic tubes. We shall see in Chapters 10, 11, and 12 that digital models of tubes are based on digital signal processing (DSP) concepts. Also, the auditory system was recognized, more than a century ago, to have properties akin to a filter bank that analyzes the spectral characteristics of the speech signal.

It therefore is desirable to include material from the DSP field, with emphasis on the filtering properties of DSP algorithms. The fundamentals of DSP are briefly reviewed and then applied to the theory and design of digital filters, with emphasis on those elements that connect to our description of speech and music coding.

## 6.2 THE *z* TRANSFORM

In this section we discuss the mathematical properties of the *z* transform. In Chapter 7 we will discuss the mathematical properties of the discrete Fourier transform. These transforms are the mathematical bridges that connect the time and frequency properties of discrete-time signals, just as the Laplace transform bridges the time–frequency properties of continuous signals. We start with a sequence $x(n)$, defined for all $n$. Define the *z* transform of $x(n)$ as

$$\tilde{X}(z) = \sum_{n=-\infty}^{\infty} x(n)z^{-n}, \tag{6.1}$$

where $z$ is a complex variable and $\tilde{X}(z)$ is a function of a complex variable.

Although Eq. 6.1 makes no explicit reference to time, in many practical cases $x(n)$ is derived by sampling a continuous signal at equally spaced time intervals.

In dealing with physical systems, it is convenient to assume that sequences begin at $n = 0$, so that $x(n)$ is undefined for negative values of $n$ and we have another definition of the $z$ transform:

$$X(z) = \sum_{n=0}^{\infty} x(n)z^{-n}. \tag{6.2}$$

We will be dealing with Eq. 6.2 unless otherwise noted. We call $\tilde{X}(z)$ the two-sided $z$ transform and refer to $X(z)$ as simply the $z$ transform, or, for the sake of clarity, the one-sided $z$ transform.

Note that the $z$ transform is a *linear* operation; that is, the $z$ transform of a weighted sum is the weighted sum of the $z$ transforms of the individual terms of the sum. This can easily be seen by inspection of Eq. 6.1 or Eq. 6.2. Also, the $z$ transform of a *delayed* sequence $x(n - m)$ is the $z$ transform of the original sequence multiplied by $z^{-m}$. (The proof is left as an exercise.) These properties will prove to be extremely useful.

## 6.3  INVERSE z TRANSFORM

Equation 6.2 is invertible; that is, we can find the sequence $x(n)$, given the function $X(z)$. To show this, multiply Eq. 6.2 by $z^{k-1}$ and perform a closed line integration on both sides of the equation. If the integration path is within the region of convergence of the infinite series, then the summation and integration can be interchanged, yielding

$$\oint X(z)z^{k-1} = \sum_{n=0}^{\infty} x(n) \oint z^{k-n-1} \, dz. \tag{6.3}$$

But (stated without proof; see Exercise 6.6),

$$\oint z^{k-n-1} \, dz = \begin{cases} 0 & \text{for } k \neq n \\ 2\pi j & \text{for } k = n \end{cases}. \tag{6.4}$$

From Eqs. 6.3 and 6.4 we have

$$x(k) = \frac{1}{2\pi j} \oint X(z)z^{k-1} \, dz. \tag{6.5}$$

In Eqs. 6.3, 6.4, and 6.5, the integration path must enclose the origin.

For many practical problems, this integration never need be explicitly done; rather, the inverse transform can often be computed by inspection, with the use of the linearity and delay properties described in Section 6.2 (see Exercise 6.3).

## 6.4  CONVOLUTION

The discrete convolution theorem is the defining equation of linear discrete systems. The mathematical statement is

$$y(n) = \sum_{m=0}^{n} x(m)h(n-m) = \sum_{m=0}^{n} x(n-m)h(m), \tag{6.6}$$

where $h(n)$ is the response of a linear system to a unit pulse, where the latter is defined as a sequence that is zero for all $n$ except $x(n) = 1$ for $n = 0$.

The unit-pulse response function $h(n)$ is associated with the time-domain behavior of the system; knowledge of $h(n)$ allows one, in principle, to find the response to any arbitrary input signal. Similarly, the system response to a discrete-time complex exponential can serve as a defining function in the frequency domain. Let the complex exponential be $e^{j\omega n}$. From Eq. 6.6, the steady state response (as $n \to \infty$) is given by

$$y(n) = \sum_{m=0}^{\infty} h(m)e^{j\omega(n-m)} = e^{j\omega n} \sum_{m=0}^{\infty} h(m)e^{-j\omega m} = e^{j\omega n} H(e^{j\omega}), \tag{6.7}$$

where $H(e^{j\omega n})$ is seen to be the $z$ transform of the system, evaluated on the unit circle. Thus, the steady state output of the system is a signal with the same frequency as the input, but with a complex gain factor that is given by the $z$ transform of the system evaluated at that frequency. This is a defining property of linear time-invariant systems.

If $X(z)$ is the $z$ transform of $x(n)$ and $H(z)$ is the $z$ transform of $h(n)$, then it can be shown that

$$Y(z) = H(z)X(z) \tag{6.8}$$

(See Exercise 6.1).

Equation 6.6 is a temporal relation between two time functions; Eq. 6.8 is the equivalent relation in the complex $z$ domain. If the value of $z$ is restricted to lie on the unit circle in the $z$ plane, then the $z$ transform reduces to the Fourier transform of the sequence.

Stated physically, convolution in the time domain leads to multiplication in the frequency domain. Furthermore, multiplication in the time domain leads to convolution in the frequency domain. We will state the result without proof.

$$y(n) = x(n)h(n),$$

$$Y(e^{j\phi}) = \frac{1}{2\pi} \int_0^{2\pi} X(e^{j\omega})H\left[e^{j(\phi-\omega)}\right] dw. \tag{6.9}$$

Integration is on the unit circle.

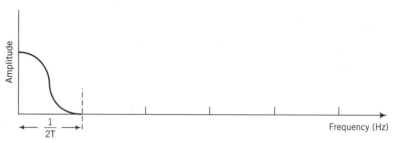

**FIGURE 6.1**   Spectrum of a continuous signal.

## 6.5 SAMPLING

If a continuous function of time $x(t)$ is sampled every $T$ seconds to produce the sequence $x(nT)$, the resultant frequency response is *periodic*, with period $1/T$. If the original signal is band limited so that it contains no frequencies greater than $1/2T$, we get the pictures of Figs. 6.1 and 6.2.

No information is lost in going from Fig. 6.1 to Fig. 6.2. In fact, the original, continuous signal can be recovered by low-pass filtering with a filter of bandwidth $1/2T$; this is a statement of the well-known sampling theorem.

It is physically clear that a sample, which has zero width in time, will have an infinite width in frequency; so, therefore, will a sequence of samples. The fact that each period in Fig. 6.2 is an exact duplicate of the original response of Fig. 6.1 can be shown in a number of ways.

First, since the samples represent the product of two signals (the original signal and a set of unity height samples $T$ seconds apart), the resulting frequency response of the product is the convolution of the two frequency responses and leads directly to Fig. 6.2.

Second, Fig. 6.1 has its counterpart in the $z$ plane. If the $z$ transform of a sequence is evaluated on the unit circle in the $z$ plane then $z = e^{j\theta}$. Let $\theta$ in Fig. 6.3 be $\omega T$; the $z$

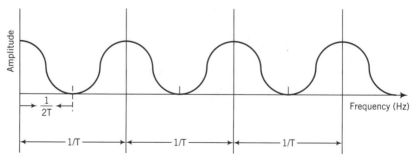

**FIGURE 6.2**   Spectrum of uniformly spaced samples.

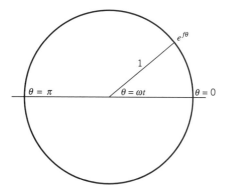

**FIGURE 6.3**   The unit circle in the complex $z$ plane.

transform becomes

$$X(e^{j\omega T}) = \sum_{n=0}^{\infty} x(n)e^{-j\omega nT}.$$ **(6.10)**

One complete path around the unit circle corresponds to $\omega T$'s traveling from 0 to $2\pi$. When $\omega T = 2\pi f_r T = 2\pi$, $f_r = 1/T$, where $T$ is the precise period of Fig. 6.3. As we keep circulating around the unit circle, subsequent periods of Fig. 6.3 are traced out.

## 6.6  LINEAR DIFFERENCE EQUATIONS

Much of the material in this book involves analysis and synthesis. Analysis often consists of studying the frequency components of signals; this is best done through filters, or through transform methods. In the analog world, filters are composed of resistors, capacitors, and inductors or their electronic equivalents. In the digital world we generally do filtering with computer programs. The signals we process are digital; they are quantized samples, or more simply, numbers.

Synthesis often consists of modeling physical devices, such as violins or human vocal tracts. In many cases the models may be approximated with linear systems. Filtering in the analog domain is mainly expressed mathematically through systems of linear differential equations. Modeling of systems in which space as well as time are parameters requires systems of partial differential equations. In the digital domain, systems of linear difference equations are needed to mathematically describe one-dimensional linear time-invariant devices. The description of more complicated systems such as acoustic tubes involves computer algorithms that have been labeled "digital wave guides" by Smith [8]. In many systems of interest, synthesis models obey some form of the wave equation, with solutions consisting of traveling or standing waves. Such models can be implemented by computer by incorporating relatively long delay elements into the algorithms. The following sections deal with both types of algorithms.

## 6.7   FIRST-ORDER LINEAR DIFFERENCE EQUATIONS

A simple example is the first-order equation:

$$y(n) = Ky(n - 1) + x(n). \tag{6.11}$$

A solution to this equation may be obtained by use of the $z$ transform. Given the previously described properties of linearity and delay, we can show the $z$-transform solution. Let $X(z)$ be the $z$ transform of $x(n)$ and let $Y(z)$ be the $z$ transform of $y(n)$; taking the $z$ transform of Eq. 6.11, we find

$$Y(z) = Kz^{-1}Y(z) + Ky(-1) + X(z). \tag{6.12}$$

Solving for $Y(z)$ yields

$$Y(z) = \frac{X(z)}{1 - Kz^{-1}} + \frac{Ky(-1)}{1 - Kz^{-1}}, \tag{6.13}$$

where $y(-1)$ can be interpreted as an initial condition of $y(n)$. For example, if $y(-1) = 0$ (the system is initially at rest), then

$$Y(z) = X(z)H(z), \tag{6.14}$$

where

$$H(z) = \frac{1}{1 - Kz^{-1}} = \frac{z}{z - K}. \tag{6.15}$$

Here $H(z)$ is the *transfer function* related to the difference equation 6.11. The transfer function is defined as the ratio of the $z$ transform of the output to the $z$ transform of the input. Since the $z$ transform of a unit impulse is unity, the transfer function can also be defined as the $z$ transform of the output when the input is a unit pulse.

The frequency response of this first-order system can be studied from the geometry in the $z$ plane.

First, note from Eq. 6.13 that knowing the input $X(z)$ anywhere in the $z$ plane, plus knowledge of the number $y(-1)$, allows determination of the $z$ transform $Y(z)$ of the output anywhere in the $z$ plane. Of special interest is the response of the system to a steady-state sinusoid. For mathematical brevity we use as the input the complex exponential $x(n) = e^{jn\theta}$. The resulting response of the system can be found by evaluating $Y(z)$ on the unit circle, as shown in Fig. 6.4. This evaluation has a simple geometric interpretation, as we can see by comparing the rightmost term of Eq. 6.15 with Fig. 6.4 for the case $z = e^{j\theta}$. The result (which can be generalized) is this: consider any point on the unit circle corresponding to an angle $\theta$. Compute the value of the vector emanating from the zero at the origin and divide by the vector emanating from the pole; this gives the (complex) frequency response at that chosen point on the unit circle. The magnitude and phase of the result is computed

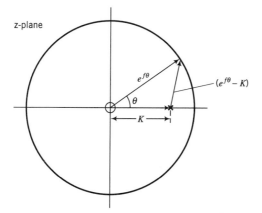

**FIGURE 6.4** Geometric interpretation of the frequency response of a first-order difference equation.

to be

$$|H| = \frac{1}{(1 + K^2 - 2K \cos \theta)^{1/2}}, \tag{6.16}$$

$$\angle H = \tan^{-1} \left( \frac{K \sin \theta}{1 - K \cos \theta} \right). \tag{6.17}$$

If $x(n)$ consists of samples from an analog signal, $\theta = \omega T$, so Eqs. 6.16 and 6.17 are direct functions of frequency.

## 6.8 RESONANCE

From Fig. 6.4 and Eqs. 6.16 and 6.17 we see that the smallest magnitude of the vector from the pole to the unit circle occurs for $\theta = 0$, so this value of $\theta$ and hence, of frequency, corresponds to the maximum value of the frequency-response magnitude. In general, as the path on the unit circle gets close to a pole, the magnitude of the frequency response increases. If we move the pole to some angle $\phi$, near the unit circle, then, as $\theta$ approaches $\phi$, the magnitude function peaks; this is resonance.

Figure 6.5 depicts such a situation.

We notice that the pole position is a complex number in the $z$ plane. What is the difference equation in the time domain that could lead to Fig. 6.5? Let us try

$$y(n) = r e^{j\theta} y(n-1) + x(n), \tag{6.18}$$

$$Y(z) = r e^{j\theta} z^{-1} Y(z) + X(z), \tag{6.19}$$

$$Y(z) = \frac{X(z)}{1 - r e^{j\theta} z^{-1}}, \tag{6.20}$$

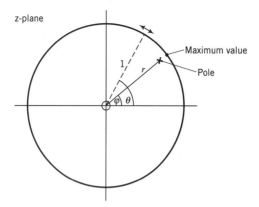

**FIGURE 6.5**   As $\theta$ approaches $\phi$, the magnitude of the $z$ transform on the unit circle peaks.

which leads to the geometry of Fig. 6.5. However, notice that $y(n)$, the inverse $z$ transform, is a sequence of *complex* numbers. The determination of resonance based on sequences of real numbers requires a second-order difference equation:

$$y(n) = Ay(n-1) + By(n-2) + x(n). \tag{6.21}$$

The $z$ transform of Eq. 6.21 is

$$Y(z) = Az^{-1}Y(z) + Bz^{-2}Y(z) + X(z). \tag{6.22}$$

Solving for $Y(z)$, we find

$$Y(z) = \frac{X(z)}{1 - Az^{-1} - Bz^{-2}} = \frac{z^2 X(z)}{z^2 - Az - B} \tag{6.23}$$

The geometry in the complex $z$ plane is shown in Fig. 6.6.

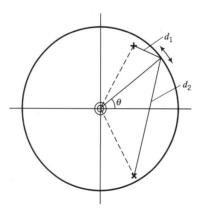

**FIGURE 6.6**   $z$-Plane depiction of resonance for a second-order system.

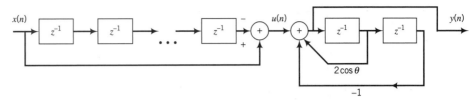

**FIGURE 6.7** Discrete-time network for Eq. 6.26 concatenated with Eq. 6.27.

If $A = 2r \cos \theta$ and $B = -r^2$, the denominator of Eq. 6.23 becomes

$$d = z^2 - (2r \cos \theta)z + r^2, \tag{6.24}$$

and this results in the roots

$$z_{1,2} = re^{\pm j\theta}. \tag{6.25}$$

Another form of resonance can be obtained from the configuration shown in Fig. 6.7.

Figure 6.8 shows the $z$-transform equivalence of Fig. 6.7.

$H_1(z)$ is the transfer function of the difference equation,

$$u(n) = x(n) \pm x(n - M), \tag{6.26}$$

and $H_2(z)$ is the transfer function of the equation

$$y(n) = 2 \cos \theta y(n - 1) - y(n - 2) + u(n). \tag{6.27}$$

Explicit expressions for $H_1(z)$ and $H_2(z)$ are easily obtained:

$$H_1(z) = 1 \pm z^{-M}, \tag{6.28}$$

$$H_2(z) = \frac{1}{1 - 2 \cos \theta z^{-1} + z^{-2}}. \tag{6.29}$$

In the $z$ plane, $H_1(z)$ is represented by $M$ zeros spaced uniformly around the unit circle. If $M$ is even (e.g., 12) and the minus sign in Eq. 6.28 is used, the zeros are as shown in Fig. 6.9. Thus, for example, the zeros at $60°$ can be cancelled by designing $H_2(z)$ to have two poles at the same angles, as shown in Fig. 6.10. Cascading the two $z$ transforms yields the pole-zero plot of Fig. 6.11.

Figure 6.12 shows the magnitude of the frequency response.

**FIGURE 6.8** $z$ Transform corresponding to the network.

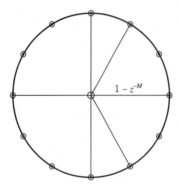

**FIGURE 6.9**   Zeros of $H_1(z)$, Eq. 6.28.

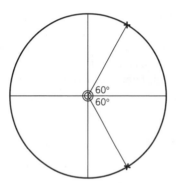

**FIGURE 6.10**   Pole-zero plot of $H_2(z)$, Eq. 6.28.

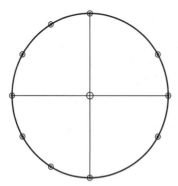

**FIGURE 6.11**   Pole-zero plot of the product.

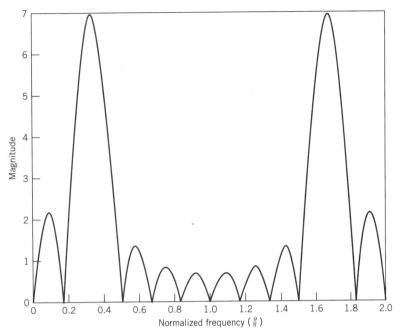

**FIGURE 6.12**  Magnitude of $H_1 H_2$ versus $\theta$ on a unit circle for $M = 12$, $H_1 = 1 - z^{-M}$.

## 6.9  CONCLUDING COMMENTS

Most new systems for the processing of speech and music are now digital, and as such are based on the fundamental mathematical tools briefly reviewed in this chapter. Filters typically perform linear convolutions, and filter responses are generally specified in terms of their $z$ transforms, which have their time-domain equivalence in terms of difference equations. First- and second-order systems form the basis of much discussion about linear discrete-time systems, and understanding the basics of the effect of transforming a continuous time signal into a sequence of numbers (sampling) is fundamental to this work.

For further reading, there is a wide range of reference texts on the subject of DSP, including [4], [1], [7], [5], and [6]. Of particular historic interest is Chapter 5, by W. Hurewicz, in [2]. Reference [3] describes the fundamental mathematics behind the $z$-transform.

## 6.10  EXERCISES

**6.1**   Prove that if $y(n)$ is the convolution of two sequences $x(n)$ and $h(n)$, then the $z$ transform $Y(z)$ is the product of the $z$ transforms $X(z)$ and $H(z)$.

**6.2**   Prove that the $z$ transform of a delayed sequence $x(n - M)$ is $z^{-M} X(z)$, where $X(z)$ is the $z$ transform of the undelayed sequence. What conditions are needed to make the proof correct?

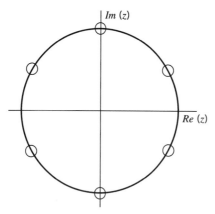

**FIGURE 6.13**   z-Plane picture of the digital filter with transfer function $H(z) = 1 + z^{-M}$, where $M = 6$. The six zeros shown are the roots of $H(z) = 0$. Note that they are three pairs of complex conjugate values. In principle there are also six poles at $z = 0$, not shown here since they have no effect on the frequency response of the filter.

**6.3**   Let $W(z) = 2X(z) + 3z^{-3}Y(z)$. Find the inverse $z$ transform of $W(z)$ by inspection of the right-hand side, that is, without using contour integrals.

**6.4**   Let $H(z) = Y(z)/X(z) = (1 - z^{-1})/(1 - 0.5z^{-1})$. Use algebra to generate an expression with $Y(z)$ on the left-hand side and then use inspection to generate a corresponding difference equation.

**6.5**   Figure 6.13 is a $z$ transform illustration of a digital filter. Choose $\theta$ in $H_2(z)$ to cancel the lowest frequency pair of $H_1(z)$. Plot the magnitude of the resulting frequency response of $H_1(z)H_2(z)$ on the unit circle.

**6.6**   Try to prove the famous Cauchy integral equation (Eq. 6.4). If you have trouble, find a book on functions of a complex variable, study the proof, and write it up.

**6.7**   Let $x(n) = K^n$, $n = 0, 1, 2, 3, \ldots$. Find the $z$ transform of $x(n)$. Given $X(z)$, use the inversion theorem Eq. 6.5 to compute $x(n)$.

**6.8**   What is the solution to the difference equation $y(n) = Ky(n-1)$ with initial condition $y(0) = 1$? Next, find the region of convergence of the $z$ transform of your solution $y(n)$. $K \leq 1$.

**6.9**   A finite sequence $h(n)$ is shown in Fig. 6.14.

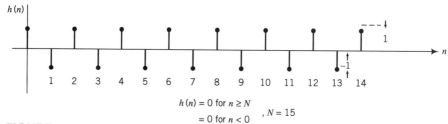

$h(n) = 0$ for $n \geq N$
$= 0$ for $n < 0$     , $N = 15$

**FIGURE 6.14**   Finite sequence.

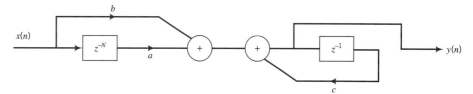

**FIGURE 6.15**  Digital filter structure.

(a) Find $H(e^{j\omega})$, and assuming that $H(e^{j\omega}) = e^{-0.5j\omega(N-1)}R(\omega)$, sketch $R(\omega)$ in the interval $0 \le \omega \le 2\pi$.

(b) This impulse response can be implemented by the structure shown in Fig. 6.15. Determine $a$, $b$, and $c$ for $N$ odd and for $N$ even.

A Hanning window $w(n)$ is applied to $h(n)$ to form the product $f(n)$, where $w(n) = 1/2 - 1/2\cos(2\pi n/N)$.

(c) Sketch the new function $f(n)$ versus $n$.

(d) Show how to implement a filter with impulse response $f(n)$ by the frequency-sampling structure shown in part (b).

**6.10**  Figure 6.16 shows the pole patterns of three $z$ transforms, $H_1(z)$, $H_2(z)$, and $H_3(z)$. Determine the conditions for which the corresponding sequences $h_1(n)$, $h_2(n)$, and $h_3(n)$ can be stable.

**6.11**  Find the magnitude of $H(z)$ on the unit circle at $\omega = 0, \pi/4, \pi/8$, and $7\pi/8$ for the network described by the difference equation $y(n) - by(n-1) = x(n) - ax(n-1)$ with initial conditions $x(n) = 0$ and $y(n) = 0$, for $a = 2$ and $b = 1/2$.

**6.12**  Find the linear convolution of the two sequences $x(n)$ and $h(n)$ shown in Fig. 6.17.

**6.13**  Consider the equation

$$y(n) = y(n-1) + x(n) - x(n-8),\tag{6.30}$$

with $y(-1) = 0$ and $x(n) = 0$ for $n \le -1$.

(a) Obtain an explicit solution for $H(z)$, the network transfer function.

(b) Draw the digital network.

(c) Does this network have all poles, all zeros, or both poles and zeros?

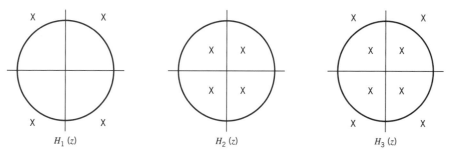

**FIGURE 6.16**  Pole patterns for three different $z$ transforms.

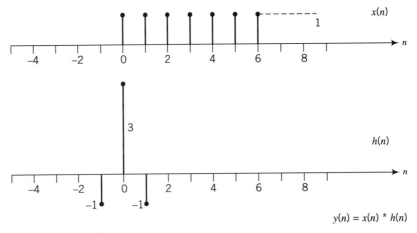

**FIGURE 6.17**  Two sequences.

**(d)** Show the positions of the network zeros, if any, in the complex $z$ plane.

**(e)** Is the network stable?

**(f)** Sketch $y(n)$ versus $n$ if $x(n)$ is a unit sample.

# BIBLIOGRAPHY

1. Gold, B., and Rader, C. M., *Digital Processing of Signals,* McGraw–Hill, New York, 1969.
2. James, H. M., Nichols, N. B., and Phillips, R. S., *Theory of Servomechanisms*, McGraw–Hill, New York, 1947.
3. Knopp, K., *Theory of Functions I*, Dover Publications, New York, 1945.
4. Kuo, F. F. and Kaiser, J. F., *System Analysis by Digital Computer*, John Wiley, New York, 1966.
5. Oppenheim, A., and Schafer, R., *Discrete-Time Signal Processing,* Prentice–Hall, Englewood Cliffs, N.J., 1989.
6. Oppenheim, A., and Schafer, R., *Digital Signal Processing,* Prentice–Hall, Englewood Cliffs, N.J., 1975.
7. Rabiner, L., and Gold, B., *Theory and Application of Digital Signal Processing,* Prentice-Hall, Englewood Cliffs, N.J., 1975.
8. Smith, J. O., "Physical Modeling Using Digital Waveguides," *Computer Music Journal*, Part I, Vol. 16, No. 4, pp. 74–91, Winter 1992.

# CHAPTER 7

# DIGITAL FILTERS AND DISCRETE FOURIER TRANSFORM

## 7.1 INTRODUCTION

If we define a filter as a device that discriminates among incoming frequencies, then filters can be implemented, for example, by mechanical, acoustical, pneumatic, or electrical elements. In the past few decades, discrete-time algorithms employing computational elements have played an increasing role in filter applications. Most commonly, these elements also incorporate discrete numerical values (as opposed to continuous circuit variables[1]) to represent signals. For simplicity we refer to these devices as digital filters. In this chapter, we deal with some of the issues that arise in the design of digital filters. Many of the ideas of digital signal processing pertain directly to digital filter design. However, in addition, many of the filtering concepts developed for the design of analog electrical filters [11] apply equally well to digital filter design. Section 7.2 is a review of these concepts. In Section 7.3 we show how simple mathematical transformations bridge the gap between the analog and digital world. The remainder of this chapter focuses on methods that have been primarily applied to discrete-time linear systems. Of particular interest is the application of the discrete Fourier transform (DFT) to filter theory and applications. Both filtering and the DFT are widely used for speech-processing applications.

The traditional analog filter design consists of two major portions; the approximation problem and the synthesis problem. Approximation refers to the development of an analytic expression that satisfies the constraints of the design problem (e.g., a ratio of polynomials, such that the frequency response is low pass with a 3-dB point at 4 kHz). Synthesis refers to the implementation of the filter, for instance as a cascade of second-order sections. In most books on analog filter design methods, the synthesis problem is the primary focus and occupies the main part of the book. This is because the analog circuits (consisting of resistors, inductors, capacitors, and operational amplifiers) have components with numerical values that are often too difficult or costly to specify with great accuracy. Thus, the trick in analog design is to create structures that are the least sensitive to component errors.

Digital filter design also requires both approximation and synthesis, but the situation is somewhat reversed; the approximation problem, how to find filter parameters to satisfy a given design criterion, predominates. Good design technique in the analog world is very important because the designer knows that his components are never perfect, so the search for a synthesis method that is least vulnerable to such imperfections is an integral part of the

---

[1]There is a class of devices in which analog values are sampled in time; an example of such a device is a switched capacitor filter.

design mystique. In the early days of digital filter design, this accuracy problem could also be critical because of the possible bad effects of the finite word lengths (in some cases only 12 bits or even fewer were available). Nowadays, the availability of 32-bit word lengths and floating point capabilities at a greatly reduced cost means that the designer of digital filters can choose highly accurate components (although low cost, high-volume components still often use shorter word widths).

A filter function has relevance in both the time and frequency domains; complete specification in the frequency domain completely specifies behavior in the time domain, and vice versa. The Fourier transform of the frequency function is precisely the time response of the filter to an impulse (a delta function) applied at time zero.

Synthesis of digital filters can be divided into FIR (finite-impulse response) and IIR (infinite-impulse response) methods. A *nonrecursive* filter has an output that is a function of the input samples and is not a function of previous output samples. Such filters have only zeros in the complex $z$ plane and are always FIR; that is, the response to a unit sample (a sequence with a value of one at time zero and zero for all other times) is zero after a finite number of samples. A *recursive* filter has an output that is a function of the input samples *and* previous output samples. Such a filter has, in general, both poles and zeros in the complex $z$ plane, and in general has an impulse response (unit sample response) that does not diminish to zero after any finite number of samples[2] – for example, a response such as $h(n) = 0.5^n$.

Our discussion will necessarily be very brief; our aim is to point the reader at the large body of knowledge that has been developed over most of this century. Thus, some of the specifics may be omitted or an appropriate reference will be mentioned.

## 7.2  FILTERING CONCEPTS

The most common application of a filter is to permit a given band of frequencies to pass through relatively undisturbed while all other frequencies are severely suppressed. However, many other applications exist. In this section we describe some of the desired characteristics of several well-known filters.

Most commonly, filters are designed to be low pass (passing frequencies below some cutoff point), high pass, bandpass, or band reject. To determine the functional relation between the filter and frequency, we introduce the filter transfer function, $H(\omega)$. In classical analog filter design, $H$ is a complex function of the complex variable $s$. (The radian frequency $\omega = 2\pi f$ is a real variable; it turns out to be very useful to define the complex variable $s$ with the stipulation $s = \alpha + j\omega$, so that the value of the radian frequency is simply the distance along the vertical axis in the complex $s$ plane.) The squared magnitude of $H$ must be real, so it is convenient to establish our design criterion in terms of this

---

[2]There are cases in which a recursive structure can lead to a FIR result. An important example of such a case is the frequency-sampling design, discussed in Section 7.3. Note also that IIR filters will typically have a response that ultimately has a low enough amplitude to be represented as zero in a discrete number system such as that used in computers.

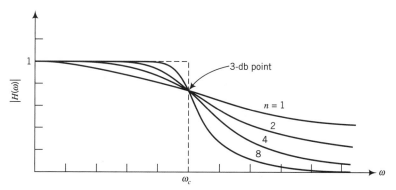

**FIGURE 7.1** Butterworth frequency response for different $n$. From [2].

squared-magnitude function. The classic solutions to this problem consist of a number of design families, most notably the Butterworth, Chebyshev, elliptic, and Bessel filters. All of these are IIR designs.

We first study the simple case of a low-pass Butterworth filter that satisfies the equation

$$|H(j\omega)|^2 = \frac{1}{1 + (\omega/\omega_c)^{2n}}. \tag{7.1}$$

Several facts can be deduced from Eq. 7.1. First, the squared-magnitude function is unity when $\omega = 0$ and approaches zero as $\omega$ approaches infinity. For all values of the parameter $n$, the function is always one-half at $\omega = \omega_c$, but for larger values of $n$, the transition is sharper, as seen in Fig. 7.1.

Figure 7.1 plots the magnitude, not the squared magnitude, of Eq. 7.1. Thus, at the cutoff frequency $\omega_c$, the function $|H|$ is 0.707, or, in logarithmic terms, $-3$ dB.

It may be desirable, in many cases, to have a sharper transition region. An alternative to a Butterworth filter is the Chebyshev filter.[3] For a given filter complexity the Chebyshev filter can fulfill this role.

The Chebyshev filter squared-magnitude function is specified by

$$|T(j\omega)|^2 = \frac{1}{1 + \epsilon^2 V_n^2(\omega/\omega_c)}, \tag{7.2}$$

where $V_n(x)$ is a Chebyshev polynomial of order $n$ that can be generated by the recursion formula

$$V_{n+1}(x) = 2x V_n(x) - V_{n-1}(x), \tag{7.3}$$

with

$$V_0 = 1, \qquad V_1 = x. \tag{7.4}$$

---

[3]Sometimes this name is spelled Tschebyscheff, corresponding to an alternate transliteration from the Cyrillic. Because of this spelling, the transfer function is often given as $T(s)$.

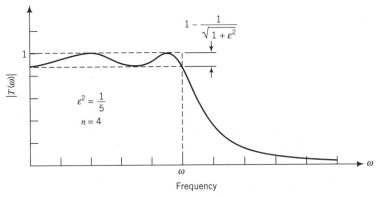

**FIGURE 7.2**   Chebyshev frequency response for $n = 4$. From [2].

The Chebyshev polynomial has the property of equal ripple over a given range of $x$, as seen in Fig. 7.2 for the case $n = 4$.

The parameter $n$, for either the Butterworth or Chebyshev filter, determines the complexity of the function. For a given value of $n$, therefore, the relative merits of the two filter types can be compared. It can be shown that, for a given complexity, the Chebyshev function has a sharper transition from passband to stop band. Comparisons are a bit tricky, since the Butterworth function is monotonic whereas the Chebyshev function oscillates.

Jacobian elliptic functions have the property of equiripple in both the passband and stop band with an even sharper transition. This function is shown in Fig. 7.3.

Thus, for a given complexity, the elliptic filter frequency response most closely resembles a rectangle (for the three filters described) whereas the Butterworth is the "worst" of the three in this respect. However, the magnitude response of a given filter is not the only criterion for comparison. In many applications, the phase response may be equally or even more important. For now it is sufficient to say that the phase response of elliptic and Chebyshev filters and the associated transient responses make these filters unsuitable,

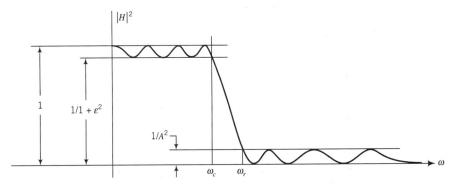

**FIGURE 7.3**   Jacobian elliptic frequency response. From [2].

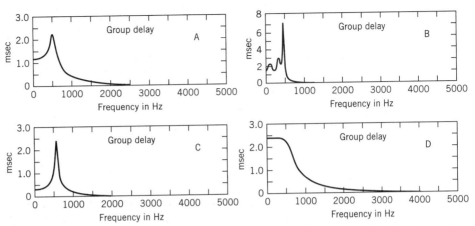

**FIGURE 7.4** Comparison of the group delay for four filter types: A = Butterworth, B = Chebyshev, C = elliptic, D = Bessel. From [2].

for example, to the design of bandpass banks for most audio applications. Of the three, the Butterworth filter is most often the designer's choice.

If phase response is overwhelmingly important, a design that focuses on the phase response might outdo all three of the above designs. Indeed, such filters have been designed and built. One design that results in a good transient response is the Bessel filter, so called because it is proportional to the inverse of the Bessel polynomial, defined by the recursion [12]

$$B_n(\omega) = (2n - 1)B_{n-1}(\omega) + \omega^2 B_{n-2}(\omega), \tag{7.5}$$

with $B_0(\omega) = 1$ and $B_1(\omega) = \omega + 1$.

A measure of the goodness of the transient response is the graph of the group delay versus the frequency of the filter function, defined as the derivative of the phase with respect to the radian frequency. Figure 7.4 shows the group-delay graphs for the four filters we have discussed.

We see from this figure that, except for the Bessel design, the group delay peaks strongly near the transition regions. This means that frequencies in that region have a markedly different delay for lower or higher frequencies. In one experiment [9], channel vocoders were simulated with each of the four types. It was found that the Chebyshev and elliptic designs caused an unacceptably high reverberation of the vocoded speech signal.

Lerner [6] showed that both a high degree of phase linearity and reasonably selective passbands could be attained. The specific function is given by Eq. 7.6:

$$L(s) = \sum_{i=1}^{m} \frac{D_i(s + a)}{(s + a)^2 + (d_i)^2}, \tag{7.6}$$

where $D_1 = 1/2$, $D_m = [(-1)^{m+1}/2]$, $D_i = (-1)^{i+1}$, and $d_i$ is the resonant frequency of the $i$th pole.

The low-pass designs of the previous section can be transformed to produce high pass, bandpass, and band stop filters. For example, we might want to transform a low-pass design that passed frequencies from 0 to 1 Hz to a bandpass filter with $\Omega_l$ as the lower cutoff frequency and $\Omega_u$ as the upper cutoff. To accomplish this we need to map the variable $s$ of the original function into the following function of $s$, $\Omega_l$, and $\Omega_u$.[4]

$$s \to \frac{s(\Omega_u - \Omega_l)}{s^2 + \Omega_u \Omega_l} \tag{7.7}$$

An analog continuous-time filter design essentially always resulted in IIR filters, given the components that were available. Discrete-time implementation has expanded the design possibilities for FIR filters. FIR filters can be designed with a perfectly linear phase by constraining the coefficients to be symmetric, since the coefficients are in one-to-one correspondence with the impulse response.

## 7.3  USEFUL FILTER FUNCTIONS

The previous chapter focused on filters that passed a band of given frequencies and attenuated other frequencies; our aim was to remove unwanted segments of the spectrum. More generally, filters can be designed to shape the spectrum. For example, filters can be designed to produce an output that approximates the derivative (in frequency space) of the input. Another design creates the Hilbert transform of the input. Still another design does not affect the amplitudes of the incoming frequencies but alters the phase, perhaps to compensate for the undesirable phase response from another part of the overall system.

Several techniques exist for designing filters with arbitrary bandpass properties. One powerful method that employs linear programming and the Remez exchange algorithm [7] is a computer-based method that leads to an optimum design in the sense that both passband and stop band are equiripple, and that the transition from passband to stop band is most rapid. We have seen that elliptic filters also have this optimum property, but this Remez method is based on a pure FIR design so that linear phase as well as good magnitude can coexist.

Frequency-sampling filters permit an arbitrary transfer function magnitude to be realized with a recursive technique that leads to a FIR design. The concept is based on the frequency-domain version of the Shannon sampling theorem for time-domain signals. If we consider the continuous function $f(t)$ and sample this function at equally spaced time intervals to obtain the samples $f(nT)$, the theorem states that the function $g(t)$ can approximate $f(t)$ as follows:

$$g(t) = \sum_{n=-\infty}^{\infty} f(nT) \frac{\sin[(2\pi/T)(t - nT)]}{(2\pi/T)(t - nT)}. \tag{7.8}$$

If the spectrum of the signal $f(t)$ is band limited, $g(t)$ is *exactly* $f(t)$, provided that the sampling interval is no greater than the inverse of twice the bandwidth. Even if the band of $f(t)$ is not limited, $g(t)$ is exactly equal to $f(t)$ at the sampling instants.

---

[4]To visualize how this transformation could have such an effect, begin with a simple low-pass function such as $1/1 + s$.

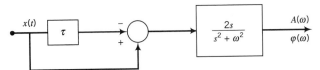

**FIGURE 7.5** Sampling network in the frequency domain.

Does the sampling theorem serve as a useful basis for filter design? The following two references give an affirmative reply: [2] and [8]. This brief passage summarizes the results.

Figure 7.5 shows how to implement an analog bandpass filter with the response (in the frequency domain) that is analogous to that of Eq. 7.8.

The transfer function for Fig. 7.5 is

$$H(s) = \frac{(1 - e^{-s\tau})2s}{s^2 + \omega_0^2}. \tag{7.9}$$

Figure 7.6 shows the pole-zero configuration of Eq. 7.9 in the complex $s$ plane. At $s = j(\omega_0)$ and $s = -j(\omega_0)$, the poles and zeros cancel. Thus $H(s)$ is like a FIR filter with only zeros. The magnitude $A(j\omega)$ and phase $\phi(\omega)$ are shown in Fig. 7.7.

The analytic expression for the magnitude along the $j\omega$ axis is

$$A(j\omega) = \left| \frac{2j\omega(1 - e^{-j\omega\tau})}{\omega_0^2 - \omega^2} \right|,$$

$$A(j\omega) = 4 \left| \frac{\omega \sin(\omega\tau/2)}{(\omega_0 - \omega)(\omega_0 + \omega)} \right|. \tag{7.10}$$

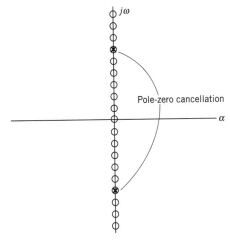

**FIGURE 7.6** Pole-zero configuration for the analog sampling network in the frequency domain.

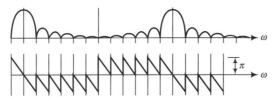

**FIGURE 7.7**   Magnitude and phase of the analog sampling network.

Equation 7.10 has the same desirable properties as a function of frequency that Eq. 7.8 has in the time domain. By adding the outputs of resonators with weights corresponding to the *samples* of a specified frequency response, we can synthesize a filter function of any desired shape.

## 7.4   TRANSFORMATIONS FOR DIGITAL FILTER DESIGN

Sections 7.2 and 7.3 emphasized useful filtering functions. How can these functions be synthesized as discrete-time filters? An approximation criterion is needed, since a discrete-time implementation can never be identical to the continuous-time filter functions. The work of Hurewicz [5] provides a technique for doing this in an impulse invariant way. By this we mean that the (discrete-time) response to a unit pulse of the derived digital filter will be samples of the impulse response of the associated analog filter. Let's begin with a simple single-pole analog function. The impulse response of this simple system is the inverse Laplace transform of this function.

$$k(t) = L^{-1}\left(\frac{A_i}{s + s_i}\right) = A_i e^{-st}. \tag{7.11}$$

If we sample $k(t)$ at equally spaced intervals $nT$ and take the $z$ transform of the resulting sequence, we obtain

$$H(z) = \frac{A_i}{1 - e^{-s_i T} z^{-1}}. \tag{7.12}$$

Thus, $H(z)$ is the impulse invariant $z$ transform corresponding to the simple single-pole system of Eq. 7.11. For more complex systems in which the $s$-plane poles are single (not multiple) poles and in which the impulse response is a real function, this leads to the correspondence

$$L(s) \rightarrow H(z) = \sum_{i=1}^{m} \frac{D_i[1 - e^{-aT}\cos(d_i T)z^{-1}]}{1 - 2e^{-aT}\cos(d_i T)z^{-1} + e^{-2aT}z^{-2}}, \tag{7.13}$$

where $L(s)$ is the function of Eq. 7.6. Thus, we have specified a digital Lerner filter in terms of the components of an analog Lerner filter.

## 7.5 DIGITAL FILTER DESIGN WITH BILINEAR TRANSFORMATION

The frequency response of an impulse invariant digital filter is an aliased version of the continuous frequency response from which it was derived. We need to examine a given design carefully to determine if this aliasing is detrimental.

Aliasing is best understood by studying Fig. 7.8. In (a) the analog response curve is almost totally contained within the Nyquist frequency range. Thus, the discrete-time version, (b), which must be periodic, is nearly identical to the analog response in that region. However, in (c) the analog response has a greater than Nyquist range and the resultant discrete-time response, (d), is significantly different than the desired analog response.

A wideband filter is more likely to cause problems when digitized by impulse invariance. An alternate approach depends on the bilinear transformation. The original $s$ plane is mapped into the $z$ plane with the $j\omega$ axis mapping into the unit circle. This mapping is

$$s \rightarrow \frac{z-1}{z+1}, \tag{7.14}$$

which leads to the mapping of $z = 1$ to $s = 0$, $z = e^{j(\pi/2)}$ to $s = j$, and $z = e^{j\pi}$ to $s = \infty$. The analog frequency $\omega_A$ maps into the digital frequency $\omega_D$.

$$\omega_A \rightarrow \tan\left(\frac{\omega_D T}{2}\right). \tag{7.15}$$

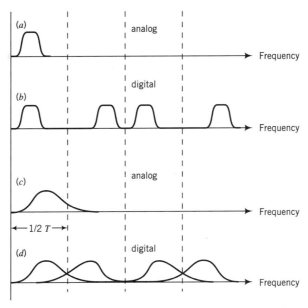

**FIGURE 7.8** Effect of aliasing. Note that the total response is the sum of the curves shown.

## 7.6   THE DISCRETE FOURIER TRANSFORM

Through the early 1960s both the Fourier transform and its mathematical cousin, the Laplace transform, were widely used as theoretical aids in signal processing, but they were not employed much in actual computation. The hardware needed to implement these functions was quite expensive and quite "klugey," but soon after it was shown how to do filtering by computer, Cooley and Tukey [1] introduced the fast Fourier transform (FFT) and sparked an explosion of theoretical and practical results in many fields. Such activity, in parallel with great advances in computers, meant that signal-processing algorithms could be implemented orders of magnitude faster with hardware orders of magnitude smaller than what was used three decades ago. Heideman et al. [4] showed that the *concept* of the FFT was understood by Gauss as well as by Good [3], but until the Cooley–Tukey paper and its digestion by signal-processing folk, no great new works resulted.

In this section we introduce the DFT and work exclusively with this digital version, trusting the interested reader to seek the many fine books and papers on the more general subject of transforms. In this section we outline some of the theoretical issues and in Section 7.7 we focus on the FFT.

The DFT of a finite duration sequence $x(n)$, $0 \leq n \leq N - 1$, is defined as

$$X(k) = \sum_{n=0}^{N-1} x(n) W^{nk}, \tag{7.16}$$

with $W = e^{-j(2\pi/N)}$. Comparing this definition with the definition (Eq. 6.2) of the $z$ transform, we can think of the DFT as a finite sum version of the $z$ transform, evaluated on the unit circle. The inverse DFT can be computed; that is, given the sequence $X(k)$, $0 \leq k \leq N - 1$, we obtain

$$x(n) = \frac{1}{N} \sum_{n=0}^{N-1} X(k) W^{-nk}. \tag{7.17}$$

The DFT computation yields the spectrum of a finite sequence; hence its great importance in signal-processing applications. Spectrum analysis can, of course, be implemented by means of a bank of analog or digital filters. There are many variations on the specific nature of spectrum analysis; these can conveniently be related to filter-bank parameters, such as the number of filters, the frequency range of the filter bank, and the amplitude and phase responses of the individual filters. In many cases, the use of a DFT can emulate (or approximately emulate) the properties of a given filter bank, and, through the FFT, perform the computation more efficiently. In Section 7.7 we illustrate how the FFT gets its speed, but it is useful to show the speed savings that can be obtained. Here we state without proof (for the present) the computational properties of a FFT.

If $N$ is a power of 2 then the DFT can be computed in $N \log_2 N$ operations. This contrasts with the "brute force" DFT computation, which takes $N^2$ operations. Thus, for example, if $N = 1024$, $\log_2 N = 10$, the savings is a factor of 100 (ignoring the details). The computational cost of a brute force DFT is in the same ball park as that of a filter-bank spectral analysis.

**TABLE 7.1  Relations between a Sequence and Its DFT[a]**

| X(n) | ⇔ | X(k) |
|------|---|------|
| even | ⇔ | even |
| odd | ⇔ | odd |
| even and real | ⇔ | even and real |
| odd and real | ⇔ | odd and imaginary |
| real | ⇔ | real part even / imaginary part odd |
| imaginary | ⇔ | real part odd / imaginary part even |
| even and imaginary | ⇔ | even and imaginary |
| odd and imaginary | ⇔ | odd and real |

[a]Recall that an odd sequence is antisymmetric and an even sequence is symmetric.

The DFT can be taken, in general, of a sequence of complex numbers. Table 7.1 relates the initial sequence $x(n)$ with its DFT $X(k)$.

In measuring the spectrum of a signal, we want to know the frequency range to be covered, the number of frequencies measured, and the resolving power of the algorithm. In general, a real-life signal such as speech or music, a radar or sonar signal, or some biological signal (e.g., an electrocardiogram) is of sufficient duration that we can say that it goes on forever. Furthermore, it usually changes character as time progresses. The researcher wanting to perform spectral analysis by DFT must decide, based on her or his insights about the signal, on the duration of the DFT, on the time between successive applications of the DFT, and on the sampling rate of the discretized signal (assuming that the data were obtained by sampling an analog signal). Additionally, the researcher must choose a window that is appropriate for the problem. For example, the parameters of a speech-production model can change over a time span of 10–20 ms. Thus, an appropriate window size could be 20 ms, with a DFT update every 10 ms. If $L$ is the assumed time duration of the window and $M$ is the number of samples in the window, then $L = MT$, where $T$ is the sampling interval.

The choice of the DFT size $N$ is dictated by the desired spectral resolution and in general can be greater than $M$. This is easily implemented by augmenting with zeros. Therefore, if it is desired that the frequency spacing between adjacent DFT samples is $\delta F$, the sampling rate $R = (\delta F)N$. The frequency range covered by the DFT is therefore $R$. However, for an input that is real (the imaginary component is zero), the DFT is symmetric about $R/2$ so that the nonredundant DFT result yields a frequency range of $R/2$.

In Chapter 6 we showed that the sampling of an analog signal in the time domain resulted in a periodic function in the frequency domain. In the DFT both time and frequency

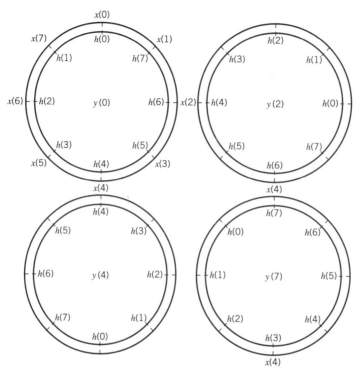

**FIGURE 7.9**  Circular convolution of two eight-point sequences. Only $y(0)$, $y(2)$, $y(4)$, and $y(7)$ are shown. All outer circles carry the same sequence as the upper-left circle.

are samples; thus, both $x(n)$ and $X(k)$ are periodic. Both Eqs. 7.16 and 7.17 are periodic with a period $N$. Consequently, if we begin with a perfectly finite-length sequence of length $N$ and compute its DFT, the inverse DFT will be periodic in $N$. As a result, the product of two DFT's results in the circular convolution of the time sequences. This is readily visualized in Fig 7.9.

For each one sample rotation of the outer circle, the convolution is the sum of all products of samples facing each other. Thus, for example, term zero of the convolution is given by

$$y(0) = x_0 h_0 + x_1 h_7 + x_2 h_6 + x_3 h_5 + x_4 h_4 + x_5 h_3 + x_6 h_2 + x_7 h_1, \tag{7.18}$$

whereas term two is

$$y(2) = x_0 h_2 + x_1 h_1 + x_2 h_0 + x_3 h_7 + x_4 h_6 + x_5 h_5 + x_6 h_4 + x_7 h_3. \tag{7.19}$$

The DFT can implement the linear convolution required for FIR filtering operations by judicious augmentation of two finite-length sequences with the proper number of zeros [10]. An example is shown in Fig. 7.10.

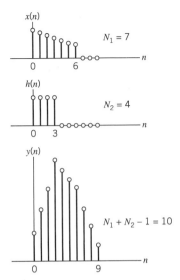

**FIGURE 7.10** Linear convolution of two finite-length sequences by DFT.

Given two sequences of length $N_1$ and $N_2$, augment each sequence with zeros so that the two augmented sequences have length $N_1 + N_2 - 1$. Then, by multiplying the DFTs of the two sequences and taking the inverse DFT of this product, one obtains the linear convolution $y(n)$. This technique is useful, for example, when one wishes to implement a FIR filter with many terms in its impulse response.

To perform filtering by DFT for a very long input waveform, one must perform sectioned convolution; the technique is described in [10] and in many other books and papers.

## 7.7 FAST FOURIER TRANSFORM METHODS

There are a great variety of FFT algorithms. They can all be derived from successive applications of a single operation, by representing a one-dimensional string of numbers as a two-dimensional array. If we have an $N$-point sequence, the integer $N$ is either a prime or a composite number. If $N$ is prime, it cannot be expressed as a product of smaller integers. If $N$ is composite, it can be expressed as the product $N_1 N_2$. If either or both $N_1$ and $N_2$ are composite, further reduction is permissible. For example, we can express the number 60 as $12 \times 5$ or $3 \times 4 \times 5$ or $2 \times 2 \times 5 \times 3$, and so on. The term radix is commonly used to describe this decomposition. If $N$ can be expressed as a product of the same integer $r$, the FFT algorithm is called a radix $r$ algorithm. The term mixed radix means that all factors of $N$ are not identical.

The computational advantage of the reductions just described comes directly from the fact that a two-dimensional DFT is more efficient than a one-dimensional DFT with the same number of input samples. This stems from the observation that a two-dimensional DFT can be implemented, for example, by performing one-dimensional DFTs on all columns to obtain a new matrix and then performing one-dimensional DFTs on all rows of the new matrix. The total computation time for $N_1$ columns by $N_2$ rows is of the order $(N_1)^2 + (N_2)^2$, and that may be appreciably smaller than the computation time $(N_1 + N_2)^2$. Be aware that the advantages of this kind of reduction are true for very special cases, such as the DFT algorithm.

Let's now derive the mathematics of this DFT trick. Let the one-dimensional index be $n$, as usual. Letting the column index be $m$ and the row index be $l$ ($M$ is the number of columns and $L$ is the number of rows), we get

$$n = Ml + m. \tag{7.20}$$

We now perform the two-dimensional DFT and choose $r$ and $s$ as the transformed variables; these can be recomposed to yield the single variable

$$k = Lr + s. \tag{7.21}$$

We are now in a position to express the DFT samples $X(k) = X(s, r)$ as the transform of $x(n) = x(l, m)$ by simply substituting Eqs. 7.20 and 7.21 into the definition of the DFT, giving

$$X(k) = X(s, r) = \sum_{m=0}^{M-1} \sum_{l=0}^{L-1} x(l, m) W^{(Ml+m)(Lr+s)}. \tag{7.22}$$

Expanding $W^{(Ml+m)(Lr+s)}$, observing that $W^{MLlr} = W^{Nlr} = 1$, and properly associating indices with summation signs, we rearrange Eq. 7.22 as

$$X(s, r) = \sum_{m=0}^{M-1} W^{Lmr} W^{ms} \sum_{l=0}^{L-1} x(l, m) W^{Msl}. \tag{7.23}$$

Notice that the $L$-fold sum is the DFT of the $m$th column of the array having the kernel $W^M$. Thus the first step in our computational procedure would be as follows.

**1.** Compute the $L$-point DFT of each column. The result is a function of $s$ and $m$; call it $q(s, m)$. Equation 7.23 can now be written as

$$X(s, r) = \sum_{m=0}^{M-1} W^{Lmr} W^{ms} q(s, m). \tag{7.24}$$

**2.** Obtain a new array $h(s, m)$ by multiplying every $q(s, m)$ by its twiddle factor $W^{ms}$. Equation 7.24 now reduces to

$$X(s, r) = \sum_{m=0}^{M-1} h(s, m) W^{Lmr}. \tag{7.25}$$

Equation 7.25 is recognized to be the $M$-point DFT of each row, with the row index $s$. Thus the final step in the procedure is as follows.

**3.** Compute the DFT of each row of the $h(s, m)$ matrix, with $W^L$ as kernel.

Several results emerge from the procedure. If $N$ is the highly composite number $2^I$ where $I$ is an integer, the DFT may be decomposed into $I$ dimensions, each being a two-point DFT. In most applications such a restriction is usually no problem, since, as we mentioned previously, augmenting with zeros simply alters the numerical meaning of the frequencies corresponding to the $k$ values.

Another result of interest is that of row–column permutation. If the computation is performed in place, the ordering of the resulting $k$ values is, in general, different than the ordering of the incoming $n$ values. The advantage of an *in place* algorithm is that no subsidiary memory is needed; when the computation is done, the result is at the same location as when the computation started. However, the result is no longer in the same order; an example is shown in Fig. 7.11.

Each two-point DFT is represented as a nodal point in the figure and the twiddle factors are shown as arrows. Each time a two-point DFT is computed, the result is stored

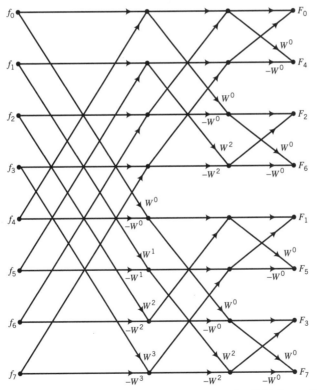

**FIGURE 7.11**  Flow chart of an eight-point FFT. From [2].

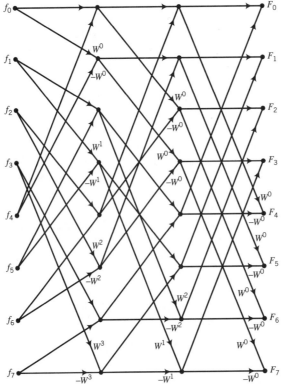

**FIGURE 7.12**   Flow chart of an eight-point FFT; no bit reversal, not in place. From [2].

back into the same memory elements. Finally, as seen in the figure, the indices of the result are bit reversed, so that, for example, register number 3, which carries the input $f_3$ (011), will wind up holding the output sample $X_6$ (110).

Bit reversal may be avoided in a radix 2 FFT by using extra memory to hold intermediate results; an example is shown in Fig. 7.12.

## 7.8   RELATION BETWEEN THE DFT AND DIGITAL FILTERS

Since both DFTs and filter banks are capable of performing spectral analysis, it is fair to inquire what the mathematical relations are between these methods. Figure 7.13 shows a filter-bank implementation of a *sliding DFT*.

Assume that the system is initially at rest. Then, it is easy to see that the initial $N$ arrivals of the signal samples will produce the required DFT output samples. Now consider the $N$th sample. Because of the $N$-sample delay, the delayed zeroth sample cancels the original contribution of the zeroth sample, so that the outputs will now register the contributions from sample 1 through sample $N$. Thus, each set of outputs at every sample

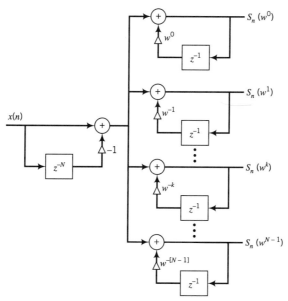

**FIGURE 7.13**  Filter-bank implementation of a sliding DFT. From [8].

point corresponds to the DFT of a sequence that has been slid over to the right by one sample.

For the filter-bank implementation of Fig. 7.13, the number of multiplications per point is $N$. Thus the filter-bank implementation seems more efficient than the sliding FFT. However, the FFT computation can be hopped. Some examples of hopping are shown in Fig. 7.14.

Notice that the hopped measurement is simply a sampling of the sliding measurement; an equivalent result is obtained by sampling the filter-bank outputs. The effects can be treated by standard aliasing arguments. Figure 7.15 shows examples of such effects.

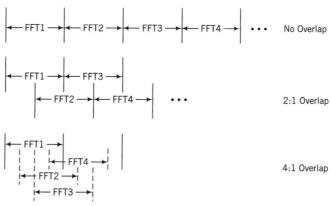

**FIGURE 7.14**  Three examples of hopped FFTs. From [8].

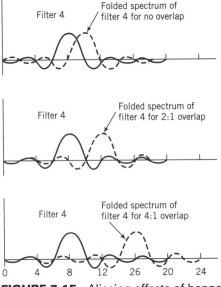

**FIGURE 7.15**   Aliasing effects of hopped FFTs. From [8].

No overlap results in severe aliasing; a 2:1 overlap significantly decreases aliasing distortion, and a 4:1 overlap (perhaps augmented with windowing) can reduce the effects to insignificance.

## 7.9  EXERCISES

**7.1**   Prove that $n = 4$ is an appropriate value for a digital Butterworth filter with a 3-dB attenuation at 1250 Hz, and more than a 20-dB attenuation at 2000 Hz. Assume a sampling rate of 10 kHz.

**7.2**   Suppose that you had the choice of a Butterworth or Chebyshev design of a low-pass filter. Discuss the criteria you might use to decide on a final design.

**7.3**   In a radix 2 FFT, prove that the number of complex multiplications is $N/2 \log_2 N$.

**7.4**   An analog signal is sampled at 10 kHz, and 23 samples are digitized and stored in computer memory. It is desired to obtain a spectrum by means of DFT that covers the range 0–5 kHz, with at a least 7-Hz resolution. Specify an algorithm to perform this task.

**7.5**   Prove that the configuration of Fig. 7.13 produces the same output as that of a sliding DFT. Which implementation (Fig. 7.13 or the sliding DFT) is more efficient?

**7.6**   Consider the sequence $x(0) = 1$, $x(1) = 2$, $x(2) = 3$, $x(3) = 4$, $x(4) = 5$, and $x(n) = 0$ for all other values of $n$.
   **(a)** Compute the sequence by linearly convolving $x(n)$ with itself.
   **(b)** Compute the sequence by circularly convolving $x(n)$ with itself.

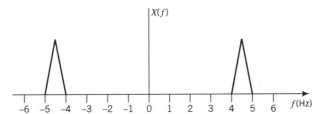

**FIGURE 7.16** Spectrum of an analog signal.

(c) Compute the circular convolution of a suitably modified version of $x(n)$ to obtain the exact result of part (a).

**7.7** Figure 7.16 shows the spectrum of an analog signal.
The signal $x(t)$ is sampled at equally spaced intervals $T = 0.5$s, thus creating the signal $x(nT)$.
(a) Draw the spectrum $X_d(f)$ of $x(nT)$.
(b) Is $x(t)$ recoverable from $x(nT)$? If so, how?

**7.8** In the analog domain, an ideal differentiator is defined by its transform $H(\omega) = j\omega$. In the discrete domain it is convenient to define a band-limited ideal differentiator as shown in Fig. 7.17.
(a) Find the unit sample response of the discrete domain differentiator.
(b) Sketch the result for $\omega_c = \pi$.
(c) Is the unit sample response causal?
(d) Is the unit sample response real?

**7.9** We are given a block of 512 samples and want to compute the DFT at 15 values of $k$. The samples were obtained by sampling an analog signal at 10,240 samples/s. The 15 values of interest uniformly cover the DFT output from $X(150)$ to $X(164)$.
    Assume that you have a choice of a radix 2 FFT or a straightforward (slow) DFT. Compare the running times of the two approaches.

**7.10** Let's take the DFT $X(k)$ of the finite sequence $x(n)$, for $N = 8$. Now, form a new sequence specified as $Y(k) = X(k)$ for even values of $k$ and zero for odd values of $k$.
(a) If $y(n)$ is the inverse DFT of $Y(k)$, find $y(n)$ for $n = 0, 1, 2, \ldots, 7$ in terms of the original sequence $x(n)$.
(b) If $x(n)$ has the values 0, 1, 2, 3, 3, 2, 1, 0, find the numerical values of $y(n)$ for $n = 0, 1, 2, 3, 4, 5, 6, 7$.

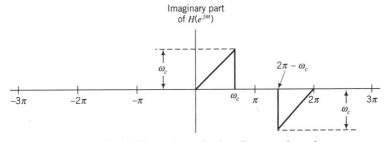

**FIGURE 7.17** Ideal differentiator in the discrete domain.

# BIBLIOGRAPHY

1. Cooley, J. W., and Tukey, J. W., "An algorithm for the machine computation of complex Fourier series," *Math. Comput.* **19**: 297–301, 1965.
2. Gold, B. and Rader, C. M., *Digital Processing of Signals*, McGraw–Hill, New York, 1969; republished by Kreiger, Fl. 1983.
3. Good, I. J., "The interaction algorithm and practical fourier series," *J. R. Statist. Soc. B* **20**: 361–372, 1960.
4. Heideman, M., Johnson, D. H., and Burrus, C. S., "Gauss and the history of the fast Fourier transform," *ASSP Mag.* 1(4): 14–21, 1984.
5. Hurewicz, W., "Filters and servo systems with Pulsed Data," in H. M. James, N. B. Nichols, and R. S. Phillips, eds., *Theory of Servomechanisms*, Chap. 5, Radiation Laboratory Series, McGraw–Hill, New York, 1947.
6. Lerner, R. M., "Band-pass filters with linear phase," *Proc. IEEE* **52**: 249–268, 1964.
7. Parks, T. W. and McClellan, J. H., "A program for the design of linear phase finite impulse response digital filters," *IEEE Trans. Audio Electroacoust.* **AU-20**: 195–199, 1972.
8. Rabiner, L. R. and Gold, B., *Theory and Application of Digital Signal Processing*, Prentice–Hall, Englewood Cliffs, N.J., 1975.
9. Rader, C. M., unpublished notes, 1964.
10. Stockham, T. J., Jr., "High speed convolution and correlation," *AFIPS Proc.* **28**: 229–233, 1966.
11. Storer, J. E., *Passive Network Synthesis*, McGraw–Hill, New York, 1957.
12. Weinberg, L., *Network Analysis and Synthesis*, McGraw–Hill, New York, 1962.

# CHAPTER 8

# *PATTERN CLASSIFICATION*

## 8.1  INTRODUCTION

Much as the discipline of digital signal processing forms the technological basis for the processing of audio signals, pattern recognition is the basic field of study that underlies application areas such as speech recognition. In this chapter and the one that follows, we present some major concepts that are essential to understanding pattern-recognition technology. We begin in this chapter with a brief survey of the basic methods of classifying simple patterns. We distinguish here between pattern recognition and pattern classification by restricting the latter to the distinction between categories for sets of observed data samples or vectors. For the more inclusive term of pattern recognition we can also mean the recognition of sequences (which may be multidimensional) that are not presegmented into the patterns to be classified. In such cases it may not even be known exactly which training examples are synchronous with the class labels. For instance, in handwriting recognition, we wish to recognize the sequence of written marks with a sequence of characters (or words), but we may not know exactly where one letter ends and another begins. Similarly, in speech recognition, we wish to associate the input speech patterns with a sequence of words. Each of these examples requires the association of an observation sequence with a labeling sequence. For now, we are going put aside the issue of temporal sequencing to concentrate on the simpler case of pattern recognition for static patterns.

A *supervised* pattern-classification system is trained with labeled examples; that is, each input pattern has a class label associated with it. Pattern classifiers can also be trained in an *unsupervised* fashion. For example, in a technique known as vector quantization, some representation of the input data is clustered by finding implicit groupings in the data. The resulting table of cluster centers is known as a codebook, which can be used to index new vectors by finding the cluster center that is closest to the new vector. This approach has several advantages for low-bit-rate coding (some of which will be discussed in later chapters), but it also can be useful for the reduction of computation in speech recognition. For the most part, the pattern classifiers referred to in this book will be trained in a supervised manner.

As noted earlier, the job of static pattern classification is to classify individuals into groups. For example, consider classifying humans with the features of height and weight. In Fig. 8.1 the circles could be Swedish basketball players, whereas the triangles might represent speech-recognition researchers. For the figure as shown, it would be very easy to devise a system to divide people into the two classes.

For the case of speech, Fig. 8.2 shows an extreme case of some vowels represented by their formant frequencies F1 and F2. The vowels represented are as pronounced in the words beet (/i/), bat (/ae/), bot (/a/), and boot (/u/). Notice that they fall into nice groupings. Unfortunately, speech data are not generally this well behaved. For instance,

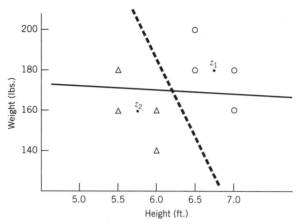

**FIGURE 8.1**    Distinct classes of Swedish basketball players (circles) and speech-recognition researchers (triangles). The points labeled $z_1$ and $z_2$ are the means for these two classes. The solid line represents a linear decision boundary derived from a minimum Euclidean distance, and the thick dashed line is a linear decision boundary that would perfectly separate classes for these data.

if more vowels were added to the chart, the classes would often intermingle in the F1–F2 space. Furthermore, there is enough variability in speech that the graphing of the F1–F2 values from a large spoken-language data base into this figure would leave little white space.

Note that in each case the individual examples were represented by some choice of variables. In the case of speech, for instance, we did not attempt to distinguish between

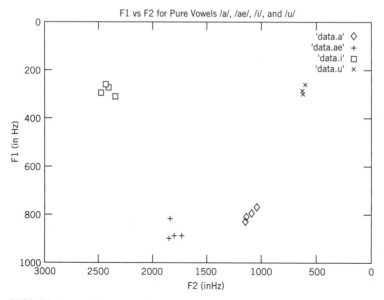

**FIGURE 8.2**    F1 and F2 for various vowels.

vowel classes by using individual waveform amplitudes as features, but rather by using variables that we believed to be relevant to classification. In general, the choice of input representation is one of the major issues in the study of pattern recognition.

## 8.2 FEATURE EXTRACTION

As shown in Fig. 8.3, the first step in any pattern-classification system is to evaluate some representation of the input pattern. Although in some cases this is the degenerate representation that consists of the raw input data, in general this can be improved with signal processing. In the case of speech, for instance, some form of power spectrum is often evaluated. Typically, these acoustic representations (vectors of relevant variables) are extracted within successive analysis windows that overlap. For example, windows of 20–30 ms overlapped by 10 ms are often used, although step sizes between frames can range from 5 to 20 ms. The features extracted are generally spectral or cepstral coefficients that condense the information in the speech signal to a vector of numbers, typically of length 5–40. For the classification of one-dimensional time series such as monaural music or speech, the input time series is transformed into a sequence of feature vectors that are sampled at a rate that is generally much lower than the original sequence (though each sample is represented by a vector rather than the original scalar).

What are the major considerations involved in the choice of a feature representation? Since the overall goal in pattern classification is to distinguish between examples of different classes, in general the goal of feature extraction is to reduce the variability for features for examples that are associated with the same class, while increasing the variability between features from examples that belong to different classes. In some cases this goal can be formulated as an explicit criterion that yields a procedure or even an analytical solution for optimal performance (such as the linear discriminant that will be briefly described later in this chapter). In general, however, we may not have good measures of the variability that will be seen during the classification process simply from looking at any given training data, and so it is desirable to normalize for irrelevant factors in the data wherever possible. For the case of speech recognition, a number of these examples will be given in a later chapter, but for now we note that such normalization schemes (for instance, to reduce the effects of overall energy or of some constant spectral factor) can be important for practical classification systems.

Despite these deep problems associated with normalizing for unseen conditions, it is still true that it is desirable to learn something about the capabilities of candidate feature sets for discrimination from the data. This should be the goal of feature extraction – in some sense, the pattern classification can be solved trivially if the features are good enough.

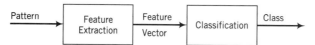

**FIGURE 8.3** Pattern classification is often divided into two major components – feature extraction and the classification of the features associated with a pattern.

In practice, for most interesting problems none of the components of the task are truly trivial.

Aside from this basic goal of feature extraction, another perspective is the determination of the best feature vector size. The best representations for discrimination on the training set have a large dimensionality. In fact, given enough dimensions, a group of patterns may be split up in any arbitrary way. However, the best representations for generalization to the test set are usually succinct. Too detailed a representation may also represent characteristics of the particular training set that will not be present for independent data sets. Thus, there is a trade-off between the training-set classification performance and generalization to the test set. For instance, in the extreme case, you could represent the training set by storing a million measurements for each example; this will permit the splitting of the training set into the labeled categories with a trivial classifier. However, if many of the dimensions are not reliable measures for distinguishing between examples of the classes in unseen data (as is almost certainly the case with so many measures), then classification on a test set will be difficult.

In general, it is desirable to find a good (and moderate-sized) set of generalizable features. There are several common methods of reducing the dimensionality of the features. Two direct analytical approaches are principal components analysis (PCA) and linear discriminant analysis (LDA). The object of PCA is to find a projection of the feature vector down to a lower dimension, such that the variance accounted for is maximized. This is the most common approach used, and it is reasonable if you don't know anything else about the data. The limitation is that the components are chosen to maximize the variance accounted for, which can be a problem if most of the variance isn't important for discrimination, for example the variance near 0 Hz within speech signals. For pattern classification, it is often better to have components that maximally account for discriminability.

Linear discriminant analysis is an approach that tries to maximize variance between classes and minimize variance within classes, given the constraint of a linear transformation of the input features. Chapter 4 of Duda and Hart [2] gives a good description of this technique. This has been usefully applied to speech recognition by a number of researchers in the 1980s and 1990s [4], [3]. In general, discriminant approaches to dimensionality reduction have tended to give preferable results to using variance-based approaches such as PCA, though this is a deep and tricky issue. For instance, features chosen to maximally discriminate between samples with a high signal-to-noise ratio may be poor choices for test examples that are corrupted by unexpected additive noise.

One can also use application-specific knowledge to reduce the dimensionality of the feature vectors, or just test different reductions by evaluating a pattern recognizer trained on the reduced features on a test set. This is often helpful, although testing any significant number of possible feature subsets can often be quite computationally intensive.

## 8.2.1   Some Opinions

Although this is a debatable point, it can be argued that it is often better to throw away bad features (or bad data) than to reduce their weight. This is true even though we can generally establish an objective criterion for error and adjust weights to minimize this criterion. However, for many applications the biggest difficulty of pattern classification can be that the training patterns are not completely representative of unseen test data, and so in practice it can be better to simply ignore dimensions or data that we believe to be unreliable.

How can one determine what data or features are bad? Often the best way to make this determination is through the analysis of errors in the pattern recognizer. We often can simplify this process by having some idea of where to look for problems. For instance, in ASR, energy can be an unreliable feature for practical usage (since talker energy can vary significantly for the same word, given different speakers, gains, etc.), and removing it has often improved results in realistic conditions. In contrast, variables that reflect the signal energy can be important for a number of discriminations, for instance between voiced and unvoiced sounds. Fortunately, the local time derivative of log energy is independent of overall (long-term) gain, since

$$\frac{d}{dt} \log C E(t) = \frac{d}{dt} [\log C + \log E(t)] = \frac{d}{dt} \log E(t), \tag{8.1}$$

where $C$ is a time-invariant (constant) quantity.

Finally, in pattern-recognition experiments, it is important to avoid two common errors during system development:

1. Testing on the training set – the training examples are used to design the classifier, but the desired system is one that will perform well on unseen examples. Therefore, it is important to assess the classifier on examples that were not seen during training. The system could implicitly or explicitly be memorizing characteristics of the specific examples in the training set that may not be present in an independent test set.

2. Training on the testing set – even when a test set is used for evaluation, features or parameters of the classifier are often modified over the course of many experiments. Thus, the system may also become tuned to the test set if it is used a number of times. It is common in speech-recognition research, for instance, to define a development set for all such adjustments; final performance is then reported on a second evaluation set for which no new parameters have been adjusted.

# 8.3 PATTERN-CLASSIFICATION METHODS

Given a feature vector choice, pattern classification primarily consists of the development or *training* of a system for classification of a large number of examples of the different classes. As of this writing, statistical methods are the dominant approach, and as such we will devote a separate chapter to their brief explanation. First, however, we describe the major methods for the training and use of deterministic pattern classifiers. It should be noted that these same methods can also be viewed statistically as well; however, the deterministic perspective may be a more intuitive starting point.

## 8.3.1 Minimum Distance Classifiers

Minimum distance classifiers are based on the notion that we can learn or store pattern examples that are representative of each class, and then we can identify the class of new examples by comparison to the stored values. For instance, if we store the heights and weights of all the Swedish basketball players and speech-recognition researchers, we can compare the height and weight of a new visitor to all of these examples and choose the class of the closest one as the class of the newcomer. One could even use some kind of a vote, in which the class of the majority of the local examples is used to determine the class of the new examples. Alternatively one could store one or more prototypes to represent each class, such as the most typical (or average) for each one, and again choose the class of the closest prototype as the class of the new example. All of these cases are types of minimum distance classifiers.

In a minimum distance classifier, a key point is the definition of a distance function that accepts as input two feature vectors, one representing a new example and one a stored vector, and returns the distance between the two vectors. The choice of this distance function is very important; it is also equivalent to implicit statistical assumptions about the data. Intuitively, however, simple distance measures can be misleading, given a different scale of values or differences in importance for the discrimination of different dimensions of the feature vector. For instance, in the case of height and weight, if weight is in pounds and height is in feet, the Euclidean distance between a new and a stored example will be dominated by weight, while in fact one might expect that height might be an important predictor of success of a career in professional basketball.

A sample minimum distance classifier might have $j$ templates $\mathbf{z}_i$, where $0 \leq i < j$. For each input vector $\mathbf{x}$, choose

$$\operatorname*{argmin}_i \left[ D_i = \sqrt{(\mathbf{x} - \mathbf{z}_i)^T (\mathbf{x} - \mathbf{z}_i)} \right] \qquad (8.2)$$

(where the symbol $T$ refers to the transpose operation). In other words, choose the $i$ which minimizes the distance $D_i$. Manipulating the algebra, we get

$$\operatorname*{argmin}_i \left[ D_i^2 = \mathbf{x}^T \mathbf{x} - 2 \left( \mathbf{x}^T \mathbf{z}_i - \frac{1}{2} \mathbf{z}_i^T \mathbf{z}_i \right) \right]. \qquad (8.3)$$

Since $\mathbf{x}^T \mathbf{x}$ is constant over $i$, we can also equivalently choose to maximize the following function:

$$\operatorname*{argmax}_i \left( D_i'^2 = \mathbf{x}^T \mathbf{z}_i - \frac{1}{2} \mathbf{z}_i^T \mathbf{z}_i \right), \qquad (8.4)$$

which is a linear function of $\mathbf{x}$. If each template $i$ is normalized so that $\mathbf{z}_i^T \mathbf{z}_i = 1$, then we can just maximize

$$\operatorname*{argmax}_i \left( D_i''^2 = \mathbf{x}^T \mathbf{z}_i \right). \qquad (8.5)$$

The minimum distance classifier in this case is equivalent to a classifier based on the maximum dot product with the stored vector.

Given this distance function, as noted earlier, we can derive a decision rule for classifying new vectors by assigning the vector the class of its nearest neighbor. This is called (appropriately) nearest-neighbor classification. In the limit of an infinite number of samples, this rule will achieve at most twice the error of the optimum classifier [2]. This can also be generalized to a $k$-nearest neighbor rule, where the class of the new vector is decided by a vote of the class of the $k$-nearest neighbors; in a variant of this, the voting weight is determined from the distance of the neighbor.

   This technique is potentially problematic in several ways. For the case of the direct use of all examples, it requires a great deal of storage for large problems; in addition, the search over these examples can also be time consuming. To reduce computation and storage, one can sometimes store only a small number of prototypes to represent the class; for instance, in Fig. 8.1, we indicate the mean of each class (written as $z_i$ for class $i$). Also, unless the feature vector is preprocessed to equalize the importance of the feature dimensions, this method is particularly prone to scaling problems (as noted earlier). Finally, for high dimensions, the space will be sparsely sampled by the training set.

## 8.3.2 Discriminant Functions

As noted earlier, finding the minimum distance to a prototype is closely related to finding the maximum dot product (essentially a correlation) between a new vector and the stored vectors. Finding the best prototypes for this measurement, then, can be viewed as determining weights for a linear recombination of the input features that will be maximum for the correct class. The term for this determination of a function that is maximum for the correct class is discriminant analysis. This function will determine decision surfaces between classes, where the value of the function of values along a surface is the same for the two classes that are bounded by the surface. For example, Fig. 8.4 shows that two discriminant functions will be evaluated for each input pattern. The input values for which these two functions are equal specify a decision boundary in the input feature space. In general, there can be many discriminant functions that can then be used by a decision-making process to determine the best class for the pattern (as in Fig. 8.4).

   Linear discriminant functions generate linear decision surfaces; in two dimensions (as in Fig. 8.1) the surface is a line, whereas in three dimensions it is a plane. In general, a linear decision surface is called a hyperplane. Returning to the case in which we use minimum distance to prototypes as the decision criterion (see Eq. 8.4), we find that each

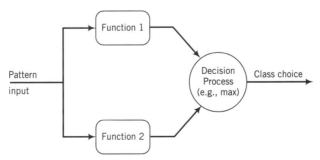

**FIGURE 8.4** A system to discriminate between two classes.

discriminant function would correspond to a constant plus a dot product between the input feature vector and the prototype. Suppose there are two prototypes (one for each class), $\mathbf{z}_1$ and $\mathbf{z}_2$. Then, for input vector variable $\mathbf{x}$, the decision boundary corresponds to

$$\mathbf{x}^T \mathbf{z}_1 - \frac{1}{2}\mathbf{z}_1^T \mathbf{z}_1 = \mathbf{x}^T \mathbf{z}_2 - \frac{1}{2}\mathbf{z}_2^T \mathbf{z}_2 \tag{8.6}$$

or

$$\mathbf{x}^T (\mathbf{z}_1 - \mathbf{z}_2) = \frac{1}{2}\left(\mathbf{z}_1^T \mathbf{z}_1 - \mathbf{z}_2^T \mathbf{z}_2\right). \tag{8.7}$$

If the prototypes (templates) are all normalized, as in Eq. 8.5, the right-hand side goes to zero.

Take the example of our Swedish basketball players and speech-recognition researchers. A number of methods can be used to learn the right hypersurface (which in two dimensions is a line) to best separate the two classes; probably the most common approach is to use the Fisher linear discriminant function, which maximizes the ratio of between-class and within-class scatter.[1] Alternatively, iterative methods such as gradient learning can be used to find the best line to divide the classes. In both cases, some criterion (such as the sum of the linear discriminant values for misclassified examples) is maximized (or minimized), in which the changes to the linear discriminant coefficients are adjusted according to the partial derivative of the criterion with respect to each coefficient.

A particularly simple approach would be to use Eq. 8.7, which was derived by using a discriminant function based on the minimum distance from a template that was the average of all of the examples of a class. The line corresponding to this equation (the thin solid line in Fig. 8.1) is not a particularly good decision boundary between the two classes. This illustrates one of the potential pitfalls with the use of minimum Euclidean distance as a criterion: in this case, the weight is represented with much larger numbers, and so it dominates the distances; this forces the decision boundary to be nearly horizontal. The dashed line in that figure shows a more sensible decision boundary, such as one might hope to get from one of the other approaches mentioned. Since this idealized line perfectly partitions the two classes, we refer to this data set as being linearly separable.

In current approaches, most often the technique for determining the linear separator is based on a statistical perspective. We will return to this point of view in the next chapter.

### 8.3.3 Generalized Discriminators

Often, one may want decision surfaces that are nonlinear. Although simple cases such as quadratic surfaces can be derived by methods similar to the ones discussed above, artificial neural networks can be utilized to get quite general surfaces. McCulloch and Pitts first modeled neurons in 1943, using a threshold logic unit (TLU). In this model, an output either

---

[1] The within-class scatter is proportional to the sample covariance matrix for the pooled data. The between-class scatter is the weighted sum $\sum_{i=1}^{C} n_i(m_i - m)(m_i - m)^T$, where $C$ is the number of classes, $m_i$ is the mean vector for the $i$th class, and $m$ is the total mean vector [2].

fires or does not; firing occurs when a linear function of the input exceeds a threshold. This model could be viewed as equivalent to a linear discriminant function. As noted in earlier chapters, Rosenblatt developed perceptrons in the late 1950s and early 1960s, based on the McCulloch–Pitts neuron. In Rosenblatt's group, systems were built based on combinations of multiple neurons and also based on a final combining layer of neurons. Originally these multilayer perceptron (MLP) systems used TLUs, and so somewhat tricky training methods had to be used since these units implemented a function that was not differentiable. In other words, for systems of this generation, it was not straightforward to adjust an internal variable by using the partial derivative of an output error criterion with respect to the variable, since the TLU implemented a binary output. Some schemes were found to work around this problem; for instance, in discriminant analysis iterative design [9], the first layer of neurons was trained by simple statistical methods (see Chapter 9), while the output neurons were trained to minimize an error criterion based on a differentiable nonlinear function of the output sum (before the threshold). In later developments [10], [7] the thresholding function was replaced with a smoother (differentiable) nonlinear function. A common choice is the sigmoidal (S-like) function

$$f(y) = \frac{1}{1 + e^{-y}}, \tag{8.8}$$

which is bounded between zero and one.

The use of the differentiable nonlinearity (as opposed to a step function) permitted uniform training procedures for all layers based on changing the weights at all layers to minimize the output error criterion, using partial differentiation and the chain rule. It can be proved that a two-layer MLP[2] with enough internal nodes (or hidden nodes) can learn any arbitrary mapping from input to output; hence, MLPs make good general discriminators. Figure 8.5 shows a typical MLP with two layers. A unit in the most common kind of modern MLP computes a nonlinear function [e.g., $f(y)$] of the weighted sum of its inputs (as shown in Fig. 8.6).

$$y_j = w_{0j} + \sum_{i=1}^{n} w_{ij} x_i. \tag{8.9}$$

No computation takes place at the input layer; it only takes place at the hidden and output layers.

The training of such networks is straightforward. An error criterion is established (e.g., a mean-squared error), and this is then differentiated with respect to the weights that were used to evaluate the MLP output in Eq. 8.9. The weights are updated proportionately to this partial derivative, and an error signal is also computed that can be propagated backward to parts of the network that contributed to the value that was an input to the neurons in the output layer. This process can continue through multiple layers by using the chain

---

[2]Unfortunately, the literature is inconsistent on the naming convention here: the MLP in the figure is sometimes called a two-layer system (since there are two layers doing the computation) and sometimes called a three-layer system (since the input features form a kind of a layer).

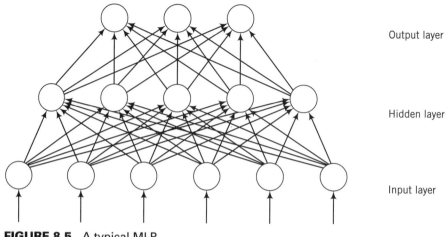

**FIGURE 8.5** A typical MLP.

rule of differentiation, since the nonlinearities that are typically used in most MLPs are differentiable functions such as the one given in Eq. 8.8. The overall algorithm is referred to as error backpropagation, and it is described in detail in such texts as [7] or [6]. A brief description and derivation are given in the Appendix at the end of this chapter.

During training, the update of the MLP parameters can be done in two different ways.

The first approach is off-line training. In this case, we accumulate the weight updates over all the training patterns and we modify the weights only when all the training patterns have been presented to the network. The gradient is then estimated for the complete set of training patterns, which guarantees, under the usual conditions of standard gradient procedures, the convergence of the algorithm.

The second way is on-line training. In this case, the MLP parameters are updated after each training pattern according to the local gradient. However, although this does

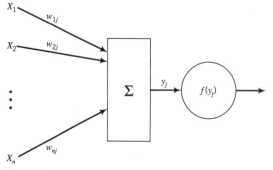

**FIGURE 8.6** A typical unit for a MLP. The unit implements a linear or nonlinear function of a weighted sum of the input variables. A typical choice for $f(y_j)$ is given by Eq. 8.8.

not actually minimize the error gradient directly, it can be shown [11] that this process will stochastically converge to the same solution.[3] In practice, the on-line training exhibits several advantages compared with the off-line procedure: it is generally acknowledged that it converges much faster and that it can more easily avoid local minima. This can be explained by the fact that the use of local gradients introduces noise in the training process, which usually improves the behavior of gradient searches because of the lowering of the risk of getting stuck in a suboptimal local minimum. Additionally, for large and varied training sets (such as are typically used for speech recognition), on-line training implies multiple passes through similar data for each single pass through the whole set.

The extension from linear to nonlinear discriminant functions permits the construction of decision surfaces that are quite complicated and that need not conform to any strong prior notion of the nature of the surface. The use of very general functions, however, makes the determination of a useful error criterion even more complicated. For this reason, as well as the availability of powerful statistical training algorithms, practical pattern-recognition problems such as speech and music have been dominated in recent years by statistical approaches (in particular for speech recognition). This is the subject of Chapter 9, but even in the statistical case it is sometimes useful for intuition's sake to map approaches back to their approximate equivalent in a deterministic framework.

## 8.4 EXERCISES

**8.1**  Suppose that you wish to classify patterns into one of two classes. (Take the case of Swedish basketball players and speech-recognition researchers, given heights and weights for each example.) Unknown to you, it turns out that the squared value of heights and of weights (in centimeters and kilograms; respectively) sum to a value greater than some constant $C$ for all basketball players, and to under this value for all the researchers. What is the shape that represents this decision surface? How would you need to modify the input features in order to make a straight line be the optimal decision surface?

**8.2**  George has designed a classifier that can perfectly distinguish between basketball players and researchers for every example in the training set. Martha has designed a classifier that makes some errors in the training set classification. However, Martha insists that hers is better. How can she be right? In other words, what are some potential pitfalls in the development of a classifier given a training set?

**8.3**  Assuming that one is willing to use an infinite number of training examples (and take an infinite amount of time for classification), would it be possible to do perfect classification of speech utterances given all possible waveforms?

**8.4**  How can one simply transform the feature values to equalize the influence that they have on a Euclidean distance? Using the data points from Fig. 8.1, do this transformation and show that the distance-based linear classifier does a better job than the original one.

**8.5**  Give some conditions under which it might be preferable to train a linear pattern classifier or a generalized nonlinear one such as a MLP.

---

[3]Since the local gradient can be viewed as a random variable whose mean is the true gradient, such an approach is sometimes called a stochastic gradient procedure.

# 8.5 APPENDIX: MULTILAYER PERCEPTRON TRAINING

There is a large body of literature on training methods for multilayer perceptrons. However, here we simply define notation and derive equations for the most common form of back-propagation training. Similar methods are useful in a large number of nonlinear systems for pattern classification. The development is largely reproduced from Chapter 4 of [1].

## 8.5.1 Definitions

We generalize the MLP of Fig. 8.5 to an $\eta$-layered perceptron consisting of $(\eta + 1)$ layers $L_\ell$ ($\ell = 0, \ldots, \eta$) of several units, where $L_0$ corresponds to the input layer, $L_\eta$ to the output layer, and $L_\ell$ ($\ell = 1, \ldots, \eta - 1$) to the hidden layers. Hidden and output units are computational units, and their output values are determined by first summing all of their inputs (as given by Eq. 8.9) and then passing the results through the sigmoid function given by Eq. 8.8. The output values of layer $L_\ell$ form a $n_\ell$ vector $h_\ell(x_n)$, which is a function of the input vector $x_n$. Here $n_\ell$ is the number of units in $L_\ell$. Input vector $h_0(x_n)$ and output vectors $h_\eta(x_n)$ are also denoted $x_n$ and $g(x_n)$ in the following. Vector $\bar{h}_\ell(x_n)$ ($\ell = 0, \ldots, \eta - 1$) stands for the $(n_\ell + 1)$ augmented vector, where the zeroth unit will be fixed to one and will account for the biases of the following layer. As the biasing unit is irrelevant for the output layer, we have $\bar{h}_\eta = h_\eta$. Layer $L_{\ell-1}$ is fully connected to layer $L_\ell$ by a $(n_{\ell-1} + 1) \times n_\ell$ weight matrix $\overline{W}_\ell$. Matrix $W_\ell$ denotes $\overline{W}_\ell$ deprived of its first row (corresponding to the biases). The state propagation is thus described by

$$h_\ell(x_n) = F\left[\overline{W}_\ell^T \bar{h}_{\ell-1}(x_n)\right], \quad \ell = 1, \ldots, \eta, \tag{8.10}$$

where $F$ is a nonlinear function, typically a sigmoid function (e.g., Eq. 8.8) that operates componentwise. Finally, we may write symbolically:

$$h_\eta(x_n) = g(x_n), \tag{8.11}$$

where $g$ is now a nonlinear function of $x_n$, depending on the parameters $\overline{W}_\ell$, $\forall \ell \in [1, \ldots, \eta]$.

The model parameters (the weight matrices $\overline{W}_\ell$) are obtained from a set of training input and associated (or desired) output pairs by minimizing, in the parameter space, the error criterion defined as

$$E = \sum_{n=1}^{N} \| g(x_n) - d(x_n) \|^2, \tag{8.12}$$

where, given each training input vector $x_n$, and $d(x_n)$ is the desired output associated with $x_n$, $g(x_n)$ represents the output vector $h_\eta(x_n)$ generated by the system. If there is at least one hidden layer, given Eqs. 8.10 and 8.11, $g(x_n)$ is a nonlinear function of the input $x_n$ (defining nonlinear decision surfaces) and contains the sigmoid function of Eq. 8.8. The total number of training patterns is denoted by $N$.

Error backpropagation training of such a system can be briefly summarized as follows. For each training iteration $t$ ($t = 1, \ldots, T$), There is presentation of all the training input vectors $x_n$, $n = 1, \ldots, N$, forward computation of the output vectors $g(x_n)$ (using Eq. 8.10), and calculation of the error function $E$. There is also backward propagation of the error (using the chain rule) to compute the partial derivative of the error criterion with respect to every weight, and update of the weights according to

$$w_{kj}(t + 1) = w_{kj}(t) - \alpha \frac{\partial E}{\partial w_{kj}(t)}, \tag{8.13}$$

in which $\alpha$ is usually referred to as the step-size parameter or learning rate and has to be small enough to guarantee the convergence of the process. In other words, we compute the error gradient for the weight vector and adjust the weights in the opposite direction.

## 8.5.2 Derivation

In the following paragraph, it is shown that the backpropagation algorithm can be derived by interpreting the problem as a constrained minimization problem. This approach, initially proposed in [5], regards MLP training as a constrained minimization problem involving the network state variables $\overline{h}_\ell(x_n)$ and the weights. We remind the reader of a method that is often used for such problems, namely that of Lagrange multipliers. Suppose that we wish to find an extreme value (e.g., minimum) for a function $f(x, y, z)$, with side constraints given by equations such as $g(x, y, z) = 0$. We then construct the auxiliary function $H(x, y, z, \lambda) = f(x, y, z) + \lambda g(x, y, z)$, and we find values of $x, y, z$, and $\lambda$ for which all the partial derivatives of $H$ are equal to zero. This gives the desired solution, and the approach can be generalized to multiple side conditions by using multiple constraint terms, each with its own Lagrange multiplier $\Lambda_i$. See standard calculus textbooks (e.g., [8]) for further explanation and derivation.

The MLP training problem may be specified by a single Lagrange function $L$ containing the objective function $E$ and constraint terms, each multiplied by a Lagrange multiplier $\Lambda_\ell$. In this particular case, the constraint terms describe the network architecture, that is, the forward equations of the network.

For each training pattern $x_n$, we want $h_\eta(x_n) = d(x_n)$ under the $\eta$ constraints represented by Eq. 8.10. By introducing $\eta$ vectorial Lagrange multipliers $\Lambda_\ell$, $\ell = 1, \ldots, \eta$, we transform the problem to one of minimizing a modified error function $L$ versus the parameters $\overline{W}_\ell$, $h_\ell(x_n)$, and $\Lambda_\ell$, for $\ell = 1, \ldots, \eta$, with

$$L = [d(x_n) - h_\eta(x_n)]^T [d(x_n) - h_\eta(x_n)]$$
$$+ \sum_{\ell=1}^{\eta} \Lambda_\ell \{ h_\ell(x_n) - F[\overline{W}_\ell^T \overline{h}_{\ell-1}(x_n)] \}. \tag{8.14}$$

A constraint is met when the corresponding term in $L$ is zero. It may be shown that

$$\nabla L[\Lambda_\ell, h_\ell(x_n), \overline{W}_\ell] = 0, \quad \ell = 1, \ldots, \eta \tag{8.15}$$

corresponds to a minimum of $E$ while meeting the constraints. We may split condition 8.15 into its constituent partials.

**Condition 1:**

$$\frac{\partial L}{\partial \Lambda_\ell} = 0, \quad \ell = 1, \ldots, \eta, \tag{8.16}$$

where the derivative is applied to each component of its argument. This leads to

$$h_\ell(x_n) = F\left[\overline{W}_\ell^T \overline{h}_{\ell-1}(x_n)\right], \tag{8.17}$$

which comprises the forward recurrences 8.10 of the error backpropagation algorithm.

**Condition 2:**

$$\frac{\partial L}{\partial \overline{h}_\ell(x_n)} = 0, \quad \ell = 1, \ldots, \eta. \tag{8.18}$$

Setting to zero the derivative with respect to $h_\eta$ leads to

$$\Lambda_\eta = 2[d(x_n) - h_\eta(x_n)].$$

Differentiation with respect to the other $h_\ell$'s yields

$$\Lambda_\ell = \overline{W}_{\ell+1} F'\left[\overline{W}_{\ell+1}^T \overline{h}_\ell(x_n)\right]\Lambda_{\ell+1}, \quad \ell = 1, \ldots, \eta - 1.$$

By defining

$$b_\ell(x_n) = F'\left[\overline{W}_\ell^T \overline{h}_{\ell-1}(x_n)\right]\Lambda_\ell, \quad \ell = 1, \ldots, \eta - 1,$$

we finally obtain

$$b_\eta(x_n) = 2F'\left[\overline{W}_\eta^T \overline{h}_{\eta-1}(x_n)\right][d(x_n) - h_\eta(x_n)], \tag{8.19}$$

and the backward recurrence

$$b_\ell = F'\left[\overline{W}_\ell^T \overline{h}_{\ell-1}\right]\overline{W}_{\ell+1} b_{\ell+1}, \quad \ell = 1, \ldots, \eta - 1, \tag{8.20}$$

which comes from substituting the expression for $\Lambda_\ell$ into the definition for $b_\ell(x_n)$ above and then replacing the expression for $b_{\ell+1}(x_n)$ in the result.

This recurrence yields a vector that is generated for each layer from the higher (later) layers. The next condition will show how this vector, which is the backward error propagation term, is used to update the weights.

**Condition 3:**

$$\frac{\partial L}{\partial \overline{W}_\ell} = 0, \quad \ell = 1, \ldots, \eta. \tag{8.21}$$

This leads to

$$F'\left[\overline{W}_\ell^T \overline{h}_{\ell-1}(x_n)\right]\Lambda_\ell \overline{h}_{\ell-1}^T(x_n) = 0 \tag{8.22}$$

or

$$b_\ell(x_n)\overline{h}_{\ell-1}^T(x_n) = 0 , \quad \ell = 1, \ldots, \eta. \tag{8.23}$$

The weight matrices $\overline{W}_\ell$ satisfying these equations can be obtained by an iterative gradient procedure making weight changes according to $\alpha(\partial L/\partial \overline{W}_\ell)$, where $\alpha$ is again the learning-rate parameter. The parameters at training step $t + 1$ are then calculated from their value at iteration $t$ by

$$\overline{W}_\ell^T(t + 1) = \overline{W}_\ell^T(t) + \alpha b_\ell(x_n)\overline{h}_{\ell-1}^T(x_n), \tag{8.24}$$

which is the standard weight update formula of the error backpropagation algorithm.

These three conditions, when met, give a complete specification of the backpropagation training of the network: optimizing with respect to the Lagrange multipliers gives the forward propagation equations; optimization with respect to the state variables gives the backward equations (the gradients); and optimization with respect to the weights gives the weight update equations.

# BIBLIOGRAPHY

1. Bourlard, H., and Morgan, N., *Connectionist Speech Recognition: A Hybrid Approach*, Kluwer, Boston, Mass., 1994.
2. Duda, D., and Hart, P., *Pattern Classification and Scene Analysis*, Wiley–Interscience, New York, 1973.
3. Haeb-Umbach, R., Geller, D., and Ney, H., "Improvements in connected digit recognition using linear discriminant analysis and mixture densities," *Proc. IEEE Int. Conf. Acoust. Speech Signal Process.*, Adelaide, Australia, pp. II-239–242, 1994.
4. Hunt, M. and Lefebvre, C., "A comparison of several acoustic representations for speech recognition with degraded and undegraded speech," *Proc. IEEE Int. Conf. Acoust. Speech Signal Process.*, Glasgow, Scotland, pp. 262–265, 1989.
5. le Cun, Y. "A theoretical framework for back-propagation," *Proc. 1988 Connectionist Models Summer School*, CMU, Pittsburgh, Pa., pp. 21–28, 1988.
6. Ripley, B., *Pattern Recognition and Neural Networks*, Cambridge Univ. Press, London/New York, 1996.
7. Rumelhart, D. E., Hinton, G. E., and Williams, R. J., "Learning internal representations by error propagation," in D. E. Rumelhart and J. L. McClelland, eds., *Parallel Distributed Processing. Exploration of the Microstructure of Cognition. vol. 1: Foundations*, MIT Press, Cambridge, Mass., 1986.
8. Thomas, G., and Finney, R., *Calculus and Analytic Geometry*, 5th ed., Addison–Wesley, Reading, Mass., 1980.

9. Viglione, S. S., "Applications of pattern recognition technology in adaptive learning and pattern recognition systems," in J. M. Mendel and K. S. Fu, eds., *Adaptive Learning and Pattern Recognition Systems*, Academic Press, New York, pp. 115–161, 1970.

10. Werbos, P. J., "Beyond regression: new tools for prediction and analysis in the behavioral sciences," Ph.D. Thesis, Harvard University, Cambridge, Mass., 1974.

11. Widrow, B., and Stearns, S., *Adaptive Signal Processing*, Prentice–Hall, Englewood Cliffs, N.J., 1985.

# CHAPTER 9

# STATISTICAL PATTERN CLASSIFICATION

## 9.1  INTRODUCTION

Audio signals such as speech or music are produced as a result of many causes. For instance, although a speech signal originates in a vocal apparatus, the received signal is generally also affected by the frequency response of the audio channel, additive noise, and so on. The latter sources of variability are unrelated to the message that is communicated. Much of the development of speech-signal analysis is motivated by the desire to deterministically account for as much of this variance as possible. For example, it would be desirable to reduce the spectral variability among multiple examples of the same word uttered with different amounts of background noise. However, with the use of common representations of speech, signals corresponding to the same linguistic message vary significantly.

For this reason, it is often desirable to model an audio signal such as speech as a random process and then to use statistical tools to analyze it. This general class of approaches has proven to be extremely useful for a range of speech-processing tasks, but most notably for speech recognition. However, before seriously discussing the applications of statistics to speech processing, we must introduce some basic concepts. In this chapter we briefly introduce statistical pattern classification (see [5] for a much more thorough treatment). As with Chapter 8, we focus on static patterns and leave the recognition of temporal sequences for a later chapter.

## 9.2  A FEW DEFINITIONS

Here we give, without proof (or detailed explanation), a few operational definitions that should prove useful for this and later chapters. More rigorous definitions as well as explanations and definitions for basic terms such as probability and random variable can be found in sources such as Papoulis [6].

**1.** A discrete random variable has a range of isolated possible values. For example, the random variable corresponding to the number of dots associated with a pair of thrown dice can take only the values 2, 3, . . . ,12. For such variables, the collection of probabilities for all possible values constitutes a discrete probability density function, whose values must sum to one.

**2.** A continuous random variable has a continuum of possible values. For example, the temperature at a particular point in space and time could be any value within some reasonable range. In reality our measurement accuracy is limited, so that in some sense we only know the temperature to the closest value with some finite resolution; however, for many problems the number of such possible values is huge, and so it is better to characterize the variable as being continuous. In this case, there is a corresponding continuous probability density function that must integrate to one; that is, for $x$, a continuous random variable, and $P(x)$, the probability density function (or pdf) for $x$,

$$\int_{-\infty}^{\infty} P(x)\,dx = 1. \tag{9.1}$$

In other words, the area under the $P(x)$ curve is one.

The integral of this function over any finite interval corresponds to a probability. For simplicity's sake, both probability densities and probabilities will be represented by $P(.)$.

    **3.** A multivariate version of a pdf is a joint pdf. For example, for random variables $x$ and $y$,

$$\int_{-\infty}^{\infty} \int_{-\infty}^{\infty} P(x, y)\,dxdy = 1. \tag{9.2}$$

In other words, the volume under the $P(x, y)$ curve is one.

    **4.** A conditional density function is equivalent to a joint density function that has been scaled down by the density function for one of the variables. In other words,

$$P(y\,|\,x) = \frac{P(x, y)}{P(x)} \tag{9.3}$$

and, consequently,

$$P(x, y) = P(y\,|\,x)P(x), \tag{9.4}$$
$$P(x, y) = P(x\,|\,y)P(y). \tag{9.5}$$

    **5.** Given Eqs. 9.4 and 9.5, we also note that

$$P(y\,|\,x) = \frac{P(x\,|\,y)P(y)}{P(x)}, \tag{9.6}$$

which is one formulation of what is often called Bayes' rule. Thus, a conditional probability density function can be expressed as a product of the opposite conditioning times a ratio of the pdf's of the individual variables. This is extremely useful for manipulating the form of pdf's.

## 9.3  CLASS-RELATED PROBABILITY FUNCTIONS

In Chapter 8 we showed that there was a close relationship between a minimum distance classifier and discriminant analysis. When speech features are viewed as random variables, their probability density functions can be used to derive distance measures, for which the

Euclidean case is a special case that implies certain statistical assumptions. Additionally, the corresponding discriminant functions have a straightforward statistical interpretation. This can be used to define (in theory) an *optimal* decision rule, that is, one that has the minimum probability of classification error.

To begin with, let us assume that there are $K$ classes. If we let $\omega_k$ be the $k$th class and let $\mathbf{x}$ be an input feature vector, then the following statements are true.

- $P(\omega_k \mid \mathbf{x})$ is the probability that the correct class is $\omega_k$ given the input feature vector. This is sometimes referred to as the posterior or *a posteriori* probability, since this probability can only be estimated after the data have been seen.
- $P(\omega_k)$ is the probability of class $\omega_k$, which is called the prior or *a priori* probability of class $k$, since this can be evaluated before $\mathbf{x}$ has been observed.
- $P(\mathbf{x} \mid \omega_k)$ is the conditional pdf of $\mathbf{x}$ (conditioned on a particular class $k$), sometimes referred to as a *likelihood* function.

Intuitively, we can think of a likelihood function as being a kind of a closeness measure (if a particular class-dependent density is closer to the new observation than other densities, it will tend to have a higher likelihood); we will see later on that the negative log of this function can often be interpreted directly as a distance.

Finally, using Bayes' rule, we can express the posterior in terms of the prior and the likelihood as follows:

$$P(\omega_k \mid \mathbf{x}) = \frac{P(\mathbf{x} \mid \omega_k)P(\omega_k)}{P(\mathbf{x})}. \tag{9.7}$$

## 9.4 MINIMUM ERROR CLASSIFICATION

We can optimally classify a feature vector into a class by using a maximum *a posteriori* (MAP) decision rule. This rule is optimum in the sense that it will give the minimum probability of error. The rule can be stated simply as follows. Assign $\mathbf{x}$ to class $k$ if

$$P(\omega_k \mid \mathbf{x}) > P(\omega_j \mid \mathbf{x}), \quad j = 1, 2, \ldots, K, \quad j \neq k. \tag{9.8}$$

In other words, choose the class with the maximum posterior probability. This should be an intuitive result; one should always choose the most probable class given the evidence available. Why is this also the optimum strategy in theory?

To illustrate this we take the example of a two-class problem. In this case the probability of error is

$$P(\text{error} \mid \mathbf{x}) = \begin{cases} P(\omega_1 \mid \mathbf{x}) & \text{if} \quad \mathbf{x} \in \omega_2 \\ P(\omega_2 \mid \mathbf{x}) & \text{if} \quad \mathbf{x} \in \omega_1 \end{cases}.$$

Clearly for any observed value of $\mathbf{x}$, this probability of error is minimized if we choose the value of $k$ corresponding to the maximum posterior probability. Finally, since the overall probability of error is

$$P(\text{error}) = \int_{-\infty}^{\infty} P(\text{error}, \mathbf{x})\,d\mathbf{x}$$

$$= \int_{-\infty}^{\infty} P(\text{error} \mid \mathbf{x})P(\mathbf{x})\,d\mathbf{x}, \tag{9.9}$$

then if $P(\text{error} \mid \mathbf{x})$ is the minimal value for each choice of $\mathbf{x}$, the overall integral is minimized.[1] This extends naturally to the case of more than two classes.

## 9.5   LIKELIHOOD-BASED MAP CLASSIFICATION

Some kinds of systems directly estimate the class posterior probabilities $P(\omega \mid \mathbf{x})$. However, often it is easier to generate estimates of likelihoods $P(\mathbf{x} \mid \omega)$. For this latter case one can typically compute statistics separately for the input vectors falling into the different classes. Given the Bayes rule formulation from earlier in this chapter, there is a straightforward (in principle) transformation between the two kinds of densities. In fact, the factor $P(\mathbf{x})$ typically does not need to be computed. Since $P(\mathbf{x})$ is a constant for all classes, finding the maximum posterior is equivalent to determining

$$\underset{k}{\text{argmax}}\ P(\omega_k \mid \mathbf{x}) = \underset{k}{\text{argmax}}\ P(\mathbf{x} \mid \omega_k)P(\omega_k). \tag{9.10}$$

Decisions between candidate classes can also be made by using a ratio of posterior probabilities:

$$\frac{P(\omega_k \mid \mathbf{x})}{P(\omega_j \mid \mathbf{x})} = \frac{P(\mathbf{x} \mid \omega_k)P(\omega_k)}{P(\mathbf{x} \mid \omega_j)P(\omega_j)}. \tag{9.11}$$

The Bayes decision rule says that if this ratio is $>1$ then we pick $\omega_k$ over $\omega_j$. This is equivalent to assigning $\mathbf{x}$ to class $k$ if

$$\frac{P(\mathbf{x} \mid \omega_k)}{P(\mathbf{x} \mid \omega_j)} > \frac{P(\omega_j)}{P(\omega_k)} \tag{9.12}$$

for all $j$ other than $k$.

The left-hand side of Eq. 9.12 is often referred to as the likelihood ratio. Taking the log of the likelihood ratio, we find that the rule states the $\omega_k$ is chosen over $\omega_j$ if

$$\log P(\mathbf{x} \mid \omega_k) + \log P(\omega_k) > \log P(\mathbf{x} \mid \omega_j) + \log P(\omega_j). \tag{9.13}$$

The MAP classification based on likelihoods is then equivalent to choosing a class to maximize a statistical discriminant function:

$$\underset{k}{\text{argmax}}[\log P(\mathbf{x} \mid \omega_k) + \log P(\omega_k)]. \tag{9.14}$$

---

[1]For simplicity's sake, we have written the integral as if $x$ were a scalar.

## 9.6   APPROXIMATING A BAYES CLASSIFIER

How do we find the Bayes classifier? Unfortunately, the actual density functions referred to earlier are essentially always unknown for real problems. Consequently, we must estimate the densities from training data. The training results in the learning of some set of parameters $\Theta$, which will then be used during classification to estimate the required probabilities. For the formulation of Eq. 9.14, this would mean that we would train estimators of $P(\mathbf{x} \mid \omega_k, \Theta)$ for each class $\omega_k$. Maximum likelihood (ML) procedures are used to learn the parameters that will give the largest possible values for these quantities. Given a good enough estimator, enough data, and perfect estimates for the prior probabilities $P(\omega_k)$, parameters $\Theta$ that are learned by the ML criterion will lead to the MAP classifier.

Some specific approaches to training the parameters $\Theta$ include the following.

**1.** For each class, count the instances for which a feature or group of features is closest to a finite set of prototypical values. When the discretization is done for a group of features (e.g., locating the nearest spectrum among one of 256 candidate spectra), it is called vector quantization, or VQ.

**2.** Assume a parametric form for the density, whose parameters can be directly estimated from the data. The most common example of this approach is the Gaussian distribution.

**3.** Assume a parametric form for the density, in which the parameters must be estimated with an iterative solution. For example, a density can be represented as a weighted sum or mixture of Gaussian densities, with the means, covariances, and mixture weights to be learned from the data. These parameters are typically learned through some variant of the expectation maximization (EM) algorithm (Section 9.8). The EM is particularly useful for modeling densities that are more complicated than simple Gaussians (e.g., multimodal distributions).

**4.** Use automatic interpolation–learning to estimate posteriors directly (e.g., with a neural network; see Chapter 27). In this case it is the posterior probabilities that are learned, rather than the likelihood densities. As with the Gaussian mixtures, complicated multimodal distributions can be represented in this way.

In practice the training of posterior estimators may lead to different results than the training of likelihood estimators. In principle direct posterior estimators will require the training of parameters that are influenced by all of the input vectors, which can greatly increase the complexity, though there can be a trade off with a smaller number of parameters. We will return to these issues in Chapter 27.

For the moment, we restrict ourselves to the case of a single Gaussian. An analysis of this case can often provide us with insight that may be useful in understanding the more complex methods of density estimation, to which we will return later in this text.

The central feature of the Gaussian assumption is that the density is completely defined by the first- and second-order statistics (means and covariances). In one dimension (one feature), the Gaussian probability density function can be expressed as

$$P(x \mid \omega_i) = \frac{1}{\sqrt{2\pi}\,\sigma_i} \exp\left[ -\frac{1}{2}\left( \frac{x - \mu_i}{\sigma_i} \right)^2 \right], \tag{9.15}$$

where $\mu_i$ is the mean and $\sigma_i$ is the standard deviation (square root of the variance) for the $i$th class.

In two or more dimensions, this takes the form

$$P(\mathbf{x} \mid \omega_i) = \frac{1}{(\sqrt{2\pi})^d |\Sigma_i|^{1/2}} \exp\left[ -\frac{1}{2}(\mathbf{x} - \mu_i)^T \Sigma_i^{-1} (\mathbf{x} - \mu_i) \right] \qquad (9.16)$$

where $\Sigma_i$ is the covariance matrix of the $i$th class,[2] $|\Sigma_i|$ is the corresponding determinant, and $d$ is the dimension of $\mathbf{x}$.

The product $(\mathbf{x} - \mu_i)^T \Sigma_i^{-1} (\mathbf{x} - \mu_i)$ is often called the (squared) Mahalanobis distance. Note that it is similar in form to the Euclidean distance, which would be $(\mathbf{x} - \mu_i)^T (\mathbf{x} - \mu_i)$. In the special case in which the covariance matrix is diagonal (all off-diagonal values equal to zero), the inclusion of the factor $\Sigma_i^{-1}$ is equivalent to scaling all the variables to a variance of one before computing the Euclidean distance. More generally, multiplying by this factor is equivalent to rotating the feature vectors so that their covariance matrix would be the identity matrix. In other words, the effect is to decorrelate the feature vectors, in addition to the scaling of each feature to have unity variance.

Recalling expression 9.13, the optimal Bayes classifier in the case of a Gaussian density would correspond to the use of a statistical discriminant function:

$$g_i(\mathbf{x}) = -\frac{1}{2}(\mathbf{x} - \mu_i)^T \Sigma_i^{-1} (\mathbf{x} - \mu_i) - \frac{d}{2} \log_e 2\pi - \frac{1}{2} \log_e |\Sigma_i| + \log_e P(\omega_i). \qquad (9.17)$$

Note that the first term of this function is the Mahalanobis distance, so that when the later terms can be ignored, choosing $i$ to maximize $g_i(\mathbf{x})$ is equivalent to choosing the class with the minimum Mahalanobis distance between the input vector and the corresponding mean vector.

Note also that the term $d/2 \log_e 2\pi$ is constant over all classes, and so it will be ignored in the rest of this discussion. More generally, maximizing $g_i(\mathbf{x})$ can be simplified by dropping terms from the discriminant function that are the same for all classes. Consider the following two cases.

1.  There is no correlation between the features for any of the classes, and all features have the same variance; that is, $\Sigma_i = \sigma^2 \mathbf{I}$

$$\Rightarrow |\Sigma_i| = \sigma^{2d}$$

$$\Rightarrow$$

$$g_i(\mathbf{x}) = -\frac{\|\mathbf{x} - \mu_i\|^2}{2\sigma^2} + \log_e P(\omega_i), \qquad (9.18)$$

where the constant and covariance determinant terms have been dropped. In the case in which the prior probabilities of the different classes are all equal, this function is a Euclidean distance. Thus, minimum distance classifiers that use a Euclidean distance are equivalent

---

[2]Reminder: the element of the $i$th covariance matrix in the $j$th row and the $k$th column is the covariance between the $j$th and $k$th features for all observations in the $i$th class. The covariance between two random variables is the expected value of their product once the means have been removed.

to Bayes' classifiers in which we assume Gaussian densities for the likelihoods, equal prior probabilities for each class, no correlation between the different features, and features that each have the same variance.

2. If the covariance matrices are the same for all classes, i.e., $\Sigma_i = \Sigma$, then an equivalent statistical discriminant function is linear in $x$. (See the Exercises for more on this.) In this case, we can estimate the optimal linear discriminant function by estimating the overall covariance and the means for each class.

The Gaussian approximation can be quite poor in some cases, for instance if the pdf is not unimodal (has more than one big bump). Density functions for real-world data can often be better approximated by more complicated models, such as a mixture (weighted summation) of Gaussians. However, the analysis above is a good place to start for understanding's sake, and in many cases a model based on a single Gaussian per class can be useful for formulating a classifier.

## 9.7 STATISTICALLY BASED LINEAR DISCRIMINANTS

As in Section 9.6, consider the Bayesian classifier with an assumption of Gaussian form for the density of each class. Further assume that the features are uncorrelated with one another so that the covariance matrix consists of variances only (no nonzero off-diagonal elements). Further assume that the covariance matrix is identical for each class. Since there are no off-diagonal elements, the net effect is to scale down each feature in the $\mathbf{x}$ and $\mu$ vectors by the corresponding standard deviation (see Exercise 1). For simplicity's sake, then, we simply remove the covariance matrix from Eq. 9.17 and assume that the random variables have been scaled to have a variance of one. Expanding out Eq. 9.17 and dropping the constant and class-independent terms, we derive a discriminant function:

$$g_i(\mathbf{x}) = \mathbf{x}^T \mu_i - \frac{1}{2}\mu_\mathbf{i}^T \mu_i + \log_e P(\omega_i). \tag{9.19}$$

Comparing this expression with the discriminant function $D_i'^2$ (Eq. 8.4) in Chapter 8, we see the following differences:

1. The prototype or template $z_i$ has been replaced by its mean (as was also done in Fig. 8.1).

2. The features are viewed as random variables and have been scaled by the within-class standard deviation (square root of variance), assumed to be the same for each class.

3. There is an additional term that corresponds to prior knowledge about how probable each class is. If this certainty is strong enough, it can outweigh evidence from the data. For instance, if 99% of the population consists of basketball players and only 1% consists of speech-recognition researchers, on the average the classification error would only be 1% if all new cases were classified as basketball players regardless of height or weight. More generally, this factor simply applies a bias.

If there is no prior information to suggest favoring one class over the other, and the features are scaled as suggested above, then the statistical discriminant function described above is equivalent to the distance-based classifier of Chapter 8. However, the statistical formulation is very general, permitting both a principled way of normalizing the amplitudes of the features (scaling down the variances to be equal to one) and a provision for incorporating prior biases about the classification. If the statistical assumptions are justified (for instance, the Gaussian form), the resulting classifier is optimal in the Bayes sense.

### 9.7.1   Discussion

As noted previously, the Bayes classifier is the optimal one that we attempt to approximate with the methods described here. The simplifying assumptions described above permit principled approaches to the approximation of this classifier. However, the class-dependent covariance matrices are not generally equal, and the distributions are not usually well described by a single Gaussian. Fortunately, approaches exist for learning more general forms of distributions. Iterative techniques can be used to find the parameters for distributions where there is no direct analytical solution, such as the means and variances for a mixture (weighted sum) of Gaussian densities. Neural networks can also be trained to determine probability density functions, which will be discussed in Chapter 27. However, for many cases (such as Gaussian mixtures) an approach can be used that guarantees finding the best possible value of the parameters given the parameters from the previous iteration. The dominant approach for this purpose is called EM.

## 9.8  ITERATIVE TRAINING:   THE EM ALGORITHM

In the earlier sections here we have shown how one can estimate the parameters for a Gaussian density corresponding to a class. Given the true densities for all classes, one may determine the optimum Bayes classifier. Given parameters trained to maximize the likelihood for the data in each class, we can approximate the Bayes classifier. However, in many problems, there is no analytical solution for the parameters of the statistical classifier given the training data. An example of such a problem is finding the means, variances, and mixture weights for a sum of Gaussian densities, given a sample of data points. We will postulate a model in which each individual Gaussian generates some of the observed data points. Given that there is no way to directly compute estimates for these quantities (since we do not know which points correspond to which Gaussian), we need an optimization technique to find the parameters for each Gaussian (and for its mixture) that will maximize the likelihood of all of the observed data.

The dominant approach to such problems is called the Expectation Maximization (EM) algorithm [4]. In this approach, the parameter-estimation problem is structured to incorporate variables representing information that is not directly observed, but that is assumed to be part of the model that generated the data. Such a variable is often called hidden or missing. For instance, in the Gaussian mixture case, a hidden variable could be the index of the Gaussian that generated a data point. The key idea of EM is to estimate the densities by taking an expectation of the logarithm of the joint density between the known and unknown components, and then to maximize this function by updating the parameters that are used in the probability estimation. This process is then iterated as required. We can show fairly easily that if this expectation is maximized, so too is the data likelihood itself, which (as noted previously) is typically the goal for statistical model training. First, we express the expectation of the joint likelihood of observed and hidden variables as a function of both old parameters $\Theta_{\text{old}}$ and new ones $\Theta$. Then for random variable $k$ (a missing or hidden variable), observed random variable $x$, and parameters $\Theta$, let

$$\begin{aligned}
\mathcal{Q}(\Theta, \Theta_{\text{old}}) &= \sum_k P(k \,|\, x, \Theta_{\text{old}}) \log P(k, x \,|\, \Theta) \\
&= \sum_k P(k \,|\, x, \Theta_{\text{old}}) \log[P(k \,|\, x, \Theta) P(x \,|\, \Theta)] \\
&= \sum_k P(k \,|\, x, \Theta_{\text{old}}) \log P(k \,|\, x, \Theta) + \log P(x \,|\, \Theta) \sum_k P(k \,|\, x, \Theta_{\text{old}}) \\
&= \sum_k P(k \,|\, x, \Theta_{\text{old}}) \log P(k \,|\, x, \Theta) + \log P(x \,|\, \Theta).
\end{aligned} \tag{9.20}$$

If $\Theta$ is chosen to be equal to $\Theta_{\text{old}}$, then

$$\mathcal{Q}(\Theta_{\text{old}}, \Theta_{\text{old}}) = \sum_k P(k \,|\, x, \Theta_{\text{old}}) \log P(k \,|\, x, \Theta_{\text{old}}) + \log P(x \,|\, \Theta_{\text{old}}). \tag{9.21}$$

Subtracting Eq. 9.21 from Eq. 9.20 and rearranging terms, we get

$$\begin{aligned}
\log P(x \,|\, \Theta) - \log P(x \,|\, \Theta_{\text{old}}) = {} & \mathcal{Q}(\Theta, \Theta_{\text{old}}) - \mathcal{Q}(\Theta_{\text{old}}, \Theta_{\text{old}}) \\
& + \sum_k P(k \,|\, x, \Theta_{\text{old}}) \log \frac{P(k \,|\, x, \Theta_{\text{old}})}{P(k \,|\, x, \Theta)}.
\end{aligned} \tag{9.22}$$

However, the last term is the relative entropy or Kullback Leibler distance, which can be shown to be nonnegative [3]. Therefore, if a change to $\Theta$ increases $\mathcal{Q}$, $\log P(x \,|\, \Theta)$ increases. In other words, changing the parameters to increase the expectation of the log likelihood of a joint distribution between the data and a hidden variable will also increase the log data likelihood itself. In principle, we could maximize the expectation for each value of $\Theta$ and then reestimate $\Theta$. Although there is no guarantee of how good the estimates would be, they are guaranteed to improve (or at least get no worse) with each iteration.

In many cases, it is possible to structure the problem so that we can analytically determine the choice for the parameters that will maximize the expectation in each iteration. Then a new expression for the expectation can be determined, followed by a new parameter estimation, and so on.

Figure 9.1 shows a histogram drawn from a bimodal distribution that is the sum of two Gaussian densities (renormalized to integrate to one). If data points could be unambiguously assigned to one of the two component densities, then one could estimate the means, variances, and weight for each component analytically. However, if the data assignments are unknown, we must evaluate these quantities by using expected values over the joint distributions (of mixture index and observed data value) and iteratively improve the estimates. This has proven to be a powerful concept and works well under many conditions. The figure shows quick convergence to a good fit for this simple example. There are many

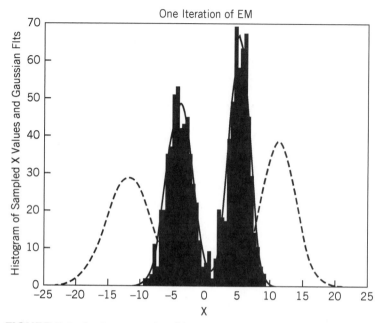

**FIGURE 9.1** In the center is a histogram for 1000 data points that were sampled from a mixture of two Gaussians. The correct mixture parameters are means of $-4.0$ and 5.0, sigmas of 2.0 and 1.5, and equal weights of 0.5. The outermost Gaussian shapes correspond to an initial (poor) guess for the mixture parameters, consisting of means of $-12$ and 11, sigmas of 3.5 and 2.6, and equal weights. The log likelihood of the data for this guess was $-5482$. After one iteration of EM, the density in the middle is obtained, which is quite a good fit to the data. The corresponding means are $-4.1$ and 5.0, sigmas are 2.0 and 1.5, and weights are 0.49 and 0.51. The log likelihood for this new estimate was $-2646$. Further iterations did not appreciably change the parameters or the likelihood for this simple example. The vertical axis for the densities is scaled up by 1000 to match the histograms, i.e., having an integral of 1000.

other useful cases for which we can find the optimum values for $\Theta$ at each stage, and in practice the optimization often converges quite quickly.

    To illustrate EM, let us take the specific example of optimizing the parameters for a density that is composed of a weighted sum (or mixture) of simpler distributions, Gaussians. This is a fairly general notion, since we can always decompose an unknown probability density $P(x \mid \Theta)$ as[3]

$$P(x \mid \Theta) = \sum_{k=1}^{K} P(x, k \mid \Theta) = \sum_{k=1}^{K} P(k \mid \Theta) P(x \mid k, \Theta), \qquad (9.23)$$

---

[3]For simplicity's sake, we will only consider the case of a univariate density; that is, the case in which the observed variable is a scalar. The more general vector case is very similar but is more intricate because of the matrix arithmetic. See [1] for a description of the vector case.

using the definitions of joint and conditional densities, and assuming $K$ disjoint (but unknown) categories. We hope to choose these categories so that the component densities can be simply computed.

As is typical for training with a ML criterion, we will attempt to approximate the true densities by maximizing the likelihood $P(x \mid \Theta)$ or its logarithm. There is no analytical expression that will generally yield the $\Theta$ to maximize either of these quantities directly. Furthermore, taking the logarithm of the rightmost expression in Eq. 9.23 would require the log of a sum, which would generally be difficult to optimize. However, we can consider the variable $k$ to be an unobserved (hidden) random variable corresponding to the mixture component from which each data sample came. Then the theorem proved in Eq. 9.22 tells us that an increase in the expectation of $\log P(x, k \mid \Theta)$ will also increase the data likelihood $P(x \mid \Theta)$. Equation 9.23 also shows that we can interpret the probability of this hidden variable as the mixture weight.

Given these interpretations, we can write an expression for the log joint density for observed and hidden variables.

$$Z = \log P(x, k \mid \Theta) = \log[P(k \mid \Theta)P(x \mid k, \Theta)], \tag{9.24}$$

with an expected value over $N$ examples and $K$ mixture components of

$$\mathcal{Q} = E(Z) = \sum_{k=1}^{K} \sum_{n=1}^{N} P(k \mid x_n, \Theta_{\text{old}}) \log[P(k \mid \Theta)P(x_n \mid k, \Theta)]. \tag{9.25}$$

Note that we have distinguished between $\Theta_{\text{old}}$, the parameters used to generate the distribution with which we will evaluate the expectation, and $\Theta$, which is a variable to be optimized (in the sense of maximizing $\mathcal{Q}$).

Finally, we can decompose this expression into two terms by using the usual properties of the log function:

$$\mathcal{Q} = \sum_{k=1}^{K} \sum_{n=1}^{N} P(k \mid x_n, \Theta_{\text{old}}) \log P(k \mid \Theta) + \sum_{k=1}^{K} \sum_{n=1}^{N} P(k \mid x_n, \Theta_{\text{old}}) \log P(x_n \mid k, \Theta). \tag{9.26}$$

We will often assume models for which the $\Theta$ parameters of the two terms are disjointed and can be optimized separately.

One way that we can train the parameters for this density is to assume that each mixture component $P(x \mid k)$ is a Gaussian with mean $\mu_k$ and standard deviation $\sigma_k$. In this case, $P(k)$ is the weight given to Gaussian $k$ in the model.

Recalling Eq. 9.15, we can express the Gaussian density for component $k$ as

$$P(x_n \mid k) = \frac{1}{\sqrt{2\pi}\,\sigma_k} \exp\left[ -\frac{1}{2}\left( \frac{x_n - \mu_k}{\sigma_k} \right)^2 \right], \tag{9.27}$$

where $\mu_k$ is the mean and $\sigma_k$ is the standard deviation for the component.

Substituting this definition into expression 9.26, we find that the expectation to be maximized becomes

$$Q = \sum_{k=1}^{K} \sum_{n=1}^{N} P(k \mid x_n, \Theta_{\text{old}}) \log P(k \mid \Theta)$$

$$+ \sum_{k=1}^{K} \sum_{n=1}^{N} P(k \mid x_n, \Theta_{\text{old}}) \left[ -\log \sigma_k - \frac{(x_n - \mu_k)^2}{2\sigma_k^2} + C \right], \tag{9.28}$$

where the $C$ term refers to constants that will disappear with the differentiation to follow.

Given this choice of parametric form for the mixture density, we can use standard optimization methods to find the best value for the parameters. Let's begin by solving for the means. Setting the partial derivative $\partial Q / \partial \mu_j$ to zero, we find that the first term (which is not dependent on the means) drops away, as do all the terms for which $k$ is not equal to $j$; thus we get

$$\sum_{n=1}^{N} P(j \mid x_n, \Theta_{\text{old}}) \left( \frac{x_n}{\sigma_j^2} - \frac{\mu_j}{\sigma_j^2} \right) = 0, \tag{9.29}$$

or (multiplying through by $\sigma_j^2$ and moving the second term to the right-hand side)

$$\sum_{n=1}^{N} P(j \mid x_n, \Theta_{\text{old}}) x_n = \sum_{n=1}^{N} P(j \mid x_n, \Theta_{\text{old}}) \mu_j. \tag{9.30}$$

Thus

$$\mu_j = \frac{\sum_{n=1}^{N} P(j \mid x_n, \Theta_{\text{old}}) x_n}{\sum_{n=1}^{N} P(j \mid x_n, \Theta_{\text{old}})}. \tag{9.31}$$

A similar procedure can be used to proceed from Eq. 9.28 to derive the optimum value for the variances (see Exercises). The result is

$$\sigma_j^2 = \frac{\sum_{n=1}^{N} P(j \mid x_n, \Theta_{\text{old}})(x_n - \mu_j)^2}{\sum_{n=1}^{N} P(j \mid x_n, \Theta_{\text{old}})}. \tag{9.32}$$

In the case of the mixture weights $P(k \mid \Theta)$, Eq. 9.28 must be supplemented with a Lagrangian term $\lambda [\sum_{k=1}^{K} P(k \mid \Theta) - 1]$, which expresses the constraint that the mixture weights must sum to one. Then let $Q^*$ be the augmented function that includes this constraint term. Taking the partial derivative $[\partial Q^* / \partial P(j \mid \Theta)]$ and setting it to be equal to zero, we find that the term involving means and variances may be disregarded, leaving the equations

$$\frac{\partial Q^*}{\partial P(j \mid \Theta)} = \frac{1}{P(j \mid \Theta)} \sum_{n=1}^{N} P(j \mid x_n, \Theta_{\text{old}}) + \lambda = 0. \tag{9.33}$$

Summing this up over all the components $j$ yields $\lambda = -N$, so that the mixture weights can be expressed as

$$P(j \mid \Theta) = \frac{1}{N} \sum_{n=1}^{N} P(j \mid x_n, \Theta_{\text{old}}) \tag{9.34}$$

Each of these results has included an expression $P(j \mid x_n, \Theta_{\text{old}})$. How do we evaluate this? This can be done using Bayes' rule, that is,

$$P(j \mid x_n, \Theta_{\text{old}}) = \frac{P(x_n \mid j, \Theta_{\text{old}}) P(j \mid \Theta_{\text{old}})}{P(x_n \mid \Theta_{\text{old}})} = \frac{P(x_n \mid j, \Theta_{\text{old}}) P(j \mid \Theta_{\text{old}})}{\sum_{k=1}^{K} P(x_n \mid k, \Theta_{\text{old}}) P(k \mid \Theta_{\text{old}})}. \tag{9.35}$$

which only includes terms that we have already shown how to evaluate. That is, $P(x_n \mid j, \Theta_{\text{old}})$ is the value of the $j$th Gaussian at $x_n$ assuming the mean and variance from the previous iteration. Similarly, $P(j \mid \Theta_{\text{old}})$ is the weight of the $j$th Gaussian from the previous optimization step.

## 9.8.1 Discussion

Equations 9.31, 9.32, and 9.34 have an interesting form. Each is an expectation over many examples of the desired parameter, with the appropriate normalization so that the densities incorporated actually sum to one. Thus, EM has led us to a sensible and satisfying result; compute the posterior distribution of the hidden variable given the observed variables and the old parameters, and use it to compute the expectations of the desired new parameters. The mean and variance for each Gaussian are estimated by weighting each instance of the sampled variable by the probability of that sample having originated from the corresponding density, and then scaling appropriately. The mean that is computed in this way is also sometimes called a center of mass for the variable. The estimate for the mixture weight, or the prior probability for each mixture, is just the average of the posterior probabilities of the mixture over all of the training samples.

An interesting special case occurs for a single Gaussian. When $M = 1$, the posterior is always equal to one, and Eqs. 9.31 and 9.32 can be reduced to the standard equations for the sample mean and covariance.

We have not discussed initialization. EM is only guaranteed to maximize the data likelihood for each step, and it is not guaranteed to find the global maximum. For simple problems such as that illustrated in Fig. 9.1, this is clearly not a problem. However, in the more general case, good initializations can be important for improving the quality of the resulting solution.

The overall EM procedure can be briefly summarized as follows.

1. Choose a form for the probability estimators (e.g., Gaussian) for the densities associated with each class.

2. Choose an initial set of parameters for the estimators.

3. Given the parameters, compute posterior estimates for the hidden variables.

**4.** Given posterior estimates for hidden variables, find the distributional parameters that maximize the expectation of the joint density for the data and the hidden variables. These will be guaranteed to also maximize the improvement to the likelihood of the data.

**5.** Assess goodness of fit (e.g., log likelihood) and compare to the stopping criterion. If not satisfied, go back to step 3.

This section constitutes a very brief and light introduction to EM. For a more complete description, the reader should consult one of the many references for EM, e.g., [2] or [4]. We will also return to EM later (in Chapter 26) in the context of hidden Markov models. In that case, we will modify the parameters for estimates of densities that are hypothesized to generate sequences.

## 9.9 EXERCISES

**9.1** Expand out expression 9.17 for the case in which only the diagonal terms of the covariance matrix are nonzero, and in which the covariance matrix is equal across classes. Reduce the expression to a simpler function by eliminating all terms that are either constant or equal for all classes.

**9.2** The heights and weights of basketball players and speech researchers are vector quantized so that there are 16 different possible reference indices $i$ corresponding to (height, weight) pairs. A training set is provided that has the height and weight and occupation of 1000 people, 500 from each occupation. How would you assign them to reference indices? Given a chosen approach for this, how would you estimate the discrete densities for the likelihoods $P(i \mid occupation)$? How would you estimate the posteriors $P(occupation \mid i)$? How would you estimate the priors $P(occupation)$ and $P(i)$?

**9.3** A Bayes decision rule is the optimum (minimum error) strategy. Suppose that Albert uses a Bayes classifier and estimates the probabilities by using a Gaussian parametric form for the estimator. Beth is working on the same problem and claims to get a better result without even using explicitly statistical methods. Can you think of any reason that Beth could be right?

**9.4** Section 9.8 derived the EM update equations for the mixture weights and means of a univariate Gaussian mixture, but it only stated the result for the variances. Derive the variance update equations.

## BIBLIOGRAPHY

1. Bilmes, J., "A gentle tutorial of the EM algorithm and its application to parameter estimation for Gaussian mixture and hidden Markov models," International Computer Science Institute Tech. Rep. ICSI TR-97-021; 1997; ftp://ftp.icsi.berkeley.edu/pub/techreports/1997/tr-97-021.ps.gz.

2. Bishop, C., *Neural Networks for Pattern Recognition*, Oxford Univ. Press, London/New York, 1996.

3. Blahut, R., *Principles and Practice of Information Theory*, Addison–Wesley, Reading, Mass., 1987.

4. Dempster, A. P., Laird, N. M., and Rubin, D. B., "Maximum likelihood from incomplete data via the EM algorithm," *J. R. Statistical Soc.* **39**: 1–38, 1977.
5. Duda, R., and Hart, P., *Pattern Classification and Scene Analysis*, Wiley–Interscience, New York, 1973.
6. Papoulis, A., *Probability, Random Variables, and Stochastic Processes*, McGraw–Hill, New York, 1965.

# ACOUSTICS

$$B_{11} = cos(M\theta)cos(L\theta) - \frac{V_k}{V_{k+1}} sin(M\theta)sin($$

$$B_{12} = jV_{k+1}cos(M\theta)sin(L\theta) + jV_k sin(M\theta)co$$

$$B_{21} = \frac{j}{V_k}sin(M\theta)cos(L\theta) + \frac{j}{V_{k+1}}cos(M\theta)sin$$

$$B_{22} = -$$

$$u_k^+ (t - x/c) \qquad z^{-M} \qquad u_{k+1}^+ (t - x/c)$$

Soun ys noght but eyr ybroken
And every speche that ys spoken
Lowd or pryvee, foul or fair,
In his substance ys but air.

—Geoffrey Chaucer

**M**UCH MODERN work in speech and audio signal processing is focused on the use of mathematical tools, such as those introduced in Part II. However, what sets this field apart from others is that sound is the signal of interest. It is essential, then, to obtain at least a cursory understanding of the nature of the acoustic signal, and in particular of the basic physics of its production, transmission, and reception. These topics are introduced in Part III. As with other areas of study in this book, acoustics is discussed in much greater detail in other sources, to which we happily direct the reader who develops a deeper interest in the subject.

Chapter 10 reviews basic material on traveling and standing acoustic waves, such as might be found in an introductory physics text. This perspective is then applied to acoustic tubes, which are commonly used as models for the production of speech and music. Chapter 11 continues in this vein for the case of speech, and Chapter 12 applies the same mathematical methods to sound production in common types of acoustic musical instruments. Finally, Chapter 13 describes a critical component between the sound source and the listener, namely, the acoustical properties of the room in which the sound is produced and received.

# CHAPTER 10

# *WAVE BASICS*

## 10.1  INTRODUCTION

Many pleasing sounds are produced by carefully designed and carefully controlled acoustic tubes or vibrating strings.[1] A brief categorization of some of these sounds follows: (a) plucked string instruments such as guitars, banjos, mandolins, and harps are set into vibration by the musician directly (i.e., with fingers); (b) bowed string instruments such as violins, violas, and cellos are set into vibration by the frictional force between the bow and string. (c) Struck string instruments like the piano are set into vibration by a complicated mechanism following the depression of the piano keys by the performer. (d) Acoustic tube resonances in the human vocal tract help produce the sounds of speech or music. (e) Acoustic tube resonances in brass instruments such as trumpets, cornets, flugelhorns, and trombones are excited by the player's lip vibrations. (f) Acoustic tube resonances in the clarinet and oboe produce sound stimulated by a vibrating reed that is set into vibration by the player's breath stream.

Many instruments can be mathematically characterized by the wave equation, a partial differential equation in both time and space. For example, the vibration of a plucked guitar string is described by knowing the motions of every point along the string. We will see that the solutions to the simplest wave equation consist of travelling waves whose properties are largely determined by boundary conditions.

The wave equation yields important insights into the properties of tubes and strings, but there are additional complications in understanding the sounds produced by practical instruments. The vocal and nasal tracts of the human voice are stimulated into sound-producing vibrations when excited by either oscillations of the vocal cords or air turbulence caused by the forcing of air through narrow openings. Interactions between these excitation sources and tube resonances may often require nonlinear modeling methods. Also, the human vocal tract is a complex acoustic tube and is lossy, so our problem will be to simplify the analysis in order to obtain mathematically tractable yet useful results. We begin with an analysis of the vibrating string as the simplest way to come to grips with the basics of wave theory as applied to the human voice and many musical instruments.

## 10.2  THE WAVE EQUATION FOR THE VIBRATING STRING

The wave equation for the vibrating string can be deduced by application of Newton's second law of motion ($F = ma$). Figure 10.1 depicts an infinitesimal section of the string.

---

[1]Although strings alone do not generate loud enough sound, their coupling to wood or metal causes the latter to vibrate to create the sounds we hear at a concert.

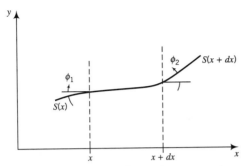

**FIGURE 10.1**   Vibrating string geometry.

To arrive at the simplest form, we need to make the following assumptions:   (a) there is no stiffness in the string; (b) tension and mass density are constant throughout the length; (c) the vertical displacement of the string from its normal position is small; and (d) there are no gravitational or frictional forces.

Let $S$ = string tension. From Fig. 10.1, we see that the total tensional force in the $y$ direction is the difference between the forces at $x$ and $x + dx$ and can be expressed in terms of the angles $\phi_1$ and $\phi_2$:

$$F_y = S(\tan(\phi_2) - \tan(\phi_1)). \tag{10.1}$$

$$\tan(\phi_1) = \left.\frac{\partial y}{\partial x}\right|_x, \quad \tan(\phi_2) = \left.\frac{\partial y}{\partial x}\right|_{x+dx}, \tag{10.2}$$

$$\tan(\phi_2) = \frac{\partial y}{\partial x} + \left(\frac{\partial^2 y}{\partial x^2}\right)dx. \tag{10.3}$$

From Eqs. 10.1, 10.2, and 10.3,

$$F_y = S\left(\frac{\partial^2 y}{\partial x^2}\right)dx. \tag{10.4}$$

If we assume that $dx = ds$ for small displacements and define $\epsilon$ to be the (uniform) mass density along the string, then the total mass in the length $dx$ is $\epsilon dx$, and Newton's second law gives

$$S\left(\frac{\partial^2 y}{\partial x^2}\right)dx = \epsilon\left(\frac{\partial^2 y}{\partial t^2}\right)dx. \tag{10.5}$$

Defining $c$ as the square root of the ratio $S/\epsilon$, we arrive at the one-dimensional wave equation:

$$c^2\left(\frac{\partial^2 y}{\partial x^2}\right) = \frac{\partial^2 y}{\partial t^2}. \tag{10.6}$$

Solutions to Eq. 10.6 can take many forms, depending on boundary conditions. If the string is plucked, traveling waves will be set up. Such a wave is described by the time-variable string displacement function $y(x, t)$. Since we know that $y(x, t)$ is a traveling wave, let's assume a solution of the form

$$y^+(x, t) = f(x - ct). \tag{10.7}$$

If the function $f(x - ct)$ has well-behaved first and second derivatives, it can be shown that $f(x - ct)$ is a solution to wave equation 10.6. The constant $c$ has the dimension of velocity, so that $f(x - ct)$ can be interpreted as a traveling wave moving in the positive $x$ direction with velocity $c$. Similarly,

$$y^-(x, t) = f(x + ct) \tag{10.8}$$

represents a wave traveling in the negative $x$ direction with the same velocity. The sum of the two traveling waves is also a solution, so we finally have a solution that will turn out to be of practical interest for the boundary conditions that we intend to impose.

$$y(x, t) = Af(x - ct) + Bf(x + ct). \tag{10.9}$$

## 10.3 DISCRETE-TIME TRAVELING WAVES

To simulate traveling waves by computer, we need to discretize both time and space. A block diagram of such a simulation is shown in Fig. 10.2.

If an external stimulus is applied to the structure of Fig. 10.2 and the system is set into motion, a wave traveling to the right is created in the upper track and a wave traveling

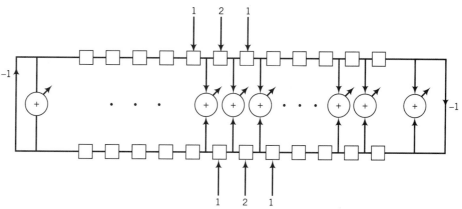

**FIGURE 10.2** Discrete-time simulation of a string fixed at both ends. At both $x = 0$ and $x = l$, $y(x, t) = 0$. The sequences 1, 2, 1 shown in the figure represent initial values of the string position.

to the left is created in the lower track; the output of the system at any time is the pattern of all the sums shown in the figure. The simulation is therefore *not* of the wave equation but of its *solutions*. Since both $x$ and $t$ of Eq. 10.9 are now discrete variables, we can replace $x$ by $mX$ and $t$ by $nT$ to obtain the discrete-time and discrete-space version of Eq. 10.9, where $n$ is a time index, $m$ is a distance index, $T$ is the time-sampling interval, and $X$ is the distance-sampling interval.

$$y(mX, nT) = Af(mX - cnT) + Bf(mX + cnT). \qquad \textbf{(10.10)}$$

Thus, for example, at any (discrete) time and (discrete) place, the state of the simulation is the weighted sum of a forward- and backward-traveling wave; the constants $A$ and $B$ are determined from the boundary conditions and initial conditions imposed on the system.

## 10.4 BOUNDARY CONDITIONS AND DISCRETE TRAVELING WAVES

As an example, consider a string fixed at both ends and given an ideal pluck at its center, as shown in the discrete-time simulation of Fig. 10.2. With these constraints the string vibration history can be followed, as in Fig. 10.3. The discrete-time version of the pluck is the upper-right pattern of Fig. 10.3. Symmetry has been invoked to decompose this initial pattern into the two patterns to the left of the equal sign. The pattern to the left of the plus sign propagates to the right while its neighbor propagates to the left; summing these produces the pattern following the equal sign. The *constraint* that the ends are fixed means that the sum is always zero at both ends; this can only happen if the two traveling waves cancel at the ends. Figure 10.3 shows the result for the first twelve clock cycles; the reader is invited to continue the sequence until the original pattern (top line) reappears.

Boundary conditions are met, as shown in Fig. 10.3, by reversing signs at both ends and then continuing the travel in the reverse direction.

## 10.5 STANDING WAVES

At any instant, each of the traveling wave patterns of Fig. 10.3 can be decomposed into a Fourier series in the $x$ dimension. Thus, we can imagine that the harmonics of the pattern are (sinusoidal) traveling waves. The sum of a left- and right-traveling sinusoid results in a *standing wave*. To see this, consider the traveling waves in Eqs. 10.11 and 10.12:

$$g = \sin(\lambda x - ct), \qquad \textbf{(10.11)}$$

$$q = \sin(\lambda x + ct). \qquad \textbf{(10.12)}$$

The sum is

$$g + q = 2 \, \cos(ct) \sin(\lambda x). \qquad \textbf{(10.13)}$$

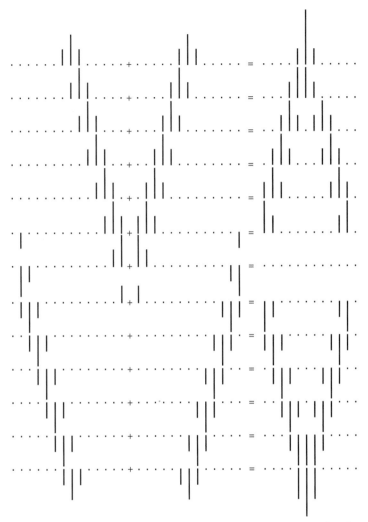

**FIGURE 10.3** Step-by-step history of traveling waves stimulated by a pluck.

Thus, the sum is a sinusoid of fixed phase in the $x$ dimension but with a time-varying amplitude.

## 10.6 DISCRETE-TIME MODELS OF ACOUSTIC TUBES

Ideal acoustic tubes obey the same wave equation as that of the vibrating string, but there are obvious differences that we discuss here in a preliminary way. The vibrating string has the single variable $y(x, t)$ of interest. An acoustic tube has two variables of interest: the pressure gradient $p(x, t)$ and the volume velocity $u(x, t)$ of a small volume of air. In a string, the vibratory motion is perpendicular, or *transverse* to the directions of wave propagation;

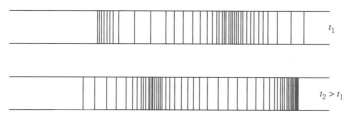

**FIGURE 10.4**  Longitudinal waves in an air column.

in a tube, the motion of the air molecules is *longitudinal* (in the same direction) as the wave motion. This motion is depicted in Fig. 10.4.

The physical law for air motion in an acoustic tube is Newton's second law, just as in the string. The pressure gradient $p(x, t)$ causes a change in $u(x, t)$, whereas the velocity gradient causes compression and rarefaction of the air, thus changing $p(x, t)$. The development of the details (left to the reader) results in the equations

$$c^2 \left( \frac{\partial^2 p}{\partial x^2} \right) = \frac{\partial^2 p}{\partial t^2}, \tag{10.14}$$

$$c^2 \left( \frac{\partial^2 u}{\partial x^2} \right) = \frac{\partial^2 u}{\partial t^2}. \tag{10.15}$$

These equations imply that both $p(x, t)$ and $u(x, t)$ can be expressed as traveling waves. A useful configuration is that of an excitation source at one end and an open tube at the other. At the interface between the open tube and the external environment, the pressure gradient will drop to zero; at the other end, the volume velocity will follow the volume velocity of the applied source.

The constants of the vibrating string are the tension $S$ and the string mass density $\epsilon$. For the air-filled tube, the constants are $\rho$, the mass density of the air and $\kappa$, the compressibility of the air. Both $p(x, t)$ and $u(x, t)$ are traveling waves with velocity $c$ given by

$$c = \frac{1}{(\rho \kappa)^{0.5}}. \tag{10.16}$$

The solutions for $u(x, t)$ and $p(x, t)$ are

$$u(x, t) = u^+(x - ct) - u^-(x + ct), \tag{10.17}$$

$$p(x, t) = Z_0[u^+(x - ct) + u^-(x + ct)], \tag{10.18}$$

where $Z_0$ has to be determined.

Application of Newton's second law leads to the relation between velocity and pressure:

$$\frac{\partial p}{\partial x} = -\frac{\rho}{A} \left( \frac{\partial u}{\partial t} \right), \tag{10.19}$$

where $A$ is the tube cross-sectional area.

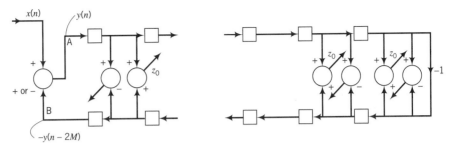

**FIGURE 10.5** Discrete-time simulation of pressure and volume velocity in an acoustic tube. Upward arrows are pressure waves; downward arrows are velocity waves; and small squares are unit delays. The plus sign on the left corresponds to a tube open at the left, and the minus sign signifies a closed tube at the left. The tube is always closed at the right.

By combining Eqs. 10.17, 10.18, and 10.19, we can show that

$$Z_0 = \frac{\rho c}{A}. \tag{10.20}$$

Here $Z_0$ can be interpreted as an impedance relationship between volume velocity and pressure. The relations of Eqs. 10.17 and 10.18 can be represented digitally by Fig. 10.5. The upper track contains the set $u(mX - cnT)$; the lower track is $u(mX + cnT)$.

## 10.7 ACOUSTIC TUBE RESONANCES

The dynamics of the propagating signals in Fig. 10.5 can be appreciated by observing how the signal at point $A$ varies with $n$ (or time). All other signals on the upper track are merely delayed versions of the signal at point $A$. Let's call this signal $y(n)$. Then the signal at point $B$ is $-y(n - 2M)$, where $M$ is the number of delays in the upper (or lower) track. Therefore

$$y(n) = x(n) \pm y(n - 2M), \tag{10.21}$$

and the corresponding $z$-transform relationship is

$$Y(z) = \frac{X(z)}{1 \mp z^{-2M}} = X(z)H(z). \tag{10.22}$$

If the minus sign is used in Eq. 10.22, the poles are distributed as shown in Fig. 10.6(a). If the plus sign is used, the poles of $H(z)$ are also distributed uniformly on the unit circle as in Fig. 10.6(b). The examples in the figure are for $M = 3$. The angle between any adjacent poles is $\pi / M$.

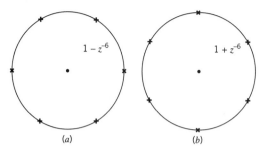

**FIGURE 10.6**   Poles of a discrete-time simulation of a lossless, uniform acoustic tube.

The distribution of poles on the unit circle immediately translates into a frequency response. To see this, replace $z$ by $e^{j\omega T}$ in Eq. 10.22 and compute the magnitude of $H(z)$ (for the plus sign in Eq. 10.22):

$$|H(e^{j\omega T})| = \frac{1}{|1 + e^{-2j\omega TM}|} = \frac{1}{|2\cos(M\omega T)|}. \tag{10.23}$$

For real acoustic tubes, of course, there will be some energy loss, so the poles will not be precisely on the unit circle and the corresponding magnitude will not be infinite for any finite sinusoidal input.

## 10.8   RELATION OF ACOUSTIC TUBE RESONANCES TO OBSERVED FORMANT FREQUENCIES

The human vocal tract can be approximately modeled as an acoustic tube excited at one end by glottal vibrations and at the lips by an open or closed tube. During articulation of the neutral vowel (the first phoneme of "above"), the vocal tract most closely resembles the uniform acoustic tube of the previous section.

In Chapter 3, the spectrogram was introduced as a speech-analysis tool, and resonances of the vocal tract were observed; for the neutral vowel, the resonant frequencies (formants) were approximately 500, 1500, 2500, and 3500 Hz, and so on. These values can be justified by noting the resonant modes that occur in a uniform acoustic tube excited at one end and open at the other end. The first three modes are shown in Fig. 10.7. One can obtain these mathematically by solving for the volume velocity, after modifying Eqs. 10.17 and 10.18 to accommodate the above constraints. The result (stated without proof and left as an exercise) when the excitation is a complex exponential $U(\omega)e^{j\omega t}$ is

$$u(x, t) = \frac{\cos\{[\omega(l - x)]/c\}}{\cos[(\omega l)/c]} U(\omega)e^{j\omega t}. \tag{10.24}$$

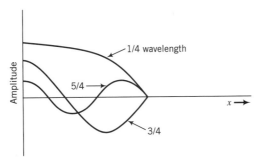

**FIGURE 10.7** First three modes of an acoustic tube open at one end.

From Eq. 10.24, we see that the poles occur when

$$\omega_n = [(2n + 1)\pi c]/2l, \quad n = 0, 1, 2, \ldots. \tag{10.25}$$

Thus the frequencies $f_n = \omega_n/2\pi$ are

$$f_n = (2n + 1)c/4l, \quad n = 0, 1, 2, \ldots. \tag{10.26}$$

For $c = 344$ m/s (in air) and $l = 0.17$ m (typical vocal tract length), $f_n = 500, 1500,$ 2500, and 3500 Hz, and so on.

We can get the same result from a purely discrete-time argument. From Eq. 10.23, and using the plus sign, the resonant radian frequencies are given by

$$\omega_n = [(2n + 1)\pi]/2MT, \quad n = 0, 1, 2\ldots. \tag{10.27}$$

Equating Eqs. 10.26 and 10.27, we get the relationship

$$MT = l/c, \tag{10.28}$$

which implies that $M$ and $T$ are functionally correlated to the actual length of the tube and the velocity of propagation. Figure 10.8 illustrates this point for the two cases $M = 3$ and $M = 8$. When the sample interval $T$ is adjusted, the same resonant frequencies are mapped out in both cases.

Since the cross-sectional area of the vocal tract varies as one moves from the glottis to the lips, a model must include this fact. One approach is to use a model that consists of relatively small acoustic tube sections; each section is a lossless tube of constant cross section over its length, but each tube may have a different length and area. Thus, each section by itself has the properties that we have just analyzed, but the model is complicated by the interfaces between the tubes. In Chapter 11, these issues will be discussed in the context of human vocal tract analyses.

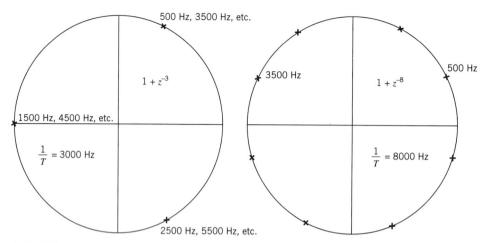

**FIGURE 10.8**  Two examples showing that the same resonant frequencies can be obtained with different sampling rates and different spatial quantization.

## 10.9  EXERCISES

**10.1**   Show that Eqs. 10.7, 10.8, and 10.9 are all solutions of wave equation 10.6.

**10.2**   Continue the sequence of Fig. 10.3 until the initial pattern reappears.

**10.3**   If the discrete-time simulation of the string (Fig. 10.2) is plucked at a position of one-fourth of its length, describe the sequence.

**10.4**   Prove that if the pluck is at $1/n$th the string length, the $n$th harmonic will be missing.

**10.5**   Find the poles of Eq. 10.22 for $M = 6$ for both the plus and minus signs.

**10.6**   Make a rough sketch of the magnitude of Eq. 10.22 for both the plus and minus cases.

**10.7**   Consider an acoustic tube excited at one end by the complex exponential $U(\omega)e^{j\omega t}$ and open at the other end. Prove that the volume velocity in the tube satisfies Eq. 10.24.

$$u(x, t) = \frac{\cos\{[\omega(l - x)]/c\}}{\cos[(\omega l)/c]} U(\omega)e^{j\omega t}. \qquad (10.29)$$

**10.8**   Consider two acoustic tubes of the same dimension. Tube 1 is open at both ends; tube 2 is open at one end and closed at the other end. Now consider the lowest modes of both tubes; which tube will have the lower of such modes? Explain.

**10.9**   As the mass density in a tube increases, does the tube impedance increase, decrease, or stay the same? Explain.

**10.10**   If the tension in a vibrating string is increased, does the vibration frequency rise, fall, or stay the same? Explain.

**10.11** An acoustic tube closed at both ends and excited at its midpoint will resonate at frequencies higher, lower, or equal to that of an acoustic tube open at one end and excited at the other end. Explain.

**10.12** Find the lowest three vibrational modes of (a) a string with a length of 6 m, (b) a string with a length of 2 cm, (c) a tube 2 ft. long, and (d) a tube 6 in. long.

**10.13** Find the three lowest modes in a helium-filled pipe closed at both ends. (Length is 0.5 m.)

# BIBLIOGRAPHY

1. Beranek, L. L., *Acoustics*, Amer. Inst. of Phys., New York, 1954.
2. Morse, P. M., and Ingard, K. U., *Theoretical Acoustics*, Princeton Univ. Press, Princeton, N.J., 1968.

# CHAPTER 11

# ACOUSTIC TUBE MODELING OF SPEECH PRODUCTION

## 11.1 INTRODUCTION

For many years, Gunnar Fant directed the Speech Transmission Laboratory in Stockholm. He performed X-ray measurements to determine the shape of the human vocal tract during phonation. In 1970 (based to a great extent on his doctoral thesis) his book *Acoustic Theory of Speech Production* [1] was published. It contained detailed information on vocal tract shapes.

For each phoneme in any spoken language there corresponds one or several sequences of vocal tract shapes. With the development of digital signal-processing concepts, these shapes can be efficiently modeled. In Chapter 10 we showed how simple acoustic tubes could be digitally modeled. In this chapter, these ideas are extended to more complicated acoustic tube structures that relate to spoken sounds.

## 11.2 ACOUSTIC TUBE MODELS OF ENGLISH PHONEMES

Fant first traced area functions from the X-ray data. An example is shown in Fig. 11.1 for the vowel /i/ as in /bid/.

On the left is the tracing and on the right we see the area of the tube as a function of the distance from the glottis. This area function is quantized as a concatenation of cylindrical tubes. This string of tubes can now be approximated by analog T networks [1] or digital waveguides [4]. Straightforward mathematical derivations for a practical system (four or more tubes) become difficult. Computer simulation using digital waveguides is a more effective method than with analog T networks. So we begin with digital waveguides and then add speech-specific attributes such as source function properties.

Our aim is to establish relationships between various acoustic tube structures and the resonant modes resulting from these structures. We will see that even a small number of tubes exhibit resonances that resemble formant measurements of the different phonemes.

Figure 11.2 shows a single section of a digital waveguide. This figure is a graphical representation of the equations describing the pressure and volume velocity at the two ends of a lossless, uniform acoustic tube, governed by Eqs. 10.17–10.20. From this figure we now derive the relationships between $u_k, p_k$ (the inputs) and $u_{k+1}, p_{k+1}$ (the outputs).

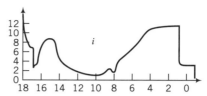

**FIGURE 11.1**  X-ray tracing and area function for phoneme /i/. From [1].

For the $k$th section we can write the equations

$$u_k = u_k^+ - u_k^-, \tag{11.1}$$

$$p_k = \frac{\rho c}{A_k}\left(u_k^+ + u_k^-\right) = V_k\left(u_k^+ + u_k^-\right), \tag{11.2}$$

where $A_k$ is the cross-sectional area and $V_k = \rho c/A_k$, $\rho$ is the density of the gas in the tube, and $c$ is the velocity of sound in the tube. We have omitted the arguments, remembering that $u^+$ always has $(t - x/c)$ as an argument and $u^-$ has $(t + x/c)$ as an argument. Notice the similarity between Eqs. 11.1 and 11.2 and Eqs. 10.17 and 10.18.

In what follows, we use the same notation for the space–time functions and their $z$ transforms. No confusion should result, since the $z$-transform versions always explicitly include $z$.

Similarly,

$$u_{k+1} = u_{k+1}^+ - u_{k+1}^-, \tag{11.3}$$

$$p_{k+1} = V_{k+1}\left(u_{k+1}^+ + u_{k+1}^-\right). \tag{11.4}$$

An inspection of Fig. 11.2 yields the equations

$$u_{k+1}^+ = z^{-M}u_k^+, \tag{11.5}$$

$$u_{k+1}^- = z^{M}u_k^-, \tag{11.6}$$

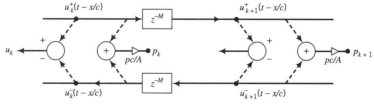

**FIGURE 11.2**  Single section of a digital waveguide. *M* is the delay in units of the sampling period.

and this leads to the following set of two equations in the two unknowns, $u_k^+$ and $u_k^-$:

$$u_{k+1} = z^{-M}u_k^+ - z^M u_k^-, \tag{11.7}$$

$$p_{k+1} = V_k\left(z^{-M}u_k^+ + z^M u_k^-\right). \tag{11.8}$$

The solutions are

$$u_k^+ = \frac{z^M}{2V_k}p_{k+1} + \frac{z^M}{2}u_{k+1}, \tag{11.9}$$

$$u_k^- = \frac{z^{-M}}{2V_k}p_{k+1} - \frac{z^{-M}}{2}u_{k+1}. \tag{11.10}$$

Substituting Eqs. 11.9 and 11.10 into Eqs. 11.1 and 11.2, we arrive at the basic chain relationship between the $k$th and $(k+1)$th stage:

$$u_k = \frac{z^M + z^{-M}}{2}u_{k+1} + \frac{1}{V_k}\frac{z^M - z^{-M}}{2}p_{k+1}, \tag{11.11}$$

$$p_k = \frac{z^M + z^{-M}}{2}p_{k+1} + V_k\frac{z^M - z^{-M}}{2}u_{k+1}. \tag{11.12}$$

Since our interest is to determine the resonances in the system, and since, for a lossless tube, the poles always appear on the unit circle, we can replace $z$ by $e^{j\theta}$ (where $\theta = \omega T$) so that Eqs. 11.11 and 11.12 become

$$u_k = \cos(M\theta)u_{k+1} + \frac{j}{V_k}\sin(M\theta)p_{k+1}, \tag{11.13}$$

$$p_k = jV_k\sin(M\theta)u_{k+1} + \cos(M\theta)p_{k+1}. \tag{11.14}$$

It is useful to express Eqs. 11.13 and 11.14 in matrix form; adding additional sections results in successive matrix multiples.

$$\begin{bmatrix} p_k \\ u_k \end{bmatrix} = \begin{bmatrix} \cos(M\theta) & jV_k\sin(M\theta) \\ \dfrac{j}{V_k}\sin(M\theta) & \cos(M\theta) \end{bmatrix} \begin{bmatrix} p_{k+1} \\ u_{k+1} \end{bmatrix}. \tag{11.15}$$

Thus, for example, if we want the relationship between the $k$th section and the $(k+2)$th section pictured in Fig. 11.3, we can write down the matrix result

$$\begin{bmatrix} p_k \\ u_k \end{bmatrix} = \begin{bmatrix} \cos(M\theta) & jV_k\sin(M\theta) \\ \dfrac{j}{V_k}\sin(M\theta) & \cos(M\theta) \end{bmatrix} \begin{bmatrix} \cos(L\theta) & jV_{k+1}\sin(L\theta) \\ \dfrac{j}{V_{k+1}}\sin(L\theta) & \cos(L\theta) \end{bmatrix} \begin{bmatrix} p_{k+2} \\ u_{k+2} \end{bmatrix}, \tag{11.16}$$

$$\begin{bmatrix} p_k \\ u_k \end{bmatrix} = \begin{bmatrix} B_{11} & B_{12} \\ B_{21} & B_{22} \end{bmatrix} \begin{bmatrix} p_{k+2} \\ u_{k+2} \end{bmatrix}, \tag{11.17}$$

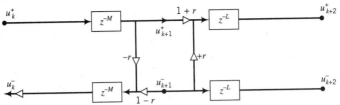

**FIGURE 11.3** Two-section digital waveguide. $M$ and $L$ are the delays of successive sections, in units of the sampling period.

with

$$B_{11} = \cos(M\theta)\cos(L\theta) - \frac{V_k}{V_{k+1}}\sin(M\theta)\sin(L\theta),$$ (11.18)

$$B_{12} = jV_{k+1}\cos(M\theta)\sin(L\theta) + jV_k\sin(M\theta)\cos(L\theta),$$ (11.19)

$$B_{21} = \frac{j}{V_k}\sin(M\theta)\cos(L\theta) + \frac{j}{V_{k+1}}\cos(M\theta)\sin(L\theta),$$ (11.20)

$$B_{22} = \frac{-V_{k+1}}{V_k}\sin(L\theta)\sin(M\theta) + \cos(M\theta)\cos(L\theta).$$ (11.21)

For many speech sounds, particularly vowels, the mouth is open so that the pressure gradient at the mouth opening is zero; setting $p_{k+2}$ to zero in Eq. 11.17, we get the simple relationship

$$u_{k+2} = \frac{u_k}{B_{22}}.$$ (11.22)

Setting $B_{22}$ to zero allows us to solve for the poles on the unit circle. $B_{22}$ can be simplified by noting from Eq. 11.1 that $V_k = \rho c/A_k$, using the trigonometric identities

$$\cos(M\theta)\cos(L\theta) = \frac{1}{2}\{\cos[(M+L)\theta] + \cos[(M-L)\theta]\},$$ (11.23)

$$\sin(M\theta)\sin(L\theta) = \frac{-1}{2}\{\cos[(M+L)\theta] - \cos[(M-L)\theta]\}.$$ (11.24)

Then, letting the factor $r = (A_{k+1} - A_k)/(A_{k+1} + A_k)$, we arrive at the relationship

$$\cos[(M+L)\theta] + r\cos[(M-L)\theta] = 0.$$ (11.25)

Given the parameters $M$, $L$, and $r$, we can find those values of $\theta$ that correspond to the resonances (formants) of the two-tube structure. Figure 11.4 shows a plot of the positions of formants 1 and 2.

Each curve in this $f_1, f_2$ plane corresponds to specific values of $M$ and $L$, and the curves trace out a trajectory that is a function of the ratio $A_2/A_1$. Also shown in the figure

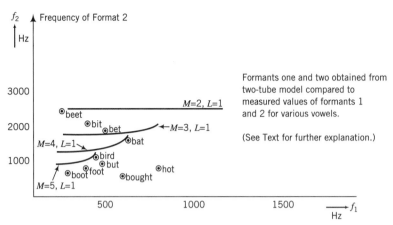

**FIGURE 11.4**  Formants 1 and 2 obtained from the two-tube model.

are the $f_1, f_2$ points for various vowels obtained from the work of Peterson and Barney [2]. Any curve passing close to a vowel implies that there exists a two-stage digital waveguide that has approximately the same $f_1, f_2$ value as that vowel. Notice that not all vowels are close to a trajectory; such vowels require a model in which the number of stages exceeds two. Also, this analysis has ignored matches to higher formants; again, a model with more stages is required. However, as is made clear in the studies by Fant [1] and Portnoff [3], *an acoustic configuration can always be found to match the measured steady-state spectrum of any speech sound.*

## 11.3  EXCITATION MECHANISMS IN SPEECH PRODUCTION

Thus far we have shown how an acoustic tube or combinations of such tubes respond to acoustic stimuli. In the human vocal system, three types of excitation exist. The speech signal is the response of the vocal tract to some combination of the three exciting signals.

During the production of vowels and vowellike sounds, the excitation is a nearly periodic sequence of pulselike pressure changes occurring at the glottal opening. Pressure changes originating in the lungs force open the vocal cords, which are then quickly closed by elastic forces, which are again forced open, and the process repeats. Neurologically controlled muscles determine the vocal cord tension and hence the degree of elasticity; thus, the frequency of this excitation signal is controlled by the speaker.

Vowels generally are excited as described above, but not always. Vowels can be whispered. In such cases the vocal cords remain open but the air stream must pass through the small glottal opening; this produces turbulence, a noiselike component in the air stream. The resonances of the vocal tract will further shape the pressure wave to produce the whispered vowel.

Turbulence can also be produced by constrictions in other parts of the vocal tract; for example, for voiceless fricatives to be generated, noise can be generated at the tongue-tip–

teeth constriction (/s/ or /th/), or, further back in the vocal tract, at the tongue–upper-palate constriction (/sh/), or at the teeth–lower-lip constriction (/f/). These excitation signals are acted on by the vocal tract complex to produce the various spectra typifying the different fricative sounds.

Such excitations can take place in concert with glottis-controlled excitations during voiced fricatives. The vocal tract configurations during these sounds are the same as the corresponding voiceless fricatives, but the vocal cords can be simultaneously vibrated, yielding sounds that contain both quasiperiodic and noise components.

Transients in the vocal tract are another source of excitation. If pressure is built up anywhere in the tract by occlusion, sudden removal of the occlusion causes a sudden pressure change that propagates throughout the vocal cavity. This occurs, for example, for (/p/),(/k/), and (/t/).

## 11.4 EXERCISES

**11.1** In Section 11.2 it is stated that for a lossless tube, the poles always appear on the unit circle. Can you justify this claim?

**11.2** Derive Eqs. 11.18–11.21 given Eqs. 11.16 and 11.17.

**11.3** What are the boundary conditions on $p_k$ and $u_k$ for an open tube and for a closed tube?

**11.4** Sketch an acoustic tube model of the voiceless fricative sounds. This sketch should be of a qualitative nature accompanied by some intuitive justification of your solution.

**11.5** Repeat the previous problem for the voiceless plosives.

**11.6** Repeat again for the voiced fricatives and plosives.

**11.7** Finally, repeat for the nasal sounds.

**11.8** If the space–time solution to the wave equation can be expressed as the separable product of a time function and a space function, derive explicit solutions for these two functions. Base your solution on the case of a single space dimension.

## BIBLIOGRAPHY

1. Fant, G., *Acoustic Theory of Speech Production*, Mouton, The Hague, 1970.
2. Peterson, G. E., and Barney, H. L., "Control methods used in the study of vowels," *J. Acoust. Soc. Am.* **24**: 175–184, 1952.
3. Portnoff, M. R., "A quasi-one-dimensional digital simulation for the time-varying vocal tract," M.S. Thesis, Massachusetts Institute of Technology, Cambridge, Mass., 1973.
4. Smith, J. O. III, "Physical modelling using digital waveguides," *Comp. Mus. J.* **16**: 74–87, 1992.

# CHAPTER *12*

## MUSIC PRODUCTION

## 12.1   INTRODUCTION

In Chapter 10, we analyzed wave motion on strings and in acoustic tubes. In Chapter 11, we applied these concepts to model speech production. The ideas of these chapters will be applied to the study of some musical instruments. Specifically, we will examine how various acoustical musical instruments work.

Many instruments use vibrating strings, whereas others use the human breath stream. In the case of speech or singing, the sound we hear is created by the resonances of the vocal tract. Thus, the breath stream represents the excitation that creates the traveling wave that is then modified by the vocal tract resonances. Similarly, the vibrating strings in a string instrument excite the body of the instrument, causing it to vibrate to produce the sounds we hear.

String instruments can be further classified as plucked (guitars, banjos, and harps), bowed (violins, violas, and cellos), or struck (piano).

The difference between the excitation functions of plucked and bowed string instruments is significant. Plucking involves a sudden perturbation of some small section of a string; the subsequent string vibrations can be represented as a straightforward source-filter model. Bowing is a more complex operation. Friction allows the bow to drag the string along horizontally (in the plane of the strings); when the restoring force of string tension becomes large enough, the string snaps back to its quiescent state, only to be again captured by the bow. Thus the string horizontal motion is a sawtoothlike motion.

Wind instruments can employ vibrating reeds (clarinet), lip vibrations (trumpet and trombone) or vocal cord vibrations (singing). Percussion instruments (drums, xylophones, and bells) are excited by striking some material. The piano, as we shall see, is sufficiently complex to be treated as a separate category.

String instruments and reed instruments have this in common:   the excitation function is the source of the overtones (harmonics), whereas the main source of sound radiation (e.g., the violin body or trumpet body) determines the overall spectral shape of the emitted sound. (Spectra are of course further modified by room or concert hall reverberations; see Chapter 13). It is convenient to consider these instruments as source-filter systems, as we do the human voice. It's worth remarking, however, that these instruments have regimes of oscillation as discussed in [3], whereby resonances from the main sound sources feed back to affect the properties of the excitation.

In this chapter we first direct our attention to the violin as (predominantly) a bowed string instrument. We examine how the strings, the bridge, and the body of the violin vibrate in response to bowing. Next we study the radiation patterns of bowed string instruments. We then discuss some aspects of piano design and construction. Finally, we include a brief section on the acoustic properties of several wind instruments.

**154**

## 12.2 SEQUENCE OF STEPS IN A PLUCKED OR BOWED STRING INSTRUMENT

Listening to a string instrument involves the following sequence of actions.

1. The string is plucked or bowed and begins to vibrate.
2. The string motion causes the wooden bridge to vibrate.
3. The bridge motion sets the upper plate of the instrument body into vibration.
4. The entire body now vibrates, as does the air inside the body.
5. An acoustic radiation pattern caused by body and air vibration is set up.
6. This radiation pattern is modified by the room or concert hall acoustic environment.
7. The listener's auditory apparatus interprets the received sound.

## 12.3 VIBRATIONS OF THE BOWED STRING

The motion of a string fixed at both ends and plucked at its center was described in Chapter 10. Figure 10.3 gives a history of the string displacement. The modes of vibration are obtained by a Fourier analysis of this displacement pattern. The string vibrations cause the bridge to vibrate, and this sets the body of the instrument into motion, effectively creating the sound we hear.

The relations between force and velocity in a bowed string can be modeled (approximately) as shown in Fig. 12.1.

The frictional force between bow and string plays a significant role in string motion. At times, this force keeps bow and string together and is independent of the amount of force;

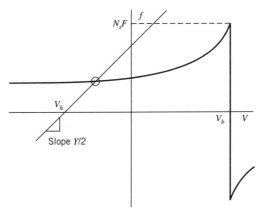

**FIGURE 12.1** Two relations between the force and velocity of a bowed string. From [11].

this situation is represented by the vertical line on the right of Fig. 12.1. At other times there is slippage between bow and string, and the force versus velocity can be represented as the curve in the figure. In addition, the performer is always applying a force to keep the string vibrating, and this can be represented by the sloping straight line. This line is displaced from the origin by the presence of all the reflected waves from previous steps. The slope of the line is $Y/2 = (Tm)^{-0.5}$, where $T$ is the string tension and $m$ is the string mass per unit length. Here $Y$ is the wave admittance of the string. This leads to Eq. 12.1.

$$v(t) = \left(\frac{Y}{2}\right) f(t) + v_h(t) \tag{12.1}$$

Thus, at each step in a computer simulation, we can compute $v_h(t)$, since its value depends only on previous history; this allows the simulator to find the next intersection between the curve and line. In this way, both the force on the string and the velocity of the string may be computed as functions of time.

If we watch the string vibration during bowing, the string seems to sweep out a parabolic path. However, high-speed photography reveals that the string is nearly straight, with a bend that races around the string and follows the curved path. The situation is depicted in Fig. 12.2, where the bottom figure shows the string displacements at the point of contact with the bow. The figures marked (a) through (h) indicate the states of the string corresponding to the same letters on the bottom figure.

Ideally, the string displacement at the bowing point should behave like a nearly perfect sawtooth wave; in practice, however, the string does not always follow this ideal. Extra slippage can be caused by variations in the boundary conditions at the nut or bridge, the two points at which the string is fastened, as well as by poor bowing. Figure 12.3 shows (in a simulation) a good result (above) and a bad result. The difference is in the method of setting up the end conditions. The two cases are the velocity waves at the bowing points. In the good case, when the rapidly moving bend gets to an end, its corner is rounded; in the bad case, no such rounding is simulated. We see that in the bad case, the string keeps slipping back under bow control during the time that it is escaping from the bow [11].

# 12.4  FREQUENCY-RESPONSE MEASUREMENTS OF THE BRIDGE OF A VIOLIN

The bridge is the transduction mechanism that translates the string vibrations into the strong body vibrations needed to produce a satisfactory audible radiation field. The outline of a normal violin bridge is shown in Fig. 12.4.

Hacklinger [6] measured bridge vibrations by employing a *mute* violin; he attached a heavy block of iron to the frame, thus decoupling the body from the bridge motions. Figures 12.5 and 12.6 are examples of the frequency responses.

Here is a summary of Hacklinger's conclusions based on his experiments:

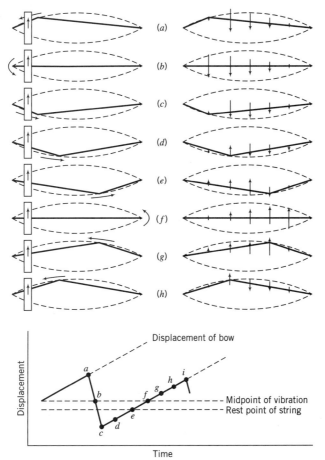

**FIGURE 12.2** String state vs. string displacement at the point of bow contact. Both columns in the figure show the string displacement at a given instant. The arrows in the left column show the directions of the motion of the maximum displacement. The arrows in the right column give an estimate of the vertical forces along the string. From [10].

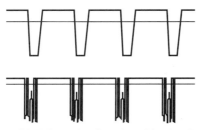

**FIGURE 12.3** Good and bad string velocity at the bowing point. From [11].

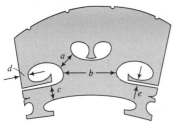

**FIGURE 12.4**   Normal violin bridge outline. From [6].

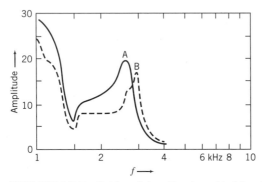

**FIGURE 12.5**   Bridge A (mellow) and bridge B (brilliant) frequency-response curves. From [6].

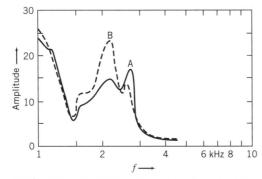

**FIGURE 12.6**   Bridge A (original) and bridge B (reduced cross section) response curves. From [6].

1. The frequency response of the bridge has a strong influence on total violin timbre.

2. The conventional bridge form is well suited to produce the desired timbre by relatively small changes in bridge dimensions.

3. Bridge influence is strongest for the two upper strings ($A_4 = 466.2$ Hz and $E_5 = 659.3$ Hz).

4. The stiffness of violin bridges should be between 1500 and 3000 N/cm.

5. The bridge is made of maple and its elasticity varies greatly, so it is advisable to measure the stiffness of the wood before constructing the bridge.

6. On the average, hard wood bridges produce darker violin sounds.

7. High bridges require thinner upper edges or a higher stiffness to obtain the same timbre.

The important consequence of all this is as follows: below 1 kHz, the bridge frequency response has little effect on the overall response. The bridge generally has a resonance between 2 and 4 kHz, and the frequency and magnitude of this resonance *does* have a significant effect.

## 12.5 VIBRATIONS OF THE BODY OF STRING INSTRUMENTS: MEASUREMENT METHODS

Here we concentrate on methods of measuring vibrations of the instrument body (item 4 in Section 12.2). Although the discussion refers to guitars, the methodology applies to the measurements of the body vibrations of other string instruments as well.

Once the upper plate of the guitar is set into vibration, the entire body plus the enclosed air is set into motion. Both the vibration of the wood body and the enclosed air radiate acoustic energy; air motion has resonances that are at lower frequencies than body resonances. Research has identified two low-frequency resonances of the guitar; one is the low-frequency resonance of the upper plate, and the other is the resonance of the body of enclosed air. Caldersmith [4] has treated the guitar as a reflex enclosure ("an externally driven plate...backed by an enclosed air volume which is vented to the external air with a hole or a tube").

Guitar body vibrations are complex. Various techniques have evolved to measure these vibrations. Among these are Chladni patterns, an external excitation method, and holography.

Chladni patterns: When sand or salt is sprinkled on a plate and then the plate is vibrated, the particles will tend to collect in the deepest regions. These regions are called nodal lines. They appear when the plate is vibrated at one of its resonant frequencies. Thus, by attaching a voltage-controlled transducer to the plate and varying the frequency of the sinusoidal voltage, one can determine the resonances of the plate.

External excitation method: Resonant modes may also be estimated by placing a microphone at some standardized distance from the plate and measuring the received signal as a function of frequency; vibrations are induced in the same manner as the production of Chladni patterns.

Holographic method:    A more accurate way of measuring resonant modes of a guitar top (or bottom) plate is through the use of optical holography. The concept of holography is explained by considering the *simultaneous* illumination of a specific object by *two* beams. We explain in the following paragraphs how this allows preservation of the phase information in the resultant wave front as well as storage of this information on photographic film. Then, by illumination of the film with one of the two beams (i.e., the reference beam), a near replica of the original wave front is perceived by the eye, and the beholder is fooled into thinking that he or she sees the original object.

Let the wave front obtained by illuminating an object of interest with a monochromatic light beam be represented as the two-dimensional vector $u(x, y)$, a complex function of the spatial coordinates $x$ and $y$. The reflected light from the object of interest is incident on photographic film. At the same time, a reference beam that reflects off the mirror $M_2$ is also incident on the film. Figure 12.7 shows how this is done.

Let the reference beam create the wave front $v(x, y)$. The total illumination on the film is the vector sum $h(x, y)$ of the two illuminations:

$$|h|^2 = hh^* = (u + v)(u^* + v^*),$$    (12.2)

$$|h|^2 = uu^* + vv^* + uv^* + vu^*,$$    (12.3)

where the asterisked vectors are complex conjugates. If we now assume that the reference illumination is constant over the recording surface, then the amplitude transmittance of the recorded film is

$$g = K + \gamma(uu^* + uv^* + vu^*),$$    (12.4)

where $K$ is a constant bias and $\gamma$ is a constant property of the film.

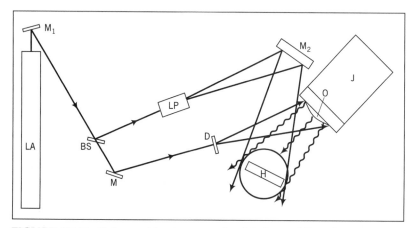

**FIGURE 12.7**   Holographic photography. LA, laser; M's, mirrors; BS, beam splitter; LP, lens–pinhole holder; D, diffuser; O, object of interest; J, jig; and H, holographic film. From [1].

Next, let's shine a beam of light *identical* to the reference illumination through the processed film; this amounts to multiplication of each term in Eq. 12.4 by $v(x, y)$; the result is

$$f = Kv + \gamma v |u|^2 + \gamma u |v|^2 + \gamma v^2 u^*. \tag{12.5}$$

The important item to notice is that the third term on the right side of Eq. 12.5 is an exact copy (except for a scale factor) of the original wave front! Thus, if we can succeed in eliminating or greatly diminishing the other three terms of Eq. 12.5, we should be able to view a *replica* of the original object by illuminating the film with the reference wave front.

Equation 12.5 is a static representation. How do holographic methods allow the determination of an object's movement? Go back to Eq. 12.3 and integrate over the exposure time $T$ of the film. Remembering that $v$, the reference, is constant, we can write

$$\int_{-T/2}^{T/2} |h|^2 \, dt \;=\; vv^* + \int_{-T/2}^{T/2} |u|^2 \, dt + v^* \int_{-T/2}^{T/2} u \, dt + v \int_{-T/2}^{T/2} u^* dt. \tag{12.6}$$

Let's focus attention on the third term on the right of Eq. 12.6 and define

$$F(x, y) = v^* \int_{-T/2}^{T/2} u(x, y, t) dt = v^* \int_{-\infty}^{\infty} p(t) u(x, y, t) dt, \tag{12.7}$$

where $p(t) = 0$ except in the interval $-T/2$ to $T/2$, during which it is unity.

Using Parseval's theorem,[1] we find

$$F(x, y) = v^* \int_{-\infty}^{\infty} \sin c(fT) U(x, y, f) df, \tag{12.8}$$

with

$$U(x, y, f) = \int_{-\infty}^{\infty} u(x, y, t) e^{-j2\pi ft} dt \tag{12.9}$$

and

$$\sin c(fT) = \frac{\sin(\pi fT)}{\pi fT}. \tag{12.10}$$

We can interpret the action of the film as performing a low-pass filtering in the frequency domain. To obtain the distribution of intensity on the film, we use the fact that the reference and object beam create an interference pattern that is time varying and dependent on the phase variation with time as the object vibrates. The geometry is shown in Fig. 12.8.

For simplicity we assume that the object is a flat plate moving in the $x$ direction. A light ray strikes the plate at coordinates $y$, $z$ at an angle $\theta_1$. It is reflected at an angle

---

[1] Parseval's theorem states that $\int_{-\infty}^{\infty} x(t) y^*(t) dt = \int_{-\infty}^{\infty} X(f) Y^*(f) df$ for the two transform pairs.

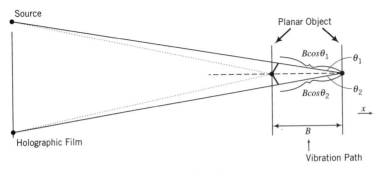

**FIGURE 12.8**  Holography of a vibrating plate.

$\theta_2$ and strikes the holographic film at the point $y_1, z_1$. If the maximum excursion of the object is $B$, we see from Fig. 12.8 that the change in path length of the incident wave over this maximum excursion is $B \cos(\theta_1)$, and the corresponding time difference is the same quantity divided by $c$, where $c$ here is the velocity of light. A similar argument holds for the reflected wave. If $\lambda$ is the wavelength of the light, and remembering that the phase $\phi$ is $2\pi f t$ and that $c = f\lambda$, we obtain the total phase excursion:

$$\Delta\phi = \frac{2\pi B[\cos(\theta_1) + \cos(\theta_2)]}{\lambda}. \tag{12.11}$$

Powell and Stetson [9] have shown that the intensity $I(y_1, z_1)$ is the product of the static intensity $I_{st}$ and the integral

$$\int e^{j\phi(y_1, z_1, t)} dt, \tag{12.12}$$

where $\phi$ is a sinusoidally varying phase with maximum value $\Delta\phi$ and the integration time depends on the film's sensitivity. This integral can be expressed as a Bessel function series, but since the film behaves like a low-pass filter, the interference pattern can be evaluated as the zeroth-order Bessel function $J_0(\Delta\phi)$:   from Eq. 12.11 we can write

$$I(y, z) = \left(J_0\left\{\frac{2\pi B[\cos(\theta_1) + \cos(\theta_2)]}{\lambda}\right\}\right) I_{st}. \tag{12.13}$$

Figure 12.9 shows $J_0(Kx)$ for three different values of $K$. As $K$ increases, the number of oscillations of $J_0(Kx)$ increases, but the value of $J_0(Kx)$ decreases more rapidly. Figure 12.10 shows the patterns of holographic reconstruction for different intensities; these can be directly related to the three plots of $J_0(Kx)$ of Fig. 12.9.

In Fig. 12.10, the dark lines correspond to the peak vibration amplitudes and the white fringes correspond to the nodes, or nulls. Figure 12.10 corresponds to the lowest mode, or resonant frequency of vibration. The evaluation of the sound at some point removed from the guitar body requires a spectral analysis, as the guitar plate is sinusoidally vibrated for different frequencies. This is a complicated procedure because the results obtained are

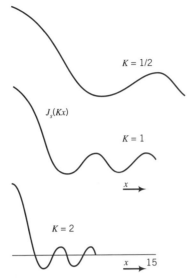

**FIGURE 12.9** Plots of a zeroth-order Bessel function for three values of $K$.

functions of the driving points, the frequencies, and the placement of the microphone, as well as the properties of the plate being tested.

## 12.6  RADIATION PATTERN OF BOWED STRING INSTRUMENTS

Meyer [7] experimentally investigated the directional characteristics of violins, violas, cellos, and contrabasses. Measurements were made in an anechoic chamber; thus, the measurements dealt with the direct sound from the instruments and did not deal with the

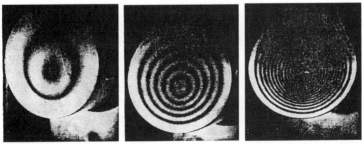

**FIGURE 12.10** Holographic reconstruction of a vibrating can bottom. Reconstructions of three holograms of a 35-mm film can bottom with a progressive increase in amplitude of excitation at the lowest resonance frequency of the can bottom. From [9].

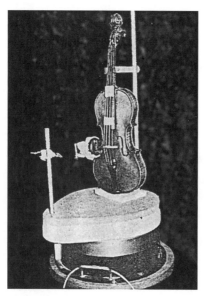

**FIGURE 12.11** Photograph of a violin mounted on a turntable. From [7].

reverberation patterns that would be generated in a concert hall or room. Reverberation is a very important component of the sound reaching the listener's ears, and it will be considered in Chapter 13.

In Meyer's experiment the instrument is mounted on a turntable, as shown in the sketch of Fig. 12.11. The string movements are damped and excitation is applied directly to the bridge by a small vibrating needle controlled by an electrodynamic oscillation system. The needle is vibrated in the same direction that is normally done by the strings. Radiated sound is measured by a microphone that is 1-m distant for violins and violas and 3.5-m distant for cellos and contrabasses. Recording is synchronized with turntable motion.

Figures 12.12–12.15 give overall results for the various instruments as a function of the frequency bands. Figures 12.12 and 12.13 show the radiation pattern of the cello in two planes. The general tendency is for the directivity to become sharper for high frequencies. For the cello frequencies 2000–5000 Hz, there are two main lobes.

Figure 12.14 shows the radiation pattern for the violin in the horizontal plane. For low frequencies (200–400 Hz), radiation is omnidirectional; between 1 and 1.5 kHz, the lobes are narrow, widening again as the frequency increases. We see from Fig. 12.15 (contrabass) that the main lobes versus frequency are very different than those of a violin.

How do anechoic measurements help determine the best arrangement of an orchestra for concert halls? We quote from Meyer's abstract: "These results suggest that different arrangements for the strings would be optimum for different concert halls and different styles of musical composition." In other words, the design of seating arrangements is a multidimensional problem, and Meyer's measurements constitute one of several considerations.

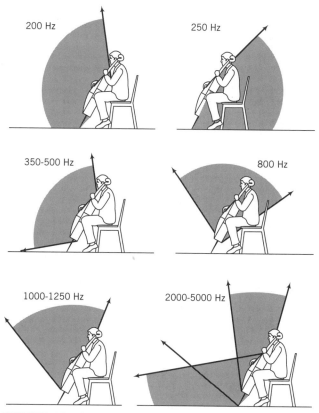

**FIGURE 12.12** Principal radiation directions of the cello in the vertical plane. From [7].

## 12.7 SOME CONSIDERATIONS IN PIANO DESIGN

The earliest piano that resembles present models was built by Bartolomeo Cristoferi (1655–1731). A version completed in 1720 still exists and is on display at the Metropolitan Museum of Art in New York City [5].

The piano strings (88 in a standard piano) vibrate when struck by the key-actuated hammers. These vibrations are transmitted via a wooden mount to the sounding board, which produces most of the sound. As Fig. 12.16 indicates, the action of a piano is quite complex.

The following conditions have to be satisfied in piano design [2]:

**1.** The hammer must rebound *immediately* after striking the string.

**2.** The string cannot be struck twice for a single note.

**3.** The pianist should be able to strike the key again before it returns to static position.

**4.** Tone should persist as long as the key is held down.

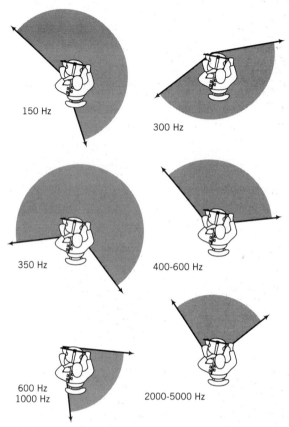

150 Hz

300 Hz

350 Hz

400-600 Hz

600 Hz
1000 Hz

2000-5000 Hz

**FIGURE 12.13** Principal radiation directions of the cello in the plane of the bridge. From [7].

From Fig. 12.16, here is a list of mechanical motions in the grand piano:

1. The performer pushes the key down.
2. The key lifts the capstan and raises the damper.
3. The wippen rotates and raises the jack.
4. The jack pushes on the roller.
5. The jack raises the hammer.
6. The jack strikes the jack regulator.
7. The jack rotates and no longer pushes on the roller.
8. The momentum of the hammer motion permits the key to be struck.
9. The hammer rebounds.
10. The hammer is caught by the back check to prevent a second striking of the string.

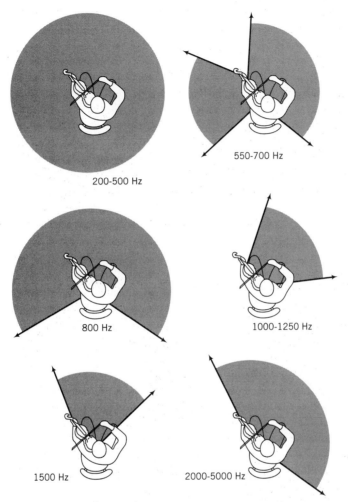

200-500 Hz

550-700 Hz

800 Hz

1000-1250 Hz

1500 Hz

2000-5000 Hz

**FIGURE 12.14** Principal radiation directions of a violin in the horizontal plane. From [7].

11. The performer allows the key to lift partially.
12. The jack upper end slopes back under the roller.
13. If the key is again depressed while it is halfway down, the hammer will strike the key again.
14. The key release allows the damper to fall back on the string.

This mechanism is repeated for each note.

Effect of the hammers: The piano hammers are made of heavy felt glued to a wooden body. If the surface of the hammer is hard, the tone is brighter (larger values of

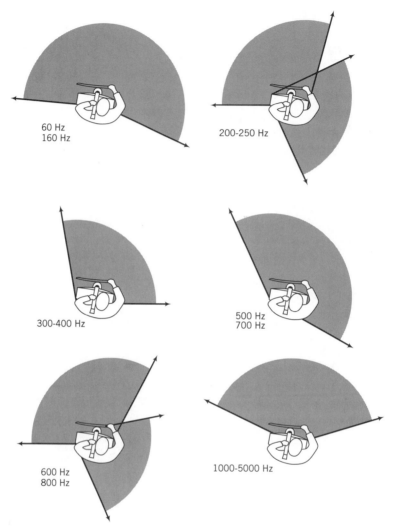

**FIGURE 12.15**  Principal radiation directions of a contrabass in the horizontal plane. From [7].

high-frequency harmonics). A soft surface creates softer tones. The piano tuner has various methods of changing the hardness of the surface.

The hammer strikes the string approximately one-seventh of the way along the string, as shown in Fig. 12.17. The precise excitation of the piano string by the hammer is a critical factor of the resulting piano tone. The choice of one-seventh is an empirically derived result.

Design of piano strings:    The frequency of a string is a function of its speaking length (the vibrating portion of the string), its tension, and its mass. Thus, the high notes of a piano use strings that are quite a bit shorter and thinner than the strings used for the bass notes. The

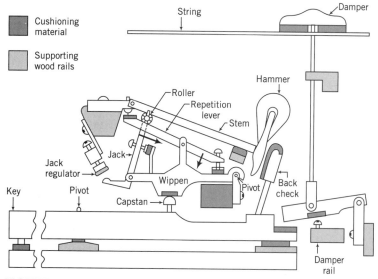

**FIGURE 12.16** Action of a grand piano. From [2].

piano designer is constrained by such factors as overall piano size (thus constraining the maximum length of low-frequency strings) and the tensile strength and fatigue resistance of the shorter strings (to prevent broken strings). Given a reasonable set of parameters for $L_{88}$ (the string length for the highest note), the following empirical formula gives a good approximation to any string length $L_n$ as long as $n$ is greater than approximately 20:

$$L_n = L_{88}(S^{88-n}), \tag{12.14}$$

with $S = 1.055$ and $L_{88} = 0.0508$ m.

Figure 12.18 shows the string length versus key number of a 2.74-m grand piano. The hexagons in the figure are based on Eq. 12.14, whereas the circles are actual string lengths of the piano.

Figure 12.19 shows a comparison of speaking lengths for three pianos. The squares in the figure are for a 0.91-m upright piano, the hexagons are for a 2.13-m grand piano, and the circles are for a 2.74-m grand piano. Notice that for keys above middle C, the larger

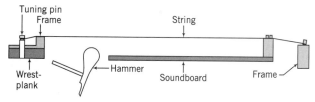

**FIGURE 12.17** Relative positions of the string and hammer. From [2].

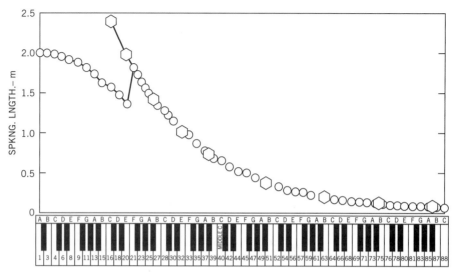

**FIGURE 12.18**  Theoretical and actual speaking lengths of a 2.74-m grand piano. From [5].

piano has longer strings. The upright piano has the least amount of discontinuity at middle C; the string lengths of the smaller grand also follow a fairly smooth trajectory.

Piano strings are usually made of steel to withstand the high tension. These strings usually have stiffness. This changes the overall response of the string; a fourth-order

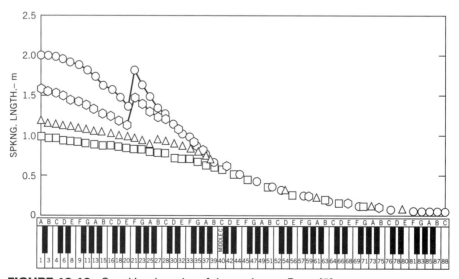

**FIGURE 12.19**  Speaking lengths of three pianos. From [5].

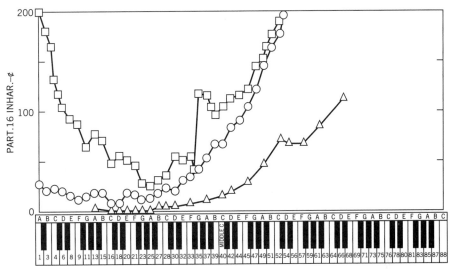

**FIGURE 12.20** Inharmonicity, in cents of 16th partial for three pianos. An octave is often given in cents; 1200 cents = 1 octave. A cent is defined logarithmically so that one cent corresponds to the ratio $2^{1/1200}$. From [5].

derivative is introduced in the differential equation [8], as follows:

$$T\frac{\partial^2 y}{\partial x^2} - QS\kappa^2 \frac{\partial^4 y}{\partial x^4} = \rho S \frac{\partial^2 y}{\partial t^2},$$ **(12.15)**

where $T$ is the tension, $S$ is the cross-sectional area, $\kappa$ is the radius of gyration, $\rho$ is the density, and $Q$ is the modulus of elasticity.

Sinusoidal waves can travel along such a wire, but the overtones are no longer exact multiples of the fundamental frequency. Although such a result may appear troublesome, it turns out that the slight inharmonicity is *liked* by most listeners. Figure 12.20 shows the deviation from a perfect harmonic series of three pianos for the 16th harmonic of notes as high as $A_5$. The vertical scale is given in cents deviation from harmonicity.

## 12.8  BRIEF DISCUSSION OF THE TRUMPET, TROMBONE, FRENCH HORN, AND TUBA

Each of these four brass instruments consists of a mouthpiece, a mouthpipe, a cylindrical section, and a bell. The reason for this design starts with the formula for the modes of a closed cylindrical pipe (See Chapter 10 on Wave Basics):

$$f_n = \frac{nv}{4L}, \quad n = 1, 3, 5, 7, \ldots,$$ **(12.16)**

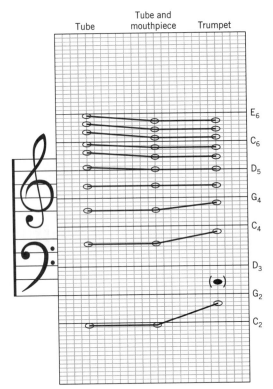

**FIGURE 12.21** The evolution of a trumpet: effects of the mouthpiece and bell. From [2].

where $v$ is the velocity of the sound wave (343 m/s in air), $L$ is the length of tubing, and $n$ refers to the $n$th mode. There are no modes for even values of $n$.

Figure 12.21 indicates how modification of the tube to construct the trumpet shifts the modes so that they are harmonically related for the trumpet.

The left column shows the modes of a simple closed tube that satisfy Eq. 12.16. Cutting off a length of pipe and replacing it by the nonuniform mouthpiece does not change the low-frequency modes but does affect the high-frequency modes, since the mouthpiece behaves like a tube closed at one end. The overall result is that the virtual length of the overall system increases with increasing frequency, with a consequent lowering of the modes, as shown in the center column of Fig. 12.21.

The bell moves the lower modes up. The tube is effectively shortened and the lower modes are shifted from the center column of Fig. 12.21 to the right column. The frequencies of the modes, beginning with the mode just below $C_4$, correspond to the second, third, and fourth (and so on) harmonics of the (fictitious) black note. This note is provided by the vibration of the player's lips at the fundamental frequency (the so-called pedal tone). The lowest mode (between $E_2$ and $F_2$) is not energized, since the player's lips provide no energy at this frequency.

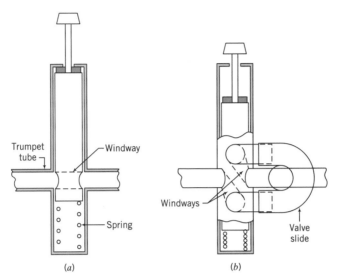

**FIGURE 12.22** Trumpet valve. From [2].

Figure 12.21 shows the result for one specific length of tubing, but the overall length can be varied by means of three valves of the trumpet. This would shift the various positions of the excited modes, but they would still be in harmonic relation to the fundamental frequency provided by the lips' vibrations.

The playing ranges of the brass instruments vary greatly; the tuba has a range from approximately 50–200 Hz, the trombone's range is approximately 80–600 Hz, the range of the French horn is approximately 70–700 Hz and the trumpet's range is approximately 150–1000 Hz.

Figure 12.22 shows a trumpet valve.

When the piston is up, air flows through the tube unmolested. Depressing the valve forces the air column to follow the arrows, thus increasing the effective length of tubing and lowering the mode frequencies.

## 12.9  EXERCISES

**12.1**  Describe a procedure whereby the complex conjugate of the original image is viewed by use of holographic techniques.

**12.2**  Prove Parseval's theorem by using both Fourier transform and $z$-transform notation.

**12.3**  Discuss the relationship between Fig. 12.9 and 12.10.

**12.4**  Most of the instruments we have discussed can be modeled as acoustic resonators excited by some form of vibration. For the human voice, a simplified representation of a vowel is that of a time-variable filter (the vocal tract), excited by glottal pulses. Devise analogous representations for the violin and trumpet.

**12.5**   Does bowing or plucking a violin string cause more high-frequency excitation? Justify your answer.

**12.6**   A piano is an example of a struck string instrument and a harpsichord is an example of a plucked string instrument. For the same note, the piano string has a higher tension and greater diameter. How does this affect the relative lengths?

**12.7**   Why are some notes on the piano produced by striking several strings at once?

**12.8**   What is the effect on the spectrum of a trumpet note when played with greater intensity?

**12.9**   Explain what is meant by regimes of oscillation in a trumpet.

# BIBLIOGRAPHY

1. Agren, C., and Stetson, K. A. "Measuring the resonances of treble viol plates by hologram interferometry and designing an improved instrument," *J. Acoust. Soc. Am.* 51 Number 6 (Part 2) 1972.
2. Backus, J., *The Acoustical Foundations of Music*, Norton, New York, 1968.
3. Benade, A. H., *Fundamentals of Musical Acoustics*, Dover, New York, pp. 394–396, 1976.
4. Caldersmith, G., "The guitar as a reflex enclosure," *J. Acoust. Soc. Am.* **63**: 1566–1575, 1978.
5. Conklin, H. A., "Design and tone in the mechanoacoustic piano," Parts I, II, and III, *J. Acoust. Soc. Am.* **99**: 3286–3296, **100**: No. 2, Part I, pp. 695–708, and **100**: No. 3, pp. 1286–1298, 1996.
6. Hacklinger, M., "Violin timbre and bridge frequency response," Acoustica **39**: 323–330, 1978.
7. Meyer, J. "Directivity of the bowed string instruments and its effect on orchestral sounds in concert halls," *J. Acoust. Soc. Am.* **51**: 1994–2009, 1972.
8. Morse, P. M., and Ingard, K. U., *Theoretical Acoustics*, Princeton Univ. Press, Princeton, N.J., 1968.
9. Powell, R. L., and Stetson, K. A., "Interferometric vibration analysis by wavefront reconstruction," *J. Opt. Soc. Am.* **55**: xx–xx, 1965.
10. Rossing, T. D., The Science of Sound, Addison–Wesley, Reading, Mass., 1990.
11. Woodhouse, J., Physical modelling of bowed strings, *Comput. Music J.* **16**: 43–56, 1992.

# CHAPTER *13*

# *ROOM ACOUSTICS*

**I**N PREVIOUS chapters, we described some of the fundamentals of the acoustics of tubes and strings, using abstractions that we showed to be relevant to the production of audio signals in the human vocal tract and in some musical instruments. Once these signals leave their sources, however, they generally encounter boundaries that are at least partially reflective. Thus, a listener or a microphone receives a multipath version of the original source signal. Therefore, room acoustics are also a fundamental concern for many audio signal-processing applications. In this chapter, we discuss the effect of room boundaries on a sound wave, the resulting phenomenon of reverberation (i.e., the smearing of a source sound over time as a result of these boundary effects), and the effect of reverberation on speech intelligibility. As with many topics discussed in this book, this chapter can only serve as a brief introduction, with a practical focus on factors that affect the goals of audio signal processing.

## 13.1 SOUND WAVES

At atmospheric pressure and standard conditions of humidity, the speed of sound is

$$c = 331.4 + 0.6T \text{ m/s,} \qquad \textbf{(13.1)}$$

where $T$ is the temperature of the air in degrees Celsius. At 20°C, the speed of sound is 343.4 m/s, or about 1127 ft/s, corresponding to a transmission time of approximately 1 ms for each foot. If a reflective boundary is 10 ft ($\sim$305 cm) from an acoustic source, it will take roughly 20 ms for the sound wave to return. This is too short for the second sound to be heard as a distinct echo. In contrast, if the boundary is 50 ft or more from the source, it will take at least 100 ms for the sound wave to return, which is long enough for the listener to hear a discrete echo. In many rooms, successive reflections are too close together to be audible as separable events.

In real acoustic environments, the speed of sound varies over time and space, as a result of factors such as temperature (as noted in the equation) and humidity. The nature of the received reflective pattern will be dependent on these sources of variability, as well as on movements of the sound source or receiver around the room. However, for a midfrequency sound wave, and for a given source and receiver position, the room acoustic effect can be seen as a linear time-invariant sum of attenuated, filtered, and delayed versions of the original signal. In other words, for this limited approximation, the room effect can be viewed as a linear convolution with an echo pattern. We will return to this perspective, but first we review acoustic wave propagation in the context of room acoustics.

### 13.1.1   One-Dimensional Wave Equation

In Chapter 10 we gave a one-dimensional equation relating space, time, and pressure for a progressive acoustic plane wave (repeated here in slightly different form):

$$\frac{\partial^2 p}{\partial x^2} = \frac{1}{c^2} \frac{\partial^2 p}{\partial t^2}, \tag{13.2}$$

with a general solution of the form

$$p(x,t) = F(ct - x) + G(ct + x), \tag{13.3}$$

where $F(ct - x)$ represents a wave moving toward increasing values of $x$, and $G(ct + x)$ represents a wave moving toward decreasing values of $x$.

For sinusoidal functions, $\lambda = c/f$ (i.e., wavelength equals the speed of sound divided by frequency). To give some physical intuition, if we assume that $c \approx 1000$ ft/s, then[1] a 20-Hz signal has a wavelength $\approx$50 ft long, a 1-KHz signal has a wavelength $\approx$1 ft long, and a 20-KHz signal has a wavelength $\approx$0.5 in. (1.27 cm) long.

In our previous examples (speech and music), we examined sound propagation in objects that were small in comparison to the wavelengths of low-frequency sounds (e.g., violins or human vocal tracts). In the case of room acoustics, we consider sound propagation in enclosures that are much larger. For wavelengths that are much smaller than room dimensions, ray paths from the source to the receiver (including reflective paths) can be traced to yield a reasonable approximation to the acoustic transmission characteristics.

It is often useful to think of sound in terms of wavelength, particularly for intuition concerning the extent to which a structure is a barrier to sound transmission. Noting the physical size (wavelength) of acoustic waves can also help provide intuition for another aspect of audio perception, known as head shadowing. The human head is a barrier of significant size to high-frequency sound waves. Thus, sound from the opposite side of the head is low-pass filtered.[2]

---

[1]Recall that the actual speed is closer to 1130 ft/s at room temperature; we assume the lower speed above for simplicity's sake only.

[2]This filtering effect is used to good advantage by the auditory system, helping to determine the direction of the arrival of sound waves. Spatial location is also assisted by the relative timing of the arrival of sound waves at each ear.

### 13.1.2 Spherical Wave Equation

Ignoring any directionality of a source, ignoring the effect of boundaries, and assuming a point source, the propagation of waves in three-dimensional space will be spherical. The wave equation in spherical coordinates is

$$\frac{\partial^2 p}{\partial r^2} + \frac{2}{r}\frac{\partial p}{\partial r} = \frac{1}{c^2}\frac{\partial^2 p}{\partial t^2}. \tag{13.4}$$

A solution is the complex exponential,

$$p(r, t) = P_0 \frac{e^{j(\omega t - kr)}}{r} = P_0 \frac{e^{j(2\pi/\lambda)(ct - r)}}{r}, \tag{13.5}$$

where $p$ is the pressure, $r$ is the distance between the source and the receiver, $P_0$ is the amplitude of the sinusoidal sum, $t$ is the time, $k = 2\pi/\lambda$, and $\omega = kc$. Note that the pressure is inversely proportional to $r$. This proportionality relationship is a reasonable approximation in free space and under many outdoor conditions.

### 13.1.3 Intensity

Intensity is defined as the amount of sound energy flowing across a unit area surface in a second. This is equivalent to the rms pressure times the rms velocity, or[3]

$$I = \overline{pv}. \tag{13.6}$$

This can also be expressed without the velocity term, which can be expressed as the ratio of pressure to impedance (as noted in Chapter 10), yielding

$$I = \frac{\overline{p^2}}{\rho_0 c}, \tag{13.7}$$

where $\rho_0$ is the medium density, $c$ is the speed of sound as before, and $\rho_0 c$ is the characteristic impedance. For a sinusoid,

$$I = \frac{P_0^2}{2\rho_0 c}. \tag{13.8}$$

So $I \propto p^2$, and for a spherical wave $p \propto 1/r$ implies $I \propto 1/r^2$.

In other words, as with many other interesting physical phenomena, sound intensity follows an inverse square law with distance – but don't forget the assumptions of a point source and lack of reflective boundaries! In addition, one must assume that the sound medium (air in the room acoustics case) does not dissipate any energy, which is particularly incorrect for high-frequency audio signals.

---

[3]For derivations of this and other relations that are simply stated here as facts, check acoustics texts such as [3].

## 13.1.4 Decibel Sound Levels

Conventionally, the difference between sound energy levels is measured in decibels:

$$L = 10 \log_{10} \left| \frac{I_1}{I_2} \right|. \tag{13.9}$$

Since the intensity is proportional to the square of the pressure,

$$L = 20 \log_{10} \left| \frac{p_1}{p_2} \right|. \tag{13.10}$$

The denominator values are often chosen to be reference values that correspond to the threshold of hearing at 1 kHz, namely

$$p_2 = 2 \times 10^{-5} \text{ N/m}^2 \quad \text{and} \quad I_2 = 10^{-12} \text{ W/m}^2.$$

For these choices, the decibel levels become the sound pressure level (SPL) and the intensity level, respectively. The two measurements are roughly equivalent for plane or spherical waves measured in the air.

## 13.1.5 Typical Power Sources

For speech in many natural situations, it is preferable to assume that the point source propagates hemispherically rather than spherically. Given this assumption, the following are typical power values for various speech sources. The number in parentheses indicates the SPL of the sound wave flowing across a 1-m$^2$ area 40 cm away from the source.

- Whispered speech:   1 nW (30-dB SPL)
- Average for speech:   10 $\mu$W (70-dB SPL)
- Loud speech:   200 $\mu$W (83-dB SPL)
- Shouted speech:   1 mW (90-dB SPL)

Ignoring boundaries, we find that the SPL would be 6 dB lower for each doubling of distance. Note that shouted and whispered speech differ by 60 dB, which is a power ratio of a million! This is certainly not comparable to the human sense of relative loudness; however, the cube root of intensity is often used as such a measure. Given this approximation, a 10-dB increase in intensity corresponds roughly to a doubling of perceived loudness. However, as we discuss in Chapter 15, loudness is frequency dependent. Perceived loudness is generally weaker for lower frequencies, though the apparent frequency response is flatter for louder sounds. The loudness control found on some stereos, as opposed to the volume, does some filtering to adjust for human sensitivity characteristics; roughly speaking, these controls usually bring up the extremes of low and high frequencies at low volumes so as to compensate for the reduced perceived bandwidth at the lower volume level.

Sound-level meters often make use of standard frequency-weighting characteristics in order to more closely approximate loudness, as opposed to acoustic energy. This certainly makes sense for measurements that will be relevant for human perception, such as the noise level in a building. One of the most common standards is the A-weighted measurement, typically used for a low to moderate loudness level, in which a large energy-loudness correction must be made for the low frequencies (reducing sensitivity to low frequencies) [6]. This is probably appropriate for applications such as describing the noise levels in a typical acoustic background, since a low-frequency rumble could have a very large amplitude before it had the same perceptual effect as a midfrequency sound. Sound-pressure-level meters commonly have an A setting for this purpose; B and C settings correspond to loudness corrections for successively higher sound pressure levels, and they apply smaller low-frequency compensations. SPL measurements for the A-weighted case are often referred to as dBA.

For machine speech-analysis purposes, these weighting curves can often be misleading, since a large low-frequency noise can sometimes have strong effects on our algorithms but will have a reduced effect on dBA measurements.

## 13.2  SOUND WAVES IN ROOMS

As suggested earlier, the mathematics used for acoustically modeling a room as a box is similar to that used for abstractions of strings and tubes. Solving the wave equation will result in characteristic resonances that will be instantiated as standing waves. For an ideal box, the lowest-frequency resonance will correspond to a wavelength that is twice the size of the room's longest dimension. Any room will actually have a large number of such resonances, and at higher frequencies these can be approximated by a continuum; as explained in [4], the number of modes (eigenfrequencies) below frequency $f$ is proportional to $f^3$.

For sound-production devices, such resonant phenomena are often the most critical aspect. However, resonances are often not the most important aspect of room acoustics for the study of the how speech and music will be perceived by listeners and machines. In particular, the biggest effect on the intelligibility of speech comes from the effects of reverberation, which is more fundamentally a time-domain phenomenon.

Thus far in this chapter we have noted that a point source emanates a spherical wavefront whose intensity is inversely proportional to the square of distance. In an enclosure, however, boundaries will reflect some of the energy, causing the receiver to get a series of delayed and attenuated versions of the original signal. The pattern of these returns establishes the audio character of the room, both for intelligibility and for sound quality (e.g., for the musical quality of a concert hall). These characteristics are determined by the geometry of the room, the positions of the source and receiver of sound, and the absorption characteristics of the boundaries. We separately describe the long-term and short-term effects of the room echo response, as they are qualitatively quite different. We ignore more complicated effects (such as diffusion from complicated surfaces).

### 13.2.1 Acoustic Reverberation

When a wave front impinges on a boundary, part of the energy is reflected, and part is absorbed; absorption includes both energy that is transmitted and energy that is dissipated into heat. A key part of architectural acoustics, then, is the use of measurements of the fraction of energy that is absorbed, called the absorption coefficient. These coefficients are typically frequency dependent; most ordinary materials absorb better at high frequencies (or, conversely, reflect better at low frequencies). For example, Table 13.1 shows that a typical coefficient for acoustic paneling is 0.16 at 125 Hz, but it is 0.80 at 2 kHz. Some materials that are less absorptive overall show less frequency sensitivity. For instance, glass typically has a coefficient of 0.04 at 125 Hz and 0.05 at 2 kHz. For nearly any real room, the effect of this frequency dependence is to shorten the time for reflections to die down for a higher-frequency energy; put another way, contributions to the current sound from previously emitted sounds are low-pass-filtered (as well as delayed and attenuated) versions of the original.

For large distances, high frequencies, or both, sound absorption of the air can cause an additional low-pass-filtering effect. In addition to the inverse square attenuation, the energy absorption of the air can be approximated by a factor of $e^{-mr}$, where $m$ is 0.0013/m at 1 kHz and 0.021 at 4 kHz (for a relative humidity of 50%).

**TABLE 13.1  Effective Absorption Coefficients of Common Building Materials[a]**

| Material | Frequency in Hz | | |
|---|---|---|---|
| | 125 | 500 | 2000 |
| Acoustic paneling | 0.16 | 0.50 | 0.80 |
| Acoustic plaster | 0.30 | 0.50 | 0.55 |
| Brick wall, unpainted | 0.02 | 0.03 | 0.05 |
| Draperies | | | |
|   light | 0.04 | 0.11 | 0.30 |
|   heavy | 0.10 | 0.50 | 0.82 |
| Felt | 0.13 | 0.56 | 0.65 |
| Floor | | | |
|   concrete | 0.01 | 0.02 | 0.02 |
|   wood | 0.06 | 0.06 | 0.06 |
|   carpeted | 0.11 | 0.37 | 0.27 |
| Glass | 0.04 | 0.05 | 0.05 |
| Marble or glazed tile | 0.01 | 0.01 | 0.02 |
| Plaster | 0.04 | 0.05 | 0.05 |
| Rock wool | 0.35 | 0.63 | 0.83 |
| Wood paneling, pine | 0.10 | 0.10 | 0.08 |

[a]From [3]. The coefficients listed here are only representative.

Whereas the early pattern of reflections can be an important cue as to the size and shape of the room (as well as the distance to the sound source), the reflections usually become quite dense within 100 or 200 ms of the first (direct) wave front. This dense pattern, which tends to have a roughly exponential decay, is typically characterized by the time that it takes a steady-state noise signal to decay by 60 dB, a value that is referred to as RT60.

There are a number of ways to derive a formula for RT60, as elaborated in such sources as [3] or [4]. Some use a stochastic approach, assuming independence of reflection, whereas others begin with a first-order differential equation to approximate the balance between absorption and generation. In either case the simple approximations lead to an exponential decay, which on the average conforms to the long-term response in regular rooms. In the 1920s a Harvard professor named Sabine showed empirically that a particularly simple approximation was a reasonable predictor of reverberation time (ignoring air absorption), namely

$$\text{RT60} = \frac{0.163V}{S\bar{\alpha}} \tag{13.11}$$

in MKS units, and

$$\text{RT60} = \frac{0.049V}{S\bar{\alpha}} \tag{13.12}$$

for English units (feet), where $S$, $V$, and $\bar{\alpha}$ are the surface area, volume, and average absorption of the room, respectively.

Including the effect of air absorption,

$$\text{RT60} = \frac{0.049V}{S\bar{\alpha} + 4mV} \tag{13.13}$$

in feet, where $m$ is the acoustic air-absorption term given above. This air term typically dominates at very high frequencies, and it is largely irrelevant for low frequencies.

For concert hall acoustics, a rule of thumb is that 75% of the absorption is contributed by the audience and orchestra, so that measurements taken in an empty hall must be interpreted with care.[4]

Reverberation has a number of effects on acoustic signals such as music and speech, most obviously to smear them out in time. When the reverberant energy is large (for instance when the distance between source and receiver is large, and when RT60 is long), syllable onsets and identities can be masked by decaying energy from previous syllables. This can hurt intelligibility, particularly when combined with noise. For music, a degree of smearing that is considered appropriate for some forms of music (such as Bach organ pieces) can cause an undesirable loss of definition for others (such as string quartets).

As noted at the start of this chapter, the net effect of reverberation on a sound signal for a *particular* source position and receiver (microphone or ear) position may be roughly expressed as a linear convolution of an echo response with the source signal. A simple approximation to this response is an exponentially decaying impulse response, such as that implied by the RT60. However, although the reverberation time is important, the

---

[4]For this rule of thumb and many other fascinating observations about 76 famous concert halls, see [1].

corresponding exponential impulse response gives a poor match to many of the important features that characterize a room's acoustics. There are, however, a number of ways of more accurately estimating the impulse response in a real room (for one specific source and receiver position). One of the most popular is to emit white noise or pseudorandom sequences from a calibrated source and correlate them with the received signal at the microphone. This can be easily shown to yield an estimate of the desired impulse response. For linear time-invariant systems in general, $R_{xy} = R_{xx} * h(t)$, where $R_{xy}$ is the correlation between processes $x$ and $y$, $R_{xx}$ the autocorrelation for $x$, and $h(t)$ is the impulse response that is convolved with $x$ to yield $y$. Then if $R_{xx} = \delta(t)$ (uncorrelated input sequence), the measured output correlation $R_{xy}$ is equivalent to the desired impulse response.

In an alternative approach, a chirp (a signal with an instantaneous frequency that rises with time) can be used as the test source. The output signal is then phase adjusted to compensate for the timing of the different sinusoidal components in the source, resulting in an estimate of the transfer function (Fourier transform of the impulse response) of the linear model for the source–receiver relationship. Figure 13.1 show an impulse response that was measured through the use of a chirp-based approach. The response was measured in an experimental chamber at Bell Labs that has walls with adjustable absorption characteristics. The adjustable characteristics were set to yield an RT60 of 0.5 s. Figure 13.2 shows two

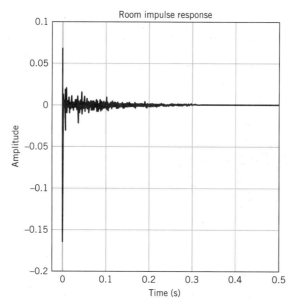

**FIGURE 13.1**   Measured impulse response from the experimental varechoic chamber at Bell Labs. This was part of a collection by Jim West and Gary Elko, along with Carlos Avendano, who kindly passed it on to us. The room reverberation time was set to be roughly 0.5 s, the distance from the source to the microphone was 2 m, and the shutters in the varechoic chamber were 43.7% open. The time of flight from the sound source to the microphone was removed from the impulse response display, so the plot doesn't illustrate that sound travels at finite speed.

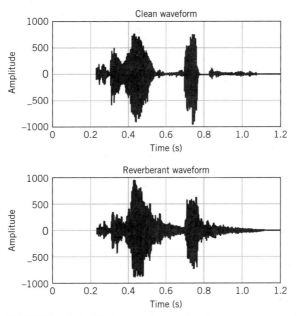

**FIGURE 13.2** Two waveforms for the continuous phrase "two oh six," uttered by a female speaker over the telephone. The first waveform is the original signal, and the second was derived from the first by convolution with the impulse response shown earlier. The phrase is part of the Numbers data base from the Oregon Graduate Institute.

versions of the waveform for a sequence of spoken numbers: one that was taken directly from the amplified and digitized microphone signal, and one that was artificially reverberated by using the measured impulse response from the Bell Labs recording setup. Note that although time smearing is evident, most of the basic energy features are still intact for this case; in fact listeners do not seem to have any trouble understanding what is said in this example. However, it is difficult to make speech-recognition models that will recognize the reverberated forms when they have been trained on the original versions.

Reverberation typically increases the loudness at a given location, both because energy generated over a range of time in the past is received in the present, and also because sound that would have radiated away in an open space is instead received by a listener in a closed space. Nonzero reflectivity of boundaries is an essential part of sound-reinforcement systems – it can be difficult to provide intelligible speech to listeners in rooms that are overly absorbent. Intelligibility is particularly aided by contributions to loudness from the early components in the impulse response.

## 13.2.2 Early Reflections

Prior to the arrival of a significant energy density, reflections are relatively sparse, and an exponential energy decay is typically not a useful description. Apparently these early

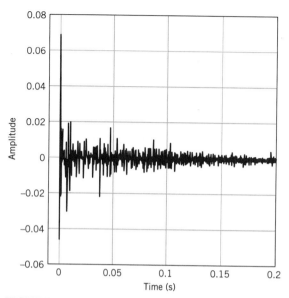

**FIGURE 13.3**   Zoom of the first 200 ms of the measured impulse response, in which the initial point (the direct sound) is omitted in order to scale up the rest of the sequence.

reflections (i.e., in the first 80–100 ms) provide critical cues for the listener's sense of intimacy or apparent room size and character [5]. A critical parameter characterizing these early reflections for concert halls is the initial time-delay gap, which is defined for concert halls as the delay between the time of receipt of the direct sound and the time of the first echo, as measured at a point midway between the orchestra and the back of the hall (or overhang of a balcony when there is one in the back). As noted in [1], the best concert halls rated by conductors are those that have an initial time-delay gap of 15–30 ms. In the case of speech, additional energy that is received from these early reflections is integrated by the listener into the apparently direct sound, which typically enhances intelligibility and quality.

Figure 13.3 shows an expanded version of the first 200 ms of the impulse response shown earlier. Note that the early response tends to look roughly like a series of discrete echos, whereas by 100 ms the sequence has a significantly more continuous character.

## 13.3   ROOM ACOUSTICS AS A COMPONENT IN SPEECH SYSTEMS

As should be clear from the preceding sections, the acoustics produced in a room undergo significant processing as a result of a multipath transmission from source to receiver. When the receiver is a human listener, moderate reverberation can improve intelligibility by increasing the signal-to-noise ratio: more signal gets to the listener. For a constant signal level received by the listener, however, reverberation tends to hurt intelligibility, particularly for large amounts of reverberant energy and for long reverberation times. This is particularly true when the signal-to-noise ratio is poor.

Even a small amount of reflected energy can effect the spectral or apparent spatial character of the sound for a human listener; but such small effects do not generally hurt intelligibility. Thus, the effects of the early reflections, as noted earlier, are largely to modify the listener's impression of the rooms size or intimacy. However, this is often not the case for machine signal processing of audio signals such as speech. Particularly for the speech-recognition application, even relatively short-term echo patterns modify the signal representations so as to be inconsistent with stored representations collected in different (i.e., nonreverberant) environments, such as speech collected from a close-talking microphone. Significant additive components from previous sounds appear to hurt machine recognition systems seriously for much smaller reverberation effects than are required to hurt intelligibility for humans.

The most common approach to handling these problems is to use a directional (noise-canceling) microphone quite close to the talker's mouth. In addition to improving the signal-to-noise ratio, such an arrangement keeps the direct-to-reverberant energy ratio high, so that the room acoustic effects are minimized. Unfortunately, there are many situations in which the microphone cannot be placed in this way. There has also been a significant amount of research on microphone arrays (see, e.g., [2]). In these studies, processing techniques such as beam forming or matched filtering have been applied to signals from as many as 200 microphones. These approaches improve spatial selectivity and thus reduce the effects of room acoustics on speech intelligibility and recognition performance.

## 13.4 EXERCISES

**13.1** On one hand, explain how a reflective ceiling in a lecture hall could potentially improve the intelligibility of speech. On the other hand, explain how reflective surfaces could hurt speech intelligibility.

**13.2** For an omnidirectional point sound source, the received sound level at a microphone is an 80-dB SPL 10 ft away in an anechoic chamber (a room with essentially no reflective boundaries). What is the SPL at a 20-ft distance? If the same source and microphone are put in a highly reverberant room for which the energy from reflections in the first 40 ms is much smaller than the energy from later reflections, what is the corresponding change in decibel SPL when the microphone distance is doubled?

**13.3** A rectangular room has a 12-ft high ceiling and is 20 ft wide and 30 ft long. The ceiling is made of acoustic paneling and the floor is wooden. The other walls are 25% glass and 75% wood paneling.

**(a)** What is the approximate RT60 at 125, 500, and 2000 Hz, ignoring air absorption?

**(b)** Let one corner of the floor be the position of the origin, and let $(x, y, z)$ be the position in feet along the 20-ft wall, 30-ft wall, and the height (respectively). For a sound source at (10, 10, 3) and a receiver at (10, 18, 3), find the propagation time for the direct sound. Also find the propagation time for the first reflection that is received after the direct sound.

# BIBLIOGRAPHY

1. Beranek, L., *Concert and Opera Halls – How They Sound*, Amer. Inst. Physics, New York, 1996.
2. Flanagan, J., Surendran, A., and Jan, E., "Selective sound capture for speech and audio processing," *Speech Commun.* **13**: 107–122, 1993.
3. Kinsler, L., and Frey, A., *Fundamentals of Acoustics*, Wiley, New York, 1962.
4. Kutruff, H., *Room Acoustics*, Applied Science, London, 1973.
5. Morgan, N., "Room acoustics simulation with discrete-time hardware," Ph.D. Thesis, University of California at Berkeley, 1980.
6. Tremaine, H., *Audio Cyclopedia*, Sams, Indianapolis, Ind., 1978.

# PART IV

# AUDITORY PERCEPTION

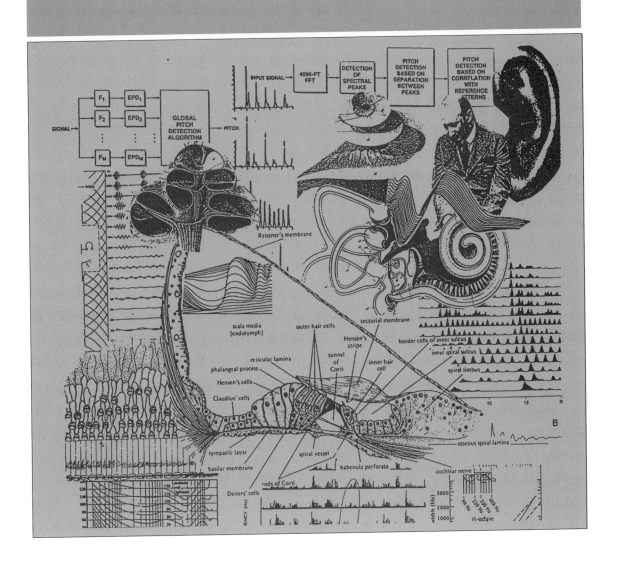

Man, if you gotta ask, you'll never know.

—Louis Armstrong (on being asked what jazz is)

**D**OES A falling tree make a sound if no one is there to hear it? By the physical definitions of Chapter 13, it most certainly does. Nevertheless, in this case the sound may not matter to us. The way in which sounds are interpreted by human hearing is obviously critical to speech communication and music appreciation. Similarly, coding and synthesis systems rely on some explicit or implicit model of how the acoustic signal is perceived by listeners. In the case of recognition, it is less obvious that human mechanisms for hearing are important, since in principle one could derive a completely artificial pattern-recognition system. However, human speech recognition is the one working example of a general and robust system for speech recognition and understanding. For all of these reasons, we believe that an introduction to the basic elements of auditory perception is critical to the goals of this text.

Chapter 14 provides a review of the peripheral auditory system, from the outer ear through the auditory nerve. Chapter 15 introduces psychoacoustics, focusing on the spectral resolution of human hearing. Such concepts are particularly important in light of the major emphasis on spectral estimation in speech and audio applications. Pitch analysis by machine has been a key component of many speech-analysis systems and has often benefited from knowledge of pitch perception, which is described in Chapter 16. All three of these chapters concern human hearing. Chapters 17 and 18 focus on perception of the particular sound signal that is speech. Chapter 17 provides an introduction to the vast literature on this topic, whereas Chapter 18 specializes further with a discussion of the recognition of speech (i.e., syllables or word sequences). As we move away from the relatively well-understood mechanisms in the auditory periphery (i.e., the ear), the description necessarily becomes more behavioral; that is, since we don't know how the brain really recognizes speech, we tend to report the results of recognition experiments rather than describe physiology. However, even in this case it can be useful to understand the comparison between human and artificial recognition performance, a topic that is introduced in Chapter 18.

# CHAPTER 14

# *EAR PHYSIOLOGY*

## 14.1  INTRODUCTION

The human ear is able to perform very useful signal processing on incoming signals. For example, there are auditory mechanisms for making sense of target signals despite noisy environments. Fine frequency and intensity differences can be measured by the ear. Von Helmholtz [4] proposed that the auditory nerve processes sound *tonotopically*; that is, by having different nerve bundles be sensitive to different frequencies. This notion of the auditory system as a sophisticated filter bank persists today and is the basis of much auditory research.

While the anatomy (i.e., the structure) of the auditory system in most animal species is fairly well understood, we still have a long way to go toward comprehending the physiology (i.e., the function of the components). There are a great many similarities among the auditory systems of many animals, including humans. Thus, in this chapter we survey the physiological knowledge garnered from animal studies. In the next chapter we survey psychophysical studies on human subjects; in succeeding chapters we use this knowledge to try to erect plausible models of how the human auditory system perceives the pitch of speech and music and how it perceives speech.

## 14.2  ANATOMICAL PATHWAYS FROM THE EAR TO THE PERCEPTION OF SOUND

The neocortex is that large part of the human brain that ultimately determines the nature of sensory input such as auditory stimuli. Here we trace the pathways that lead from the outer ear to this cortical percept. Figure 14.1 is a diagrammatic sketch of this pathway.[1]

Figure 14.1 shows the ascending pathways from the right cochlea to the cortex. Notice that there are both right and left versions of the intermediate neural nuclei. The right cochlea shown at the bottom right is part of the peripheral auditory system and its fibers innervate (excite) the right cochlear nucleus, but we see that there are pathways from there to both the right and left way stations. The left ear follows a comparable path (not shown to avoid confusion). Fibers that follow these ascending pathways are called afferent (feedforward).

There are also feedback mechanisms in the auditory system. Figure 14.2 shows half of the descending paths:  those ending at the right cochlea. We see that there are efferent (feedback) neurons from most of the way stations that eventually terminate in the periphery.

---

[1]Figures 14.1 and 14.2 omit many details of the auditory pathways. Only the most studied nucleii are shown.

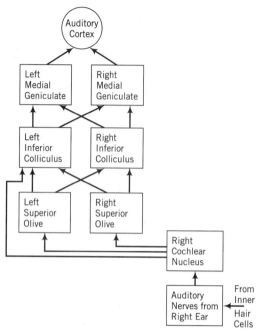

**FIGURE 14.1**   Auditory pathways linking the right ear to the brain.

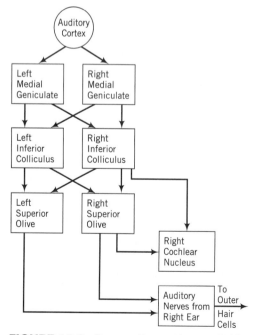

**FIGURE 14.2**   Descending pathways linking the brain to the right ear.

**TABLE 14.1   Cells in the Auditory Nuclei of the Monkey[a]**

| Central Auditory Nucleus | Number of Cells |
|---|---|
| Cochlear nuclei | 88,000 |
| Superior olivary complex | 34,000 |
| Nuclei of lateral lemniscus | 38,000 |
| Inferior colliculus | 392,000 |
| Medial geniculate body (pars principalis) | 364,000 |
| Auditory cortex | 10,000,000 |

[a]From [12].

There are approximately 30,000 auditory neurons associated with each cochlea in humans. Approximately 1000 of these neurons connect to approximately 20,000 outer hair cells. The 3500 inner hair cells connect to the nearly 30,000 remaining neurons. Table 14.1 is a list of cells in the auditory nucleii of the monkey. From the numbers in this table, it seems that the knowledge still to be obtained vastly exceeds the knowledge presently known.

Although the anatomy of the auditory path is fairly well understood, the physiology is still only partly understood. Many measurements on cats and other mammals have been made at the periphery, the nerve bundle leading from the inner ear into the cochlear nucleus. The cochlear nucleus (the next stage of neural processing beyond the peripheral auditory nerve) is only partially understood. There is a much greater variety of functions in this neural nucleus than there is in the auditory nerve, and if this pattern continues as we work our way up through all the pathways, it will be many years before sufficient physiological information is available so that scientists can propose a plausible model of auditory function. For these reasons, most of this chapter focuses on the peripheral auditory system; in particular, it focuses on the inner ear, which contains the cochlea.

## 14.3   THE PERIPHERAL AUDITORY SYSTEM

Figure 14.3 shows the three components of the peripheral auditory system: the outer, middle, and inner ears. The input is an acoustic signal and the output is a collection of neural spikes that enter the brain, as indicated in Figs. 14.1 and 14.2.

The auditory canal is an acoustic tube that transmits sound to the eardrum, where acoustic energy is transduced to vibrational mechanical energy of the middle ear. The middle ear consists of three very small bones (ossicles); the malleus is attached to the other side of the eardrum and its vibration is transmitted through the incus to the stapes. The stapes motion impinges on the oval window of the inner ear. The oval window is a flexible membrane, and its motion sets the fluid within the cochlea in motion. This motion is transmitted to the basilar membrane within the cochlea. The final transducing medium is the collection of hair cells sitting atop the basilar membrane that implement the transformation to the neural spikes of the auditory nerve bundle. The semicircular canals (not shown in the

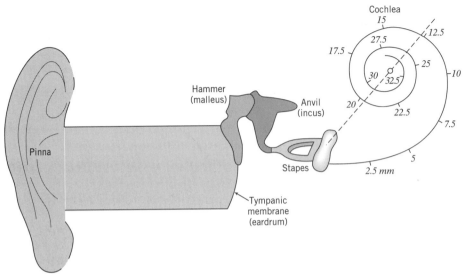

**FIGURE 14.3**  The peripheral auditory system. Boundaries between outer, middle, and inner ears are approximate.

figure) are part of the vestibular system that controls the sense of balance and are not part of the organs of hearing.

As seen in Fig. 14.3, the shape of the cochlea resembles that of a snail, but we can better picture what happens by looking at Fig. 14.4. In this figure we have unwound the snail. Where the stapes impinges on the oval window is called the base; the far end (deep inside the snail) is the apex. Near the base, the basilar membrane (BM) is relatively narrow and stiff; near the apex it is wider and less stiff, with the result that high frequencies excite the basal portion and vibrations die out as they approach the apex. At low frequencies, vibrations begin at the base but reach peak amplitude further down, as seen in the figure. It is important to realize that high- and low-frequency disturbances created by stapes motion arrive at their respective peak basilar membrane points nearly simultaneously, because wave propagation · is predominantly a fluid phenomenon and these traveling waves are appreciably faster than wave propagation on an isolated basilar membrane. This leads to the supposition that the BM action is akin to a filter bank, and in Chapter 19 we shall see that much research activity has centered around this concept. Figure 14.5 is a pictorial representation of BM activity.

The importance of the BM derives from the location, on the BM, of the auditory transducers, the hair cells. Motion of the hairs, or  stereocilia, causes firing of the auditory nerves that  innervate (connect to) the hair cells, and it is the spikes produced by the auditory neurons that relay *all* auditory information to higher brain centers.

## 14.4  HAIR CELL AND AUDITORY NERVE FUNCTIONS

Figure 14.6 shows how stereocilia motion (caused by BM motion) leads to neural spiking of the auditory nerve that is connected to the corresponding hair cell. To follow this activity,

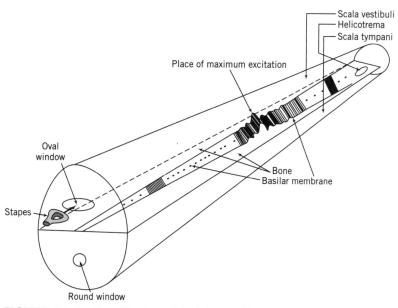

**FIGURE 14.4** Simplified model of the cochlea.

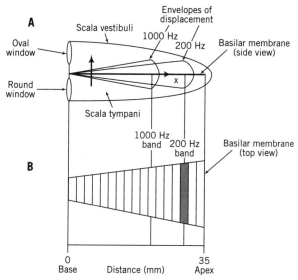

**FIGURE 14.5** Pictorial representation of activity along the basilar membrane. From [3].

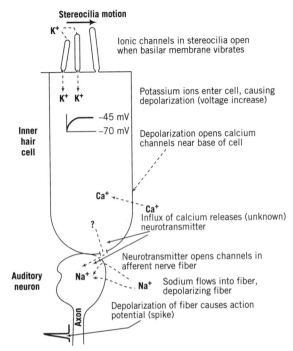

**FIGURE 14.6**  Neural spiking produced by hair-cell stereocilia motion.

we need to make a slight detour in our discussion. All cells are inside a membrane that can permit molecules to flow from inside to outside of the cell (and vice versa). Control of this chemical flow is through molecules that are embedded in the cell membrane. Many of the chemicals both inside and outside the cell are charged so that this molecular flow can cause changes in the voltage potentials of the cells. Thus, in the case shown in Fig. 14.6, it is the flow of potassium and sodium ions that create the voltages needed to trigger neural spikes (action potentials).

In Fig. 14.6, depolarization is equivalent to making the voltage difference from inside to outside the cell more positive, because of the flow of positively charged sodium ions from outside to inside. For this flow to occur requires the presence of a neurotransmitter that opens channels in the auditory fiber. The precise mechanism of spike generation involves an understanding of the Hodgkin–Huxley model of neural firing [11], which is beyond the scope of this brief synopsis.

## 14.5  PROPERTIES OF THE AUDITORY NERVE

Many experiments have been performed on the auditory nerves of small mammals. Cats have been the mammal of choice in most cases because the auditory system of the cat closely

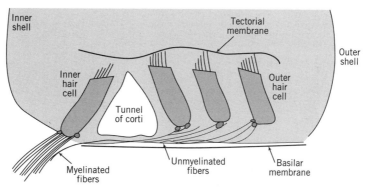

**FIGURE 14.7** Sketch of the inner and outer hair cells in a cross section of the cochlea.

resembles that of the human. The properties that have been evaluated include spontaneous spike generation, adaptation, tuning, synchrony, and various nonlinear effects.

Figure 14.7 is a drawing of some of the anatomy inside the cochlea. We see that three rows of outer hair cells and a single row of inner hair cells sit on the basilar membrane. The stereocilia of the hair cells impinge on the tectorial membrane, and the resultant forces open channels in the hair cells that eventually can cause spiking in the cochlear (auditory) nerve bundle shown in the lower left of the figure. Figure 14.8 is a schematic diagram of the innervation patterns; notice the three rows of outer hair cells (OHCs) and the one row of inner hair cells (IHCs), but also notice that approximately 90% of the afferent (ascending) auditory neurons come from this inner single row, whereas most of the efferent (descending) neurons go to the three outer rows.

Physiological measurements have uncovered some general properties of auditory nerves, including adaptation, tuning, synchrony, and nonlinearity (including masking).

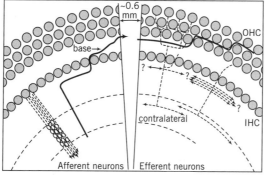

**FIGURE 14.8** Schematic diagram of inner (IHC) and outer (OHC) hair cells. From [13].

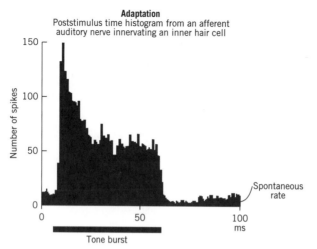

**FIGURE 14.9**   Adaptation by an auditory nerve. From [7].

Adaptation:    When a stimulus is suddenly applied, the spike rate of an auditory nerve fiber rapidly increases. If the stimulus remains (such as a steady tone), the rate decreases in an exponential manner to a steady value. Figure 14.9 shows a poststimulus time histogram for an afferent auditory nerve innervating an inner hair cell.

The poststimulus time histogram is obtained by the use of the following procedure: the tone is applied to the ear and the time of each spike referred to the initiation of the tone burst is measured. The tone is applied many times so that a histogram of spike-arrival times is obtained. We see from the figure that in the absence of the stimulus the neuron produces spikes at a small but discernable rate. This is called the spontaneous rate. At the beginning of the tone burst the neuron responds with many spikes, but after approximately 20 ms a steady rate is achieved. When the tone is removed, the spike rate decreases to slightly lower than the spontaneous rate for a short time before resuming its normal spontaneous rate. Therefore, Fig. 14.9 shows us that the neuron is more responsive to *changes* than to steady inputs.

Tuning:    We have mentioned that the action of the basilar membrane resembles that of a bank of tuned filters and that the tuning frequency is a function of the position on the BM. We also know that different hair cells lie on different parts of the BM. Thus, it is no surprise that auditory nerves have tuning properties. Figure 14.10 shows an example.

Figure 14.10 is obtained by the use of the following procedure:    for a given frequency, a 50-ms tone burst is applied every 100 ms. The sound level is gradually increased until the spike discharge rate is increased by one spike/s, at which time the sound pressure level (or SPL, as defined in Chapter 13) is recorded. This procedure is repeated for all frequencies to create the curves shown in the figure. The lowest values of these curves correspond to the frequency at which the nerve is most sensitive and therefore to its resonant frequency. Thus, these neural tuning curves look like the inverse of a bandpass filter tuning curve.

The results obtained here can be contrasted with von Bekesy's [2] measurements on basilar membranes. Treated as bandpass filters, the neural measurements result in appreciably narrower filters than von Bekesy's BM filters. For some time, physiologists

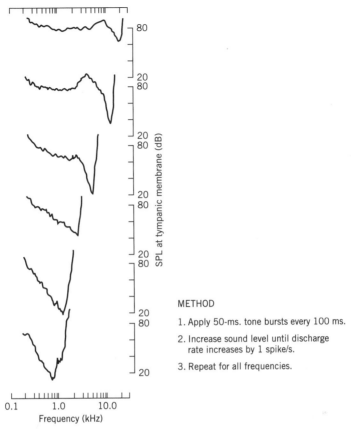

**FIGURE 14.10**  Tuning curves of six auditory nerve fibers. From [6].

tried to erect models to explain this discrepancy, but recent, more refined measurements on BM motion indicate agreement between BM and neuron. This allows us to advance the hypothesis that the primary component of an auditory fiber's tuning properties is the basilar membrane motion and that the curves of Fig. 14.10 are determined mainly by the BM motion and the position of the hair cell that innervates the neuron. There is, however, evidence that the hair bundles (stereocilia) of the hair cells also contribute to the tuning curves [5].

Also notice the shape of the tuning curves. The left side of the resonance curve is a relatively slow function of frequency, whereas the right side has a much steeper slope. This fact also correlates well with the BM response to tones. For example, Fig. 14.5A shows the BM response to a 1000-Hz tone; the envelope of BM displacement is seen to be the approximate inverse to the tuning curves of Fig. 14.10.

Synchrony:   Again we apply a tone and perform a measurement on a cat's auditory neuron. This time we measure the histogram of time intervals between adjacent spikes. When this measurement is repeated many times, interval histograms such as those shown in Fig. 14.11 are created.

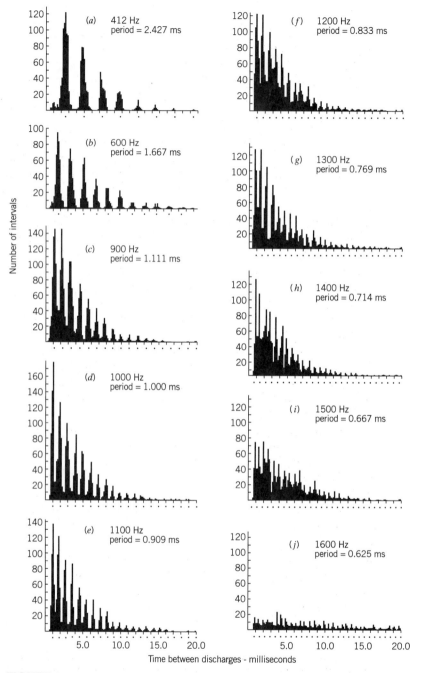

**FIGURE 14.11** Interval histograms showing periodic distributions of interspike intervals to pure tones of different frequencies. The stimulus frequency is indicated in each graph. The intensity of all stimuli is an 80-dB SPL, and the tone duration is 21 s. The responses to 10 stimuli constitute the sample on which each histogram is based. (The abscissas plot time in milliseconds between successive neural discharges. Dots below the time axes indicate integral values of the period of each frequency employed.) From [9].

Notice that the time between peaks in both histograms is the inverse of the frequency. This indicates that spikes tend to occur in synchrony with the applied stimulus. There is still a probabilistic component to this phenomenon, as indicated by the finite width of the peaks and the small but observable noise floor – some small percentage of the spikes do *not* occur in synchrony with the signal.

Phase locking is another way to describe synchrony. Experimentally, it has been shown that neurons fire in phase with the stimulus primarily at low frequencies. Phase locking does not exist in the cat's auditory nerve for frequencies beyond 5 kHz and gradually diminishes above 1 kHz.

Nonlinearities:    There are several phenomena that can be traced to the nonlinear behavior of the cochlea:    saturation, two-tone suppression, masking by noise, and combination tones.

Saturation:    The number of spikes that a nerve fiber can generate in a given time is limited by the biology of the fiber. It is also true that different fibers have varying properties. For example, the spontaneous rate can vary over many decibels. This can lead to an interesting result, shown in Fig. 14.12. Part D is the input power spectrum of the vowel in the word bet. Each point in parts A, B, and C is the normalized rate of a fiber at a given characteristic frequency (CF); these are high spontaneous rate fibers (and therefore low threshold fibers). As the input is increased, more fibers saturate so that as one progresses from A to C, the system loses its ability to represent the spectrum. In contrast, the low spontaneous fibers shown in parts E and F remain unsaturated for the same intensity inputs and thus continue to create a plausible image of the spectrum. It is currently of interest to auditory scientists to study the way in which more central auditory neurons make use of this diversity to yield a large dynamic range (100 dB) from the firings of individual neurons with dynamic ranges varying from 20 dB to 40 dB.

Two-tone suppression:    Figure 14.9 shows adaptation to a tone by a nerve fiber. We see that after approximately 20 ms, the firing rate is at a steady state. If, now, a new tone is applied *without removing the old tone*, then, depending on the parameters, the old tone can be suppressed. Figure 14.13 shows this effect. After a short interval the old tone is strongly suppressed, but then the firing rate increases to a new steady state that is lower than the previous steady state. When the new tone is removed, the firing rate increases suddenly and then adapts to the steady state. The specific result depends on the parameters, including frequency of the masker (suppressing tone), amplitude of the masker, frequency and amplitude of the signal, the time relation between the signal and masker, the characteristic frequency of the fiber, and the threshold of the fiber. Figure 14.14 shows the shaded areas where a tone of a frequency different than the probe tone suppresses the probe tone by 20% or more. Outside the shaded areas but inside the tuning curve, the addition of a second tone causes an increase in the firing rate.

Masking of a tone by noise:    If a tone plus noise is presented to a fiber, the noise has a suppressing effect on the response of the fiber to the tone. Figure 14.15 shows the firing rate of an auditory nerve as a function of the tone intensity, both with and without additive wideband masking noise. As the tone intensity increases toward its saturation, we see that the tone in noise fires at a lower rate when noise is present. The tone frequency and fiber CF were both 2.9 kHz. The noise band was 2.5–4 kHz.

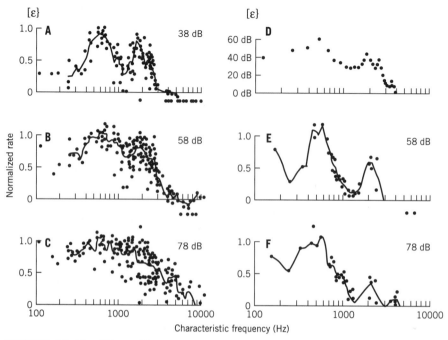

**FIGURE 14.12** Effect of input intensity on the rates of high spontaneous rate fibers (A,B,C) and low spontaneous rate fibers (D,E,F). From [10].

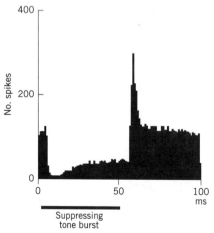

**FIGURE 14.13** Suppression of a continuous tone by a tone burst. From [7].

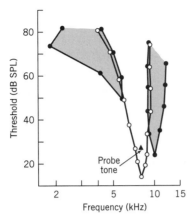

**FIGURE 14.14**  Suppression regions of an auditory nerve fiber. From [1].

Combination tones:   If a fiber is excited by two tones, a combination tone may appear that was not present in the stimulus. Thus, it is possible to excite a fiber with the combination tone $2f_1-f_2$ when both $f_1$ and $f_2$ are far from the CF of the fiber. For example, if the two applied tones are $f_1 = 1.0$ kHz and $f_2 = 1.1$ kHz, then the combination tones 0.7 kHz ($4f_1-3f_2$), 0.8 kHz ($3f_1-2f_2$), and 0.9 kHz ($2f_1-f_2$) will also be able to excite the appropriately tuned fiber.

# 14.6  SUMMARY AND BLOCK DIAGRAM OF THE PERIPHERAL AUDITORY SYSTEM

1. The complete mammalian auditory system contains many millions of neurons, but the *peripheral auditory system* of humans contains approximately 30,000 neurons and is the best understood component.

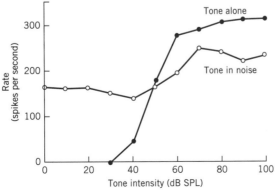

**FIGURE 14.15**  Rate vs. intensity with and without additive noise. From [8].

2. The auditory system consists of both ascending and descending fibers. Thus, there is feedback at most levels.

3. The outer ear terminates at the eardrum and affects the acoustics much like that of an acoustic tube.

4. The middle ear performs mechanical impedance transformation, from the malleus (driven by the eardrum) to the stapes (which drives the inner ear fluids).

5. The basilar membrane behaves like a bank of mechanical tuned circuits, over the complete range of auditory signals.

6. BM motion is transmitted to the stereocilia of the hair cells, and this leads to the firings of peripheral auditory neurons.

7. Auditory nerves adapt to stimuli, spiking vigorously at the beginning of a new input and then continuing to spike in the steady state at a reduced rate.

8. Each auditory nerve has a best or characteristic frequency, which is a function of its position on the basilar membrane.

9. For low frequencies (below 5 kHz) spikes tend to synchronize with periodic stimuli. This is called phase locking.

10. Various nonlinearities exist in the auditory system, leading to such phenomena as the limited dynamic range of a nerve, masking effects, and combination tones.

Figure 14.16 is a summary block diagram of many of the connections in the system.

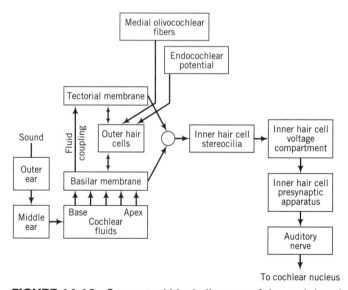

**FIGURE 14.16**   Conceptual block diagram of the peripheral auditory system. From [3].

## 14.7 EXERCISES

**14.1** Assume that the ear canal in a typical person is 3 cm long and of cylindrical shape with a diameter of 0.8 cm. Also, assume that the ear drum behaves like a solid, inflexible wall. Describe how the acoustical properties of the eardrum affect the frequency response of the overall system. How do these properties affect neural spiking for a 100-Hz pure tone stimulus and for a 3-kHz stimulus?

**14.2** The dynamic range of a normal human ear is approximately 100 dB, but the measured dynamic range of many neurons is approximately 20–30 dB. How does the auditory system manage such a high dynamic range with such restricted elements?

**14.3** In response to pure sinusoidal tones, a specific auditory nerve will spike at different rates, depending on the stimulus frequency. Various theories of the mechanism have been advanced. Discuss these theories and present empirical justifying evidence.

**14.4** Describe the sequence of events leading to auditory nerve spiking when an acoustic pressure wave appears on the outer ear.

**14.5** The frequency response in the peripheral auditory system is tonotopic (center frequency changes with place). Explain how this comes about.

**14.6** Present a heuristic justification for the statement that the basilar membrane behaves like a bank of bandpass filters.

**14.7** Refractory times in auditory nerves (intervals immediately following a spike during which the nerve is incapable of firing again) are at least several milliseconds. Explain how the auditory system is capable of responding to high-frequency stimuli (5 kHz or higher).

**14.8** Many people are deaf because their hair cells have been destroyed. Can you think of a way whereby some hearing is restored?

**14.9** Devise one or more thought experiments that demonstrate that listeners can hear frequencies that are not present in the stimulus.

## BIBLIOGRAPHY

1. Arthur, R. M., Pfeifer, R. R., and Suga, N., "Properties of 'two-tone inhibition' in primary auditory neurones," *J. Physiol.* **212**: 593–609, 1971.
2. von Bekesy, G., *Experiments in Hearing*, McGraw–Hill, New York, 1960.
3. Frishkopf, L. S., Class notes, Quantitative Physiology II–Sensory Systems, Massachusetts Institute of Technology, 1989.
4. von Helmholtz, H., *On the Sensation of Tone as a Physiological Basis for the Study of Music*, 4th. ed., A. J. Ellis, trans., Dover, New York, 1954; orig. German, 1862.
5. Hudspeth, A. J., "The cellular basis of hearing: the biophysics of hair cells," *Science* **230**: 745–752, 1985.
6. Kiang, N. Y.-S., and Moxon, E. C., "Tails of tuning curves of auditory nerve fibers," *J. Acoust. Soc. Am.* **55**: 620–630, 1974.
7. Pickles, J., *An Introduction to the Physiology of Hearing*, 2nd ed., Academic Press, New York/London, 1988.

8. Rhode, W. S., Geisler, C. D., and Kennedy, D. T., "Auditory nerve fiber responses to wide-band noise and tone combinations," *J. Neurophysiol.* **41**: 692–704, 1978.

9. Rose, J. E., Brugge, J. F., Anderson, D. J., and Hind, J. E., "Phase-locked response to low-frequency tones in single auditory-nerve fibers of the squirrel monkey," *Journal of Neurophysiology* **30**: 262–286, 1967.

10. Sachs, M. B., and Young, E. D., "Encoding of steady state vowels in the auditory nerve:   representation in terms of discharge rate," *J. Acoust. Soc. Am.* **66**: 470–479, 1979.

11. Shepherd, G. M., "Neurobiology," *Oxford Univ. Press*, London/New York, pp. 101–119, 1988.

12. Tobias, J. V., ed., *Foundations of Modern Auditory Theory II*, Academic Press, New York/London, 1972.

13. Yost, W. A., and Nielsen, D. W., *Fundamentals of Hearing – An Introduction*, Holt, Rinehart & Winston, New York, 1977.

# CHAPTER 15

# PSYCHOACOUSTICS

## 15.1 INTRODUCTION

Psychoacoustics is the science in which we quantify the human perception of sounds. The ultimate aim is to derive a quantitative model that matches the results of all auditory experiments that we can contrive. This is quite a tall order, since, to a great extent, the human auditory system remains a "black box," despite many years of physiological research. In this chapter we survey some of the results that have the most obvious relevance to speech and audio applications. Where possible, we correlate psychoacoustical phenomena with physiological measurements.

We can establish some objective variables that will be adjusted in order to assess human perception of sounds. For frequency and intensity, standardized instruments can produce outputs that are linearly proportional to the stimulus. For example, a device that counts the number of zero crossings of a sinusoid over a prescribed time interval can be calibrated to read what we can define as the frequency of the signal. A measure of the spectrum of a sound can also be defined, for instance by a particular form of spectrogram. Duration is another objective property of a sound.

Each of these sound characteristics has a corresponding perceptual variable. The perception of frequency is called pitch, the perception of intensity is called loudness, and the perception of spectrum is called timbre. These human response variables are not linearly proportional to the value of the corresponding stimulus variables. Thus, if a person hears a pure tone at some given frequency $f$, followed by another tone at $2f$, the perception will *not* be that the frequency of the second tone is double that of the first tone. Similarly, if the intensity of the tone is doubled, the human subject will not describe it as twice as loud. The same argument holds for perception of duration. Furthermore, the response variables are often dependent on more than one of the stimulus variables. For instance, the subjective impression of pitch, although primarily dependent on frequency, can vary with other parameters, such as intensity or spectrum. The same holds for loudness and timbre.

Some issues of interest in this area include the following.

1. How sensitive is human hearing? How does the ear[1] respond to different intensities?
2. How does the ear respond to different frequencies?
3. How well does the ear focus on a given sound of interest in the presence of interfering sounds?

---

[1] In common parlance the ear actually refers to the entire apparatus for hearing, up to and including the brain.

Such questions can be quantitatively addressed by conducting psychoacoustic tests. In the following sections we review some of the classic experiments that have been performed to try to answer a few of these types of questions. In particular, we discuss experiments to demonstrate the dependence of perceived loudness on objective parameters of intensity, duration, and spectrum. Pitch will be discussed in Chapter 16.

The reader's understanding can be further enhanced by listening to a set of demonstrations that were released on a commercial compact disk, *Auditory Demonstrations* [7]. A review paper by Hartmann [6] contains evaluations based on class listening tests for these demonstrations.

## 15.2  SOUND-PRESSURE LEVEL AND LOUDNESS

Sensations (hearing, seeing, smelling, etc.) increase logarithmically as the intensity of the stimulus increases. Many experiments have certainly verified (at least approximately) this law and have led to the use of the decibel scale. To define a subjective measure of loudness, we introduce the sone. Based on the work of Stevens [11] and others, an empirical relation between the sound pressure $p$ and the loudness $S$ in sones gives the result

$$S \propto p^{0.6}, \tag{15.1}$$

where a sone value of one is set to be the loudness of a 1000-Hz tone at an intensity of a 40-dB SPL. Recalling from Chapter 13 that the intensity is proportional to the square of the pressure, we obtain

$$S \propto I^{0.3}. \tag{15.2}$$

Roughly speaking, we can say that the loudness is proportional to the cube root of the intensity.[2] From Eq. 13.9, we can derive that a 10-dB increase in sound intensity corresponds to an increase in intensity by a factor of 10. According to Eq. 15.2, then, this 10-dB increase roughly corresponds to a doubling in sone value.

The constant of proportionality implied by Eq. 15.1 is frequency dependent. In general, the subjective listener response of loudness is a function of the intensity, frequency, and quality of the sound. Figure 15.1 shows a standardized set of curves for pure tones. Each curve denotes the measurement of equal loudness as a function of tone frequency. Typically, the listener adjusts the intensity at a given frequency until it is judged to be of equal intensity to a standard 1000-Hz tone. We see from these curves that the ear is most sensitive to sounds of approximately 4 kHz. Each contour of loudness level corresponds to units called phons. The phon level is set to be equal to the SPL in decibels at 1000 Hz. Thus, for example, we see that for a SPL of 40 dB, the loudness of a tone at 100 Hz is 20 phons less than the loudness at 1000 Hz.

This frequency-dependent sensitivity leads to the differing standards for sound-pressure-level meters, as were briefly described in Chapter 13, Section 13.1.5 in particular;

---

[2]Stevens' empirical formula is not very different than a logarithmic relation over the intensity values of interest.

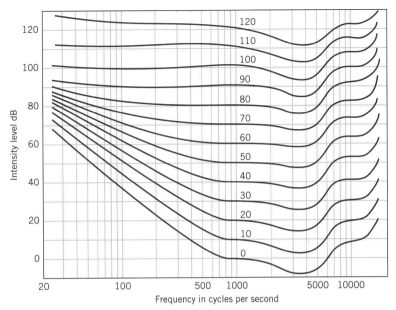

**FIGURE 15.1** Equal loudness curves for pure tones. From [4].

the relative insensitivity to low-frequency sounds led to the development of A weightings, which deemphasize these sounds. Note also that this variation in sensitivity is reduced for louder sounds, which led to the development of B-weighted and C-weighted SPL measurements.

These relationships are illustrated in [7] by demonstrations 4 and 6. In demonstration 4, tones are played in intensity steps of 1, 3, and 6 dB. In demonstration 6, tones are played in intensity steps of 5 dB for frequencies ranging from 125 to 8000 Hz.

Finally, there are many experiments that have been performed to demonstrate the dependence of loudness on the duration of a sound. Such experiments have shown that if the duration of a sound is smaller than approximately 200 ms, it will be less loud than a sound of the same intensity with a duration greater than 200 ms. In demonstration 8 of [7], for instance, pulses are presented with decreasing sound-pressure levels (0, −16, −20, −24, −28, −32, −36, and −40 dB). The listener responds by noting at which step the sound became inaudible. The test is repeated for durations of 1000, 300, 100, 30, 10, 3, and 1 ms. Figure 15.2 shows the result averaged over 103 listeners in a reasonably reverberant classroom, as reported in [6].

The scale on the left of Fig. 15.2 shows the pulse level, relative to the first pulse, at which audibility was lost. The dashed line is a reference with a slope of −10 dB per decade of duration. We see that the result asymptotes at approximately 100 ms, at which point further duration increases change the perceived threshold very little. This time would probably be longer if the experiment had been conducted with headphones, as the reverberation in the classroom undoubtedly extended the length of each pulse as received by the listeners.

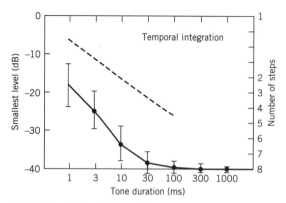

**FIGURE 15.2**    Effect of duration of a pulse in noise on loudness. For each tone duration the ordinate on the solid curve gives the smallest audible tone level (on the left) and the corresponding step number (on the right). Error bars of 2 standard deviations are shown. The broken line is a reference to show a slope of −10 dB per decade of duration, which corresponds to a simple integration of signal power. From [6], Fig. 5.

## 15.3  FREQUENCY ANALYSIS AND CRITICAL BANDS

In Section 15.2, we described the frequency dependence of a listener's sensitivity to pure tones. However, most signals of interest are more complex. This suggests other types of experiments to learn about the perception of multiple tones. Some of the most important of these are based on the notion that some kind of frequency analysis is central to human hearing.

As we discussed in the previous chapter, auditory neurons are tuned to specific characteristic frequencies. Thus, we surmise that the auditory system behaves like some sort of a filter bank. Experiments have been performed for many years to determine the characteristics of these auditory filters. Fletcher [2] performed some of the early experiments of this kind. In one such series of experiments, he based his measurements on the ear's response to a pure tone in band-limited white noise. Initially, the tone level was set to be heard by normal ears. The tone level was then decreased in discrete 5-dB steps until the listener did not hear it, and the number of steps was registered. The noise bandwidth (still with the same flat spectrum) was decreased and the experiment was repeated. Until the noise bandwidth was decreased to some critical value, the listener's ability to hear the tone remained the same, despite the decrease in noise power. However, for noise bandwidths *lower* than this critical value, the listener's response was monotonically enhanced.

The experiment could then be repeated with a tone of a different frequency, and in this way these critical values could be plotted over the total audible band.

The implication is this:  in listening to a tone of a given frequency, the listener applies a psychological filter of width approximately equal to this critical value. The filter ignores noise outside this bandwidth. Thus, the decision as to the absence or presence of the tone is based on the signal-to-noise ratio *within this band*. Figure 15.3 is a caricature

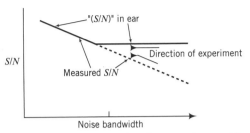

**FIGURE 15.3** Critical ratio experiment. The solid line shows the apparent signal-to-noise ratio of the psychological filter; the dashed line shows the signal-to-noise ratio of the stimulus.

of the apparent and measured signal-to-noise (S/N) ratios for this type of experiment. Note that the break point in the solid line in this figure can be interpreted as showing the noise bandwidth corresponding to a critical band for the particular tone frequency used.

Fletcher called these critical filter widths the critical bands. Later researchers (see Chapter 19) have developed methods of estimating the shapes of auditory filters.

From an engineering standpoint, probably the most important result that emerges from critical band research is that auditory filters with higher center frequencies have greater bandwidths. The Bark scale and the scale proposed by Greenwood, as shown in Fig. 15.4, are reasonable approximations to critical bands obtained from psychoacoustic measurements.

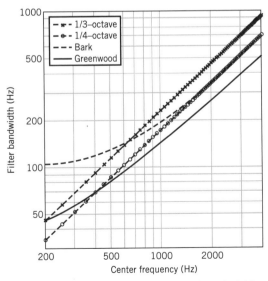

**FIGURE 15.4** Plots of estimated bandwidth as a function of center frequency for the Bark scale, and Greenwood's cochlear frequency-position function, with two constant-Q scales also shown for comparison. From [8].

**FIGURE 15.5** Critical bands by loudness comparison. Solid curves represent the spectrum of the test noise burst; dashed curves represent the spectrum of the reference noise burst. The loudnesses of reference and test bursts are compared by listeners. From [7].

Similar results can be obtained by having listeners compare the loudness for a pair of noise bursts. In one experiment demonstrated in [7], the reference noise burst had a bandwidth that was a fixed 15% of the center frequency, whereas the test noise burst maintained a constant power by lowering the amplitude and widening the bandwidth as shown in Fig. 15.5. As long as the test-signal bandwidth was smaller than the critical band, the loudness of the two stimuli remained equal, but when the test stimulus exceeded the critical band, it was typically judged to be louder than the reference.

These two critical band experiments are presented in demonstrations 2 and 3 in [7].

## 15.4  MASKING

In Section 15.3, we described an experiment in which the ability to hear a tone was masked by the presence of noise within the same spectral region. This led to models for the spectral resolution of human hearing. Similarly, many experiments are based on the effect of one tone on another, which it is hoped will lead to a better understanding of the perception of complex sounds. When two tones are presented simultaneously, the weaker tone may, in some cases, not be heard. A number of results have commonly been observed for this type of experiment. Closer tones have a greater effect, and louder tones affect tones that are further away in frequency. It has also been observed that a tone more easily masks a tone of higher frequency than one of lower frequency.

In masking experiments, there is typically a target signal, which is the one to be detected, and a masker signal, which is the one being manipulated to affect the listener's perception of the target. In one such experiment, a 2000-Hz target signal and a 1200-Hz masker are applied simultaneously. The masker consists of eight bursts and the signal consists of four bursts over the same interval, as shown in Fig. 15.6. The sequence is repeated 10 times, with the target tone set at a decreased intensity level for each presentation (down by 15 dB for the second sequence, and reduced by 5 dB more for each new sequence). After all presentations, the masker and target signals are reversed, as shown in the figure.

Some of the results quoted in [6] for experiments of this kind are ambiguous. They are affected by other factors, such as the overall intensities and the amount of room reverberation present. However, the asymmetry of the masking affect, namely the greater spread upward in frequency than downward, is illustrated in the histogram shown in Fig. 15.7. The figure plots a histogram of the number of listeners who heard more $n$ pulse streams with the 2000-Hz masker than with the 1200-Hz masker. The open bars are for the 60-dB masker level and the dark bars are for the 75-dB masker level. It is clear, at least for this test, that for

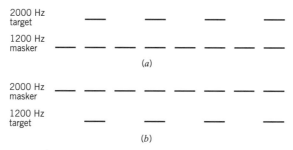

**FIGURE 15.6** Signals and maskers for a simultaneous masking experiment. The length of each line represents the duration of a tone. From [6].

a sufficiently intense masker, the asymmetry is decidedly in favor of the lower-frequency masker.

Demonstration 9 of [7] presents this experiment for the listener.

Masking can also occur when the signal and masker are nonsimultaneous. In experiments of this sort, a short signal (often called a probe) is presented at various times as a target signal, and the effect of the masker is measured. When the masker precedes the probe, the effect on perception is referred to as forward masking. When the masker follows the probe, the effect is referred to as backward masking. Many results have been observed, although many of them are ambiguous (particularly for the backward-masking experiments). We only note here that the amount of forward masking appears to increase for masker

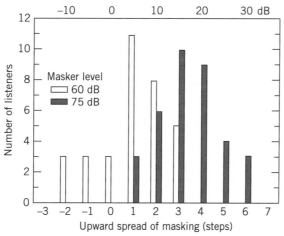

**FIGURE 15.7** Asymmetry of the masking of one tone by another. For each level (upper horizontal axis), the vertical bars show the number of listeners for whom $n$ more streams are masked by the lower-frequency masker than are masked by the higher-frequency masker, where $n$ is marked on the lower horizontal axis. From [6].

signals that are closer in time to the probe. Thus, sound signals at one time will affect the hearing of later signals. The effect is strongest over 10–50 ms, but forward masking has been observed for masker–probe delays of at least 100 ms [9]. Temporal (nonsimultaneous) masking experiments are presented in demonstration 10 of [7].

## 15.5  SUMMARY

In this chapter, we have presented a brief introduction to experiments that have demonstrated the dependence of loudness on a number of stimulus characteristics. In choosing these experiments, we have focused on those characteristics that seemed fundamental to our overall goals, namely the signal processing of speech and audio. Some of the key results of this chapter are as follows.

1. Loudness is roughly proportional to the cube root of sound intensity, with a doubling of loudness being observed for a 10-dB increase in the SPL.

2. Loudness is frequency dependent, and this frequency dependence is itself amplitude dependent. The sensitivity is greatest at approximately 4 kHz, with large deviations for low frequencies at low signal levels (up to 80 dB less sensitivity at 20 Hz). Smaller deviations are observed for low frequencies at high signal levels.

3. Longer sounds of a given intensity sound louder, up to a duration of 200 ms. This implies some kind of temporal integration, as do the results of forward-masking experiments.

4. Experiments indicate the existence of critical band filters. The bands are wider at high frequencies than at low frequencies. Further details on these filter characteristics will be described in Chapter 19 in the context of filter-bank models that can be used in speech and audio applications.

5. Pure tones that are close in frequency mask one another, with the lower-frequency tone masking the higher-frequency tone more than vice versa. Higher-amplitude tones also mask neighboring tones more than lower-amplitude tones. This set of results will also affect the filter-bank models described in Chapter 19.

As with many of the topics of this book, we can only provide an introduction to a major area of study such as psychoacoustics (for further reading, see [3], [5], [9]). In both the previous chapter and this one, for instance, we have avoided discussion of binaural (two ear) hearing, even though this aspect of hearing is fundamental to the ability to locate sound sources spatially; the use of multiple receivers by the auditory system also improves intelligibility under noisy or reverberant conditions. Similarly, there have been many studies that explore the ability of humans to identify and separate auditory streams; a comparable visual phenomenon would be the ability to recognize a car on the other side of a white picket fence. Many experiments of this latter type are described in [1]. Although we do not go into further detail on these topics, it is essential to the understanding of speech and audio signal-processing algorithms that we study the perception of pitch. This is the subject of Chapter 16.

## 15.6 EXERCISES

**15.1**    Critical band experiments result in psychoacoustical tuning curves that resemble bandpass filter frequency-response curves. Theory says that these curves are related to the tuning curves of Fig. 14.10, but these latter curves seem to be the *inverse* of critical band filters. How do you explain this apparent discrepancy?

**15.2**    Having explained the previous problem, we are still faced with the fact that physiological tuning curves and psychoacoustic tuning curves are different. Can you provide a plausible explanation for this difference?

**15.3**    As the intensity of a pure tone is increased, it sounds louder. What do you think happens neurophysiologically to make this happen?

**15.4**    What reason can you give to explain why a low-frequency tone is better able to mask a tone of higher frequency than vice versa?

**15.5**    One of the difficulties with two-tone masking is the beating effect. Thus, although the weaker tone is effectively masked, the listener hears the beats. Devise an experiment that overcomes this difficulty.

**15.6**    Three tones (100 Hz, 2000 Hz, and 7000 Hz) are presented monaurally over wideband headphones (40 Hz–16 kHz) to a young adult subject with normal hearing. In each case, the sound-pressure level at the subject's ear is 40 dB. What would be the expected loudness for the tones, going from the most loud to the least loud?

## BIBLIOGRAPHY

1.  Bregman, A. S., *Auditory Scene Analysis*, MIT Press, Cambridge, Mass., 1990.
2.  Fletcher, H., "Auditory patterns," *Rev. Mod. Phys.* **12**: 47–65, 1940.
3.  Fletcher, H., *Speech and Hearing in Communication*, Van Nostrand, Princeton, N.J., 1953.
4.  Fletcher, H., and Munson, W. J., "Loudness, its definition, measurement and calculation," *J. Acoust. Soc. Am.* **5**: 82–108, 1933.
5.  Green, D. M., *An Introduction to Hearing*, Wiley, New York, 1976.
6.  Hartmann, W. M., "Auditory demonstrations on compact disk for large N," *J. Acoust. Soc. Am.* **93**: 1–16, 1993.
7.  Houtsma, A. J. M., Rossing, T. D., and Wagenaars, W. M., "Auditory demonstrations," Philips compact disk, Inst. Perceptual Research and Acoustical Soc. Am., Eindhoven, 1987.
8.  Kingsbury, B. E. D., Perceptually Inspired Signal Processing Strategies for Robust Speech Recognition in Reverberant Environments, PhD Thesis, U.C. Berkeley, 1998.
9.  Moore, B. C. J., *An Introduction to the Psychology of Hearing*, 3rd ed., Academic Press, New York/London, 1989.
10. Rossing, T. D., *The Science of Sound,* Addison–Wesley, Reading, Mass., 1990.
11. Stevens, S. S., "The direct estimation of sensory magnitudes:    loudness," *Am. J. Psych.* **69**: 1–25, 1956.

# CHAPTER 16

# MODELS OF PITCH PERCEPTION

## 16.1 INTRODUCTION

Human pitch perception is performed by the complete auditory system. Aside from the periphery, our knowledge of this system is still so fragmentary that the task of modeling human pitch perception depends primarily on the interpretation of many psychoacoustics experiments, abetted somewhat by continuing physiological explorations. In this chapter, we first review some proposals (and the accompanying controversies) about the nature of this remarkable facility of ours. We then elaborate on some of these ideas by comparing performances of models with experimental results. The reader should be aware that there is a long and rich history associated with this problem. For greater detail than we can provide here, consult the excellent review by de Boer [2].

As with Chapter 15, the understanding of some of these concepts can be improved by listening to the relevant demonstrations from [10].

## 16.2 HISTORICAL REVIEW OF PITCH-PERCEPTION MODELS

As noted in Chapter 14, von Helmholtz [8] conceived of the auditory system as a bank of many overlapping bandpass filters. The relationship between this model and the known physiology of the periphery can be seen from Figs. 16.1 and 16.2.

As noted in Chapter 14, the basilar membrane and associated hair cells respond more to high frequencies at the entrance to the cochlea. As the vibrations penetrate more deeply into the cochlea, the basilar membrane (BM) response becomes more sluggish, corresponding to filters with lower center frequencies. A pure tone would cause greatest vibration at a specific *place* on the BM, and this would ultimately lead to perception of that tone. An important fact to note is that the ultimate perception of the tone is dependent on the activities of specific cortical neurons. We will return to this point later.

An engineering model of the auditory system is shown in Fig. 16.2. Note that the components of the inner ear (BM, hair cells) are represented by a filter bank.

Although von Helmholtz's place theory supplies a credible explanation for the pitch of pure tones, it runs into some difficulty accommodating the perception of complex tones. There are many instances when pitch can be readily identified when the fundamental frequency is completely absent. Many years before the electronic age, Seebeck [18] demonstrated this result by using a siren, as shown in Fig. 16.3.

214

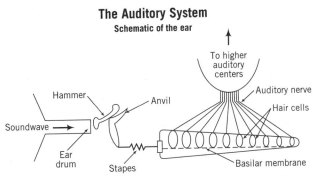

**FIGURE 16.1**  Schematic of the outer, middle, and inner ears.

The rotating disk with a single opening, subjected to a wideband acoustic field, produced pulses of sound at the frequency $1/T$. When the disk had two openings at opposite ends, the repetition rate was doubled. When one of the openings was slightly displaced, the frequency was again $1/T$, but the first harmonic could be made extremely small by reducing the displacement. Thus, Seebeck created the original version of the *missing fundamental* experiment. In such experiments, listeners are presented with complex tones with essentially no energy at the fundamental frequency in order to determine the pitch perception for such a stimulus.

Figure 16.4 illustrates a modern version of Seebeck's experiment, presented in demonstration 20 of [10]. The listener hears a complex tone at a fundamental frequency of 200 Hz. Successive harmonics, beginning with the first, are removed. Many listeners perceive the same pitch as would be heard for a sequence that included the fundamental.

How can the place explanation for pitch perception be maintained, given experiments such as Seebeck's? von Helmholtz defended his hypothesis by arguing that an (unspecified) nonlinear operation at the basilar membrane caused the BM to vibrate at the place corresponding to the fundamental frequency, even when there was no energy at that frequency in the original signal. Many years later this argument was proved false by Licklider [12]. He

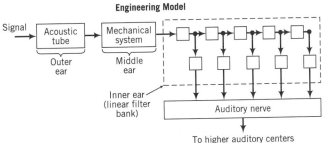

**FIGURE 16.2**  Engineering model of the outer, middle, and inner ears. The dashed curve encloses a model of the inner ear as a linear filter bank; this model will be specified further in Chapter 19.

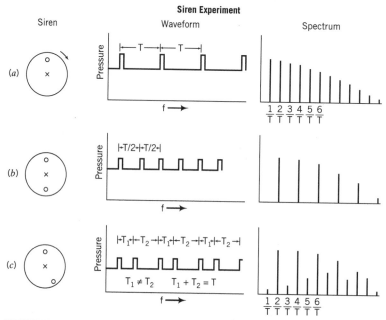

**FIGURE 16.3** Pulse train and resultant spectrum in Seebeck's experiment. The first column represents the physical arrangement for the production of each sound; the other columns show the corresponding time waveform and power spectrum. Note that for the sound labeled c, the spectral component corresponding to the disk rotation speed ($1/T$) has a magnitude of zero.

alternately played a pure tone (at the fundamental) and a harmonic series of the same fundamental frequency but with the actual first harmonic physically absent. The listener then perceived two sounds of equal pitch but different timbre. Then noise with a band centered at the fundamental was added to this sequence. It was found that the pure tone was completely masked, whereas the harmonic series was still heard. If perception of the harmonic series

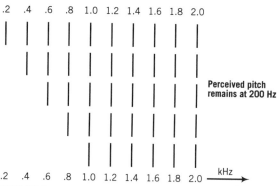

**FIGURE 16.4** Virtual pitch demonstration.

were dependent on the combination tone appearing at the fundamental frequency place in the BM, this combination tone should also have been masked. Licklider's experiment is reproduced in demonstration 22 of [10].

Licklider's result reinforced the so-called periodicity model of Schouten [17], who proposed that the auditory system perceived pitch by somehow measuring the periods of the signals as they traveled from BM (by means of hair cells) onto auditory fibers. Schouten's model is depicted in Fig. 16.5.

From this figure we observe that for a 200-Hz pulse train, the lower-frequency filters (below 1000 Hz) resolve the lower harmonics of the pulse train. For the high-frequency

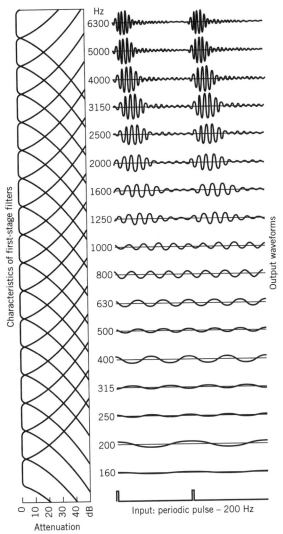

**FIGURE 16.5** Schouten's depiction of BM responses to a periodic 200-Hz pulse train. From [15].

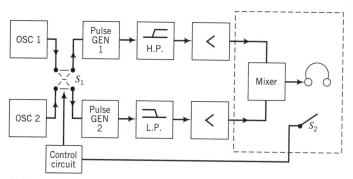

**FIGURE 16.6**  Ritsma's setup to determine dominant frequencies for pitch perception. From [16].

channels, however, a carrier signal that is within band for each filter appears to be modulated by a periodic signal so that the period of the input signal is easily recognized. Schouten thus hypothesized that pitch perception would be more salient at these higher harmonics. He asked his student Ritsma to set up an experiment to try to verify this hypothesis [16]. The experimental setup is shown in Fig. 16.6. Both the low-pass and high-pass filters in the figure are subjected to pulse trains having repetition rates that depend on the oscillators. Ritsma showed that the listener always chose the output of the low-pass filter as the preferred pitch. (The precise design of an experiment to arrive at the result is deferred to the exercises).

Thus, it does not appear to be the case that pitch perception is primarily a high-frequency phenomenon. More specifically, the results of testing the pitch perception of many listeners showed that the dominant frequencies for perception were in the vicinity of the third, fourth, and fifth harmonic. This result destroyed Schouten's hypothesis but still left open the question of a suitable model.

Other important insights were obtained from the experiments of Houtsma and Goldstein [9]. In one experiment, musically trained listeners were asked to recognize the interval between two successively played signals.[1] Each signal contained two successive harmonics of a given fundamental. When the two signals were presented to both ears, the trained subjects had no trouble identifying the intervals. They then repeated this experiment with one major modification:   for each signal, only one harmonic was presented to one ear and the other harmonic to the opposite ear. Again, pitch intervals were correctly identified. This result implied that the perception of pitch is centrally located. That is, perception appears to have taken place after the auditory signals from the two ears had been combined. From Fig. 14.1, this means that processing to determine pitch must take place at the superior olivary complex or higher. Also, significantly, von Helmholtz's place theory again plays some role in these more recent perspectives on pitch. In particular, the model proposed by Goldstein [7] assumes that the BM resolves the low-frequency harmonics and that the auditory system recognizes the pattern of excitations on the BM.

---

[1]An interval is the ratio between the fundamental frequencies of two signals.

# 16.3 PHYSIOLOGICAL EXPLORATION OF PLACE VERSUS PERIODICITY

Miller and Sachs [14] collected poststimulus time histograms of spiking intervals over a wide range of fiber characteristic frequencies (CFs) from cats. Results for the synthetic speech stimulus "da" are shown in Fig. 16.7.

Their results show that some fibers yield patterns that support the Goldstein place concept, whereas others support the periodicity concept. Eight fiber discharge patterns are shown in Fig. 16.7, with CFs ranging from 250 Hz to 3620 Hz. Also shown on the figure are formants 1, 2, and 3. For the fibers with CF = 250 Hz and CF = 400 Hz, responses are synchronized to individual harmonics of the fundamental frequency of approximately 120 Hz. For the fibers with CF = 330 Hz and CF = 970 Hz, responses are synchronized to the fundamental frequency. It is clear that the auditory system may use more than a single mechanism in arriving at a pitch estimate. A more complete explanation for these results is left as an exercise for the reader (see Exercise 6).

In recent decades, many deaf patients have been given cochlear implants; this operation sometimes provides new auditory capabilities. In this procedure, the dysfunctional hair cells are bypassed, and implanted electrodes excite the auditory nerve bundle directly. Since the electrodes are distributed throughout a specific region of the cochlea, it is possible to observe patient responses when individual electrodes, placed at specific places on the BM, are stimulated. Results show [5] that responses to the same periodic stimuli vary as a

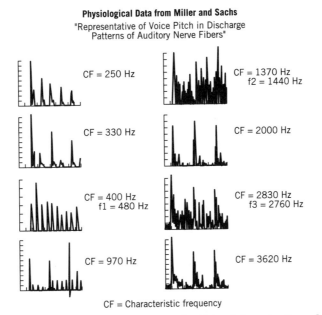

**Physiological Data from Miller and Sachs**
"Representative of Voice Pitch in Discharge Patterns of Auditory Nerve Fibers"

CF = 250 Hz

CF = 1370 Hz
f2 = 1440 Hz

CF = 330 Hz

CF = 2000 Hz

CF = 400 Hz
f1 = 480 Hz

CF = 2830 Hz
f3 = 2760 Hz

CF = 970 Hz

CF = 3620 Hz

CF = Characteristic frequency

**FIGURE 16.7** Histograms of spiking intervals. From [14].

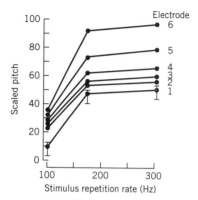

**FIGURE 16.8** Scaled pitch response of a single implanted patient. Each solid curve corresponds to the response for a different electrode. From [5].

function of the place of excitation. Figure 16.8 gives the results for a six-electrode implant as a function of a pulselike stimulus repetition rate. For repetition rates between 100 and 200 Hz, the perception of pitch increases almost linearly with the rate. Beyond 200 Hz, pitch is quite constant, which is quite different than for normal ears. We also notice that when the electrode is closer to the base (i.e., electrodes 5 and 6), pitch is higher than for the electrodes closer to the apex. This observation is reminiscent of the early place hypothesis of von Helmholtz.

## 16.4 RESULTS FROM PSYCHOACOUSTIC TESTING AND MODELS

As we noted earlier, the two major categories of pitch-perception models are those based on BM place and those based on the periodicity of the outputs from the BM. Although modern perspectives often include aspects of each theory, we have found it instructive to compare the response of these models to different stimuli.

A periodicity model is shown in Fig. 16.9. A correspondence is assumed between the basilar membrane and the filter bank. Further, the hair cell–auditory nerve complex is modeled as the elementary pitch detectors (EPDs). As in the research by Ritsma, the filters cover the low-frequency portion of the speech spectrum (100–2000 Hz). The ability

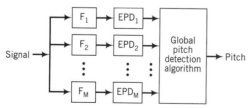

**FIGURE 16.9** Block diagram of the periodicity model.

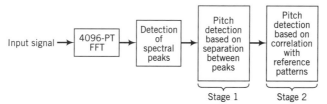

**FIGURE 16.10** Block diagram of the place model.

of these filters to resolve harmonics is a function of the pitch and spectrum of the incoming signal.

Neural spiking tends to follow the peaks of the signal. Given an auditory nerve spike, that same nerve cannot respond to further stimulation during the refractory period. Following this period, the voltage difference between the interior and exterior of the neuron gradually returns to normal, thus increasing the probability of subsequent firings. The global algorithm shown in the figure generates a histogram of the intervals between successive spikes and spikes two, three, or four intervals apart. Pitch period is determined by choosing the interval corresponding to the maximum value of the histogram.

A place model is shown in Fig. 16.10. The underlying hypothesis of this model is the ability of the auditory system to resolve harmonic peaks of the stimulus. This resolution probably takes place at higher auditory levels above the periphery. Stage 1 of the figure is a version of the Seneff algorithm [19] that performs a statistical separation of the frequency spacing between spectral peaks. Stage 2 is related to the harmonic sieve algorithm of the Goldstein model, as implemented by Duifhuis [4]. The spectral peaks are correlated with sets of harmonically spaced narrow windows. The nominated sets are based on the winning pitch of stage 1. The final winner corresponds to the set of maximum correlation.

We can refer to two examples from [10] in order to test the fidelity of each of these models to human pitch perception. In the first of these, we present listeners with signals that are not harmonic that result from the shifting of each of several harmonics upward by the same amount; this is presented in demonstration 21 of [10]. Figure 16.11 shows the harmonic structure of two signals. The top signal clearly leads to a pitch of 200 Hz,

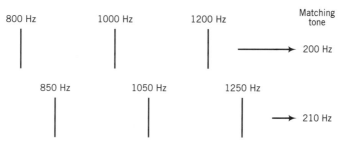

**FIGURE 16.11** Shift of virtual pitch.

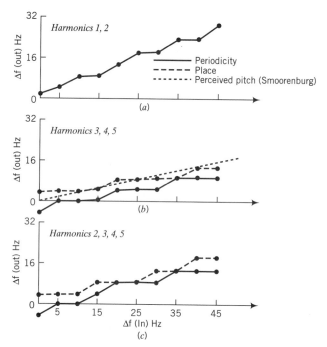

**FIGURE 16.12**   Responses of both models to virtual pitch shifts.

either for humans or for either model. However, the bottom signal is a shifted version of the top signal and leads to a perceived pitch of 210 Hz. Many experiments, using different parameters, verify that shifted versions of harmonic signals result in reliable pitch perception. At issue is how to explain the *amount* of pitch shift as a function of the shift of the stimulus frequencies. Figure 16.12 shows the results for both place and periodicity models for a variety of conditions. Both models respond more or less correctly to the stimuli.

In another case, we can present an entirely different type of stimuli. In demonstration 26 from [10], a five-octave diatonic scale is played with pulse pairs. This is followed by a four-octave diatonic scale built from samples of a Poisson process. Finally, a four-octave scale is played with bursts of comb-filtered white noise.[2] Here we tested the two pitch perception models for the last of these stimuli. Figure 16.13 shows how the noise is comb filtered. By sequentially changing the delay, one can control the pitch (which is inverse to the delays) to produce the four octaves shown in Fig. 16.14.

Figure 16.15 shows how the models respond to the comb-filtered noise. Instead of the use of the diatonic scale of Fig. 16.14, a slightly different sequence of delays was presented to the models, consisting of ten delays ranging from 12.0 to 2.1 ms. Each of the signals was

[2]Comb filtering refers to multiplying the input spectrum by some simple periodic function, for instance by placing equidistant zeros around the unit circle in the $z$ plane. Since the result is a filtering out of periodic chunks of the spectrum, the filtering action appears like a comb.

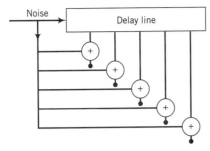

**FIGURE 16.13** Comb-filtered noise with different delays.

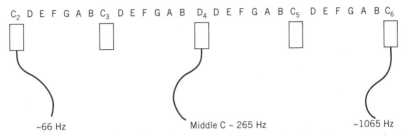

**FIGURE 16.14** Diatonic scale for a comb-filtered noise demonstration.

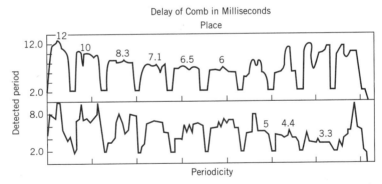

**FIGURE 16.15** Tracks of the detected periods for both place and periodicity models of pitch, given comb-filtered noise for different delays.

processed by both models for 160 ms, followed by a pause of 55 ms. Figure 16.15 shows the resulting pitch-period estimates; the dips correspond to the off times.

The place model follows the psychoacoustic results for the initial six of the 10 cases. The periodicity model more or less follows the same pattern but very erratically. Interestingly, the higher pitches are better represented by this model.

More details on listener response to various noise signals can be found in [1].

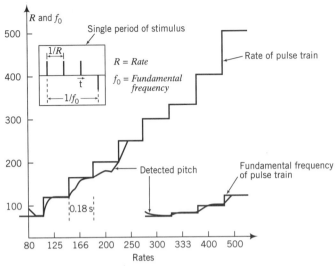

**FIGURE 16.16**   Response of periodicity model to Flanagan–Guttman pulse train.

The results from comb-filtered noise seemed to suggest that place-based models were a better match to human perception. However, in other cases the periodicity model performs quite well. An example in point is an experiment by Flanagan and Guttman [6] that demonstrated two modes of pitch perception. The listener was presented with a sequence of periodic pulses of pulse rate $R$. Each period consisted of three positive pulses followed by one negative pulse; thus the fundamental frequency was $R/4$. When $R$ was less than approximately 150 Hz, listeners perceived pitch to be $R$; for $R$ greater than 200 Hz, listeners perceived pitch to be $R/4$. Figure 16.16 shows the result obtained with the periodicity model. In this case, this model gives a good match to the behavior of listeners. The place model that we used gave erratic results and is not reproduced here.

# 16.5  SUMMARY

The reader may feel at this point that there are no conclusions that can be made at all. Certainly our understanding of pitch perception is incomplete, and the many experiments on the topic often seem to point in different directions. However, the two models discussed do follow, to some extent, the psychoacoustics data. One can argue, as do Meddis et al. [13] and Delgutte and Cariani [3], that the interval histograms of fiber firings supply all the required pitch information. Yet, it is still necessary to explain how this periodicity information is translated by the auditory system into the necessary place information in the cortex. One can subscribe to the modified place model of Goldstein, but we still need to understand how the various resolved harmonics get translated into a single answer. In either case, it is likely that the brain makes use of all available information, particularly since some aspects of the signal may be obscured under different acoustics conditions. In any event, these models have provided much food for thought for engineers who have designed automatic

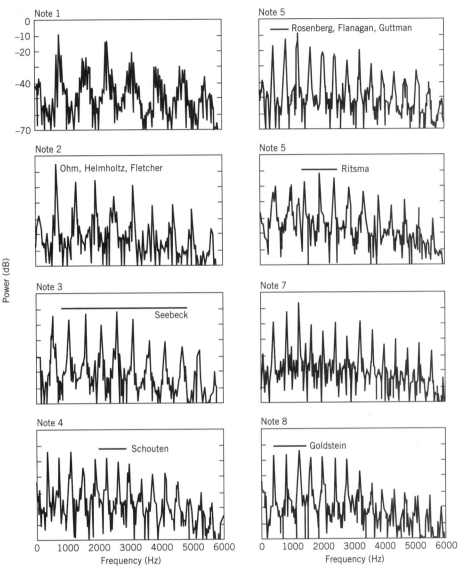

**FIGURE 16.17** Spectra from "West End Blues" and proposed important spectral sections for pitch perception.

pitch-detection systems for use in speech communications or for diagnostic purposes in detecting vocal cord or vocal tract illnesses.

We close the chapter by giving a musical example to illustrate the portions of the spectrum that are the proposed critical inputs for the auditory system according to some of the model builders referenced in this chapter. Figure 16.17 shows the spectra of the first eight notes of the great trumpet cadenza by Louis Armstrong in "West End Blues."

## 16.6 EXERCISES

**16.1** Experiments have demonstrated that nerve fibers have a *refractory* period; that is, for several milliseconds following the production of a spike, the neuron is incapable of producing a new spike. This implies that the spike frequency of a single neuron cannot exceed several hundred hertz. Given this, how can one justify the periodicity model of perception?

**16.2** Combination tones produced by the ear's nonlinearity have been verified from psychoacoustic data. Why doesn't this fact nullify the validity of Licklider's demonstration that the missing fundamental was not re-created in the cochlea by nonlinearities?

**16.3** Design an experiment using Ritsma's setup (Fig. 16.6) to show that the auditory system perceives pitch based on the low-frequency portion of the spectrum.

**16.4** Ritsma showed that the dominant frequencies for pitch perception were harmonics 3, 4, and 5. Do you think that this relationship holds for very low ($\approx$50 Hz) or very high ($\approx$1000 Hz) fundamental frequencies? Discuss.

**16.5** **(a)** Miller and Sachs collected Poststimulus time histograms to produce Fig. 16.7. Explain how this is done.

**(b)** Instead of poststimulus time histograms, interval histograms are often collected. Explain the difference. Would interval histograms for the same stimuli have been as useful?

**(c)** Licklider [11] and Meddis et al. [13] both presented a pitch-perception model based on a sum of autocorrelation functions across frequency channels. Derive a relationship between this function and the interval histogram.

**16.6** Explain the results of Fig. 16.7. Why are some fibers synchronized to a harmonic and others to the fundamental frequency?

**16.7** Since most of us don't have perfect pitch, we can't simply assign a number or a note to the pitch of a sound. Think of one or more ways of determining pitch perception experimentally.

**16.8** Both place and periodicity models performed quite well for the stimulus of Fig. 16.11. Explain how the models managed to emulate human performance for this task.

**16.9** What is the property of comb filtered noise that makes it plausible for the place model to emulate human perception for at least the lower pitches?

**16.10** What is the pitch of a harmonic series with missing fundamental frequency? Is it (a) the fundamental frequency, (b) twice the fundamental frequency, (c) halfway between (a) and (b), or (d) none of the above? (In that case, what is your guess?)

## BIBLIOGRAPHY

1. Bilsen, F. A., "Pitch of noise signals: evidence for a central spectrum," *J. Acoust. Soc. Am.* **61**: 150–161, 1977.
2. de Boer, E., "On the 'residue' and auditory pitch perception," in W. Keidel, ed., *Handbook of Sensory Physiology 5*, Springer-Verlag, New York/Berlin, pp. 479–583, 1976.
3. Delgutte, B., and Cariani, P., "Coding of the pitch of harmonic and inharmonic sounds in the discharge patterns of auditory nerve fibers," in M. E. H. Shouten, ed., *The Processing of Speech*, Mouton–DeGruyer, s-Gravenhage, The Netherlands, 1992.

4. Duifhuis, H., Willems, L. F., and Sluyter, R. J., "Measurement of pitch in speech: an implementation of Goldstein's theory of pitch perception," *J. Acoust. Soc. Am.* **71**: 1568, 1982.

5. Eddington, D. K., "Speech discrimination in deaf subjects with cochlear implants," *J. Acoust. Soc. Am.* **68**: 885–891, 1980.

6. Flanagan, J. L., and Guttman, N., "On the pitch of periodic pulses," *J. Acoust. Soc. Am.* **32**: 1308–1319, 1960.

7. Goldstein, J. S., "An optimum processor theory for the central formation of the pitch of complex tones," *J. Acoust. Soc. Am.* **54**: 1496–1516, 1973.

8. von Helmholtz, H., *On the Sensation of Tone as a Physiological Basis for the Study of Music,* 4th ed., A. J. Ellis, trans., Dover, New York, 1954; orig. German, 1862.

9. Houtsma, A. J. M., and Goldstein, J. L., "The central origin of the pitch of complex tones: evidence from musical interval recognition," *J. Acoust. Soc. Am.* **51**: 520–529, 1972.

10. Houtsma, A. J. M., Rossing, T. D., and Wagenaars, W. M., "Auditory demonstrations," Philips compact disk, Inst. Perceptual Research and Acoustical Soc. Am., Eindhoven, 1987.

11. Licklider, J. C. R., "A duplex theory of pitch perception," *Experientia* **7**: 128–133, 1951.

12. Licklider, J. C. R., "Periodicity pitch" and "place pitch," *J. Acoust. Soc. Am.* **26**: 945 (A) (1954).

13. Meddis, R., and Hewitt, M. J., "Virtual pitch and phase sensitivity of a computer model of the auditory periphery," *J. Acoust. Soc. Am.* **91**: 233–245, 1991.

14. Miller, M. I., and Sachs, M. B., "Representation of voice pitch in discharge patterns of auditory-nerve fibers," *Hearing Res.* **14**: 257–279, 1984.

15. Moore, B. C. J., *An Introduction to the Psychology of Hearing*, 2nd ed., Academic Press, New York/London, 1982.

16. Ritsma, R. J., "Frequencies dominant in the perception of the pitch of complex sounds," *J. Acoust. Soc. Am.* **42**: 191–198, 1967.

17. Schouten, J. F., "The residue, a new component in subjective sound analysis," *Proc. Kon. Acad. Wetensch (Neth.)* **43**: 356–365, 1940.

18. Seebeck, A., "Ueber die Sirene," *Ann. Phys. Chem.* **60**: 449–481, 1843.

19. Seneff, S., "Real-time harmonic pitch detector," *IEEE Trans. Acoust. Speech Signal Proc.* **ASSP-26**: 358–365, 1978.

# CHAPTER *17*

## SPEECH PERCEPTION

## 17.1  INTRODUCTION

How do people perceive speech? This question, in addition to being of great scientific interest, is also of central concern to those concerned with building systems for the processing of speech. For instance, engineering choices in the development of speech coding or synthesis systems should incorporate knowledge of what distortions affect intelligibility or voice quality for the human listener. It could also be argued that the human system for speech recognition and understanding is the one known example of a *robust* speech recognizer, that is, a recognizer whose performance is insensitive to variability over the range of nonlinguistic factors in the speech signal. Although simple mimicry of human mechanisms may not be a good approach to engineering design, it is still likely to be useful to study the functional characteristics of human speech perception.

What can we say about the physiological response of the human auditory system to a speech stimulus? How might this be related to the psychology of what listeners hear for such stimuli? These are the kinds of issues we intend to introduce in this chapter.

## 17.2  VOWEL PERCEPTION:  PSYCHOACOUSTICS AND PHYSIOLOGY

In Chapter 11, we saw that vowel articulation depended greatly on resonances (formants) of the vocal tract configuration. What is the relation between formants and the perception of vowels? This can be asked both at the level of the neural response to vowels and at the level of the human perception of the vowel identity. The answers to such questions are complex and controversial. Nonetheless, it is apparent that there is *some* relationship between formants and vowel perception. We have already observed in Chapters 14 and 15 that the effective bandwidth of auditory filters widens considerably for higher center frequencies. For low-formant frequencies (e.g., below 1 kHz), $F_1$ and $F_2$ would be resolved by the narrow bandwidth region of the auditory system, but for higher-formant frequencies, adjacent formants would not be resolved. This led to the idea that perhaps the perception of vowels by the ear really depended on just two effective formants. Carlson et al. [2] ran a psychoacoustic test to study this conjecture. The listener controlled the $F_2$ parameter of a two-formant vowel synthesizer by trying to match it to a four-formant vowel synthesizer. In general, the best matches were for the adjustable $F_2$ to be appreciably higher in frequency than the reference $F_2$ for the front vowels (high $F_2$) but equal to the reference $F_2$ for the back vowels (low $F_2$). This result strengthened the two-formant conjecture, at least at the level of the full human system for speech perception.

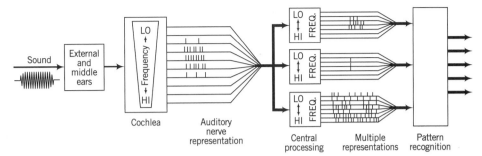

**FIGURE 17.1** Block diagram of sound representation in the auditory system. From [16].

At the finer, more physiological level, the story is less clear. Sachs et al. [16] studied the response of auditory neurons and cochlear nucleus neurons in the cat to steady-state synthetic vowel stimuli. Their concept of representation is illustrated in Fig. 17.1.

This figure implies that auditory physiologists have obtained a large amount of data on patterns of spiking in the tonotopically[1] organized peripheral auditory system. In the central processing region, tonotopic organization is maintained, but unlike the periphery, different nuclei contain neurons with diverse characteristics, so that the same input will create different pathways to the auditory cortex where recognition takes place.[2]

The physiological measurement of mammalian auditory systems demands great skill by the investigator and is a time-consuming and laborious procedure. Sachs et al. have performed a physiological coup by computing simultaneous responses of hundreds of peripheral auditory nerves to the same stimulus; their valuable data have been used by other researchers (e.g., [19] and [18]). In one of their experiments they studied the responses to the steady-state vowel /e/ (as in bet) by computing the poststimulus-time histogram (PSTH)[3] and its Fourier transform. Figure 17.2 shows the PSTH and its Fourier transform for a single neuron with CF = 400 Hz. These graphs show synchrony to the fourth harmonic of the stimulus.

If we now look at the results for many neurons, as in Fig. 17.3, we see that different neurons will synchronize to different harmonics of the stimulus. The measure used is the synchronization index, defined as the Fourier transform magnitude at a specific frequency divided by the average rate of the unit. Thus, in Fig. 17.3a, the fibers in the vicinity of 400–500 Hz are in synchrony to the fourth harmonic, whereas in Fig. 17.3c fibers in the vicinity of 1700–2000 Hz are in synchrony to the fourteenth harmonic (1793 Hz). If we now look at Fig. 17.3d, we see neurons that are in synchrony to harmonic 19 (2432 Hz). Therefore, in this figure, the neural ensemble appears to be sensitive to formant 3, and

---

[1]Recall from Chapter 14 that tonotopic organization refers to a structure in which the different parts have different frequency sensitivities.

[2]Note that the final stage in Fig. 17.1 is blank, reflecting our relative ignorance of the mechanisms for pattern recognition in the brain. However, it is true that in recent years there have been more experiments that have attempted to explore the function of the auditory cortex, as described, e.g., in [7].

[3]See Chapter 14.

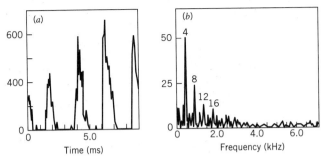

**FIGURE 17.2**　Period histogram and Fourier transform of single 400-Hz fiber. From [16].

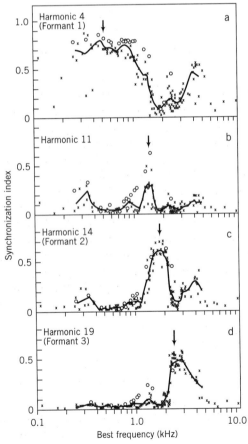

**FIGURE 17.3**　Synchronization indices for many peripheral auditory neurons. From [16].

these data are in contrast to the two-formant conjecture arising from the modeling results. Of course this measurement arises from the auditory periphery only, and the two-formant results correspond to the complete auditory system. From this evidence we might infer that the periphery retains fairly fine spectral information that may or may not be used later on in the auditory chain.

## 17.3  THE CONFUSION MATRIX

Early work on speech perception (in the 1950s) focused on the properties of vowels. At the time, it was fairly well understood that vowellike sounds were produced through the resonances (formants) of the vocal tract; it was natural to ask how the perceptual mechanism behaved toward this well-understood input. Subsequent research has verified that steady-state vowel perception is one of many interesting aspects of speech perception in general.[4] Formant transitions, at the boundaries between consonants and the following vowel, play a significant role in the correct perception of consonants [12]. Furthermore, consonants typically play a much greater role than vowels in the human understanding of speech. (In written Hebrew, for example, vowels are not explicitly displayed.)

Research by Miller and Nicely [13] examined the degree of confusion by listeners presented with different consonant transitions into the vowel /a/ as in "father". Sixteen consonants were chosen:  p, t, k, b, d, g, f, s, sh (as in shoe), th (as in thin), z, v, th (as in then), zh (as in azure), m, and n. The listeners were forced to make a decision and the results were compiled into a *confusion matrix*. Two examples of the results are shown in Figs. 17.4 and 17.5.

|     | *p* | *t* | *k* | *f* | *θ* | *s* | *ʃ* | *b* | *d* | *g* | *v* | *ð* | *z* | *ʒ* | *m* | *n* |
|-----|-----|-----|-----|-----|-----|-----|-----|-----|-----|-----|-----|-----|-----|-----|-----|-----|
| *p* | 240 |     | 41  | 2   | 1   |     |     |     |     |     |     |     |     |     |     |     |
| *t* | 1   | 252 | 1   | 1   |     |     |     |     |     | 1   |     |     |     |     |     |     |
| *k* | 18  | 3   | 219 |     |     |     |     |     |     |     |     |     |     |     |     |     |
| *f* |     |     |     | 225 | 24  |     |     | 5   |     |     | 2   |     |     |     |     |     |
| *θ* | 9   |     | 1   | 69  | 185 |     |     | 3   |     |     |     | 1   |     |     |     |     |
| *s* |     |     |     |     |     | 232 |     |     |     |     |     |     |     |     |     |     |
| *ʃ* |     |     |     |     |     |     | 236 |     |     |     |     |     |     |     |     |     |
| *b* |     |     |     |     |     | 1   |     | 242 |     |     | 24  | 12  | 1   | 1   |     |     |
| *d* |     |     |     |     |     |     |     |     | 213 | 22  |     |     |     |     |     |     |
| *g* |     |     |     |     |     | 1   |     |     | 33  | 203 |     |     | 3   |     |     |     |
| *v* |     |     |     |     |     |     |     | 6   |     |     | 171 | 30  |     |     | 1   |     |
| *ð* |     |     |     |     |     | 1   |     | 1   |     |     | 22  | 208 | 4   |     |     | 1   |
| *z* |     |     |     |     |     |     |     |     | 2   | 4   | 1   | 7   | 238 |     |     |     |
| *ʒ* |     |     |     |     |     |     |     |     |     |     |     |     |     | 244 |     |     |
| *m* |     |     |     |     |     |     |     |     |     |     |     | 1   |     |     | 274 | 1   |
| *n* |     |     |     |     |     |     |     |     |     |     |     |     |     |     |     | 252 |

**FIGURE 17.4**  Confusion matrix for S/N = +12 dB and a frequency response of 200–6500 Hz. From [13].

[4]In recent years it has become increasing clear that in natural speech, there are few regions with truly steady-state spectral characteristics; rather, it is much more common to have spectral characteristics constantly in flux. Thus, the steady-state vowel should be viewed as an abstraction rather than a universal physical phenomenon.

| | p | t | k | f | θ | s | ∫ | b | d | g | τ | δ | z | 3 | m | n |
|---|---|---|---|---|---|---|---|---|---|---|---|---|---|---|---|---|
| p | 80 | 43 | 64 | 17 | 14 | 6 | 2 | 1 | 1 | | | 1 | | | | |
| t | 71 | 84 | 55 | 5 | 9 | 3 | 8 | 1 | | | | 1 | 2 | | | |
| k | 66 | 76 | 107 | 12 | 8 | 9 | 4 | | | | | 1 | | | | |
| f | 18 | 12 | 9 | 175 | 48 | 11 | 1 | 7 | 2 | 1 | 2 | 2 | | | | |
| θ | 19 | 17 | 16 | 104 | 65 | 32 | 7 | 5 | 4 | 5 | 6 | 4 | 5 | | | |
| s | 8 | 5 | 4 | 23 | 39 | 107 | 45 | 4 | 2 | 3 | 1 | 1 | 3 | 2 | | |
| ∫ | 1 | 6 | 3 | 4 | 6 | 29 | 195 | | 3 | | | | | | | |
| b | 1 | | | 5 | 4 | 4 | | 136 | 10 | 9 | 47 | 16 | 6 | 1 | | |
| d | | | | | | | 8 | 5 | 80 | 45 | 11 | 20 | 20 | 26 | | |
| g | | | | | 2 | | | 3 | 63 | 66 | 3 | 19 | 37 | 56 | | |
| τ | | | | 2 | | | 2 | 48 | 5 | 5 | 145 | 45 | 12 | | | |
| δ | | | | | 6 | | | 31 | 6 | 17 | 86 | 58 | 21 | 5 | | |
| z | | | | | 1 | 1 | 1 | 7 | 20 | 27 | 16 | 28 | 94 | 44 | | |
| 3 | | | | | | | | 1 | 26 | 18 | 3 | 8 | 45 | 129 | | |
| m | 1 | | | | | 4 | | 4 | | | 4 | 1 | 3 | | 17 | |
| n | | | | 4 | | | | 1 | 5 | 2 | 7 | 1 | 6 | | | 4 |

**FIGURE 17.5** Confusion matrix for S/N = −6 dB and a frequency response of 200–6500 Hz. From [13].

Each figure is a 16 × 16 matrix. The spoken syllables are listed vertically; the transcribed results for each consonant are listed horizontally. For example, the top row of Fig. 17.4 shows that the consonant "p" was heard correctly 240 times and incorrectly 44 times (distributed among the four consonants shown). For a lower signal-to-noise ratio (Fig. 17.5), confusion in spotting the "p" has considerably increased. Miller and Nicely collected a total of 17 such matrices for different signal-to-noise ratios and different filters. In the figures, standard symbols are used for the phonemes; thus, row (and column) 5 corresponds to th (as in thin), row 7 corresponds to sh (as in shoe), row 12 corresponds to th (as in then) and row 14 corresponds to zh (as in azure).

The 17 conditions are listed in Fig. 17.6.

| Condition | S/N | Band |
|---|---|---|
| 1 | −18 | 200–6500 |
| 2 | −12 | 200–6500 |
| 3 | −6 | 200–6500 |
| 4 | 0 | 200–6500 |
| 5 | 6 | 200–6500 |
| 6 | 12 | 200–6500 |
| 7 | 12 | 200–300 |
| 8 | 12 | 200–400 |
| 9 | 12 | 200–600 |
| 10 | 12 | 200–1200 |
| 11 | 12 | 200–2500 |
| 12 | 12 | 200–5000 |
| 13 | 12 | 1000–5000 |
| 14 | 12 | 2000–5000 |
| 15 | 12 | 2500–5000 |
| 16 | 12 | 3000–5000 |
| 17 | 12 | 4500–5000 |

**FIGURE 17.6** Seventeen conditions of S/N and filtering. From [13].

The sounds of any language can further be categorized by *features*.[5] The features used by Miller and Nicely were voicing, nasality, affrication, duration, and place of articulation, and they are defined as follows.

Voicing:   As noted earlier in this book, vocal cord vibration is present for certain phonemes such as vowels. Other sounds, such as voiceless fricatives, are made with an open glottis and no vocal cord vibration. Exceptions occur; for example, when speech is whispered, the vocal cords stay open.

Nasality:   In English, the three phonemes, m, n, and ng (as in sing), are articulated with an open velum, allowing the breath stream to go through the nasal passageway as well as through the oral cavity. In French, many vowels are nasalized. In English, vowels preceding or following a nasal are also often nasalized.

Affrication:   The fricative sounds, such as s, z, f, and v, are produced with open vocal cords but with a constriction somewhere in the vocal tract (depending on the particular sound), which causes turbulence in the breath stream.

Duration:   This feature is self-explanatory.

Place of articulation:   In the plosive sounds, the phoneme p is characterized by a sudden opening of the lips, whereas the phoneme k is characterized by the sudden opening toward the back of the vocal tract. In these and other sounds, the position of a critical spot in the vocal tract changes as a function of the sound.

Many interesting questions can be explored with confusion matrices. For example, one can ask which features are more vulnerable to noise.

The feature concept was created by Jakobson et al. [10]. Their intent was to establish a feature set that encompassed all languages, so it was more extensive than the Miller–Nicely set and other sets that have been proposed. Figure 17.7 is the Jakobson et al. version of the distinctive feature set. Notice that all features can be classified as either place or manner of articulation. Furthermore, as implied in the figure, the features of the left-hand column can

| Place | p | k | t | b | d | g | f | thin | s | sh | v | the | z | azure | m | n | ng | l | r | w | h |
|---|---|---|---|---|---|---|---|---|---|---|---|---|---|---|---|---|---|---|---|---|---|
| bilabial | + | − | − | + | − | − | − | − | − | − | − | − | − | − | + | − | − | − | − | + | − |
| labiodental | − | − | − | − | − | − | + | − | − | − | + | − | − | − | − | − | − | − | − | − | − |
| dental | − | − | − | − | − | − | − | + | − | − | − | + | − | − | − | − | − | − | − | − | − |
| alveolar | − | − | + | − | + | − | − | − | + | − | − | − | + | − | − | + | − | + | − | − | − |
| palatal | − | − | − | − | − | − | − | − | − | + | − | − | − | + | − | − | − | − | + | − | − |
| velar | − | + | − | − | − | + | − | − | − | − | − | − | − | − | − | − | + | − | − | − | − |
| pharyngeal | − | − | − | − | − | − | − | − | − | − | − | − | − | − | − | − | − | − | − | − | + |
| Manner | | | | | | | | | | | | | | | | | | | | | |
| glide | − | − | − | − | − | − | − | − | − | − | − | − | − | − | − | − | − | + | + | + | − |
| nasal | − | − | − | − | − | − | − | − | − | − | − | − | − | − | + | + | + | − | − | − | − |
| stop | + | + | + | + | + | + | − | − | − | − | − | − | − | − | − | − | − | − | − | − | − |
| fricative | − | − | − | − | − | − | + | + | + | + | + | + | + | + | − | − | − | − | − | − | − |
| voicing | − | − | − | + | + | + | − | − | − | − | + | + | + | + | + | + | + | + | + | + | + |

**FIGURE 17.7**   Binary distinctive feature set of Jakobson et al. From [10].

[5]Note that the term *features* here has a different meaning than we used in our earlier chapters on pattern recognition; in those, features just referred to the observed variables. In the phonetics literature the word typically refers to standard binary variables that characterize speech sounds.

| Place of articulation | Glide | Nasal | Manner of articulation Stop Voiced | Stop Unvoiced | Fricative Voiced | Fricative Unvoiced |
|---|---|---|---|---|---|---|
| **Front** | | | | | | |
| Bilabial | w,ʍ | m | b | p | | |
| Labiadental | | | | | v | f |
| **Middle** | | | | | | |
| Dental | | | | | δ | θ |
| Alveolar | j,l | n | d | t | z | s |
| Palatal | r | | | | 3 | ʃ |
| **Back** | | | | | | |
| Velar | w,ʍ | ŋ | g | k | | |
| Pharyngeal | | | | | | h |
| Glottal | | | ʔ | | | |

**FIGURE 17.8** Articulatory classification of consonants. From [15].

be represented as binary entities. Finally, the set includes the semivowels or glides, plus /h/ (last five columns in Fig. 17.7).

The set of consonants in Fig. 17.7 can also be classified into articulatory categories, as shown in Fig. 17.8. This figure defines the consonants in terms of the required vocal tract shape (including the role of the glottis) and the place in the vocal tract of greatest constriction [15].

Voiers [22] used six features in his diagnostic rhyme test (DRT). He called his features voicing, nasality, sustention, compactness, sibilation, and graveness. The DRT has been used to test a great variety of speech-communication systems, most often a vocoder system. In assessing the results of a DRT, he made use of these features to produce a score of the complete test.

Ghitza [9] extended the DRT to include tiling. In this paradigm, word pairs in the DRT were modified by interchanging selected time–frequency acoustic regions (called tiles) and the errors induced by these changes were evaluated. In this way, perceptual distance metrics were derived, allowing the experimenter to determine the critical spectral–temporal regions for the perception of a given consonant.

## 17.4 PERCEPTUAL CUES FOR PLOSIVES

English plosive sounds or stop consonants (p, t, k, b, d, and g) can contain four successive acoustically distinct intervals. Typically a period of silence (commonly called a closure) occurs to allow for the pressure buildup in the closed vocal tract. This occlusion is suddenly opened, resulting in an acoustic burst. For voiceless plosives, the burst is followed by an interval of aspiration and then by a formant transition into the vowel. For the voiced plosives, the aspirate interval is missing, so these contain three acoustically distinct intervals. This description excludes other cases, for example, a plosive followed by a fricative. Note also that unvoiced plosives sometimes have reduced aspiration when preceded by a fricative sound (such as s).

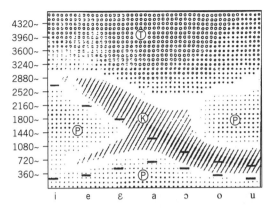

**FIGURE 17.9** Perceptual responses to different burst frequencies and vowels for the voiceless plosives. From [3].

The length of time between the release of closure and the start of voicing for a following vowel is referred to as the voice onset time (VOT). The VOT is an important perceptual cue; in the extremes, a very long VOT nearly always leads to a percept of an unvoiced stop (p, t, or k), whereas a very short VOT leads to the percept of a voiced stop (b, d, or g).

Cooper et al. [3], using the Haskins Pattern Playback (see Chapter 29), studied the perceptual roles played by the release burst and the following vowel, for voiceless stop consonants. Figure 17.9 shows results for seven vowels and 12 burst frequencies. It can be seen that high-frequency bursts (greater than 2500 Hz) were always perceived as /t/, whereas low-frequency bursts (below 1000 Hz) were always perceived as /p/. For intermediate burst frequencies, listeners chose /p/ for vowels with low first formants and /k/ for vowels in which $F_1$ and $F_2$ were relatively close together. The tentative conclusion: cues (i.e., perceptual hints) for voiceless plosives were based on *both* burst frequency and the following vowel, at least for /k/ and /p/.

## 17.5 PHYSIOLOGICAL STUDIES OF TWO VOICED PLOSIVES

Miller and Sachs [14] have studied the response of cats' ears to the sounds /da/ and /ba/. The stimuli were generated with the Klatt synthesizer (see Chapter 29). Figure 17.10 shows the stimulus properties for /da/.

For each 20-ms interval, PSTHs of the signals were taken for a large ensemble of fibers over a wide range of characteristic frequencies. Fourier transforms of these histograms were computed, and a synchronization measure called the average localized synchronized rate (ALSR) was introduced. The ALSR is computed as a function of the $k$th index of the discrete Fourier transform of the PSTH of the $L$th fiber, divided by the fiber's average rate ($R_{kL}$), so that

$$\text{ALSR}(k) = \frac{1}{M_k} \sum_{L \in C_k} R_{kL}, \qquad \textbf{(17.1)}$$

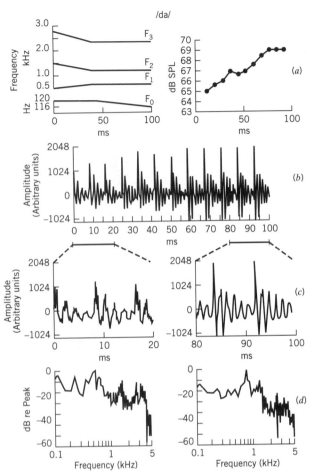

**FIGURE 17.10** Stimulus for the syllable /da/. The left plot of (a) shows the formants and fundamental frequency of the applied synthesizer parameters; the right plot shows the overall SPL time variation. (b) shows the time waveform, (c) shows the first 20 ms (left) and last 20 ms, and (d) shows the corresponding spectra. From [14].

where $C_k$ is the set of fibers whose CFs are within $\pm 0.25$ octave of that frequency, and $M_k$ is the number of fibers in $C_k$. The ALSR shows how new results may be obtained as a result of the ability to obtain data from many fibers based on the *same* stimulus. Figure 17.11 shows one example.

The dashed curves shown in the right column correspond to the formants used in the synthesizer. We can see that these formant trajectories are well represented by the ALSRs. Thus the temporal aspects of the spike patterns in the auditory nerve prove to be a good representation of the actual stimulus spectrum.

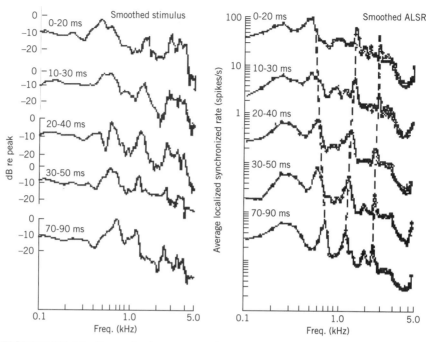

**FIGURE 17.11**  Smoothed spectra and corresponding ALSRs for the five 20-ms intervals at the beginning of the /da/ stimulus. From [14].

## 17.6  MOTOR THEORIES OF SPEECH PERCEPTION

As of this writing, our knowledge of the representation of speech in the brain is very primitive. When incoming speech is transformed into a neural representation, what does the brain match this representation to in order to decide on what was said? Although we don't know the answer to this, there have been a number of theories advanced. One prominent view (though probably not a dominant one) is referred to as the motor theory of speech perception. This theory is based on the hypothesis that our brains interpret the received speech patterns in terms of the neural pattern production that we need to *articulate* the same incoming speech. To quote Pickett [15]:

> *It derives from the 'functional' school of psychology. . . . The functionalists insisted that all perception is organized to serve as a basis for behavioral action [1].*

An engineering model of the motor hypothesis was proposed by Stevens and Halle [20]; a diagram is shown in Fig. 17.12.

In this model, called analysis by synthesis, a tentative analysis is first carried out as shown in box A in the figure. To quote Pickett (pp. 192–193) again [15]:

> *The speech sound input on the left is first analyzed by the auditory mechanism, A, which may provide an analysis in terms of distinctive features. The auditory analysis is*

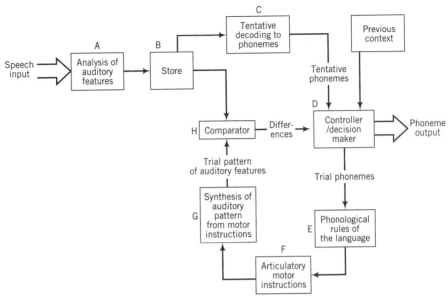

**FIGURE 17.12** Analysis by synthesis motor theory of speech perception. From [15].

*fed to a store, B, which retains the auditory features for subsequent use in the comparator, H, and in a preliminary phoneme decoder, C. This decoder converts the auditory features into a tentative sequence of phonemes, which is combined with the previous context by the controller/decision mechanism, D, to give a trial sequence of phonemes. The trial sequence is then converted by phonological rules, E, to a set of motor articulatory instructions, F. These are fed to a synthesizer, G, which converts the instructions into a set of auditory features, a trial auditory pattern that can be compared with the stored auditory pattern. The difference, or 'error,' seen by the comparator, H, are assessed by the decision mechanism; if the error is low the decision confirms and puts out the preliminary phoneme sequence as the final decision; if the error is high the differences are used to generate a new, different sequence of trial phonemes, and the synthesis process is repeated for the new phonemes. The process reiterates until a final, best decision is made.*

Several variations on the motor hypothesis have been proposed. Work by Galunov and Chistovich [8] led to a proposal for a syllabic motor theory. Studdert-Kennedy [21] pointed out that although consonants are processed categorically (quantal response on continuously controlled parameters) and vowels are *not* processed categorically, this reflects a basic structure of the acoustic syllable. He then proposed a motor theory of speech acquisition in infants that postulates the existence of inherited articulatory and auditory templates that make it possible for the growing child to learn to speak and to recognize speech.

In 1965, Lane [11] showed experimentally that categorical perception is not restricted to speech sounds but works for certain nonspeech stimuli; he thus cast doubt on the motor hypothesis and the notion that the human brain has special speech-processing neural entities. As Pickett (p. 206) remarks, the motor theory, at this stage, may be more appealing than compelling.

## 17.7 NEURAL FIRING PATTERNS FOR CONNECTED SPEECH STIMULI

Delgutte [6] has recorded neural firing patterns in the cat in response to a spoken English sentence of several seconds. As did Sachs et al., he measured PSTHs, but his focus was primarily on the phoneme transitions. Figure 17.13 shows a comparison between the pattern of neural spiking and the sound spectrogram below. It can be seen that the transitions as represented by the spectrogram are also well represented by the firing patterns.

Figure 17.13 indicates how neural spike patterns respond to transients in speech. Delgutte [5] proposed a model for neural detection of the VOT. The model hypothesizes a measure of the difference between onset times of low-CF and high-CF neurons. An example is seen in Fig. 17.13 for the /J/ in "Joe" and the /t/ in "took"; one can find the VOTs by measuring the time intervals between the 400-Hz and 4000-Hz CF onset times.

Speech contains relatively steady-state as well as transient intervals (although fluent speech is essentially in constant flux, as noted earlier). Chistovich et al. [4] proposed a speech perception system that included a tonic and a phasic component; we have already described the phasic as the system response to transients; the tonic is the response to steady stimuli that frequently occur for vowels and fricatives. Figure 17.14 shows the PSTHs for neurons with CFs ranging from several hundred hertz to 3000 Hz for two periods of the steady-state vowel /ae/ (as in hat).

These measurements are the same ones shown in Fig. 17.13 but with a greatly expanded time scale. It can be seen, for instance, that intervals between peaks in the histograms of the lower-CF fibers are inverse to $F_1$, whereas CF fibers between 1300 and 1800 Hz show peaks at intervals of 0.7 ms, the inverse of $F_2$. Notice that the lowest CF fiber resolves the fundamental frequency $F_0$ of 100 Hz. Interestingly, the high-CF fibers do *not* resolve the harmonics and the histograms also reflect $F_0$.

In Figure 17.15, interval histograms rather than PSTHs are measured for the same stimulus as that of Fig. 17.14. (Notice that the scales for the two figures differ.) Peaks appear at roughly the same intervals, but, in addition, $F_0$ is represented by a major peak in *all* fibers so that the pooled distribution (B in the figure) yields a very strong representation of the fundamental period. In Section 17.3 on Confusion Matrices, we plotted the confusions for each of 16 consonants preceding the vowel /a/ (as in father). Delgutte [6] has obtained response patterns of a single fiber for several cases. These are shown in Fig. 17.16.

The spectrograms of the synthesized consonant–vowel sounds are shown in the left column of the figure. The middle column gives the response of an auditory filter with CF = 1800 Hz. The rightmost row gives the response of a fiber in the inferior colliculus with CF = 1100 Hz. Note that this final row shows a sparser response to the stimulus, a tendency that is often seen farther up the auditory chain.

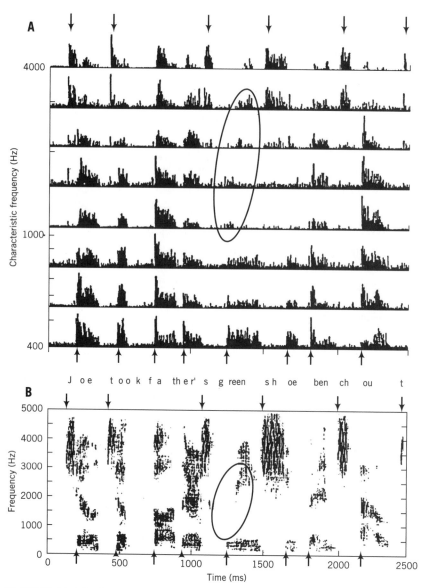

**FIGURE 17.13** Neural firing patterns for different CFs and spectrograms of the same sentence. From [6].

## 17.8 CONCLUDING THOUGHTS

It would be easier to present a coherent picture of the physiological basis of speech perception if it was very well understood – but it is not. There have been many experiments showing the response of neurons to simple stimuli (particularly in the periphery, i.e., at the auditory nerve), but the results of these experiments provide us with more detail than

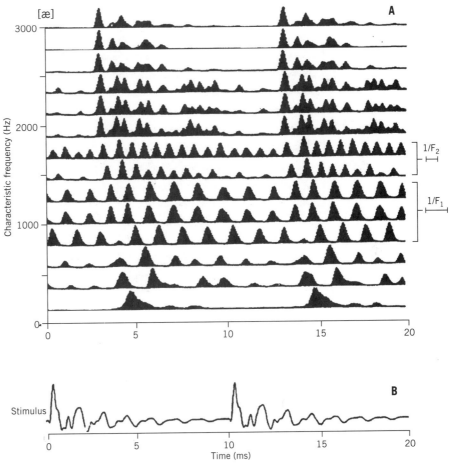

**FIGURE 17.14** PSTHs for two periods of a steady vowel. Plot B shows the vowel waveform. From [6].

understanding of the overall operation of the system. It is apparent that the auditory nerve provides rich representations to the higher centers, and that both spectral information and temporal information are available for these later systems to use. Some of more central auditory subsystems (such as the cochlear nuclei) contain cells that react in a number of diverse ways to the stimuli; as noted in Chapter 14, the auditory nerve has both high-threshold (low spontaneous rate) fibers and low-threshold (high spontaneous rate) fibers, so that even the notion of a spectral representation at the auditory nerve is made complicated. As noted in Chapter 16, there is also significant evidence that some temporal information is passed on to higher centers, and this information may be used for more than pitch perception. However, key features of formant trajectories for short-syllable stimuli do appear to be well represented at the auditory nerve (although higher-pitch speech appears to generate representations with less of a clear correspondence with formant structure.) It is also apparent

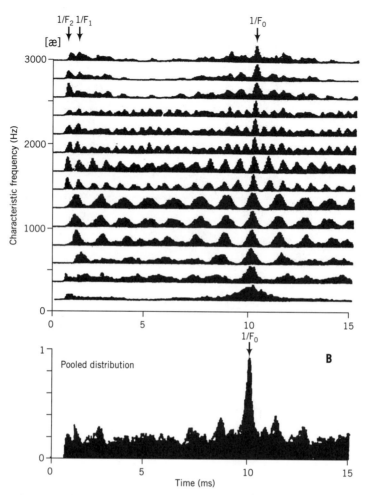

**FIGURE 17.15** Interval histograms for two periods of a steady vowel. Plot B shows the pooled distribution, summing up over the characteristic frequencies. From [6].

that at least some of the sounds that are commonly confused by listeners generate somewhat similar representations at the auditory nerve.

In this chapter, we also presented some classic results of psychophysical experiments with simple speech stimuli – in particular, with individual phonetic elements and with simple plosive-vowel syllables. Such experiments were often done with synthesized sounds or with careful talker presentations of isolated elements. Some of the basic ideas that are now used in discussions of speech perception came from these studies. In particular, these include the notion of common confusions, the notion of phonetic features that characterize individual sounds, and the apparent existence of some internal representation that seems to be more coarse than what is suggested by a pure formant structure. We also noted some of the aspects of stop-consonant perception, including voice onset time, burst frequency, and formant transitions. We have necessarily skipped many important areas of study, such

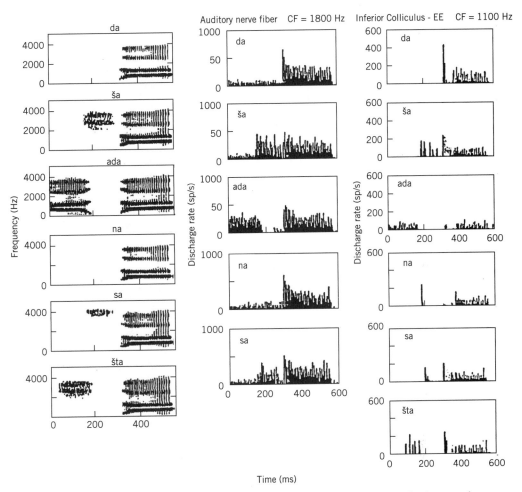

**FIGURE 17.16** Response patterns of two fibers to various synthetic speech consonant–vowel transitions. From [6].

as prosody (patterns of pitch, intensity, and durations), effects of accent and dialect or pronunciation in general, speech as part of multiple auditory streams in real environments, and so on. In Chapter 18, we will extend the discussion to the problem of human speech recognition.

## 17.9 EXERCISES

**17.1** In Section 17.2, it is stated that "the best matches were for the adjustable $F_2$ to be appreciably higher in frequency than the reference $F_2$ for the front vowels but equal to the reference $F_2$ for the back vowels." Why did this result strengthen the two-formant conjecture?

**17.2**   What phonemes would you think would be most difficult to perceive over a quiet telephone channel? What about a noisy one? Give an explanation.

**17.3**   Three measures of auditory neural activity are based on the times of occurrence of neural spikes; these are (a) PSTHs, (b) interval histograms, and (c) period histograms. In (a), a histogram of the time of occurrence of spikes following stimulus onset is accumulated. This involves repeated presentations of the same stimulus; the stimulus is typically 10–20 ms in duration. In (b), the same stimulus may be presented to the animal's ear, but the measurement is the time between adjacent spikes. In (c), the times are measured relative to the reference time in a periodic stimulus; for example, the time between the positive-going zero crossing of a sinusoidal stimulus and the occurrence of spikes within the subsequent stimulus period. Make diagrammatic sketches of how a single neuron might respond to a single presentation for each of these measurements. Then, discuss the relationships between the three methods, using the responses to a tone and a click as examples. Can you think of cases in which one of the measurements is superior to the others?

**17.4**   Jakobson et al. [10], Miller et al. [13], and Voiers [22] each developed a distinctive feature set. The concept was based on the notion that phonemes could be described in terms of more elementary entities. As the text indicates, the work of Jakobson et al. was more extensive. Discuss how the feature systems of Voiers and Miller et al. can be derived from the Jakobson et al. system. Compare Voiers and Miller and give your opinion as to the relative merits and demerits of each for assessing speech-communication systems. Why did Miller and Voiers find it necessary to simplify the Jakobson system?

**17.5**   The ability of Sachs and Young [17] to measure the responses of many neurons to the same stimulus created new opportunities for physiological research on auditory systems. Discuss how the ALSR is a byproduct of these technical developments. Can you cite other measurements that became available, given this new capability?

**17.6**   The sound spectrogram (described in Chapters 2 and 3) has served for many years as a standard way of visualizing the properties of a speech signal. More recently, researchers have developed the neurogram to describe speech in the time–frequency domain. Discuss the relationship between these two methods and give your opinion as to the pros and cons of each method. (Review Section 17.7, plus any other material you know about as a preliminary to your discussion).

**17.7**   What is meant by categorical perception? Do you know of any examples of this in domains other than speech?

**17.8**   Discuss the motor theory of speech, giving arguments for and against it.

# BIBLIOGRAPHY

1. Boring, E. G., *A History of Experimental Psychology*, 2nd ed., Appleton–Century–Crofts, New York, 1950.
2. Carlson, R., Fant, G., and Granstrom, B., "Two-formant models, pitch, and vowel perception," in G. Fant and M. Tatham, eds., *Auditory Analysis and Perception of Speech*, Academic Press, New York, 1975.
3. Cooper, F. S., Delattre, P. C., Liberman, A. M., Borst, J. M., and Gerstman, L. J., "Some experiments on the perception of synthetic speech stimuli," *J. Acoust. Soc. Am.* **24**: 597–606, 1952.

4. Chistovich, L. A., Lublinskaya, V. V., Malinnikova, T. G., Ogorodnikova, E. A., Stoljarova, E. I., and Zhukov, S. J. S., "Temporal processing of peripheral auditory patterns of speech," in R. Carlson and B. Granstrom, eds., *The Representation of Speech in the Peripheral Auditory System*, Elsevier, Amsterdam/New York, pp. 165–180, 1982.

5. Delgutte, B., "Analysis of French stop consonants with a model of the peripheral auditory system," in J. S. Perkell and D. H. Klatt, eds., *Invariance and Variability of Speech Processes*, Erlbaum, Hillsdale, N.J., 1986.

6. Delgutte, B., "Auditory neural processing of speech," in W. J. Hardcastle and J. Laver, eds., *Handbook of Phonetic Science*, Blackwell, Oxford, 1997.

7. Dinse, H., and Schreiner, C., "Dynamic frequency tuning of cat auditory cortical neurons: specific adaptations to the processing of complex sounds," in W. Ainsworth and S. Greenberg, eds., *Proc. Workshop Auditory Basis Speech Perception*, Worth Printing, Keele, U.K., pp. 45–48, 1996.

8. Galunov, V. I., and Chistovich, L. A., "Relationship of motor theory to the general problem of speech recognition," *Sov. Phys. Acoust.* **11**: 357–365, 1966.

9. Ghitza, O., "Adequacy of auditory models to predict internal human representation of speech sounds," *J. Acoust. Soc. Am.* **93**: 2160–2171, 1993.

10. Jakobson, R., Fant, C. G. M., and Halle, M., "Preliminaries to speech analysis," Tech. Rep. No. 13, Acoustics Laboratory, Massachusetts Institute of Technology, 1952.

11. Lane, H., "The motor theory of speech perception: "a critical review" *Psychol. Rev.* **72**: 275–309, 1965.

12. Liberman, A. M., Cooper, F. S., Shankweiler, D. P., and Studdert-Kennedy, M., "Perception of the speech code," *Psychol. Rev.* **74**: 341, 1967.

13. Miller, G. A., and Nicely, P. E., "An analysis of perceptual confusions among some English consonants," *J. Acoust. Soc. Am.* **27**: 338–352, 1955.

14. Miller, M. I., and Sachs, M. B., "Representation of stop consonants in the discharge patterns of auditory-nerve fibers," *J. Acoust. Soc. Am.* **74**: 502–517, 1983.

15. Pickett, J. M., *The Sounds of Speech Communication*, Pro-ed, Austin, Texas, 1980.

16. Sachs, M. B., Blackburn, C. C., and Young, E. D., "Rate-place and temporal-place representations of vowels in the auditory nerve and anteroventral cochlear nucleus," *J. Phonet.* **16**: 37–53, 1988.

17. Sachs, M. B., and Young, E. D., "Encoding of steady-state vowels in the auditory nerve: representation in terms of discharge rate," *J. Acoust. Soc. Am.* **66**: 470–479, 1979.

18. Secker-Walker, H. E., and Searle, C. L., "Time-domain analysis of auditory-nerve fiber firing rates," *J. Acoust. Soc. Am.* **88**: 1427–36, 1989.

19. Shamma, S. A., "Speech processing in the auditory system I: the representation of speech sounds in the responses of the auditory nerve," *J. Acoust. Soc. Am.* **78**: 1612–1621, 1985.

20. Stevens, K. N., and Halle, M., "Remarks on analysis by synthesis and distinctive features," in W. Wathen-Dunn, ed., *Models for the Perception of Speech and Visual Forms*, MIT Press, Cambridge, Mass., pp. 88–102, 1967.

21. Studdert-Kennedy, M., "Speech perception," in *Contemporary Issues in Experimental Phonetics*, N. J. Lass, ed., Academic Press, New York, pp. 243–293, 1976.

22. Voiers, W. D., "Evaluating processed speech using the diagnostic rhyme test," *Speech Technol.* **1**: 30–39, 1983.

# *HUMAN SPEECH RECOGNITION*

## 18.1 INTRODUCTION

How do people recognize and understand speech? As with other aspects of perception that we have touched on, this is an area for which many books and articles have been devoted. Our task is further complicated by the fact that, despite the profusion of articles on the subject, very little is understood in this area; at least there is very little that experts agree on.

Here we can only hope to introduce a few key concepts and in particular to lay the groundwork for the reader to think about aspects of human recognition that are different from the common approaches to artificial speech recognizers. For this purpose, we focus on two particular studies:  the perception of consonant–vowel–consonant (CVC) syllables in decades-long studies, directed by Harvey Fletcher of Bell Labs (and reexamined recently by Jont Allen [1]); and the direct comparison of human and machine "listeners" on tasks of current interest for speech-recognition research, as described by Richard Lippmann of Lincoln Labs [6].

## 18.2 THE ARTICULATION INDEX AND HUMAN RECOGNITION

Recently Jont Allen from AT&T revived interest in a body of work done at Bell Labs in the 1920s by a group headed by Harvey Fletcher; [1] is an insightful summary of Allen's perspective on this work. Here we describe only a few key points from that paper.

### 18.2.1 The Big Idea

A principal proposal of this paper is that humans do not appear to use spectral templates (e.g., one set of local spectral features every 10 ms), but rather they do partial recognition of phonetic units across time, independently in different frequency ranges. In other words, Allen's model suggests a subband analysis for speech recognition, in which partial decisions are developed independently and then combined at the level of phonetic categorization.

This suggestion is notable for a number of major reasons.

**1.** It is based on decades of measurements with human listeners; in other words, even if it does not turn out to be entirely correct, it is not a casual suggestion, and it is likely to at least be related to the character of human hearing.

**2.** It is also based on a theory of human hearing that was developed by Fletcher to model the experimental results. In other words, there is a theoretical structure that can be used. In particular, Fletcher defined a measure called the articulation index (called AI long before the term was ever used for artificial intelligence). In this context he intended the word articulation to refer to the probability of identifying nonsense speech sounds. This was in contrast to the notion of intelligibility, which refers to the identification of meaningful speech sounds such as words or sentences.

**3.** Finally, it is at odds with virtually every speech-recognition system that engineers have built; as of 1997, nearly all of them use framewise features that are based on short-term spectral estimates.

## 18.2.2 The Experiments

Starting around 1918, Bell Labs researchers designed data bases from CVC, consonant–vowel (CV), and vowel–consonant (VC) nonsense syllables. Their estimate at the time was that these types comprised approximately 74% of the syllables used over the telephone, and as such provided a good idealized testbed for the recognition of speech without the more complex factors of multisyllabic acoustic context or syntactic or semantic factors. Over the following years, listening tests were conducted with differing signal-to-noise ratios and frequency ranges; the latter were established by using high-pass or low-pass filters. A fundamental motivation was to determine the bandwidth that was required for the telephone system; in fact it was these experiments that led to the frequency range that is in use today.

There were many results from these experiments, but two that Allen focused on were as follows.

**1.** The probability of getting a CVC syllable correct (determined by counting the number of times that listeners correctly identified the syllable for a given condition) was roughly the product of the probabilities of having the initial C, the V, or the final C phone correct in the syllable identification. This meant that, as far as this measure and experiment were concerned, the phone identifications could be treated as being independent.

**2.** For speech low-pass filtered and high-pass filtered at the same point, the phone error probability for the total spectrum was equal to the product of the error probabilities for each of the two bands. In other words, the probability of error for the band from 100 Hz to 3000 Hz would be the product of the error probabilities for bands from 100 to 1000 Hz and from 1000 to 3000 Hz. More formally, let $s(a, b)$ be the articulation (probability of correct phone classification) for speech with a lower spectral limit of $f_a$ and a higher spectral limit of $f_b$. Then Fletcher's experiments seemed to show that

$$[1 - s(a, c)] = [1 - s(a, b)][1 - s(b, c)]. \tag{18.1}$$

Equivalently,

$$\log_{10}[1 - s(a, c)] = \log_{10}[1 - s(a, b)] + \log_{10}[1 - s(b, c)]. \tag{18.2}$$

Fletcher then defined an AI as

$$\mathrm{AI}(s) = \frac{\log_{10}(1 - s)}{\log_{10}(1 - s_{\max})}, \tag{18.3}$$

where $s_{max}$ is the maximum articulation (that is, the best that people could achieve with a high signal-to-noise ratio and wide bandwidth), measured by Fletcher to be 0.985.

Given the experimental result of Eq. 18.1, dividing Eq. 18.2 by $\log_{10}(1 - s_{max})$, we see that the AI has the property that

$$\text{AI}[s(a, c)] = \text{AI}[s(a, b)] + \text{AI}[s(b, c)] \quad f_a \Leftarrow f_b \Leftarrow f_c. \tag{18.4}$$

Thus, a relatively simple nonlinear transformation of the probability of phone correctness converted it to a measure that would be roughly additive over frequency.

### 18.2.3   Discussion

The articulation index measure also implied an underlying density, for which the integral over some contiguous range corresponded to the AI for that range. Allen notes that Fletcher and Stewart generalized from these results to a multi-independent channel model of phone perception. As noted from Eq. 18.1, the error for a wideband signal is equal to the product of errors for two individual bands. If correct, this would be an astonishing kind of independence – essentially it says that if any single band leads to perfect phone identification, errors in any of the other bands don't matter! Such a system is in some sense optimal, though the form of optimality is one that we currently don't know how to implement in engineering systems. However, Allen's interpretation of the Fletcher data is that this was indeed what was observed.

In recent years, there have been some objections to the AI theory as a literal truth; for instance, Lippmann showed human performance for speech that had been filtered to essentially remove components between 800 and 3000 Hz to in fact be significantly *better* than that which would be predicted from AI analyses [8]. The choice of phones as the fundamental unit of speech recognition is also controversial, as others have suggested transition-based units, half-syllables, or even complete syllables as being more primary (particularly during natural speech, for instance during human-to-human conversations) [4]. However, it seems likely that there is a significant amount of analysis that is performed in the human auditory system on limited bands of the speech over time, and that this information is later integrated into some kind of incomplete decision about sound unit identity. A number of speech researchers have begun to investigate the feasibility of incorporating ideas such as this into ASR systems [3], [2], [5]. More generally, it is likely that human hearing incorporates many maps for decisions about what was said; see Fig. 17.1 in the previous chapter, for instance, for the Sachs et al. perspective, in which multiple maps that are *each* tonotopically organized are used in order to make phonetic distinctions.

## 18.3   COMPARISONS BETWEEN HUMAN AND MACHINE SPEECH RECOGNIZERS

Although the automatic speech recognition (ASR) research of the past few decades has resulted in great advances, much more remains to be done to achieve the oft-stated goal of

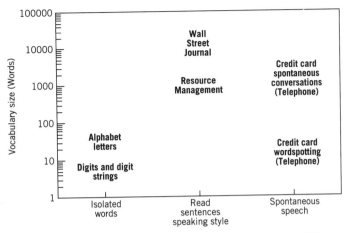

**FIGURE 18.1**  Six speech-recognition corpora. From [6].

devices that equal or exceed human performance. Lippmann [6] has compared recognition accuracy for machines and people on a range of tasks. Figure 18.1 shows six recognition corpora with vocabularies ranging from 10 to 20,000 words, and this includes isolated words, read sentences, and spontaneous speech. All cases are speaker independent (trained on one set of speakers but used on a different set). Table 18.1 lists the characteristics of six speech corpora.

**TABLE 18.1  Characteristics of Six Talker-Independent Recognition Corpora[a]**

| Corpus | Description | No. of Talkers | Vocabulary Size | No. of Utterances | Total Duration (h) | Recognition Perplexity |
|---|---|---|---|---|---|---|
| TI digits | Read digits | 326 | 10 | 25,102 | 4 | 11 |
| Alphabet letters | Read alphabet letters | 150 | 26 | 7800 | 1 | 26 |
| Resource Management | Read sentences | 109 | 1000 | 4000 | 4 | 60–1000 |
| Wall Street Journal | Read sentences | 84–284 | 5000–20,000 | 7200–37,200 | 12–62 | 45–160 |
| Credit-card continuous speech recognition | Spontaneous telephone conversations | 70 | 2000 | 35 conversations, 1600 segments | 2 | 100 |
| Credit-card wordspotting | Spontaneous telephone conversations | 70 | 20 keywords | 2000 keyword occurrences | 2 | |

[a]From [6].

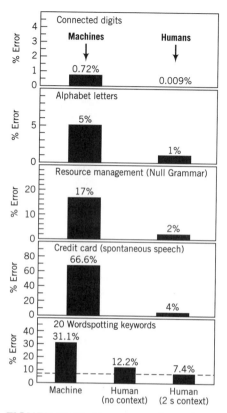

**FIGURE 18.2**　Five comparisons between human and ASR devices. From [6].

The column marked Recognition Perplexity is a measure of the average number of words that can occur at any point in an utterance, assuming a particular grammar, as defined in Chapter 5. Given the ability of many recognizers to use word-sequence constraints, perplexity is a good measure of the linguistic uncertainty in the grammar, and as such tends to correlate with the recognition difficulty for a task. It does not account for the acoustic difficulty of a task; for instance, an unconstrained digit sequence has roughly the same perplexity as an unconstrained sequence of letter names that rhyme with "e," but the latter has words that are much more acoustically similar to one another, so error rates tend to be higher. The perplexity for the last row is left blank, as the row refers to a wordspotting task, which does not use a constraining grammar in the same sense as the other tasks.

The five boxes in Fig. 18.2 show comparisons between humans and the best ASR devices for all table entries but the Wall Street Journal case.

These results are for the cases of clean speech; in other words, there are essentially no environmental effects, such as a poor signal-to-noise ratio (SNR). As the top 3 boxes show, both ASR and human error rates increase with perplexity. The bottom two boxes are for very different conditions so that perplexity is no longer a reasonable measure of difficulty.

**TABLE 18.2  Word Error Rate for a 5000-Word Wall Street Journal Task, Using Additive Automotive Noise**

| System | 10-dB SNR (%) | 16-dB SNR (%) | Quiet (%) |
|---|---|---|---|
| Baseline HMM ASR | 77.4 | 42.2 | 7.2 |
| ASR with noise compensation | 12.8 | 10.0 | |
| Human listener | 1.1 | 1.0 | 0.9 |

These boxes indicate that for more normal (less formal) human discourse, *huge* increases in ASR error rates occur.

Another comparison indicator is given in Table 18.2; namely, the effect of adding noise to a test set from the Wall Street Journal corpus.

Discussing these data, Lippmann notes [7]:

> *. . . the error rate of a conventional high performance HMM (Hidden Markov Model) recognizer increases dramatically from 7.2% in quiet to 77.4% at a SNR of 10 dB at noise levels that do not affect human performance. This enormous increase in error rate occurs for all high-performance recognizers tested in this noise and trained using quiet speech. A noise adaptation algorithm reduces this dramatic drop in performance and provides an error rate of 8.4% at a SNR of 22 dB and 12.8% at a SNR of 10 dB.*

We note that although the noise adaptation algorithm helps a great deal, the error rate still almost doubles between the quiet and 10-dB SNR case, while increasing only slightly for human listeners over this range. Further, even with noise adaptation there is still an order of magnitude greater error with ASR than for the human experiment.

Lippmann summarizes his results as follows.

> *Results comparing human and machine speech recognition demonstrate that human word error rates are roughly an order of magnitude lower than those of recognizers in quiet environments. The superiority of human performance increases in noise and for more difficult speech material such as spontaneous speech. Human listeners do not rely as heavily on grammars and speech context . . . . Humans do not require retraining for every new situation . . . . These results and other results on human perception of distorted speech suggest that humans are using a process for speech recognition that is fundamentally different from the simple types of template matching that are performed in modern hidden Markov speech recognizers.*

In [6] Lippmann further suggests that examining more narrow spectral (and temporal) regions, as well as learning how to ignore or focus on different kinds of phonetic evidence, will be key problems for future ASR research.[1]

---

[1]Lippmann later refined this material and described it in a journal paper [9].

## 18.4  CONCLUDING THOUGHTS

As we noted in Chapters 14–17, the peripheral auditory system has been explored for many years. Although it still would be presumptuous of us to assume that the physiology up to the auditory nerve is well understood, there is a moderate amount of agreement among scientists about the basics in this area. Farther up the auditory chain, our knowledge is certainly much more limited. Overall, as of this writing it would be fair to say that the internals of human speech recognition are little understood.

However, as noted in this chapter, we do know that people generally recognize speech very well, even under conditions that appear to pose great difficulty for our ASR systems. What do we know about human speech recognition that differs from our best artificial systems?

Signal processing:   Although we don't know exactly what signal processing occurs in the auditory system, we do know that processing occurs with a range of time constants and bandwidths. Given the robustness of human listening to many signal degradations, each of which would severely degrade an individual representation, it is likely that many maps of the input signal are available to the brain. ASR's current use of simple functions of short-term spectra measured every 10 ms may be a significant limitation. However, even within this constraint, a number of the characteristics of auditory perception have been incorporated into speech-processing systems, and these will be discussed in the next few chapters.

Subword recognition:   humans seem to be able to adapt their use of the multiple signal representations according to the requirements of the moment. In [6] Lippmann calls for the addition of "active analysis in the front ends of speech recognizers to determine when a feature is present and when it is a component of a desired speech signal. This supplementary information can be used by classifiers that can compensate for missing features." As noted earlier in this chapter, Allen has focused on the use of subband information over time, and he suggests some representation incorporating a correlation between subbands as a measure; combining these threads of information is an open problem, though the current leading contenders for a solution are all based on ideas from statistical pattern recognition.

Temporal integration:   humans are able to understand utterances with a wide range of speaking rates, implying some kind of time normalization. In contrast, durations are often key components in phonetic discriminations. For example, as noted in Chapter 17, the VOT (the time between the burst and the following vowel) is the key discriminating factor between "pa" and "ba." In ASR, the most common form of temporal normalization is a crude compromise that does not sufficiently reduce variability (ASR systems often do much worse on fast speech) but also eliminates critical information about internal timing.

Integration of higher level information:   in many cases in which the acoustic evidence is equivocal, the utterance identification can still be made based on the expectations from syntax, semantics, and pragmatics (where the latter refers to facts from the particular situation). Additionally, for many tasks it is not really necessary to recognize all the words, but only to get the relevant point. This is made obvious by examining written transcriptions of the spoken word. Particularly if there is no second corrective pass, there are often many differences between what is said and what was written. Essentially, people are trained to recognize the gist of what was said, and usually not the precise word sequence. In our ASR systems, we tend to focus equally on all words, both informative and noninformative; furthermore, our current capabilities for the integration of higher-level information are quite primitive, except in specialized systems that depend on an extremely restricted application domain.

This concludes the first half of our text. In the remaining chapters, we will focus on the engineering approaches that are currently the basis of audio processing systems, with a particular focus on speech recognition, synthesis, and vocoding.

## 18.5 EXERCISES

**18.1** Imagine a tonotopically organized (subband-based) phoneme recognition system such as the one discussed by Allen. What might be the potential advantages or disadvantages of such a system?

**18.2** Lippmann points out many ways in which 1996 speech-recognition technology is inferior to the capabilities of human speech recognition. Suggest some situations in which human speech recognition could potentially be worse than an artificial implementation.

## BIBLIOGRAPHY

1. Allen, J. B., "How do humans process and recognize speech?," *IEEE Trans. Speech Audio Proc.* **2**: 567–577, 1994.
2. Bourlard, H., and Dupont, S., "ASR based on independent processing and recombination of partial frequency bands," in *Proc. Int. Conf. Spoken Lang. Process.* Philadelphia, Pa., 1996.
3. Bourlard, H., Hermansky, H., and Morgan, N., "Towards increasing speech recognition error rates," *Speech Commun.* **18**: 205–231, 1996.
4. Greenberg, S., "Understanding speech understanding:   towards a unified theory of speech perception," in W. Ainsworth and S. Greenberg, eds., *Workshop Auditory Basis Speech Percept.*, Keele, U.K., pp. 1–8, 1996.
5. Hermansky, H., Pavel, M., and Tibrewala, S., "Towards ASR on partially corrupted speech," in *Proc. Int. Conf. Spoken Lang. Process.* Philadelphia, Pa., 1996.
6. Lippmann, R., "Speech recognition by humans and machines," in W. Ainsworth and S. Greenberg, eds., *Workshop Auditory Basis Speech Percept.*, Keele, U.K., pp. 309–316, 1996.
7. Lippmann, R., "Speech recognition by machines and humans," Lincoln Laboratories, Lexington, MA, 1996.
8. Lippmann, R., "Accurate consonant perception without mid-frequency speech energy," *IEEE Trans. Speech Audio Process.* **4**: 66–69, 1996.
9. Lippmann, R., "Speech recognition by machines and humans," *Speech Communication*, **22**: 1–15, 1997.

# SPEECH FEATURES

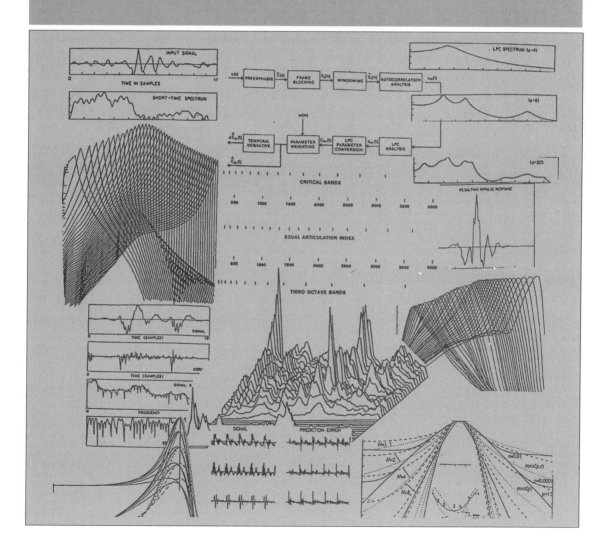

My sources are unreliable, but their information is fascinating.
—Ashleigh Brilliant

**B**EGINNING WITH  Part V, we illustrate how signal processing and pattern-recognition techniques can be applied to problems in speech and audio. In Parts V–VIII we will incorporate the ideas hinted at in Part I, using the mathematics from Part II, assuming the underlying structure of the signal suggested in Part III, and with a perspective derived from our view of human hearing as described in Part IV.

In every speech-processing system, signal processing is used to extract relevant features. In the case of analysis for synthesis, for instance, representations of the short-term spectrum are computed in order to preserve perceptually relevant features for later reproduction. In the case of speech recognition, features are typically computed so that sounds associated with different linguistic units can be distinguished. In each of these cases, researchers have developed representations to optimize for the specific goals of the application. However, much as the colors in a painter's palette are used to compose a range of possible hues, basic approaches to the representation of short-term spectral information are used as components in more complex types of feature extraction that are described later in this text. The three chapters in Part V describe archetypes for speech feature extraction; these are the filter bank, cepstral analysis, and linear prediction.

# CHAPTER 19

# *THE AUDITORY SYSTEM AS A FILTER BANK*

## 19.1 INTRODUCTION

As noted in [10], one of the key measurements used in speech processing is the short-term spectrum. In all of its many forms, this measure consists of some kind of local spectral estimate, typically measured over a relatively short region of speech (e.g., 20 or 30 ms). This measure has been shown to be useful for a range of speech applications, including speech coding and recognition. In each case, the basic notion is that of capturing the time-varying spectral envelope for the speech, and in each case it is desirable to reduce the effects of pitch on this estimate; either pitch is used separately (as with a vocoder or a tone language speech-recognition system), or it is generally discarded as irrelevant to the discrimination (as in most English language speech-recognition systems). Therefore, in speech applications, the short-term spectral algorithm is usually designed to estimate a spectral envelope that has a reduced influence from the pitch harmonics in voiced speech.

In this chapter and the following two, we will describe three basic approaches to the estimation of the short-term spectral envelope: filter banks, cepstral processing, and linear predictive coding (LPC). The first and oldest approach is that of temporally smoothed power estimates from a bank of bandpass filters. Since much of the inspiration for such an approach originates from models of the human auditory system, we will begin with a discussion of the interpretation of the auditory system as a filter bank.

In Chapter 14 we displayed tuning curves of individual auditory nerve fibers and showed that the bandwidth increased with the CF (characteristic frequency) of the fiber. In Chapter 15 we discussed psychological tuning. Here we discuss filter-bank designs that can be used to model these aspects of the human auditory system. We review Fletcher's early experiments on critical bands. We then move to more recent experiments, in particular to Patterson's results, that lead to specifying shape as well as to the bandwidth of auditory filters. Following this we discuss versions of the so-called gammatone filters, which are attempts at physical realizations of the tuning curves of Chapter 14. We conclude with a more informal discussion of some of the filter-bank designs that have been at least partially influenced by auditory system research.

## 19.2 REVIEW OF FLETCHER'S CRITICAL BAND EXPERIMENTS

As noted in Chapter 18, Harvey Fletcher and his collaborators at Bell Laboratories experimented extensively on human hearing [1]. In addition to the experiments on CVC syllable perception, they did extensive work on masking phenomena. In a test of what is called simultaneous masking, the listener was presented with a tone plus wideband noise. Initially, the tone was of low enough intensity so that it was not perceived. The intensity of the tone was then gradually increased until it was just barely perceived; this intensity was called the threshold intensity. As the noise bandwidth was decreased, no change took place in the threshold until a *critical band* was reached. As bandwidths decreased still further, the threshold of detection decreased.

Such experiments suggested the existence of an auditory filter in the vicinity of the tone that effectively blocks extraneous information from interfering with the detection of the tone. This vicinity is called a critical band and can be viewed as the bandwidth of each auditory filter. The experimental results showed that the width of a critical band increases with the higher frequency of the tone being masked. Thus, the results yielded important information about the bandwidth of the auditory filter, though not about its shape.

The quantitative result is shown in Fig. 19.1 (a repeat of Fig. 15.4, added here for convenience). The Bark scale of Fig. 19.1 (a good approximation to psychoacoustic critical band measurements) yields bandwidths that are below 200 Hz until the center frequency exceeds 1000 Hz. As center frequency increases above 1000 Hz, the Bark scale adheres closely to bandwidths that are logarithmic functions of the center frequencies. Thus, for frequencies above 1000 Hz, the data is another example of Weber's law, which states that our peripheral senses tend to follow a logarithmic law of sensation as a response to a stimulus.

Below approximately 800 Hz the bandwidths measured by Fletcher were fairly constant, with a bandwidth of approximately 100 Hz. More recently, Moore et al. [4] performed

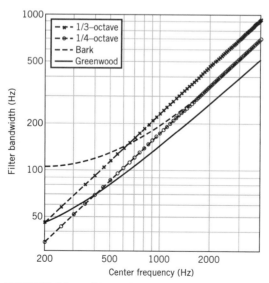

**FIGURE 19.1** Plots of estimated bandwidth as a function of center frequency for two constant-Q scales, the Bark scale and Greenwood's cochlear frequency-position function. From [2].

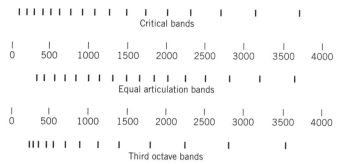

**FIGURE 19.2** AI vs. frequency. The short lines show the spacing of regions of equal AI along the frequency axis. The spacings of the critical band and third octave band are shown for comparison.

a different set of psychoacoustic measurements to estimate bandwidths for these low frequencies. These measurements seemed to show that the auditory filter bandwidths for low-frequency tones increased significantly between 100 and 800 Hz. The degree to which they increase is still a matter of some controversy among experts in the field.

Another result from Fletcher, based on very different psychoacoustic data (discussed in Chapter 18), related the articulation index (AI) to auditory bandwidths. As noted previously, the AI is a simple function (given by Eq. 18.3) of the average phone accuracy associated with a listener's response to CVC (consonant–vowel–consonant) nonsense syllables. Measurements were made by gradually increasing the bandwidth of a low-pass filtered version of these spoken syllables. As the bandwidth increased, the AI increased; in this way the AI could be directly associated with the speech bandwidth. As shown in Fig. 19.2, each mark indicated an equal increment in the AI, and we see from that figure that higher frequencies require a greater bandwidth to achieve the same AI increase as that of the lower frequencies. Thus, this test also leads to a model for which the auditory filters increase in bandwidth for higher frequencies.

## 19.3 RELATION BETWEEN THRESHOLD MEASUREMENTS AND HYPOTHESIZED FILTER SHAPES

Figure 19.3 shows two noise bands. The low-pass noise ranges from zero to 600 Hz, and the high-pass noise has a cutoff at 1200 Hz.

The listener's task is to detect a tone as its frequency varies from 400 to 1400 Hz. The small circles in the figure show the psychoacoustic results. The three curves shown correspond to three different hypotheses as to the auditory filter shape. Presumably, the assumed auditory filter shape can be varied until the computed threshold for any frequency is equal to the measured threshold. In this experiment, curves were found for three specific filter shapes: the rectangular filter (i.e., a filter with very steep transitions, approximating a rectangular frequency response) used by Fletcher, a standard resonance (a pole pair with

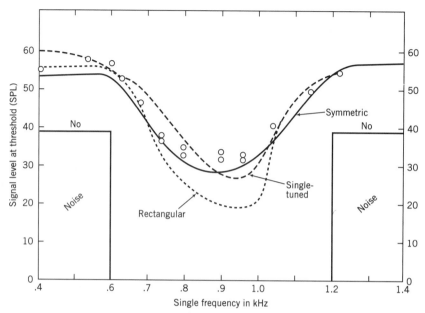

**FIGURE 19.3**   Predicted vs. measured threshold signal power in the region of a noise notch. The three curves correspond to three hypothetical filter shapes. The trajectory of small circles shows psychoacoustic results. The height of each noise rectangle is $N_0$ as in Eq. 19.2. From [5].

resonance frequency of the tone), and a symmetric filter (to be discussed later in the chapter). We note that the rectangular filter assumption results in the greatest error relative to the psychoacoustic results. The data for this experiment are from [11].

In this experiment, the noise was kept fixed while thresholds were computed for different tone frequencies. This permitted fitting of the auditory filter shape given a fixed filter paradigm, but it did not allow for an arbitrary filter transfer function to be designed directly from the psychoacoustic measurements. Patterson [5] developed a somewhat more complex method by varying the width of the rectangular noise band shown in Fig. 19.4 and keeping the tone frequency fixed. For each choice of a noise bandwidth, he measured the signal threshold or the SPL that was required for the tone to just barely be heard. He began with the following mathematical representation:

$$P = K \int_0^\infty N(f)|H(f)|^2 \, df, \tag{19.1}$$

where $P$ is the tone power at the threshold, $N(f)$ is the power spectrum of the noise, $H(f)$ is the transfer function of the auditory filter, and $K$ is the proportionality constant relating tone power at the threshold to the noise leaking through the filter, which is represented by the integral of the product of the noise spectrum and the filter shape.

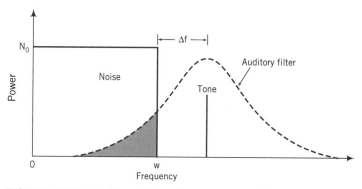

**FIGURE 19.4** Auditory filter shape computed by measuring the threshold as a function of the low-pass filter noise bandwidth. $W$ is the cutoff frequency for the noise; $\Delta f$ is the difference between the tone frequency and $W$. From [5].

If the noise spectrum is very close to rectangular, it can be removed from the integral, and Eq. 19.1 reduces to

$$P = KN_0 \int_0^W |H(f)|^2 \, df, \qquad (19.2)$$

where $W$ is the cutoff frequency of the low-passed noise, and $N_0$ is the constant noise power spectral level; $P$ is thus a function of $W$. By differentiating Eq. 19.2, we obtain an explicit result for the auditory filter magnitude function:

$$|H(W)|^2 = \frac{1}{KN_0} \frac{dP}{dW}. \qquad (19.3)$$

Note that this method can yield the magnitude function of the auditory filter but not its phase.

We can see from Fig. 19.4 that for a value of $W$ that is appreciably lower than the tone frequency, Eq. 19.3 results in a sensitive measure of $H(f)$. However, when the noise bandwidth $W$ is in the vicinity of the tone frequency, sensitivity is poor. To combat this problem, Patterson introduced a variable high-pass noise and varied the cutoff of the noise from well above the tone to just below it.

The results, normalized about the tone frequency, are displayed in Fig. 19.5; tone sensitivities are displayed in terms of signal levels that were discernible by 75% of the subjects. The top display shows the skirts of the filter at the low-frequency side; the bottom displays the high-pass noise case. The tone frequencies used are shown as parameters, ranging from 0.5 to 8 kHz. The abscissa is also in kilohertz, and we clearly see the large spread of the skirts as the tone frequency increases.

The implicit assumption in all results discussed thus far is that the auditory filter whose shape is to be found is centered around the tone to be detected. Figure 19.6 illustrates a shortcoming of this method. What if the observer's auditory filter is off center, as shown in

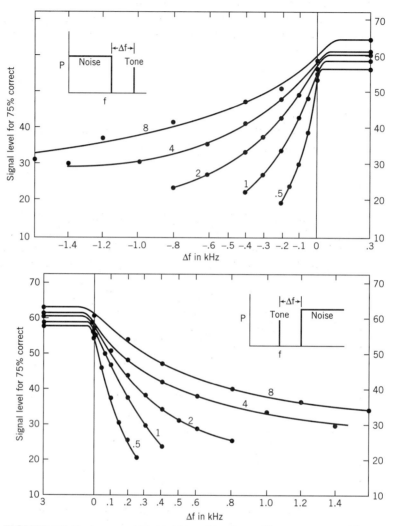

**FIGURE 19.5** Low and high skirts of auditory filters computed by varying the bandwidth of both low- and high-pass noise. The five curves show the data obtained at five different signal frequencies:   0.5, 1.0, 2.0, 4.0, and 8.0 kHz. From [5].

Fig. 19.6(b)? With this, the noise is lowered and this should lower the resultant threshold. A similar argument holds for high-pass noise. However, if off-center listening is really taking place, it becomes very difficult to use Eqs. 19.1–19.3 to compute $H(f)$.

Patterson [6] recognized this difficulty and devised a way to avoid it; the idea is illustrated in Fig. 19.6(c). Instead of separate low-pass and high-pass noise spectra, the listener is presented with notched wideband noise. As seen in the figure, off-frequency listening simply shifts the noise from one side of the noise source to the other, leaving the

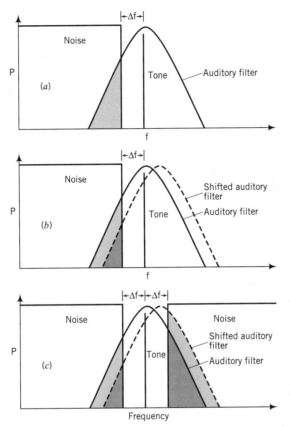

**FIGURE 19.6** Bandpass notched noise to minimize effects of off-frequency listening. (a) Noise presented to a hypothetical auditory filter centered around the tone. (b) Hypothetical auditory filter that is shifted. (c) Noise spectrum at both low and high frequencies, so the shaded noise area is the same for both hypothetical filters. From [6].

total masking noise the same. Patterson [5] was able to show that the derived auditory filters that led to the responses of Fig. 19.5 could be quite accurately represented by the so-called symmetric filter:

$$|H(f)|^2 = \frac{1}{[(\Delta f/\alpha)^2 + 1]^2}. \qquad (19.4)$$

The parameter $\alpha$ is a measure of the filter selectivity; $1.29\alpha$ is the 3-dB filter bandwidth. The function is symmetric on a linear frequency scale. If we define the bandwidth as BW and assume the symmetric filter response, and if we also maintain the previous assumption that the noise has a constant spectrum over its bandwidth, then we can show that

$$\frac{P}{N} = 1.22K\text{BW}, \tag{19.5}$$

where $K$ is the constant of proportionality given in Eq. 19.1; Patterson takes $K$ to be 1.0 in the discussion in [5].

Thus, the symmetric filter predicts that when a tone is masked by wideband noise, the signal-to-noise (power) ratio at the threshold will be proportional to the bandwidth of the auditory filter centered at the tone frequency.

## 19.4 GAMMA-TONE FILTERS, ROEX FILTERS, AND AUDITORY MODELS

Figure 19.7 defines the gamma-tone filter in terms of its impulse response in the analog (continuous) domain. The name derives from the form of the envelope, which is an $N$th-order gamma function. Notice that there are four parameters in this formula. In particular, when $\omega_r$ and $b$ are varied, these impulse response functions can implement filters of different center frequencies and bandwidths.

The Laplace transform of the gamma-tone filter impulse response is a function with both poles and zeros. Its frequency response (magnitude for $s = j\omega$) is displayed as GTF

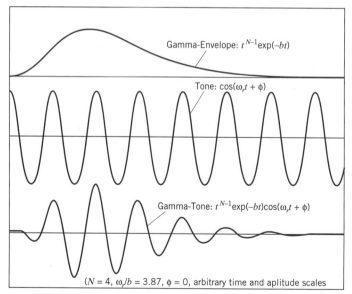

**FIGURE 19.7** Impulse response of gammatone filter. The first curve shows the gamma envelope as a function of time, and the second shows a reference cosine at the peak frequency of the filter. The final curve shows their product, which is the impulse response. From [3].

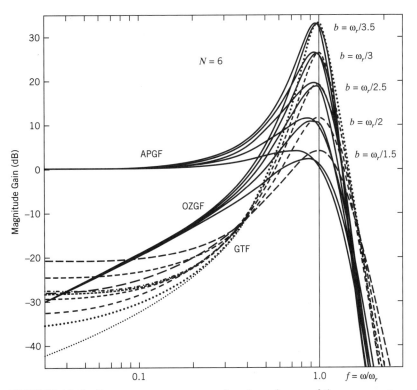

**FIGURE 19.8** Frequency responses of various forms of the gamma-tone filter. From [3].

in Fig. 19.8 for five different ratios of $b$ to $\omega_r$. GTF stands for gamma-tone filter. Also shown in the figure are results for APGF (all-pole gamma-tone filter) and OZGF (one-zero gamma-tone filter). All three cases are displayed for the same five values of the parameters.

By discarding the zeros of the original GTF to produce the APGF, Lyon [3] claims that an improvement in auditory modeling is obtained:

1. The APGF is simpler and more well behaved.

2. The APGF provides a more robust foundation for modeling auditory data.

3. The low-frequency tail of the APGF is unaffected by the bandwidth parameter, unlike the awkward behavior of the GTF.

4. The APGF has a very simple implementation; in the digital (or analog) domain, implementation consists of a cascade of second-order sections.

The APGF has the following Laplace transform:

$$H(s) = \frac{K}{[(s-p)(s-p^*)]^N} \tag{19.6}$$

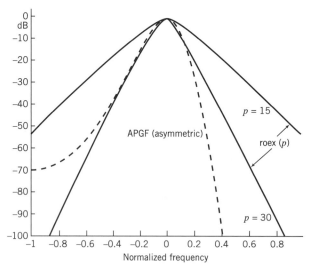

**FIGURE 19.9**  Approximations to auditory filters:   the APGF for $N = 16$, and two roex filters for $p = 30$ and $p = 15$. Adapted from [3].

As seen in Fig. 19.8, the APGF has a flat unity gain at very low frequencies. Lyon [3] states "...it is not necessarily desirable. A sloped but otherwise linear tail can be obtained by adding ... a zero at $s = 0$." This is called the OZGF; the linear tail can be observed in the figure.

It's worth remarking that the Patterson *symmetric* filter is an APGF for $N = 2$.

Roex (rounded exponential) filters were introduced by Patterson and Nimmo-Smith [7]. One of their versions is described by the following equation:

$$\text{roex}(p) = (1 + p|g|)e^{-p|g|}, \tag{19.7}$$

where $p$ is a parameter defining the filter and $g$ is a normalized frequency value, where $g = 0$ for the center (peak) frequency.

Figure 19.9 (adapted from Lyon's paper) compares the responses of an APGF and two roex approximations to auditory filters. We can see that the roex filters are more symmetric, whereas the APGF has a slower rise and a steeper fall. Thus, a well-designed APGF tends to more closely resemble neural and psychoacoustic tuning curves.

## 19.5   OTHER CONSIDERATIONS IN FILTER-BANK DESIGN

Aside from the precise filter transfer function, other factors must be included in designing a set of filters that is, in some sense, a reasonable emulation of an auditory filter bank. A simple question to ask (though difficult to answer) is, How many filters should be used? Historically, this question has been answered by trial and error. In his choice of the Voder filter-bank design (see Chapter 2), Dudley was influenced by several factors. From psychoacoustic experiments by Fletcher and others, he decided that the filter bank should extend over the frequency range 300–3000 Hz. Then, since the Voder was to be controlled by a keyboard to switch on various spectral shapes, a suitable number of filters was 10, corresponding to the number of available fingers. This led to a bank of 10 bandpass filters, each with a width of 300 Hz.

At the time that Dudley conceived of and designed the first vocoder (the late 1930s), implementation required large and relatively expensive components, so there was a natural desire to keep the parts count low. Like the Voder, the first vocoder built had only 10 channels, each including a filter with 300-Hz bandwidth. Results, however, were not satisfactory. A decade later, Vaderson gave a lecture demonstration of a 30-channel vocoder that had excellent quality.

These early vocoders did not take advantage of the variable-frequency resolution of the ear. Later vocoders built at Bell Laboratories did have wider bandwidths for higher center frequencies, and this resulted in a reduction to 16 channels covering the same frequency range.

In a vocoder, the purpose of the analysis filter bank is to generate a reasonable estimate of the speech spectrum. The fulfillment of this goal is complicated by two important properties of speech: (a) its quasi-periodicity during voiced segments, and (b) its variation with time.

During voiced speech, the spectrum is very close to periodic; an idealized example is shown in Fig. 19.10. The top figure shows an example with many spectral lines. We see that the bank of narrow-band filters leads to a spectral estimate that follows the comblike

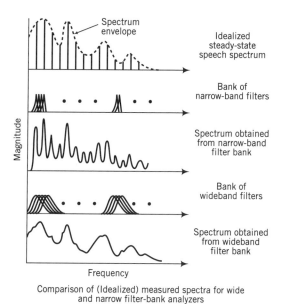

Comparison of (Idealized) measured spectra for wide
and narrow filter-bank analyzers

**FIGURE 19.10** Narrow-band and wideband spectral analyses for an idealized speech sound.

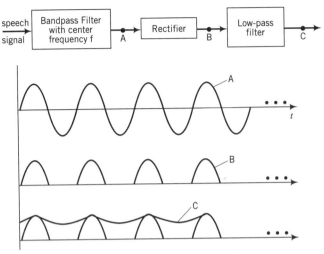

**FIGURE 19.11**   Bandpass power estimation.

properties shown, whereas a bank of wideband filters yields a spectral estimate that tends to follow the spectral envelope.

The spectrum of speech is time varying, but the *rate* of variation is based on articulator movements and is thus slow (of the order of 10–20 Hz). Thus, the spectral envelope at a given instant and frequency can be estimated from the output shown in Fig. 19.11. What should be the bandwidth of the low-pass filter? If made too narrow (e.g., 10 Hz), the output may not be able to follow spectral variations at that frequency. If made too wide, pitch ripple will appear at the low-pass filter output.

## 19.6  SPEECH SPECTRUM ANALYSIS USING THE FFT

The incorporation of fast Fourier transform (FFT) programs as spectrum analyzers created a different set of design options. With the FFT it was easy to generate a high-resolution spectrum analysis, since the computation time only increased logarithmically with increases in resolution. However, the same issues that complicated filter-bank design still held for FFT analysis. For example, if we want the equivalent of 1024 bandpass filters, this can surely be implemented with a 1024-point FFT. However, if we implement this by choosing 1024 samples to analyze a rectangular window and then invoking the FFT program, results are not too good. Assume, for example, that the speech was originally sampled at 10 kHz. This means that 1024 samples corresponds to approximately 100 ms of speech. During this time, the spectrum could have changed greatly, which means that this type of spectrum analysis will not track the natural spectral change in the speech. There are many tricks to overcome this problem. First, windowing only 20 ms of speech (200 samples) and then augmenting the input to the FFT with 824 zeros will still produce a spectrum with 1024 samples, but since the result is based on only 20 ms of speech, the result is close to a snapshot of the

20-ms segment. Also, multiplying the 20 ms by a suitable window, e.g., Hamming or Kaiser, removes most of the artifacts produced by the abruptness of a rectangular window (see [8] for a more extended discussion of windowing).

## 19.7 CONCLUSIONS

Psychoacoustic and physiological research have made it possible to estimate the frequency resolution properties of the human auditory system. This research has helped influence the design and implementation of many kinds of spectral analysis methods that look carefully at the speech spectrum. In addition to frequency resolution issues, there are also highly significant temporal issues. Designers need to consider both time and frequency design issues in the context of the specific applications.

## 19.8 EXERCISES

**19.1**  Explain why the auditory bandwidths obtained by psychological measurements do not necessarily agree with the tuning curves of auditory neurons (as described in Chapter 14).

**19.2**  Explain why the bandwidths determined by the equal articulation index differ from the critical bands of Fig. 19.2.

**19.3**  Explain why Fletcher's critical band experiments were not able to predict the shape of the auditory filters.

**19.4**  Can you present one or more physiological explanations of why auditory filter bandwidths increase with frequency? Your answer can be speculative but should be buttressed with some facts.

**19.5**  High-frequency hearing loss in normal hearing adults increases with age. Give one or more explanations.

**19.6**  The AI devised by Fletcher was based on listeners' responses to nonsense CVC syllables. Why did Fletcher choose nonsense CVC's instead of CVC's from a natural language such as English?

**19.7**  Develop a mathematical model of Patterson's notched noise experiment (Fig. 19.6). Derive an expression for the auditory filter as a function of the measured threshold of detection of a tone when (a) there is no off-frequency listening and (b) when the listener performs off-frequency listening with an increment $\delta f$.

## BIBLIOGRAPHY

1.  Fletcher, "Auditory patterns," *Rev. Modern Phys.* **22**: 47, 1940.
2.  Kingsbury, B. E. D., Perceptually Inspired Signal Processing Strategies for Robust Speech Recognition in Reverberant Environments, PhD Thesis, U.C. Berkeley, 1998.
3.  Lyon, R. F., "The all-pole gammatone filter and auditory models," from *Computational Models Signal Process. Audit. Syst.*, Forum Acusticum '96, Antwerp, Belgium, 1996.

4. Moore, B. C. J., Peters, R. W., and Glasberg, B. R., "Auditory filter shapes at low center frequencies," *J. Acoust. Soc. Am.* **88**: 132–140, 1990.

5. Patterson, R. D., "Auditory filter shape," *J. Acoust. Soc. Am.* **55**: 802–809, 1974.

6. Patterson, R. D., "Auditory filter shapes derived with noise stimuli," *J. Acoust. Soc. Am.* **59**: 640–654, 1976.

7. Patterson, R. D., and Nimmo-Smith, I., "Off-frequency listening and auditory filter asymmetry," *J. Acoust. Soc. Am.* **67**: 229–245, 1980.

8. Rabiner, L. R., and Gold, B., *Theory and Applications of Digital Signal Processing*, Prentice–Hall, Englewood Cliffs, N.J., 1975.

9. Rossing, T. D., *The Science of Sound,* Addison–Wesley, Reading, Mass., 1990.

10. Schafer, R. W., and Rabiner, L. R., "Digital representations of speech signals," *Proc. IEEE* **63**: 662–667, 1975.

11. Webster, J. C., Miller, P. H., Thompson, P. O., and Davenport, E. W., "The masking and pitch shifts of pure tones near abrupt changes in a thermal noise spectrum," *J. Acoust. Soc. Am.* **24**: 147–152, 1952.

# CHAPTER *20*

# THE CEPSTRUM AS A SPECTRAL ANALYZER

## 20.1  INTRODUCTION

In Chapters 11 and 12, models of speech and music production were introduced. The basic structure of these models could be identified as an excitation that was input to a system of resonators; the convolution of the former with the impulse response of the latter component produced the approximation to the modeled speech or music signal. It is therefore natural to contemplate an analysis of the signal as first, a separation of the source, that is, excitation, and filter, that is, resonators. This is an example of a process that is often called deconvolution, or the separation out of a signal from an impulse response that has been convolved with it. In the channel vocoder, for example, the excitation is modeled as either a quasi-periodic pulse train (caused by vocal cord vibration) or a noise signal caused by turbulence. In Chapter 16 we studied how the auditory system perceives the pulse train component of the excitation; in Chapter 30, methods of detecting both the periodic and noisy components will be reviewed. In Chapter 19 there was a brief discussion of how a vocoder analyzer models the spectral envelope, which is a function of the vocal tract articulator positions. In summary, a channel vocoder separates excitation from the filter, and therefore in some sense performs deconvolution. Similarly, it is generally desirable in recognition applications (particularly if speaker independence is desired) to separate the excitation characteristics, which for American English contain only limited information about the linguistic content (other than the voiced–unvoiced categorization), from the filter, which provides the major cues for phone classification.

Cepstral analysis performs deconvolution of the speech signal by use of techniques that are quite different from those incorporated in a channel vocoder. In order to understand how cepstral analysis performs deconvolution, we first need to delve into a little relevant theory.

## 20.2  A HISTORICAL NOTE

Bogert et al. [1] may have been the first to use cepstral processing; in this case it was used for seismic analysis. While Bogert was performing his research, Alan Oppenheim, then an MIT graduate student, was working on a fairly complete mathematical theory that he called homomorphic processing. During a visit to Bell Labs by Oppenheim and one of this book's authors in the early 1960s, Bogert and Oppenheim exchanged ideas on the subject. Subsequently, Oppenheim became convinced that his concepts could be usefully applied to vocoder design. He later spent a 2-year sabbatical at the MIT Lincoln Laboratory (in the late 1960s) and developed a complete analysis–synthesis system based on what he called homomorphic (i.e., cepstral) processing. Further important work along these lines was carried out by Oppenheim et al. [3], [4] and also by Schafer [6], [5] and Stockham [7].

## 20.3   THE REAL CEPSTRUM

It is convenient to assume that the signal consists of a discrete time sequence, so that the spectrum consists of a $z$ transform evaluated on the unit circle. Let us consider a speech example, with $X$ referring to the spectrum of the observed speech signal, $E$ to the excitation component (for instance, the glottal pulse train), and $V$ to the vocal tract shaping of the excitation spectrum. We begin with a multiplicative model of the two spectra (the excitation and the vocal tract). Thus, the spectral magnitude of the speech signal can be written as

$$|X(\omega)| = |E(\omega)||V(\omega)|. \tag{20.1}$$

Taking the logarithm of Eq. 20.1 yields

$$\log|X(\omega)| = \log|E(\omega)| + \log|V(\omega)|. \tag{20.2}$$

Particularly for voiced sounds, it can be observed that the $E$ term corresponds to an event that is relatively extended in time (e.g., a pulse train with pulses every 10 ms), and thus it yields a spectrum that should be characterized by a relatively rapidly varying function of $\omega$; in comparison, because of the relatively short impulse response of the vocal tract, the $V$ term varies more slowly with $\omega$. With the use of this knowledge, the left-hand side of Eq. 20.2 can be separated into the two right-hand-side components by a kind of a filter that separates the log spectral components that vary rapidly with $\omega$ (the so-called high-time components) from those that vary slowly with $\omega$ (the low-time components). Such an operation would essentially be performing deconvolution.

Equation 20.2 has transformed the multiplicative formula 20.1 into a linear operation and thus can be subjected to linear operations such as filtering. Since the variable is frequency rather than time, notations must be changed. Thus, for example, rather than filtering (for time), we have liftering (for frequency); instead of a frequency response, we have a quefrency response; and the DFT (or $z$ transform or Fourier transform) of the $\log|X(\omega)|$ is called the cepstrum. The cepstrum is computed by taking the inverse $z$ transform of Eq. 20.2 on the unit circle, yielding

$$c(n) = \frac{1}{2\pi} \int_{-\pi}^{\pi} \log|X(\omega)|e^{j\omega n}\, d\omega, \tag{20.3}$$

where $c(n)$ is called the $n$th cepstral coefficient. The deconvolutional properties of the cepstrum for speech can be visualized by using Fig. 20.1 depicting the sequence of operations from the speech wave to the cepstrum. Figure 20.1(d) shows the cepstrum, as defined by Eq. 20.3. The spectral envelope, which varies slowly with respect to frequency, yields large-valued cepstral coefficients for low values of $n$, but it dies out for high $n$. The spectral fine

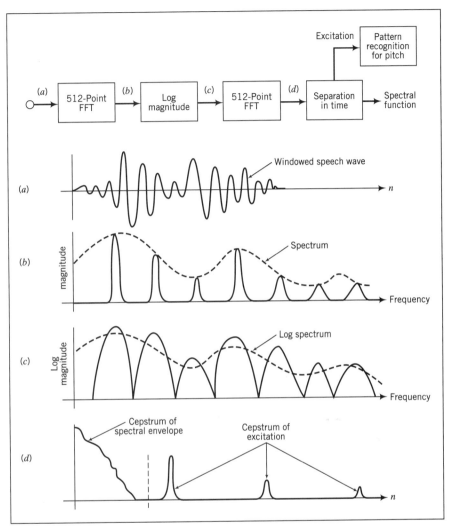

**FIGURE 20.1**  Cepstral analysis.

structure is more rapidly varying with $\omega$, and it yields small-valued cepstral coefficients for small $n$, but large values beyond the crossover point shown in the figure. Thus, the contribution of the excitation and the vocal tract filter can (in principle) be separated in the cepstral domain. Both components can be *inverted* to generate the original spectral magnitudes.

## 20.4  THE COMPLEX CEPSTRUM

Thus far, all our equations have used real functions. It is also possible to define a complex cepstrum that gives useful insight into properties of actual systems.

Let's start with a sequence $x(n)$ that can be of finite or infinite duration. We assume that this sequence is the impulse response of a well-behaved linear system that can be described in terms of a rational fraction of $z$ transforms. Then

$$X(z) = \frac{A \prod_{k=1}^{M_i}(1 - a_k z^{-1}) \prod_{k=1}^{M_o}(1 - b_k z)}{\prod_{k=1}^{N}(1 - c_k z^{-1})}. \tag{20.4}$$

Given that $a_k$, $b_k$, and $c_k$ are all less than unity, Eq. 20.4 represents a digital network with poles inside the unit circle (to ensure stability) and zeros inside the unit circle (first product in the numerator) and zeros outside the unit circle (second product in the numerator). $A$ is simply a multiplicative factor.

The logarithm of Eq. 20.4 is represented as

$$\log X(z) = \log A + \sum_{k=1}^{M_i}\log(1 - a_k z^{-1}) + \sum_{k=1}^{M_o}\log(1 - b_k z) - \sum_{k=1}^{N}\log(1 - c_k z^{-1}), \tag{20.5}$$

and the complex cepstrum $\hat{x}(n)$ of Eq. 20.5 is determined from

$$\hat{x}(n) = \frac{1}{2\pi}\int_{-\pi}^{\pi} \log X(\omega) e^{j\omega n}\, d\omega. \tag{20.6}$$

At this point, we state without proof (left as an exercise) that the complex cepstrum can be evaluated to be

$$\hat{x}(n) = \begin{cases} \log A, & n = 0 \\ \sum_{k=1}^{N}\frac{(c_k)^n}{n} - \sum_{k=1}^{M_i}\frac{(a_k)^n}{n}, & n > 0. \\ \sum_{k=1}^{M_o}\frac{(b_k)^{-n}}{n}, & n < 0 \end{cases} \tag{20.7}$$

These equations lead to some interesting relations between pole-zero positions and complex cepstral values. For example,

- If $\hat{x}(n) = 0$ for $n \geq 0$, this must correspond to an all-zero (FIR) filter with all the zeros outside the unit circle.

- If $\hat{x}(n) = 0$ for $n < 0$, this must correspond to a filter with all the poles and zeros inside the unit circle. This defines a minimum phase filter.

An example of a practical application of such results is the development of a filter bank based on physiological measurements of the tuning curves of cat auditory neurons. Delgutte [2] has measured and documented the results of these measurements. Only the magnitude of the cat's neural tuning curve was directly measured (not the phase). In this application (as in many others) it would be desirable to estimate the complete transfer function, including the phase.

There is some physical evidence that the basilar membrane vibrations (and therefore the auditory tuning curves) can be represented as minimum phase filters [2]. The application of Eq. 20.7 makes it possible to estimate the phase under this assumption, employing the following procedure.

1. Measure the auditory nerve tuning curves over a variety of neurons with different CFs; these responses resemble the tuning curves shown in Chapter 14 (Fig. 14.10).

2. These tuning curves can be inverted to produce magnitude functions of auditory bandpass filters. We now have an approximation to the function $X(z)$ of Eq. 20.4.

3. Using a discrete-time version of Eq. 20.6, compute the complex cepstrum. Instead of evaluating the integral, a DFT-based version gives good results.

4. Set $\hat{x}(n)$ to zero for $n < 0$. This means that the truncated version of $\hat{x}(n)$, which we will denote $\hat{\hat{x}}(n)$, must correspond to a log spectrum that is minimum phase.

5. The final step is the inversion of $\hat{\hat{x}}(n)$.

Step 3 can be implemented with a DFT, using the formula

$$\hat{x}_p(n) = \frac{1}{N} \sum_{k=0}^{N-1} \log[X_p(k)] W^{-nk}, \tag{20.8}$$

where, as in Chapter 7, $W$ is a shorthand for $e^{-j(2\pi/N)}$, and where the subscript $p$ is a notation to indicate that all the processing is performed in discrete time.

Step 5 can also be implemented with a DFT as

$$\log[\hat{X}_p(k)] = \sum_{k=0}^{N-1} \hat{\hat{x}}(n) W^{nk}. \tag{20.9}$$

The result that we are seeking, namely, the value of the complex (minimum phase) spectrum $\hat{X}_p(k)$, can be obtained by simply exponentiating the left side of Eq. 20.9.[1] In this way, each neural tuning curve can be well approximated by a minimum phase filter in which both the magnitude and phase of each filter is specified.

It should be mentioned that these techniques can, in exactly the same manner, be applied to obtain filters corresponding to the psychoacoustic tuning curves obtained by methods such as those described in Chapter 19. Notice, by the way, that the all-pole gamma-tone filter (APGF) of Chapter 19 is indeed a minimum phase filter. An interesting exercise would be to try to find an APGF that approximates, *in both magnitude and phase*, the filter response functions obtained by the method described above.

The complex cepstrum must thus be distinguished from the traditional cepstrum, which deals entirely with real functions. Figure 20.2 shows the basic difference between the complex cepstrum and the cepstrum.

# 20.5  APPLICATION OF CEPSTRAL ANALYSIS TO SPEECH SIGNALS

Figure 20.3 shows the result of various operations on a windowed speech signal to produce both the complex cepstrum and the cepstrum. Figure 20.3(b) shows the result of computing the log magnitude of the DFT of the signal shown in Fig. 20.3(a). Since we also want to

---

[1] All logs in this chapter are assumed to be natural logarithms.

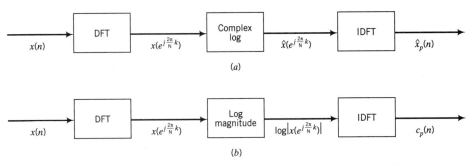

**FIGURE 20.2**   Practical implementations of systems for obtaining (a) the complex cepstrum and (b) the cepstrum.

obtain the complex cepstrum, we need to save the phase component of the DFT. Phase computation is a tricky operation, usually producing a value between $-\pi$ and $\pi$, as in Fig. 20.3(c). However, for this situation, it is necessary to unwrap the phase, as shown in Fig. 20.3(d). The complex cepstrum can be obtained by an inverse DFT (IDFT), from the log magnitude of Fig. 20.3(b) and the phase of 20.3(d). The cepstrum can be obtained by simply computing an IDFT of the function shown in Fig. 20.3(b).

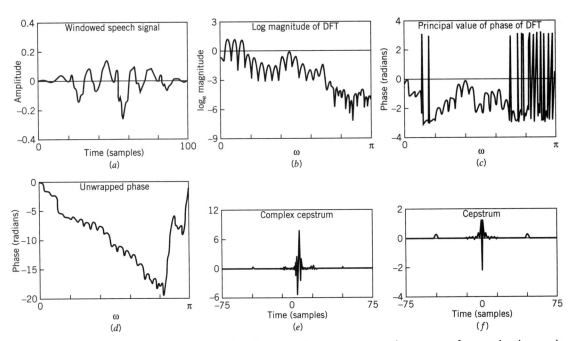

**FIGURE 20.3**   Computing the complex cepstrum and cepstrum for a voiced speech segment. From [5].

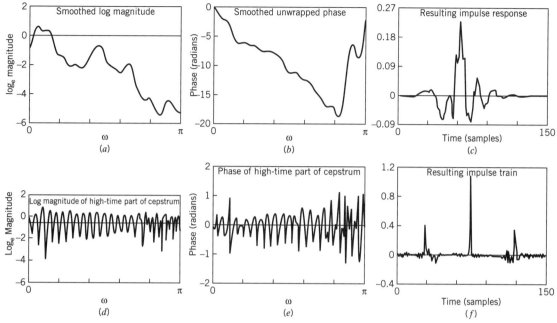

**FIGURE 20.4** Cepstrum "liftering" of voiced speech. From [5].

Figure 20.3 shows how both the complex cepstrum and the cepstrum can be computed from the original speech segment. Figure 20.4 now shows the steps involved in cepstral filtering, also called homomorphic filtering, of the same speech segment. Begin by multiplying the complex cepstrum, Fig. 20.3(e), by a rectangular window that includes the major energy centered about zero but excluding the small peaks at ±45 samples. When this windowed complex cepstrum is inverted, the smoothed log magnitude and the smooth unwrapped phase of Figs. 20.4(a) and 20.4(b) are obtained. Exponentiation of the complex spectrum of these two figures results in Fig. 20.4(c), which can be labeled as the impulse response of the vocal tract. If we now perform comparable operations on the high-time or high-quefrency part of the complex cepstrum of Fig. 20.3(e) (the part not in the rectangular window), we obtain the results shown in Figs. 20.4(d), 20.4(e), and 20.4(f). Therefore, Figs. 20.4(c) and 20.4(f) show how cepstral analysis and filtering perform deconvolution, generating distinct functions of the excitation function [Fig. 20.4(f)] and the vocal tract impulse response.

## 20.6 CONCLUDING THOUGHTS

When it is necessary to return to a time waveform without making minimum phase assumptions, it is necessary to use the complex cepstrum. However, in many practical applications, processing of the phase is not a useful operation, and in any case adds complexity. In these cases, cepstral processing can be implemented by dealing entirely with real functions; such a situation is displayed in Fig. 20.1, where separation of the excitation and filter can be seen to occur at the cepstral level. This separation has been shown to be useful in many applications, including

- Pitch estimation for vocoding (see Chapter 30).
- Spectral envelope estimation for vocoding (see Chapter 31).
- Cepstral computations for other kinds of of spectral estimates; for example, see Chapter 21 for LPC, and Chapter 22 for critical band filter banks. These representations are used for a range of applications, including speech recognition.

In the latter two cases, a moderate number of cepstral coefficients (typically 10–14) are used to represent the short-term spectral envelope. The choice of a small number of coefficients provides a further smoothing of the spectral estimate beyond what might be necessary for separation of the excitation alone. This is often beneficial to pattern-recognition tasks, in which we wish to suppress minor spectral differences between examples of the same sound.

## 20.7  EXERCISES

**20.1**   Prove that the cepstrum $c(n)$ is the even part of the complex cepstrum $\hat{x}_n$.

**20.2**   Find the cepstrum of an all-pole model of the vocal tract. Assume that the vocal tract transfer function can be expressed as

$$V(z) = \frac{1}{\prod_{k=1}^{N}(1 - c_k z^{-1})(1 - c_k^* z^{-1})}, \tag{20.10}$$

where $c_k = r_k e^{j\Theta_k}$.

**20.3**   Consider the FIR filter

$$y(n) = x(n) - \frac{3}{2}x(n - 1) - x(n - 2). \tag{20.11}$$

(a) Find the complex cepstrum of the filter impulse response.
(b) Is the filter a minimum phase filter? If not, show how it can be transformed into a minimum phase filter with the same spectral magnitude.

**20.4**   Assume that a speech or music signal can be represented as the product of two signals; one rapidly varying and the other slowly varying. You are asked to design a gain control to minimize the effects of the slowly varying signal on the quality of the perceived sound. Sketch the design of a homomorphic system that accomplishes this.

# BIBLIOGRAPHY

1. Bogert, B., Healy, M., and Tukey, J., "The quefrency analysis of time series for echos," in M. Rosenblatt, ed., *Proc. Symp. on Time Series Analysis*, Chap. 15, Wiley, New York, pp. 209–243, 1963.
2. Delgutte, B., Cambridge, MA, personal communication, 1990.
3. Oppenheim, A. V., "Generalized Linear Filtering," Chapter 8 in B. Gold and C. M. Rader, *Digital Processing of Signals*, McGraw–Hill, New York, pp. 233–264, 1969.
4. Oppenheim, A. V., Schafer, R. W., and Stockham, T. G. Jr., "Nonlinear filtering of multiplied and convolved signals," *Proc. IEEE* **56**: 1264–1291, 1968.
5. Rabiner, L. R., and Schafer, R. M., *Digital Processing of Speech Signals*, Prentice–Hall, Englewood Cliffs, N.J., 1978.
6. Schafer, R. M., "Echo removal by discrete, generalized linear filtering," Tech. Rep. 466, Res. Lab. of Electronics, Massachusetts Institute of Technology, Cambridge, Mass., 1969.
7. Stockham, T. G., "The application of generalized linearity to automatic gain control," *IEEE Trans. Audio Electroacoust.* **AU-16**: 267–270, 1968.

# CHAPTER *21*

# *LINEAR PREDICTION*

## 21.1  INTRODUCTION

In Chapter 19, we described spectral representations that are based on the signal and (to some extent) some of the properties of human hearing, in particular the property of requiring less frequency resolution at high frequencies. In Chapter 20, we showed that cepstral processing could provide a smoothed spectral representation that is useful for many speech applications. In both cases, however, we made no explicit use of our knowledge of how the excitation spectrum is shaped by the vocal tract. As noted in Chapters 10 and 11, speech can be modeled as being produced by a periodic or noiselike source that is driving a nonuniform tube. It can be shown that basing the analysis (in a very general way) on such a production model leads to a spectral estimate that is both succinct and smooth, and for which the nature of the smoothness has a number of desirable properties. This is the main topic of this chapter.[1]

## 21.2  THE PREDICTIVE MODEL

In Chapter 10, we showed that a discrete model of a lossless uniform tube led to an input–output relationship for an excitation at one end and the other end closed (see Eqs. 10.21 and 10.22, and Figs. 10.5 and 10.6). For the case in which the far end of the tube is open, we noted that the complex poles of the tube transfer function would be on the unit circle at frequencies given by

$$f_n = (2n + 1)c/4l, \quad n = 0, 1, 2, \ldots. \tag{21.1}$$

We further noted that, for the average-length (17 cm) vocal tract and the speed of sound at room temperature (344 m/s), this meant that such a tube would have one resonance/kHz. In the more realistic case with energy loss at the boundaries, the poles will be inside the unit circle, but still at the angles implied by the resonance frequencies.

Of course, the real vocal tract is far more complicated than a uniform tube, often being represented by a nonuniform tube consisting of multiple shorter concatenated tubes of differing cross-sectional areas but having the same length. This could be viewed as an approximation to a continuous vocal tract shape. The resulting tube would have a set of

---

[1]Many of the figures in this chapter were taken from [4].

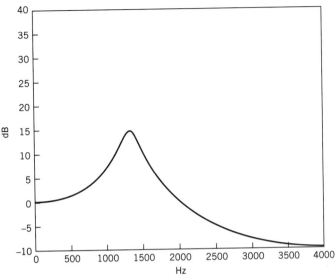

**FIGURE 21.1** Resonator frequency response. The resonator consisted of a pair of complex poles with a radius of 0.9 and an angle of 60°; the sampling rate is assumed to be 8 kHz.

resonances that (one would hope) would be similar to those for an actual vocal tract (in the shape required to produce a particular sound). However, for our current purposes it is sufficient to note that real vocal tracts do generate resonances whose number can be predicted reasonably well by tube models. Experiments in speech perception, such as those described in Chapter 17, have long suggested the fundamental importance of the formants for human listeners. Therefore, we will assume for now that we only need a model that can represent a sufficient number of resonances.

Suppose, then, each formant can be represented by a pole-only transfer function of the form

$$H_i(z) = \frac{1}{1 - b_i z^{-1} - c_i z^{-2}} \tag{21.2}$$

(where for the moment we ignore the filter gain). A typical frequency response for such a filter is shown in Fig. 21.1. We note in passing that the values of $c_i$ are always less than one for a stable filter (left as an exercise for the reader).

Assuming a 5-kHz bandwidth, one would typically need five such resonators in cascade to represent the five formants that would be expected on the average. Ordinarily one would also expect to require one or two more poles (possibly real) to represent the nonflat spectrum of the driving waveform, so a complete vowel spectrum could be represented reasonably well by six such sections.

Although this cascaded approach has been used in many synthesis applications, it is useful for our current purposes to imagine multiplying through all of these sections to get

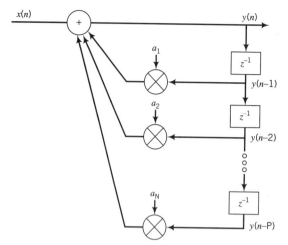

**FIGURE 21.2** All-pole model for the generation of a discrete-time sequence.

a direct-form implementation of the spectral model:

$$H(z) = \frac{1}{1 - \sum_{j=1}^{P} a_j z^{-j}},$$  **(21.3)**

where $P$ is twice the number of second-order sections going into the product ($P = 12$ in the example above), and the $a$ coefficients are the coefficients of the resulting $P$th-order polynomial.

Figure 21.2 is a diagram of the complete model. In later chapters such a system will be used as a starting point to describe linear predictive approaches to speech synthesis, but in the current context it will be used as a model to represent the signal spectrum. Thus, the short-term spectrum of a speech signal can be represented by a filter that can be specified by $P = 2 * (\text{BW} + 1)$ coefficients, where BW is the speech bandwidth in kilohertz. Note that since the driving-signal spectrum is folded into the filter, the model excitations are considered to be white.

For the system shown in Fig. 21.2, the discrete-time response $y(n)$ to an excitation signal $x(n)$ would be

$$y(n) = x(n) + \sum_{j=1}^{P} a_j y(n - j).$$  **(21.4)**

The coefficients for the second term of this expression are generally computed to give an approximation to the original sequence, which will yield a spectrum for $H(z)$ that is an approximation to the original speech spectrum. Thus, we attempt to predict the speech signal by a weighted sum of its previous values. That is,

$$\tilde{y}(n) = \sum_{j=1}^{P} a_j y(n - j)$$  **(21.5)**

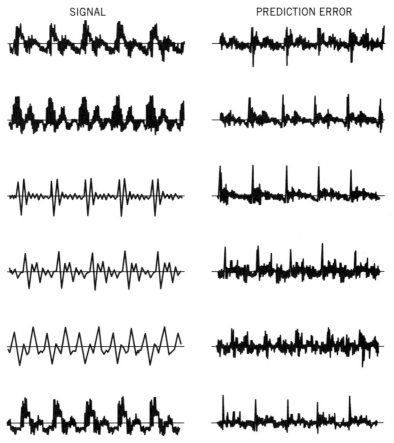

**FIGURE 21.3**  Residual error waveforms for several vowels. From [4].

is the linear predictor. Note that this has the form of a FIR filter, but that when it is included in the model of Fig. 21.2 the resulting production model is IIR. The coefficients that yield the best approximation of $\tilde{y}(n)$ to $y(n)$ (usually in the mean squared sense) are called the linear prediction coefficients. In the statistical literature, the overall model is sometimes called an autoregressive model.

The difference between the predictor and the original signal is referred to as the error signal, also sometimes called the residual error, the LPC residual, or the prediction error. When the coefficients are chosen to to minimize this signal energy, the resulting error signal can be viewed as an approximation to the excitation function. The residual signal $e(n) = y(n) - \tilde{y}(n)$ consists of the components of $y(n)$ that are not linearly predictable from its own previous samples, which is the case for a periodic excitation in this model, assuming that the number of samples between excitation pulses is much larger than the order of the filter. Figure 21.3 shows several examples of such error signals (preceded by the original waveforms) for steady-state vowels; note that the prediction error has large peaks that occur once per pitch period.

## 21.3   PROPERTIES OF THE REPRESENTATION

We have shown that a simple model for speech production[2] leads to a spectral representation that is the minimum required to potentially represent the vocal tract resonances that shape the speech spectrum, particularly for voiced sounds. However, this still leaves us with (at least) two remaining questions:

1. What error criterion should we minimize between the model spectrum and the observed spectrum?

2. What is the best (not the minimum) number of coefficients to put into the representation?

We do not know if current answers to these questions are optimal. However, in the basic form of LPC that has been traditionally incorporated in audio signal-processing systems, coefficients have been chosen to minimize the squared error between the observed and predicted signals. As shown below, this leads to a solution that is represented in terms of its autocorrelation function. We state without proof (see, for instance, [3]) that minimizing the squared error criterion is also equivalent to minimizing the integrated quotient between the speech power spectrum and the model power spectrum, or

$$ D = \int_{-\pi}^{\pi} \frac{|Y(\omega)|^2}{|H(\omega)|^2} \frac{d\omega}{2\pi}. \tag{21.6} $$

where for simplicity we have ignored a gain term corresponding to the error power. Thus, minimizing the mean squared difference between $y(n)$ and its linear predictor $\tilde{y}(n)$ is equivalent to minimizing a kind of distortion between the signal spectrum and the model filter spectrum.

What are the characteristics of this particular spectral distortion criterion? Note that for this measure, the portions of the spectrum for which $|Y(\omega)|^2$ is smaller than $|H(\omega)|^2$ will make small contributions to the integral. For a harmonic signal that is being modeled with LPC, this means that the model spectrum will tend to hug the harmonic peaks, but not the valleys between. Therefore, for model orders that are not too large, the error criterion given in Eq. 21.6 will lead to a spectrum that is an estimate of the envelope of the signal spectrum. See Fig. 21.4(c) for a short-term speech spectrum and the corresponding linear predictive spectrum.

Since, as noted earlier, it is a common goal to model spectral resonances, the squared error criterion seems reasonable. Additionally, moderate amounts of additive noise will have only small effects on the estimate, since the largest change to the spectrum will occur in the spectral valleys. However, a squared error criterion is not necessarily ideal; for instance, portions of the spectrum that have large magnitudes will tend to dominate the error, which

---

[2]Although the discussion has been oriented toward speech, similar arguments could also be made for acoustic instruments with a simple source-resonator structure.

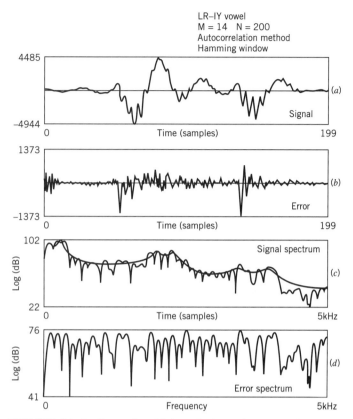

LR–IY vowel
M = 14   N = 200
Autocorrelation method
Hamming window

**FIGURE 21.4** Example of (a) a windowed speech signal, (b) the LPC error signal, (c) the signal spectrum with the LPC spectral envelope superimposed, and (d) the LPC error spectrum. From [4].

may not necessarily be those parts that are most relevant to either speech intelligibility (for coding or synthesis applications) or phone discrimination (for recognition applications). This weakness is often ameliorated by pre-emphasizing the data with a fixed first-order FIR filter to help to flatten the spectrum.

Given this model structure and error criterion, what should be the number of coefficients used? In the previous section we gave an approximation to the number of coefficients that are required in order to represent the spectral resonances. However, since the model only provides an approximate fit to the short-term signal spectrum, clearly using a greater number of coefficients (a higher-order discrete-time model) will yield model spectra that are a better match; in general, using more parameters in a least-squares fit to a sequence will provide a better fit. However, having the greater detail is not always an advantage. In particular, the typical goal for the short-term spectral analysis of speech is to compute a spectral envelope that is relatively unaffected by pitch; thus, in general the details of harmonic structure should not be modeled at this stage. Additionally, a larger number of parameters will tend to make the estimate more susceptible to errors that are due to additive noise in the observed signal.

Therefore, the model order used depends critically on the goal for the spectral analysis. In cases in which an extremely accurate spectral representation is required, higher model orders may be used. For applications such as speech recognition, though, the model order is typically kept very close to the rule of thumb described earlier (though sometimes a slightly higher model order is found to be helpful). Although this may sound somewhat ad hoc, more formal approaches are not always good enough to accurately determine the best model order for a given application. Probably the best known of these is the use of the Akaike information criterion, or AIC [1]. In this approach, the error variance is replaced by a new distortion in order to penalize the model complexity. In particular, the AIC is given by $AIC = \log \sigma^2 + 2p$, where $\sigma^2$ is the error variance and $p$ is the autoregressive model order. This measure has the right property, namely penalizing model size in addition to the error variance; however, the choice of the weight of the model order is based on assumptions that may not be correct in practice. Makhoul [3] applied this measure to a speech sample and got a reasonable result, with tenth-order predictors having the lowest AIC; however, in practice the best value would tend to be found by trial and error.

Figure 21.5 shows the spectral envelope for LPC models of different orders given the same speech sample. It is apparent that the low model orders do not adequately capture the formants, whereas the very high model orders begin to track the harmonic content.

Figure 21.6 shows the reduction in prediction error energy for increased model orders. Note that in general the error is larger for unvoiced sounds, which are inherently less predictable.

## 21.4  GETTING THE COEFFICIENTS

Let us assume that in the situation of interest, $x(n)$ of Eq. 21.4 is unknown. In this case, as noted earlier, we choose the $a$ parameters in order to minimize the squared error between $y(n)$ and $\sum_{j=1}^{P} a_j y(n-j)$ over the sequence. For speech applications, the $y(n)$ of interest is typically a locally windowed form of the original speech sample sequence; that is,

$$y(n) = w(n)s(n), \tag{21.7}$$

where $w(n)$ is a local window (e.g., a 20-ms-long Hamming window) that we assume to be $N$ points long, and $s(n)$ is the sampled speech data. The error signal between the model output and the signal then is

$$e(n) = y(n) - \sum_{j=1}^{P} a_j y(n-j). \tag{21.8}$$

Defining a distortion metric over the window, we find

$$D = \sum_{n=0}^{N-1} e^2(n) = \sum_{n=0}^{N-1} \left[ y(n) - \sum_{j=1}^{P} a_j y(n-j) \right]^2. \tag{21.9}$$

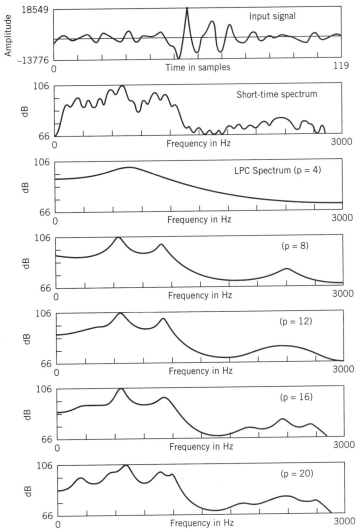

**FIGURE 21.5** LPC speech spectra for different model orders. From [4].

If we take partial derivatives with respect to each $a$, we get $P$ equations of the form

$$\sum_{j=1}^{P} a_j \phi(i, j) = \phi(i, 0) \quad \text{for } i = 1, 2, \ldots, P, \tag{21.10}$$

where $\phi(i, j)$ is a correlation sum between versions of the speech signal delayed by $i$ and $j$ points.

For the case of the windowed signal for which no points outside of the window are used for the estimate, $\phi(i, j)$ is only a function of the absolute value of the difference between $i$

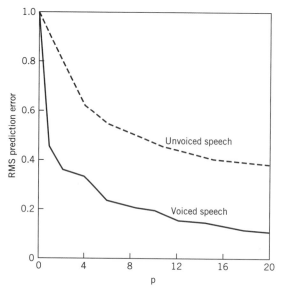

**FIGURE 21.6**    Root-mean-square prediction error for different model orders. From [4].

and $j$, and the resulting correlation matrix at the left-hand side of Eq. 21.10 is Toeplitz; that is, not only is it symmetric, but all the values along each left-to-right diagonal are equal. For this special case, the system of equations can be solved by efficient procedures $[O(P^2)]$ known as the Levinson or the Durbin recursions. Although efficient, they do require significant numerical precision. These procedures are described in detail in a number of other sources, including [3], and they are not be repeated here. For what is called a covariance analysis, the correlations are computed by sliding the signal window along outside of the blocked area, so that the resulting matrix is not Toeplitz (though it is symmetric). The resulting system of equations is typically solved by a Cholesky decomposition (sometimes called the square-root method), which takes more computation $[O(P^3)]$ but which is numerically quite stable (see, for instance, [2]).

## 21.5    RELATED REPRESENTATIONS

In practice, the prediction coefficients are often not a good representation to use for most applications. In cases in which the digital word length is critical, the polynomial coefficients tend to be too sensitive to numerical precision. When a covariance analysis is used, the stability of the resulting filter is not guaranteed, and it is not checked easily with a predictor polynomial. The coefficients are not orthogonal or normalized, which potentially creates other difficulties for classifiers that might use these features.

For all of these reasons, LPC coefficients are generally transformed into one of a number of other representations, including the following.

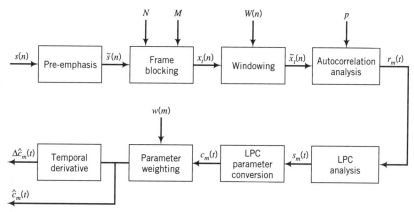

**FIGURE 21.7** Block diagram of LPC processing for speech recognition. Frame blocking simply refers to the choice of points for analysis, with *N* points that are stepped along every *M* points; an overlap of at least 50% is typical. Parameter weighting refers to a cepstral lifter that is often used in practice. From [4].

**1.** Root pairs: the polynomial can be factored (using commonly available iterative procedures) into complex pairs, thus finding something like the resonances discussed in this chapter. Each of these is implemented by a second-order filter, which has simple stability properties and good numerical behavior. This can be useful for synthesis but has not tended to be used for recognition.

**2.** Reflection coefficients: the polynomial can be transformed into a set of coefficients that represent the fraction of energy reflected at each section of a nonuniform tube (with as many sections as the order of the polynomial). The reflection coefficients also can be used directly as the coefficients for a lattice filter for synthesis. The first few coefficients must be represented with more precision than the later ones, and all of the values are bounded by $-1$ and 1 for stable filters. This is probably the most common representation for LPC synthesis.

**3.** Cepstrum: there is a recursion that can be used to generate a set of cepstral coefficients corresponding to the LPC spectrum; it is efficient, as it does not require any explicit spectral computations. The resulting variables are orthogonal (as are the DFT-based cepstral coefficients) and well behaved numerically. They are the most common form of LPC-based variables used for speech or speaker recognition.

Figure 21.7 shows the complete LPC process, including the transformation into cepstral coefficients for speech recognition. In this case temporal derivatives are also commonly used to augment the feature vector.

## 21.6 CONCLUDING DISCUSSION

We have shown that linear prediction can be used to generate estimates of the spectral envelope for the short-term spectrum of speech that is commonly of interest for applications in synthesis and recognition. It is based on a model of the vocal tract shaping of an excitation signal. When the model is a good one (for instance, for steady-state vowels), the approach yields a good match to the spectral envelope; in other words, using an eighth- or tenth-order model to represent four formants works pretty well under good conditions. When the model is a poor match to the physical generation, for instance, when the sound is nasalized (the extra side chain creates significant zeros), the results tend not to be so good; additionally, unvoiced sounds tend to have a different (usually simpler) spectral shape and may be overparameterized by a model order that is appropriate for the vowels. Still, even in these cases LPC often provides a reasonable spectral estimate.

Another issue of concern for linear prediction modeling is the use of sharp antialiasing filters to precede sampling. In many signal-processing applications we tend to think of aliasing distortion as evil, and so we wish to prevent it at all costs – it is common to set the corner frequency for such filters at 40% or so of the sampling frequency rather than at the ideal of 50% required by the Nyquist theorem (for a filter with a perfect rectangular frequency response). However, if the analysis used is LPC, this choice will put a steep low-pass filter characteristic within the band modeled by LPC analysis. Essentially, the analysis only has a fixed number of poles to place in order to model the spectrum, and having such a steep filter will reduce the degrees of freedom available to model the speech. From another perspective, having a large range of spectral values to model tends to make the correlation matrices ill conditioned, with a corresponding large range in the matrix eigenvalues [3]. Practically speaking, when LPC is used, it is common to set the antialiasing corner frequency at roughly the half-sampling frequency, particularly for applications in speech and speaker recognition, even though the aliasing distortion becomes significant.

LPC analysis focuses on the spectrum as a product of resonances. As we have noted in previous chapters, resonant (formant) frequencies have often been associated with phonetic identity, particularly for vowels. However, there is significant variation in these frequencies among people; in particular, the vocal tract length tends to scale these frequencies, leading to very different average values for adult males, adult females, and children. Thus, there tends to be a built-in speaker dependence for LPC-based features. In contrast, filter banks and cepstral analysis are less tied to the specific resonant frequencies, but they consequently lack some of the previously noted advantages of basing the envelope estimate on these resonances. Table 21.1 summarizes a number of other comparative points between these spectral envelope estimates; note that the LPC column refers to the predictor polynomial only, as some of the weaknesses indicated there for LPC can be softened by using cepstral parameters from the LPC analysis. In Chapter 22 we will discuss approaches that have been developed to benefit from all three spectral envelope estimation methods.

We note that this chapter has skimmed over the implementation mathematics very lightly; please refer to references [3] and [4] for much more complete discussions of the autocorrelation and covariance methods, including the efficient recursions used. Lattice implementations of LPC filters will be briefly discussed in Chapter 31. Finally, reference [5] is a good source for a completely different perspective based on maximizing the entropy of the spectral estimate but leading to similar solutions.

**TABLE 21.1  Summary of Characteristics of Basic Methods for Spectral Envelope Estimation in Speech[a]**

| Characteristic | Filter Banks | Cepstral Analysis | LPC |
|---|---|---|---|
| Reduced pitch effects | × | × | × |
| Excitation estimate | | × | × |
| Direct access to spectra | × | | |
| Less resolution at HF | × | | |
| Orthogonal outputs | | × | |
| Peak-hugging property | | | × |
| Reduced computation | | | × |

[a]An × in a column means that the analysis method indicated by the column label is a particularly good match to the characteristic indicated by the row label. The last column refers to predictor coefficients; many of the limitations indicated are softened by transformation to cepstral coefficients. The excitation estimate row refers to getting *some* estimate of the excitation, not necessarily the one that is commonly used in the application. HF refers to high frequency.

## 21.7  EXERCISES

**21.1**  We have indicated that the squared error criterion leads to a spectral ratio error criterion (between the power spectra associated with the speech and the model). Use Parseval's theorem to show this.

**21.2**  A signal includes a very noisy spectral slice between 500 and 600 Hz. An engineer proposes implementing a steep notch filter (band reject) to remove this noise. If this signal is going to be analyzed with linear prediction, what is a potential difficulty with this plan?

## BIBLIOGRAPHY

1. Akaike, H., "A new look at statistical model identification," *IEEE Trans. Autom. Control* **AC-19**: 716–723, 1974.
2. Golub, G., and van Loan, C., *Matrix Computations*, Johns Hopkins Univ. Press, Baltimore, 1983.
3. Makhoul, J., "Linear prediction: a tutorial review," *Proc. IEEE* **63**: 561–580, 1975.
4. Rabiner, L., and Juang, B.-H., *Fundamentals of Speech Recognition,* Prentice–Hall, Englewood Cliffs, N.J., 1993.
5. Schroeder, M., "Linear prediction, entropy, and signal analysis," *IEEE ASSP Mag.* **1**: 3–11, 1984.

# AUTOMATIC SPEECH RECOGNITION

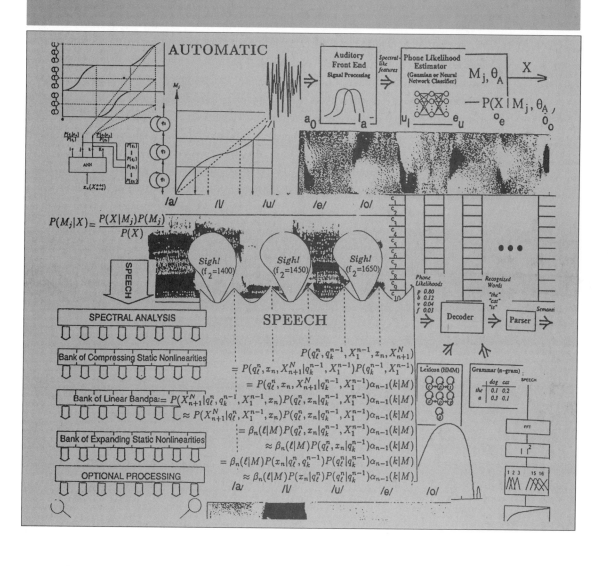

I always heard it couldn't be done, but sometimes it don't always work.
—Casey Stengel

**B**UILDING ON both the mathematical techniques of Part II and the feature extraction methods of Part V, we now focus on a single important application area: speech recognition. We describe a major aspect of speech-recognition systems in each of the seven chapters in Part VI. In Chapter 22, we extend the archetypal processing paradigms of Part V to the style of signal-processing features that are most commonly used for 1998 ASR systems. Chapter 23 introduces the linguistic categories that are most frequently used in such systems, such as phones and phonemes. The next two chapters describe methods for determining a sequence of words from measures of the similarity or dissimilarity between training examples of words and new test data. In both cases, the primary technique described is a search method known as dynamic programming. In Chapter 24, distance between sounds in training versus test examples is used as the measure of dissimilarity. Although this approach per se is rarely used today, its description can often be a useful introduction to more advanced techniques. In Chapter 25, the statistical generalization of this approach is developed, and its relation to the mathematical abstraction known as hidden Markov models is described. In this case, the distances used in the dynamic programming are probabilistic. These values are estimated by a system that has previously been trained, a process that is described in Chapter 26. Chapter 27 extends these approaches to discriminant training, a class of algorithms that are designed to improve the ability to separate classes for a linguistic unit (e.g., phonemes or words). Finally, Chapter 28 outlines the elements of a complete system, including language models. Although the estimation and training of these models are not typically viewed as signal processing per se, they are essential for any real recognition system.

# CHAPTER 22

# FEATURE EXTRACTION FOR ASR

## 22.1 INTRODUCTION

In previous chapters, we have introduced some general classes of feature extraction that researchers and system developers have found useful for the representation of speech. Filter banks, cepstral analysis, and LPC are indeed the generic representations of choice for a range of applications in speech and audio processing. However, for each application area, there are specific representations that have been developed, and they often have some of the characteristics of more than one of these archetypes.

For current ASR systems, the goal has generally been to find a representation that is relatively stable for different examples of the same speech sound, despite differences in the speaker or environmental characteristics. In this chapter, we briefly discuss a few of the common approaches. For most of these, the representation will be computed roughly once every 10 ms over a window of 20 or 30 ms. We also briefly describe some of the common techniques used to further process the feature vectors in order to make the overall system robust to simple linear distortions of the input signal (that is, to produce the same recognition results despite these deviations). Finally, we briefly discuss a few of the many research approaches that are being explored in the area of improved feature extraction.

## 22.2 COMMON FEATURE VECTORS

Over the past few decades, many variants of filter banks, LPC, and cepstral vectors have been used for speech recognition. More recently, the majority of systems have converged to the use of a cepstral vector derived from a filter bank that has been designed according to some model of the auditory system. Although there are a number of variants, we describe here the two that are used in most systems: mel cepstrum [3], [5] and perceptual linear prediction (PLP)[13]. In fact, these approaches are quite similar, and in recent years researchers have experimented with hybrids of the two.

Here we give the basic steps of such analyses, noting the differences between the mel cepstral and PLP approaches for each. We use Fig. 22.1, which gives the procedure for PLP, as a point of reference for these steps; we also refer to Fig. 22.3 for a prototypical example of the effects of each step.

**1.** Compute a power spectral estimate for the analysis window; typically for both mel cepstrum and PLP this is done by windowing the analysis region (e.g., with a Hamming window), calculating the FFT, and computing its squared magnitude. This step corresponds

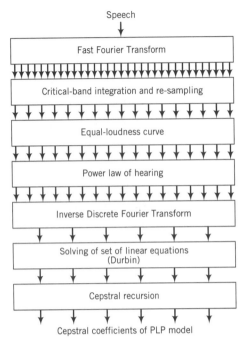

Speech

Fast Fourier Transform

Critical-band integration and re-sampling

Equal-loudness curve

Power law of hearing

Inverse Discrete Fourier Transform

Solving of set of linear equations
(Durbin)

Cepstral recursion

Cepstral coefficients of PLP model

**FIGURE 22.1**    Steps in the computation of PLP.

to the first rectangle of Fig. 22.1, and it yields a power spectrum such as the one shown in (1) of Fig. 22.3.

**2.** Integrate the power spectrum within overlapping critical band filter responses (second box of Fig. 22.1). There are a number of forms used for these filters, but all of them are based on a frequency scale that is roughly linear below 1 kHz and roughly logarithmic above this point, as discussed in earlier chapters. The mel scale is based on pitch perception and is used in the filter bank for the mel-cepstral approach (hence the name). Since it is based on human experimental data, there are a number of approximations and models that have been used. In the mel case, the integration step is done with a triangular window applied to the log of the power spectrum, as shown in Fig. 22.2. For the case of PLP, trapezoidally shaped filters are applied at roughly 1-Bark intervals, where the Bark axis is derived from the frequency axis by using a warping function from Schroeder:

$$\Omega(\omega) = 6 \ln \left\{ \frac{\omega}{1200\pi} + \left[ \left( \frac{\omega}{1200\pi} \right)^2 + 1 \right]^{0.5} \right\}. \tag{22.1}$$

Here $\omega$ is the frequency in radians/second. The trapezoidal window is an approximation to the power spectrum of the critical band masking curve from Fletcher [8]. In both cases, the net effect is to reduce frequency sensitivity over the original spectral estimate, particularly

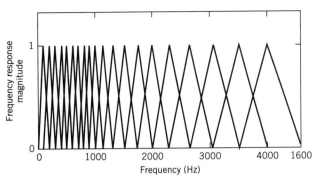

**FIGURE 22.2** Mel-scale filter bank. From [18], in turn adapted from [5].

at high frequencies. The higher frequencies are also somewhat emphasized given the wider filter bandwidths. The results are illustrated in (2) and (3) of Fig. 22.3.

**3.** Pre-emphasize the spectrum to approximate the unequal sensitivity of human hearing at different frequencies. In most mel-cepstrum analyses, this is actually done before the original spectral analysis, and an important side effect is to eliminate the effect of dc offsets in the speech signal. In PLP analysis, this step is implemented as an explicit weighting of the elements of the critical band spectrum.[1] This step corresponds to the equal-loudness curve box in Fig. 22.1 and also to (4) in Fig. 22.3.

**4.** Compress the spectral amplitudes (power law of hearing in Fig. 22.1). Typically, the log is applied after the integration. In PLP, the cube root is taken rather than the log, which is an approximation to the power-law relationship between intensity and loudness [21]. Aside from matching this property of human hearing, the effect of this step is to reduce amplitude variations for the spectral resonances, as can be seen in (5) of Fig. 22.3.

**5.** Perform an inverse DFT (third box from the bottom, Fig. 22.1). Although this step is not illustrated in Fig. 22.3 (since that figure solely describes the spectral effects), it is a critical step for both mel-cepstral analysis and for PLP. In the former case, it is the step that yields cepstral coefficients. For PLP, since the log has not been computed, the results are more like autocorrelation coefficients (though they are still from a compressed spectrum). Since the power spectral values are real and even, only the cosine components of the inverse DFT need be computed.

**6.** Perform spectral smoothing. Although the critical band spectrum suppresses some detail, another level of integration has been shown to be useful for reducing the effects of nonlinguistic sources of variance in the speech signal. In mel-cepstral processing, this step is accomplished by cepstral truncation; typically the lower 12 or 14 components are computed from 20 or more filter magnitudes. Thus, the higher Fourier components in the compressed spectrum are ignored, and the resulting representation corresponds to a smoothed spectrum. In the case of PLP, an autoregressive model (derived by the solution of linear

[1]In recent software implementations of PLP, there is sometimes an explicit high-pass filter at the input to handle dc offsets, in addition to the frequency-domain pre-emphasis.

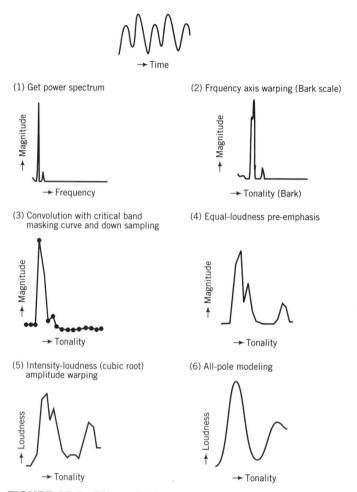

**FIGURE 22.3**  Effect of PLP steps on the spectrum.

equations constructed from the autocorrelations of the previous step) is used to smooth the compressed critical band spectrum; as with conventional LPC, the resulting smoothed spectrum is a better fit to the spectral peaks than the valleys. Many researchers have found that this approach leads to better noise robustness and speaker independence than the cepstral truncation. Part (6) in Fig. 22.3 shows a smoothed representation of the peaks that is suggestive of the kind of results one gets with PLP processing.

**7.** Use orthogonal representation. For mel-cepstral analysis, no further step is necessary to get orthogonal features – the elements of the truncated cepstral vector have this property, which typically simplifies the pattern recognition that follows. For PLP, the autoregressive coefficients are converted to cepstral variables.

**8.** Perform liftering. Not shown in either diagram is an additional step that is often taken:   the cepstral parameters are often multiplied by some simple function (such as $n^{\alpha}$, where $n$ is the cepstral index and $\alpha$ is a parameter between zero and one). The purpose of this function is to modify the distances that could be computed with these features to be more or less sensitive to the amplitude of resonant peaks in the spectrum. When $\alpha$ is set to one, the cepstrum is said to be index weighted, and the resulting lifter typically has the effect of roughly equalizing the variances of the different cepstral coefficients.

Both mel-cepstral analysis and PLP provide a representation corresponding to a smoothed short-term spectrum that has been compressed and equalized much as is done in human hearing. As noted earlier, in recent years a number of researchers have also experimented with blending the two approaches; in particular, some have found it advantageous to use PLP but with the triangular integration windows from the typical mel-cepstral analysis.

Thus, the features of mel cepstrum and PLP are extremely similar. Each has the reduced resolution at high frequencies that is indicative of auditory filter-bank-based methods, yet provides the orthogonal outputs that typify cepstral analysis. The principal difference between the two lies in the nature of the spectral smoothing (cepstral or LPC based). Since PLP is LPC based, it is interesting to compare the analysis to the more standard LPC approach described in the Chapter 21. Figure 22.4 shows this comparison. Note that the autocorrelation step of LPC is expanded out to its frequency-domain equivalent (the inverse DFT of the squared magnitude spectrum) for this comparison. Essentially, the pre-emphasis

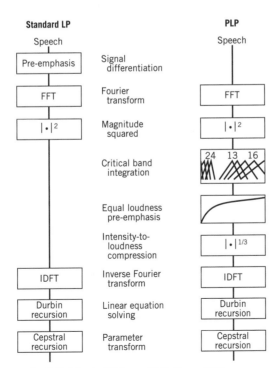

**FIGURE 22.4** LPC vs. PLP. From [12].

often done in LPC analysis is equivalent to the equal-loudness filtering in PLP. The major remaining difference lies in the computation of the compressed critical band spectrum for PLP, which is not done in the LPC analysis. Thus, PLP can be either be viewed as a mel-cepstral analysis with LPC-like spectral smoothing, or as a LPC analysis for an implicit version of the speech that has been warped according to auditory properties (in particular, critical band and power-law models of hearing). This warping has the effect of altering the error criterion of the autoregressive modeling. In practice it has often been found that the result of this change has been that lower-order PLP models are required (in comparison with standard LPC), and that a somewhat greater speaker independence is achieved (particularly for low order PLP).

## 22.3   DYNAMIC FEATURES

Feature vectors computed from mel-cepstral or PLP analyses appear to provide good smooth estimates of local spectra. However, it could be argued that a key characteristic of speech is its dynamic behavior. Because of this, many researchers have made use of estimates of the local time derivatives of the short-term spectrum or cepstrum.

One of the most common forms of this measure is the so-called delta cepstrum [9]. This is typically implemented as a least-squares approximation to the local slope, and as such is a smoother estimate of the local derivative than a simple difference between cepstra for neighboring frames. This can be expressed as

$$\Delta C_i(n) = \frac{\sum_{k=-N}^{N} k C_i(n+k)}{\sum_{k=-N}^{N} k^2}. \tag{22.2}$$

Thus, each stream of delta cepstral values is computed by correlating the corresponding stream of cepstral values with a straight line that has a slope of one.

The second derivative (commonly referred to as the delta-delta cepstrum) is also often useful, and it corresponds to a similar correlation, but with a parabolic function.

Many speech-recognition systems have incorporated features such as these. They tend to emphasize the dynamic aspects of the speech spectrum over time and to be relatively insensitive to constant spectral characteristics that might be unrelated to the linguistic content in speech, such as the long-term average spectral slope. However, the resulting feature vectors miss some of the gross characteristics that are salient in static spectral representations, and typically they are not sufficient for good recognition performance. In practice most systems that incorporate delta features use them as an add-on to static measures such as mel cepstra or PLP cepstra.

## 22.4   STRATEGIES FOR ROBUSTNESS

### 22.4.1   Robustness to Convolutional Error

In the previous section, we suggested that simple transformations, such as estimates of local time derivatives of cepstral parameters, could be robust to constant spectral components in the data. Let us be somewhat more precise. Suppose that a speech signal with short-term spectrum $S(\omega, t)$ is processed by a linear time-invariant filter with transfer function $H(\omega, t)$. Then, if $X(\omega, t)$ is the short-term spectrum of the observed signal, we may say

$$X(\omega, t) = S(\omega, t)H(\omega, t). \qquad (22.3)$$

Then the corresponding short-term log power spectrum would be[2]

$$\log |X(\omega, t)|^2 = \log |S(\omega, t)|^2 + \log |H(\omega, t)|^2. \qquad (22.4)$$

Thus, a convolutional effect in the time domain (as caused by the filter) corresponds to a multiplication in the frequency domain, and to a sum in the log power domain. If these two additive components have different properties over time then they can be separated fairly simply. For instance, if $H$ is constant over time, and if constant components of $S$ are not useful, one can simply estimate the constant component of the sum by computing the mean of the log spectrum. Alternatively, one may compute the Fourier transform of the above components, yielding cepstra, and remove the means in this domain. This operation is a standard one in many speech-recognition systems, and it is referred to as cepstral mean subtraction, or CMS.

Let us dwell on this notion momentarily. A disturbance has affected the speech, and the disturbance might be unknown – a change in telephone channel, a switch in microphones, or perhaps just a turn of the speaker's head so that the overall spectral characteristic is changed. The above analysis suggests that distinguishing between the signal components on the basis of how quickly the log spectrum or cepstrum changes with time can separate out the speech from the convolutional disturbance. In other words, disturbances that were convolutional in the time domain become additive in the log spectral domain. If such additive components have different spectral characteristics, linear filters can be used to separate them out.

Viewed in this more general framework, CMS can be seen as a specific example of a more general notion of filtering in the domain of the time trajectories of log power spectral or cepstral coefficients. Another specific example of such a principle is the approach referred to as RASTA-PLP. This is a modification to PLP analysis that is an on-line approach to achieving robustness to convolutional disturbances [15], [16], [14].[3] In this approach, the log of each critical band trajectory is filtered with a bandpass filter; typically there is a zero at dc, and the restriction at the higher frequencies establishes the passband in the domain of critical band modulations to a range that appears to be required for speech intelligibility. The resulting filtered trajectory is then exponentiated to yield a modified critical band power spectrum for analysis in the later steps of PLP.

Figure 22.5 shows the basic steps of RASTA processing. A common form for this processing is to use the first steps of PLP or mel-cepstral analysis for the spectral analysis

---

[2]Conventionally, a base of 10 would be used for the logarithm, and the result is often multiplied by 10 for graphical displays of log power spectra so that the axis can be labeled in terms of decibels.

[3]The more general idea of filtering temporal trajectories of subband energies, or simple transformations such as cepstral trajectories, is sometimes also called RASTA filtering. RASTA has also been applied to analysis approaches other than PLP; for instance, it has been applied to mel cepstra.

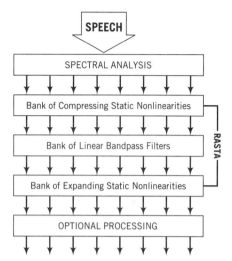

**FIGURE 22.5** RASTA processing. In the most common implementation, the compressing nonlinearity is a log, and the expanding function is an exponential.

block shown in the figure. Figure 22.6(a) shows a bandpass filter characteristic that was determined experimentally, that is, by modifying a pole position in order to maximize the recognition accuracy for a particular task (in this case, isolated digit recognition). Interestingly enough, this characteristic bears a strong resemblance to an independent measurement made from human sensitivity to frequency modulation, as shown in Fig. 22.6(b). Thus, it appears that a basic property of human hearing, namely the sensitivity to the modulation rate in audio, has a strong similarity to the bandpass characteristic of spectral envelope

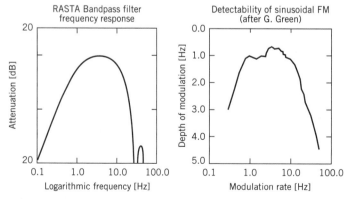

**FIGURE 22.6** (a) RASTA filter characteristic, optimized for a recognition task. (b) Human sensitivity to frequency modulation. Both figures courtesy of Hynek Hermansky. Original data for (b) from [11].

modulation rates that we find useful to pass in our recognition system. We speculate that, given this characteristic of hearing, speakers tend to produce spoken language that can be most easily heard.

Aside from the details of PLP, RASTA can be seen as comparable to a short-time version of CMS, or as a reintegration of the delta coefficients referred to earlier. The short-time characteristic, though, results in a nontrivial modification of the spectrum. Figures 22.7(a)–22.7(c) show time--frequency plots for three different representations of five Czech vowels:

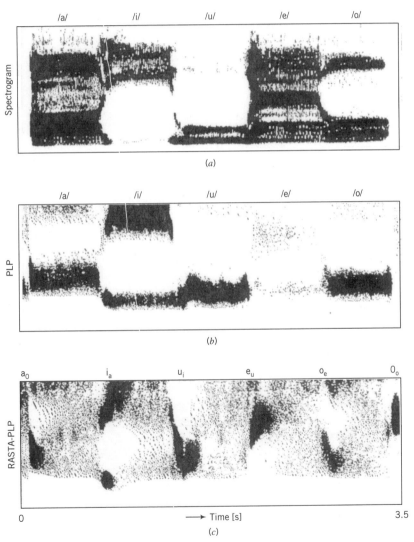

**FIGURE 22.7** Spectrograms for five Czech vowels: (a) conventional, (b) after PLP processing, and (c) after RASTA-PLP processing. From [14].

the conventional spectrogram, a spectrogram computed from low-order PLP cepstra, and a spectrogram computed from RASTA-PLP cepstra, respectively. The major effect of low-order PLP (fifth order) on the spectrum is to emphasize a small number of gross spectral features. The addition of RASTA causes the signal to decay after the onset of each steady-state region. The result is much more sensitive to transitions, which are smeared into the formerly steady-state regions. Although RASTA does have an improved independence from convolutional error, users must be aware of the effects of initial conditions (abrupt starts in artificial data segmentations, for instance); additionally, the statistical models used must include some representation of context from the left, since there is an increased statistical dependence on earlier signals because of the filter characteristic.

### 22.4.2   Robustness to Additive Noise

Generally speaking, the approaches of the previous section used some period of speech to determine an average value in a log spectral or cepstral domain and subtracted it off to eliminate components of the log spectrum that were essentially unchanging. For the case of additive noise, similar arguments can be used to suggest subtracting off something like an average power spectrum (see Exercises). In a common version of this approach, generally referred to as spectral subtraction, estimates of the noise power spectrum are literally subtracted from the observed short-term power spectrum. When a second channel is available to assist with the noise estimate (e.g., a second microphone), the estimation of the noise spectrum is simplified (see Section 22.6). Otherwise, estimates of the noise can be accumulated over time from histograms of the signal; typically the probability distributions of noise and signal energy are quite different, and the components can be estimated over time by a variety of means that will not be discussed here. Spectral subtraction is discussed extensively in [6].

Additionally, RASTA can be adapted to the noise problem by changing the compressive nonlinearity of Fig. 22.5. In this on-line approach, called J-RASTA (sometimes called lin-log RASTA), a noise power estimate is used to modify the nonlinear processing function prior to the RASTA filtering. RASTA approaches in general are discussed in much greater detail in [14].

There are many other compensation algorithms that modify the cepstral vector based on estimates of the noise or other linear modification of the signal, in general based on some computation of the difference between training and testing environments. Many of these (such as SNR-dependent cepstral normalization) have been developed at CMU and are described in [20].

### 22.4.3   Caveats

The preceding discussion was restricted to certain classes of signal disturbance:   linear and additive disturbances that have different temporal characteristics from speech. Further, it was assumed that spectral modifications did not completely obliterate the signal – an extremely sharp low-pass filter at 2 kHz will destroy high-frequency components in a way that RASTA or CMS will not be able to handle. Many real disturbances are nonlinear, phone dependent, or both. Nonetheless, many systems have found simple linear normalization schemes to be very important for acceptable performance under realistic conditions.

## 22.5  AUDITORY MODELS

In some sense many of the standard ASR features described earlier make use of insights and perspectives concerning the auditory system. The emphasis on power spectrum (ignoring the short-term phase), the integration into critical bands, the use of mel or bark scales, compression with log or cube-root functions, and spectral smoothing all are at least consistent with current perspectives on auditory function. For a particular example, that of the IBM Tangora system, an approximation to neural firing rates was used, incorporating a model of short-term adaptation [4]. Even RASTA processing appears to have some relation to models of forward temporal masking. Thus, many of the common feature computations already in use are in some sense related to auditory models.

Many other auditory-inspired models have been proposed. In work at MIT, a model was developed that computed auditory-inspired parameters called mean rate and synchrony [19]. At Bell Labs, another representation was developed called the ensemble interval histogram (EIH). The EIH computed a kind of a spectrum based on interval statistics [10]. For both the MIT and Bell Labs models, researchers have sometimes observed useful properties (in particular, robustness to some kinds of additive noise), but the techniques have not achieved widespread use.

Finally, as noted in Chapter 18, a number of experiments have been recently performed in which features (such as cepstral parameters) are separately computed for different bands of the spectrum (e.g., 4–7 for the telephone bandwidth) [2]. These were inspired by models of speech perception that incorporated within-band measurements across time, as described in [1].

## 22.6  MULTICHANNEL INPUT

Nearly all ASR systems incorporate a single channel of input. However, improved robustness to noise, reverberation, and nonlinguistic disturbances in general can often be obtained by using several microphones and computing features that take advantage of the multiple sources of acoustic information. In two-microphone schemes, for instance, differences in the SNR between the two microphones (which can be achieved by placing one microphone relatively near the talker) can be used to improve speech-signal estimation. Figure 22.8 shows a simple noise-reduction scenario, in which it is assumed that the second input signal contains noise only. If $y_1(n)$ is related by a linear transformation to the isolated noise signal $y_2(n)$, then adaptive least-squares approaches (on-line regression between the two signals) will attempt to find the best signal to subtract off from the input. In practice, such techniques can work quite well; in one demonstration of this technique, given at an ICASSP in the early 1980s, helicopter noise that was significantly louder than the speech was successfully subtracted to yield a very intelligible signal.

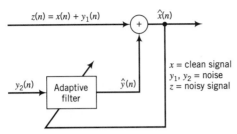

**FIGURE 22.8**  Two-microphone noise reduction. From [18].

When practical, microphone arrays can use beam-forming approaches to attempt to focus on the sound source; thus, the spatial location of the desired or undesired signals can be incorporated for their separation [7]. In some cases cross-correlation approaches are used, which are at least inspired by processing that is often presumed to occur in human binaural hearing [22].

## 22.7  DISCUSSION

Since the topic of this chapter was feature extraction for ASR, we have ignored characteristics of ASR systems that are more typically associated with the pattern-matching functions, for instance the statistical estimation. However, optimal functioning of these different components is necessarily linked. In particular, one of the most effective approaches to noise compensation has been the modification of statistical models to permit them to better represent the corrupted speech. Some methods (such as CMU's codeword-dependent cepstral normalization) modify the statistical parameter estimation for the case of cepstral input variables that have been degraded by linear and additive spectral effects. In the case of representations such as RASTA-PLP, it has been shown that some kinds of statistical estimators are much better able to handle these features than others (in particular, RASTA incurs a strong dependence on the left context, and the model must be able to represent this). Other kinds of feature vectors can have quite different properties, for instance a high dimensionality, components that are very correlated, and an added emphasis on transition regions. Other features (such as EIH) are computed over longer stretches of time than the more common feature sets, which strongly affects the corresponding statistics. These may be difficult to accommodate properly with traditional structures, and future research in ASR may need to simultaneously consider changes to the feature extraction and the statistical components.

## 22.8  EXERCISES

**22.1**   Recall that an octave is a frequency range of a factor of 2, and that fractions of an octave are computed on a log scale (e.g., the frequency that is one-half an octave above $f_1$ is $\sqrt{2}f_1$).

(a) Compute a set of frequencies that are one-third an octave apart starting at 100 Hz and going up to 3200 Hz.

(b) Use Eq. 22.1 to warp these numbers to the Bark scale.

(c) Using these results, characterize the difference between filters centered at these frequencies from filters centered at 1-Bark intervals.

**22.2** Show that the slope of the best line approximating cepstral feature $C_i$ from time $n - N$ to time $n + N$ (in the sense of the total squared error) is given by Eq. 22.2.

**22.3** Describe some circumstances in which autoregressive smoothing of the spectrum, as done in PLP, could be preferable to smoothing by cepstral truncation, as is done for mel cepstra. Describe a situation in which mel cepstra could be preferable.

# BIBLIOGRAPHY

1. Allen, J. B., "How do humans process and recognize speech?," *IEEE Trans. Speech Audio Process.* **2**: 567–577, 1994.
2. Bourlard, H., Hermansky, H., and Morgan, N., "Towards increasing speech recognition error rates," *Speech Commun.* **18**: 205–231, 1996.
3. Bridle, J. S., and Brown, M. D., "An experimental automatic word recognition system," JSRU Rep. No. 1003, Joint Speech Research Unit, Ruislip, England, 1974.
4. Cohen, J. R., "Application of an auditory model to speech recognition," *J. Acoust. Soc. Am.* **85**: 2623–2629, 1989.
5. Davis, S., and Mermelstein, P., "Comparison of parametric representations of monosyllabic word recognition in continuously spoken sentences," *IEEE Trans. Acoust. Speech Signal Process.* **28**: 357–366, 1980.
6. Deller, J. R., Proakis, J. G., and Hansen, J. H., *Discrete-Time Processing of Speech Signals*, MacMillan Co., New York, 1993.
7. Farrell, K., Mammone, R., and Flanagan, J., "Beamforming microphone arrays for speech enhancement," in *Proc. IEEE Int. Conf. Acoust. Speech Signal Process.* San Francisco, pp. 285–288, 1992.
8. Fletcher, H., *Speech and Hearing in Communication*, Krieger, New York, 1953.
9. Furui, S., "Speaker independent isolated word recognizer using dynamic features of speech spectrum," *IEEE Trans. Acoust. Speech Signal Process.* **34**: 52–59, 1986.
10. Ghitza, O., "Auditory models and human performance in tasks related to speech coding and speech recognition," *IEEE Trans. Speech Audio Process.* **2**: 115–132, 1994.
11. Green, G., "Temporal aspects of audition," PhD Thesis, Oxford University, London, 1976.
12. Hermansky, H., "An efficient speaker–independent automatic speech recognition by simulation of some properties of human auditory processing," in *Proc. IEEE Int. Conf. Acoust. Speech Signal Processing*, Dallas, pp. 1159–1162, 1987.
13. Hermansky, H., "Perceptual linear predictive (PLP) analysis of speech," *J. Acoust. Soc. Am.* **87**: 1738–1752, 1990.
14. Hermansky, H., and Morgan, N., "RASTA processing of speech," *IEEE Trans. Speech Audio Process.* **2**: 578–589, 1994.
15. Hermansky, H, Morgan, N., Bayya, A., and Kohn, P., "Compensation for the effect of the communication channel in auditory-like analysis of speech (RASTA-PLP)," in *Proc. Eurospeech '91*, Genova, Italy, pp. 1367–1371, 1991.

16. Hirsch, H., Meyer, P., and Ruehl, H., "Improved speech recognition using high-pass filtering of subband envelopes," in *Proc. Eurospeech '91*, Genova, Italy, 1991.
17. Junqua, J.-C., and Haton, J.-P., *Robustness in Automatic Speech Recognition,* Kluwer, Boston, Mass., 1996.
18. Rabiner, L., and Juang, B.-H., *Fundamentals of Speech Recognition,* Prentice–Hall, Englewood Cliffs, N.J., 1993.
19. Seneff, S., "A joint synchrony/mean-rate model of auditory speech processing," *J. Phonet.* **16**: 55–76, 1988.
20. Stern, R., Acero, A., Liu, F.-H., and Oshima, Y., "Signal processing for robust speech recognition," in C. H. Lee, F. Soong, and K. Paliwal, eds., *Automatic Speech and Speaker Recognition*, Kluwer, Boston, Mass., 1996.
21. Stevens, S. S., "On the psychophysical law," *Psychol. Rev.* **64**: 153–181, 1957.
22. Sullivan, T., and Stern, R., "Multi-microphone correlation-based processing for robust speech recognition," in *Proc. IEEE Int. Conf. Acoust. Speech Signal Process.* Albuquerque, New Mexico, pp. 845–848, 1990.

# CHAPTER 23

# LINGUISTIC CATEGORIES FOR SPEECH RECOGNITION

## 23.1 INTRODUCTION

In the past few chapters, we have introduced the fundamentals of feature extraction for ASR. The resulting features are gathered together into feature vectors that are associated with linguistic categories during training, and then during recognition they are integrated over time to find the best linguistic sequence to assign to the observed sequence of feature vectors.

Here,[1] we discuss the linguistic categories that have been used or proposed for use in ASR. Many of these have been alluded to previously (articulatory features, phones, phonemes, syllables, words, phrases, sentences, etc.); here we attempt to define terms more rigorously, particularly with regard to their application in speech engineering. We also highlight some of the research areas in terms of the representation of these linguistic categories in ASR, particularly in the context of fluent speech.

## 23.2 PHONES AND PHONEMES

### 23.2.1 Overview

Words are a natural unit for modeling in ASR, particularly since there are many applications for which isolated words are an adequate form of input. Even for continuous speech, using complete words as the fundamental linguistic unit permits acoustic modeling of the word-specific context of the sounds used. However, it is a wasteful use of training data to ignore any commonality between sounds within different words. Therefore, subword units are nearly always employed in large-vocabulary ASR. Additionally, for more natural speech (for instance, continuous multiword utterances), pronunciations can vary considerably, making it useful to define some smaller linguistic unit for ASR modeling. For instance, consider the pronunciations of "You did eat" and "Did you eat?" The former sentence is likely to be used in a clarifying context, with emphasis on the "did" (e.g., as a response to "I haven't eaten yet"). The latter, however, can be used in a more informal context (such as when your friend is ready to gnaw her own arm off and is looking for dining company); consequently, the pronunciations of "you" and "did" are often quite different, leading to pronunciations such as "Dijaeet?" When models are built on a word-by-word basis, one would have to build

---

[1]This chapter was largely written by Eric Fosler-Lussier.

many different models of words like "did" and "you," which can vary in different contexts. Moreover, we would like to have some relationship between "Dijaeet?" and "Woujaeet?" (Would you eat?) – a task difficult to achieve in word-by-word models. We would like to have a notion of what makes these two words ("did" and "would") similar.

Borrowing from the field of linguistics, many ASR systems divide words up into units called phones or phonemes (depending on the system). Although phones and phonemes are often confused in the ASR literature, each of these terms has a distinct meaning in linguistics, and we explore these differences further in this section. We also discuss what makes phone(me)s similar to each other, and how linguists and speech researchers write down phones and phonemes. Finally, we focus on how these linguistic concepts are implemented in an ASR system.

## 23.2.2   What Makes a Phone?

Large acoustic variability for the same word is typical of spoken language. In Chapter 17, we discussed the concept of categorical perception – the fact that small acoustic variations (such as shifts in the spectrum for stop consonants) often do not change the listener's perceptions of the segment in question.

Linguists have categorized many of the sounds of the languages of the world into segments called phones. Although not all linguists agree on the identity of these phones, phoneticians in general do have some system for codifying them. Phones are not even necessarily the smallest units used to describe sounds – linguists will often use modifiers to phones, to refer, for instance, to fronted or raised vowels – but they represent a base set of sounds that can be used to describe most languages.

Phones are usually written in brackets (e.g., [m]). The IPA[2] phonetic transcription for spat, for instance, would be [spætʰ]. This indicates that the word is made up with an s, followed by an unaspirated p, the short vowel a, and an aspirated t. Note that in a phonetic description, features such as aspiration will be important. To determine whether a consonant such as p or t is aspirated, hold your hand in front of your mouth. If you feel a breath of air as you say the consonant in the word, then it is aspirated. Try saying spat versus pat; in the latter case, with the larger amount of air produced, the p is said to be aspirated, as opposed to the unaspirated[3] p in spat.

## 23.2.3   What Makes a Phoneme?

Since the set of phones is designed to cover the set of all languages, the inventory of these can be quite large. Not surprisingly, every language will choose to use only a subset of them. The set of unique sound categories that a language uses are called the phonemes of the language. Two sounds are considered to be parts of different phonemes if they make a distinction between two words; these words are called minimal pairs. The words mat and pat are lexically distinct; from this we can conclude that in English, /m/ and /p/ are

---

[2]International Phonetic Alphabet – see Section 23.3 for a description.

[3]The linguistic term unaspirated means that there is only a little bit of air. This is different from the term unreleased, which means that no air is produced whatsoever.

**TABLE 23.1  Korean Distinction between Aspirated and Unaspirated [p]**

| Korean | English Gloss |
| --- | --- |
| [pʰul] | 'grass' |
| [pul] | 'fire' |

different phonemes. In general, we write phonemes between slashes, to distinguish them from phones.

There will be some cases in which a sound in one language is just not used in another. For example, the velar fricative[4] *ch* in German (e.g., in the word sprach, represented by the IPA symbol [x]) is not a sound used in English. Conversely, the English voiced *th* sound (as in this and that, represented by IPA [ð]) doesn't appear in German; speakers who have trouble with this sound will often substitute [z], or less frequently, [d].

In other cases, different phones will be possible given the same phoneme. In this case, the phones are called the allophones of the phoneme. In English, for example, the aspirated [pʰ] and unaspirated [p] correspond to the same phoneme, /p/. If you try to say spit with an aspirated [pʰ], or pit with an unaspirated [p], the result is not some different word, but rather just an unusual pronunciation of these words (perhaps sounding like they came from a foreign speaker). This is not the case in Korean,[5] however – there is a minimal pair which shows a distinction between the two (Table 23.1).

The fact that adding aspiration changes the meaning of the word indicates that /p/ and /pʰ/ are really two separate phonemes in Korean.

## 23.3  PHONETIC AND PHONEMIC ALPHABETS

In trying to write about phones and phonemes, linguists have found that the alphabets of English and other languages are not an optimal choice for linguistic descriptions. Consider the words thing and that. In these words, the sounds made by the letters th are different from each other. This indicates that we need some way to distinguish between them other than English orthography. Also, the fact that English requires two letters to represent this one sound is suboptimal, particularly when the sounds th make are only minimally related to the individual sounds of [t] and [h]; we want a system that represents one sound with one symbol.

The system that phoneticians have devised for this purpose is called the International Phonetic Alphabet (or IPA). This alphabet has a base of approximately 75 consonants and 25 vowels, which covers most languages, and a large inventory (50 or so) of diacritics that can be used to modify the base phones in order to achieve finer phonetic distinctions.

[4]See the next section on articulatory features for the definition of a velar fricative.

[5]This example is due to Finegan and Besnier [4].

In the late 1980s, a data base of read sentences called TIMIT[6][1], [7] was phonetically labeled (using a combination of automatic and manual methods) for use in automatic speech recognition. So that the data base could be machine readable, an ASCII symbol set was used, where the requirement of one symbol per sound was relaxed. The TIMIT phoneset features a slightly smaller inventory of phones than IPA (61 phones), but it is English specific and still makes relatively fine phonetic distinctions. Table 23.2 shows the relationship between the these two symbol sets.

# 23.4 ARTICULATORY FEATURES

## 23.4.1 Overview

Linguists have tried to characterize what makes some phones and phonemes similar to each other, and what makes them different. In Chapter 17, we looked at articulatory features, which are designed to capture some similarities between the way different segments are produced by the vocal tract.[7] Here we look at some of the correlates across phones that can be used in the categorization of these sounds.

## 23.4.2 Consonants

Consonants are made by constricting the tube of the vocal tract in various ways, usually with the tongue. Two main categorizations used in determining similarities between phones are the place and manner of articulation. The place of articulation refers to the point of closest constriction in the oral cavity. Figure 23.1 gives a cartoonish picture of the oral cavity, with labelings of different constriction sites. The places of articulation found in English and other languages are as follows.

**Bilabial:** These segments are made with a constriction at the lips. Examples of this are [b] and [m].

**Labiodental:** Labiodental segments require speakers to place their lower lip against their upper teeth, as in [f] and [v].

**Interdental:** (also known as just *dental* segments) For these segments, the constriction is placed between the teeth. The only sounds in this category for English are [θ] (*th*ing) and [ð] (*th*at).

---

[6]This term is an acronym for Texas Instruments, where the data base was collected, and the Massachusetts Institute of Technology, where it was transcribed.

[7]It is interesting to note that linguistics has latched onto the idea of characterization by the production method, rather than by some features of perception. Partially, this is because it is much easier to introspect how we produce sounds than how we perceive sounds. There is some debate over whether trying to use a production-based model of phones in a recognition scheme is the proper thing to do. However, there is not currently an adequate model of the features of categorical perception that can be used.

## TABLE 23.2  TIMIT Phone Types

| | | Phones in the TIMIT Database | | | |
|---|---|---|---|---|---|
| TIMIT | IPA | Example | TIMIT | IPA | Example |
| pcl | p̚ | (p closure) | bcl | b̚ | (b closure) |
| tcl | t̚ | (t closure) | dcl | d̚ | (d closure) |
| kcl | k̚ | (k closure) | gcl | g̚ | (g closure) |
| p | p | **pea** | b | b | **bee** |
| t | t | **tea** | d | d | **day** |
| k | k | **key** | g | g | **gay** |
| q | ʔ | bat | dx | ɾ | dirty |
| ch | t͡ʃ | **choke** | jh | d͡ʒ | **joke** |
| f | f | **fish** | v | v | **vote** |
| th | θ | **thin** | dh | ð | **then** |
| s | s | **sound** | z | z | **zoo** |
| sh | ʃ | **shout** | zh | ʒ | **azure** |
| m | m | **moon** | n | n | **noon** |
| em | m̩ | bottom | en | n̩ | button |
| ng | ŋ | sing | eng | ŋ̩ | Washington |
| nx | ɾ̃ | winner | el | l̩ | bottle |
| l | l | like | r | r | right |
| w | w | wire | y | j | yes |
| hh | h | hay | hv | ɦ | ahead |
| er | ɝ | bird | axr | ɚ | butter |
| iy | i | beet | ih | ɪ | bit |
| ey | e | bait | eh | ɛ | bet |
| ae | æ | bat | aa | ɑ | father |
| ao | ɔ | bought | ah | ʌ | but |
| ow | o | boat | uh | ʊ | book |
| uw | u | boot | ux | ü | toot |
| aw | ɑʷ | about | ay | ɑʸ | bite |
| oy | ɔʸ | boy | ax-h | ə̥ | suspect |
| ax | ə | about | ix | ɨ | debit |
| epi | | (epenthetic sil.) | pau | | (pause) |
| h# | | (silence) | | | |

***Alveolar:***   Behind the teeth sits the alveolar ridge; constrictions here will result in segments such as [t], [n], and [z].

***Palatal-alveolar:***   If the constriction is made slightly behind the alveolar ridge, at the junction of the ridge and the hard palate, sounds like [ʃ] (as in *sh*erry) and [ʒ] (measure) can be made.

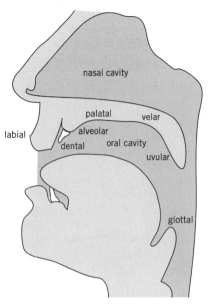

**FIGURE 23.1**   Places of the oral cavity used for articulations.

**Palatal:**   The only English phone made by a constriction on the hard palate is [d͡ʒ] (*j*udge), although this sound can also be palatal-alveolar. In German, the sound of the letters "ch" after some vowels like "i" will result in a palatal segment: for example, "ich" would be [iç].

**Velar:**   These segments are made by constrictions even further back in the mouth, closer to the soft palate (velum). Sounds in English that are velar include [k] and [ŋ] (as in si*ng*).

**Labiovelar:**   Some sounds are actually made by constrictions in two distinct places; closures at the lips and velum are a common example, which describes the English phone [w].

**Uvular:**   The uvula is the small flap of skin which hangs down at the back of the mouth. There are no segments in the English inventory which are actually uvular; in French, however, the phoneme /r/ is often realized with a uvular phone [ʁ], as in rouge.

**Glottal:**   A closure can sometimes take place as far back the oral cavity as the glottis; a little-known about segment that occurs in English is the glottal stop [ʔ], which will often precede word-initial vowels (as in the negative utterance uh-uh [ʔəʔə]), or replace segments like [t] in some dialects of British English (for example, bo*tt*le [bɑʔəl]).

The manner of articulation refers to the amount of constriction that is made in a consonantal gesture. The main categories of these seen in English are as follows.

**Stops:**   These segments are produced by a complete blockage of the airstream. Often, when stops are generated, two acoustic events result: first, the airstream is completely stopped (often called the closure), and then, after a short duration, the closure is released. Examples of stops are [t$^h$] or [p]. Another type of stop is the flap or tap, the alveolar version of which (IPA: [ɾ]) is often seen in American English replacing [t] or [d], in words like butter (compare with deter, where the /t/ sound is a full [t]). Flaps are often shorter than full stops; the tongue just taps for a short time against the roof of the mouth before returning to a less constricting position.

*manner*

**Fricatives:**   Fricatives can be thought of as "almost stops" – the tongue comes very close to a complete closure, enough to create turbulence in the airstream. Examples include [f] and [ʃ] (as in sherry).

**Affricates:**   An affricate is really a combination between a stop and a fricative. Affricates start with a stop closure, but instead of a normal stop release, a fricative is produced. In the IPA, this is notated by the combination of two symbols; for example, the English word cherry is spelled [t͡ʃɛri] in IPA.

**Nasals:**   Related to the stops are the nasals. Nasals such as [n] or [ŋ] (sing) close off the oral cavity in the same way that stops do. However, the nasal passage is opened at the velum, allowing air to escape through the nose, rather than building up behind the closure as in a stop.

**Liquids and Glides:**   These classes are often called *approximants*; these segments, although still consonantal, are more vowel-like in nature. They will often combine with vowels, blurring the boundaries between the vowel and the consonant. The English liquids include [l] and [r], which tend to color preceding vowels with their sound. The glides [y] and [w] often lengthen vowels to create diphthongs (such as boy).

**Other Articulations:**   Other articulations are also possible that are not commonly used in English. *Trills* are made by loosely placing the tongue in the place of articulation, and allowing the airstream to vibrate the tongue. Spanish uses the [r] trill in some words such as perro, "dog." Other types of stops are possible as well; *ejective* stops build up a large amount of pressure behind the stop before releasing with a poplike sound; *implosive* stops are created while breathing in, rather than out. Some African and Native American languages also utilize *clicks*, in which the tongue creates a small vacuum in the oral cavity, which is then released with a clicklike sound.

   One other major distinction in English consonants is voicing:   two phones can be made with exactly the same articulations, but with the vocal folds vibrating in one case and not in the other. When the vocal folds vibrate, a consonant is called voiced; it is voiceless otherwise. The difference between voiceless and voiced consonants can be seen in Sue versus zoo, or pat versus bat.

### 23.4.3  Vowels

Vowels can be categorized in a similar manner as consonants; often the manner of articulation is just considered to be vowel, and the place can vary over the range that consonants range. However, linguists generally tend to use three major parameters to describe vowels: frontness, height, and roundedness.

The frontness (or backness) of a vowel is similar to the place of articulation, in that it gives a general indication of the place of greatest constriction (i.e., highest tongue height) in the mouth. In English, we tend to use three different positions:   front, as in b*eat* [i], b*it* [ɪ], b*et* [ɛ], and b*at* [æ]; central, as in "high schwa" inhib*it* [ɨ], "schwa" *about* [ə]; and back, as in b*oot* [u], b*oat* [o], b*ought* [ɔ], b*ut* [ʌ], and f*ather* [ɑ].

The frontness correlates with higher second-formant frequency ($F_2$) values, although some use the difference $F_2 - F_1$ to describe frontness.

The height of a vowel refers to how far the lower jaw is from the upper when making the vowel. High vowels, such as [i] and [u], have the lower and upper jaws close together. Low vowels, such as [æ] and [ɑ], have a more open oral cavity. The height of a vowel correlates well with the first-formant frequency – high vowels have a low $F_1$, low vowels have a higher $F_1$. The actual values of the formant ranges depends on the speaker's vocal tract; for instance, men tend to have lower resonances than women, because they have a longer vocal tract. English speakers use approximately four different vowel heights in the inventory of vowels. This is difficult to pinpoint, since the variability in vowel formant frequency is actually rather large.

Roundedness refers to whether the lips have been rounded (rather than spread) in vowel production. In English, the front vowels are often unrounded, whereas the back vowels are rounded – as shown by b*it* versus b*oot*. In other languages, such as German and French, front vowels may also occur in rounded form, such as in the German word for vegetable, Gem*ü*se [gəmüzə]. Rounding tends to lower the first- and second-formant frequencies.

There are other features that can also affect vowel quality. In English, vowels often become rhoticised (r-colored) when followed by an /r/ sound, as in b*ir*d, or st*ar*. Nasalization happens when vowels are adjacent to nasals – as seen in the difference between the vowels in the words in and it, and can and cat.

### 23.4.4  Why Use Features?

Why are linguists (and speech researchers) interested in categorizing phones by using articulatory features? One reason is that phones with similar articulatory features are similar acoustically. Features also allow generalizations to be captured over classes of phonemes; for example, [t] and [d] segments can often be realized as a flap ([ɾ]), in words like but*t*er and la*dd*er. We can create a rewrite rule, called a phonological rule, which can describe this effect:

$$t \rightarrow ɾ \quad / \quad \begin{bmatrix} +\text{vowel} \\ +\text{stress} \end{bmatrix} \underline{\quad\quad} \begin{bmatrix} +\text{vowel} \\ -\text{stress} \end{bmatrix}.$$

The rule above says "change [t] to [ɾ] when the phone on the left is a stressed vowel and the phone on the right is an unstressed vowel." However, we would have to specify two rules to completely describe this phenomenon – one for [t] and one for [d]. Using articulatory features, we can only specify one rule to cover both instances:

$$\begin{bmatrix} +\text{stop} \\ +\text{alveolar} \end{bmatrix} \rightarrow \text{ɾ} \quad / \quad \begin{bmatrix} +\text{vowel} \\ +\text{stress} \end{bmatrix} \underline{\quad\quad} \begin{bmatrix} +\text{vowel} \\ -\text{stress} \end{bmatrix}.$$

In this section, we have barely scratched the surface of phonetics and phonology. For a further explanation of phonemes, phones, and the way phones are produced by the vocal tract, a good introductory text is *A Course in Phonetics*, by Peter Ladefoged [13].

## 23.5  SUBWORD UNITS AS CATEGORIES FOR ASR

Ultimately, in ASR we would like to be able to define a set of categories for statistical pattern recognition; an obvious set would be some type of phonetic or phonemic system. One of the open questions in ASR is what the appropriate set of phones is for classification. The choice of phoneset generally depends on the dictionary used; a dictionary provides a lookup table from words to their phonetic or phonemic pronunciations.

The TIMIT phoneset uses an inventory of 61 phones – a rather large set in ASR terms. Some dictionaries, such as the CMU dictionary [29], use a more phonemic representation, with only 40 classes. The trade-off is that having fewer classes often makes it easier to discriminate between classes since there will be more samples per class on average, but more classes will allow finer phonetic distinctions that may include some contextual information.

Some systems (such as IBM's) use completely data-driven categories for subword units; in IBM's case these categories were called fenones, which were models derived from statistical clustering techniques. Typically there were a somewhat larger number of fenone units than the number of phones used in such systems – a number such as 200 was used. Other ASR systems use self-organizing categories, splitting phones into three or so subunits; that is, while the initial (bootstrap) definition of the subword unit may come from human definition as established in a data base such as TIMIT (i.e., phones), the ultimate class associations for the ASR system will still be automatically determined in the training of the system. Still, the use of linguistic units such as phones or phonemes, even if only for bootstrap models, permits the incorporation of structure in other ways, such as the representation of alternate pronunciations for spoken language.

## 23.6  PHONOLOGICAL MODELS FOR ASR

Often, phonological rules such as the one just mentioned are used to map between phonemic and phonetic category sets. Different pronunciation dictionaries often use different phonesets. In order to convert pronunciations that use a set of 40 phonemes into a phoneset that has 61 phones, a set of rules, such as the flap rule just described, can be used [27], [14]. These rules provide a mapping between different phonesets.

Another use for phonological rules is to expand the set of pronunciations that can represent a word. For example, the word else [ɛls] can sometimes be pronounced [ɛlts], as is pointed out by Tom Lehrer's rhyme scheme in song:[8]

> *You give me welts,*
> *like nobody else,*
> *as we dance to the masochism tango.*

The [t] that gets inserted into else is called an epenthetic stop. In order to allow for the possibility of this pronunciation of else (and false, pulse, and Alzheimer's), we can write another phonological rule:

$$\emptyset \rightarrow t \quad / \quad l \,\underline{\quad}\, s \, \$.$$

This rule says "insert a [t] (i.e., replace nothing with [t]) between an [l] and an [s] before a syllable boundary ($)." With a large set of rules, one can generate many pronunciations – in fact, overgeneration is possible. Consider a rule of final alveolar stop deletion, which deletes word-final alveolar stops after [n] (and other segments). This rule would make the words an and and pronounced the same, potentially increasing the confusability between words for the recognizer. Usually, ASR systems either use a small number of pronunciations or have probabilities attached to the pronunciations in order to rank order them.

## 23.7   CONTEXT-DEPENDENT PHONES

Previously, we said that phonemic categorizations often use less contextual information than phonetic ones. However, some ASR systems get around this by explicitly modeling the surrounding context of a phone. For example, in Fig. 23.2 we can replace the single phone (or monophone) models with ones that have the left and right contexts. This means that the [k] and [t] models, which were previously shared by cat and kit, are now different for each word, depending on the central vowel.

One can use increasing amounts of context to make quadraphones, quintaphones, and so on. However, this increases the number of models by quite a bit. For a 40-phone set, the

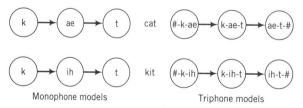

**FIGURE 23.2**   Phone models of cat and kit.

---

[8]Thanks to John Ohala for this example. Really.

number of triphones is already of the order of $40^3 = 64,000$, some of which are very unlikely to be seen in a moderate size of training data.[9] One way to reduce the dimensionality is to use acoustic features to cluster them. If the clustering is done by manner or place of the left and right context phones, this is called generalized triphones. If the models are just clustered together by using learning techniques such as decision trees, they are usually called clustered triphones. Some researchers have learned contextual relationships for several neighboring phones (e.g., two on either side), typically using decision trees.

## 23.8 OTHER SUBWORD UNITS

Phonemes, phones, clustered subphones, and context-dependent versions of these units have been the dominant structures for ASR over the past decade. However, other units have been considered and potentially have a number of significant advantages. Some researchers have suggested that the key element in speech intelligibility is not the classification of steady-state speech sounds (which are actually infrequent in natural fluent speech), but rather the classification of the *transitions* between phones. A unit that is a good match to this perspective is the diphone, which is typically defined as extending from the middle of one steady-state region to the middle of the next. Diphones are actually used quite commonly in commercial speech synthesis, but they have been used only occasionally for speech recognition.

Another unit of interest to the speech community has been the syllable (and sometimes a half-syllable unit called a demisyllable). The onset of syllables appear to be easier to detect acoustically than for phones, and all syllables seem to have structure constraints that can potentially be used in an ASR system. Speech-recognition researchers who work with languages other than American English have often incorporated the use of syllable or syllable-related units. Examples include considerable work in Japanese [18], Chinese [16], German [21], [11], [12], Hungarian [28], and Spanish [17].

A few research projects are currently exploring syllables for English, though the work is controversial; detractors view English as being a poor case for syllable-based structure, since there are a range of complex syllable types in English, and syllable timing is complicated by stress patterns. However, a recent examination of the patterns of conversational speech has shown that extremely simple syllabic types and timing patterns represent the lion's share of fluent conversational speech [8]. In general, syllables are associated with contours in energy and pitch that are 150–250 ms long.

For phonological purposes, syllables are often divided into three parts: the onset, nucleus, and coda. The nucleus is the minimal component, and, as the name indicates, is the center of the syllable. Nuclei are usually vocalic segments. The onset is typically consonantal material that precedes the nucleus (in a vowel-initial syllable there is no onset other than the beginning of the nucleus), whereas the coda is the material following. In the word spat, the onset is [sp], the nucleus is [æ], and the coda is [t$^h$].

---

[9]The actual number is probably smaller as a result of phonotactic constraints; for instance, the triphone [b-f-z] is not one that will be found in English.

### 23.8.1 Properties in Fluent Speech

A well-known rule of thumb is that onsets are preferred over codas in English. This means that syllabifications will often change in fluent speech. For instance, consider the word five ([fɑ$^y$v]), which only has one syllable in isolation. In the phrase five eight, the possibility of resyllabification ([fɑ$^y$] [vet]) exists, where five would occupy part of two syllables. This means that in fluent speech, word boundaries and syllable boundaries do not have to coincide.

## 23.9  PHRASES

In most current ASR systems, the acoustic models are structured as words (which are typically composed of subword units as noted earlier), and the words are grouped together into complete utterances by using models of language (typically simple statistical models, as will be discussed in later chapters). Ordinarily there is no other acoustic unit larger than a word. However, some work is beginning on incorporating phrase structure in speech recognition [15], [2]. Acoustically, phrases appear to have some coherence in terms of contours of energy and pitch, essentially corresponding to a sequence of syllables that can include multiple words; they are also commonly distinguished by breaks of lower energy (i.e., silence) between the phrases. Phrase boundaries also often serve to indicate a change in topic or a syntactic boundary (such as the beginning of a relative clause). However, currently the main application for phrase structure is as part of a natural language syntax that is sometimes used in a spoken language system for some limited domain (see the brief mention of the Air Travel Information System in the following paragraph).

## 23.10  SOME ISSUES IN PHONOLOGICAL MODELING

In the late 1980s and early 1990s, many speech researchers focused on problems in the read-speech domain, or human–computer interaction. In terms of U.S. government-funded tasks, read-speech problems included the Resource Management task (recognizing scripted commands to a data base of naval warships) and the North American Business News task (recognizing sentences read from the Wall Street Journal and other newspapers); the Air Travel Information System task used recordings of spontaneous interactions between a human and a computer system, in which the human was trying to get airline information.

Recently, there has been a move toward harder tasks, including several human–human interactions in recorded telephone calls (Switchboard, CallHome data bases), online computer language translation in human-to-human conversations (Verbmobil), or recorded spontaneous radio interviews (parts of the Broadcast News data base). Researchers have found that the pronunciation variability increases quite dramatically in these situations. The big challenge is to find ways to predict the variability that occurs in these settings.

Siegler and Stern [25] and Mirghafori et al. [19] have shown that abnormal speaking rates correlated well with word error in the 1993 Wall Street Journal speech-recognition evaluation; a higher or lower than average speaking rate was a good indicator that the speech recognizer would make more errors. This has also been seen in spontaneous corpora such as the OGI Numbers corpus [20] and Switchboard data base [10]. It is likely (although yet to be proven) that some of the errors are due to changes in pronunciation, such as the reduction or deletion of some phones. Researchers are looking into ways of incorporating speaking-rate information into the pronunciation model.

Another factor that affects pronunciation models is function words (high-frequency words including some prepositions and pronouns), which often contain little semantic content and have a wide variation of pronunciations within conversational speech. These function words often do not receive the same word stress as other words [24] and are often quite reduced,[10] as their pronunciation depends heavily on the surrounding context. For example, in a recent survey of pronunciations of the ten most frequent words[11] within a transcribed portion of the Switchboard corpus, Greenberg [9] cites an average of over 60 pronunciations per word. Words that have been repeated in a conversation also tend to have reduced pronunciations compared to the initial mention of the word [6].

Another issue currently under research is how to build pronunciation alternatives to be used in an ASR system. Using phonological rules [27], [14] allows the speech researcher to take advantage of the work of linguists, but one must search out and write in all of the rules. Riley [22] and Chen [3] used decision trees as statistical models for learning phonological rules from data. This use saves the work of trying to find all of the relevant phonological rules but may miss some generalizations that have very little data, and it requires hand-transcribed data, which are expensive to obtain. Schmid et al. [23] and Sloboda [26] use machine alignments instead of hand transcriptions in order to get enough data to build word models. Fosler et al. [5] combine the two approaches above and allow phonological processes to occur across words (to handle "dijaeat" vs. "did you eat" phrases) in a more dynamic approach: in their system, the pronunciations of words change depending on what the surrounding words are.

## 23.11 EXERCISES

**23.1**  Why would lip rounding tend to lower the first- and second-formant frequencies of a vowel? (Hint: consider the acoustic tube model discussed in Chapter 11.)

**23.2**  What would the phonological rule for final alveolar deletion after [n] described earlier look like? Use the symbol # for word boundaries.

**23.3**  Can you explain why the word think is more similar to the word thing than to the word thin?

---

[10]That is, pronunciations are transformed from their canonical form, often with vowels centralized, or phones shortened or deleted.

[11]This list consists of I, and, the, you, that, a, to, know, of, and it.

# BIBLIOGRAPHY

1. "TIMIT Acoustic-Phonetic Continuous Speech Corpus," Speech Disc 1-1.1, NIST Order No. PB91-505065, National Institute of Standards and Technology, 1990.
2. Bates, B., Stolcke, A., and Shriberg, E., Palo Alto, Calif., personal communication, 1997.
3. Chen, F., "Identification of contextual factors for pronounciation networks," in *Proc. IEEE Int. Conf. Acoust. Speech Signal Process.*, Albuquerque, NM, pp. 753–756, 1990.
4. Finegan, E., and Besnier, N., *Language Its Structure and Use*, Harcovrt Brace Jovanovich, San Diego, Calif.; 1989.
5. Fosler, E., Weintraub, M., Wegmann, S., Kao, Y-H., Khudanpur, S., Galles, C., and Saraclar, M., "Automatic learning of word pronunciation from data," *Proc. Int. Conf. Spoken Lang. Process.*, Philadelphia, Addendum, pp. 28–29, 1996.
6. Fowler, C., and Housum, J., "Talker's signaling of 'new' and 'old' words in speech and listeners' perception and use of the distinction," *J. Memory Lang.* **26**: 489–504, 1987.
7. Garofolo, J., Lamel, L., Fisher, W., Fiscus, J., Pallett, D., and Dahlgren, N., "DARPA TIMIT acoustic-phonetic continuous speech corpus," Rep. NISTIR 4930, U.S. Department of Commerce, National Institute of Standards and Technology, 1993.
8. Greenberg, S., "Understanding speech understanding:  towards a unified theory of speech perception," in *Proc. ESCA Tutorial Adv. Res. Workshop Auditory Basis Speech Percep.*, Keele University, England, pp. 1–8, 1996.
9. Greenberg, S., "On the origins of speech intelligibility in the real world," in *Proc. ESCA Workshop Robust Speech Recog. Unknown Commun. Channels*, Germany, 1997.
10. Ostendorf, M., Byrne, B., Bacchiani, M., Finke, M., Gunawardana, A., Ross, K., Roweis, S., Shriberg, E., Talkin, D., Waibel, A., Wheatley, B., and Zeppenfield, T., "Modeling systematic variations in pronunciation via a language-dependent hidden speaking mode," Tech. Rep. 24, *Fourth LVCSR Summer Res. Workshop*, Johns Hopkins University, Baltimore, 1996.
11. Kirchhoff, K., "Phonologically structured HMMs for speech recognition," in *Proc. Second Meet. Assoc. Comput. Linguistics SIG Comput. Phonology,* Santa Cruz., pp. 45–50, 1996.
12. Kirchhoff, K., "Syllable-level desynchronization of phonetic features for speech recognition," in *Proc. Int. Conf. Spoken Lang. Process.*, Philadelphia, pp. 2274–2276, 1996.
13. Ladefoged, P., A Course in Phonetics, San Diego: Harcourt Brace Jovanovich College Publishers, 1993.
14. Lamel, L., and Adda, G., "On designing pronunciation lexicons for large vocabulary, continuous speech recognition," in *Proc. Int. Conf. Spoken Lang. Process.*, Philadelphia, pp. 6–9, 1996.
15. Lee, C.-H., "Key-phrase detection and verification for flexible speech understanding," Lecture, Bell Labs' dialog systems research Dept., at ICSI, March 1997.
16. Lin, S. C., Chien, L. F., Chen, K. J., and Lee, L. S., "A syllable-based very-large-vocabulary voice retrieval system for Chinese databases with textual attributes," in *Proc. Eurospeech '95*, Madrid, Spain, pp. 203–206, 1995.
17. Lleida, E., Mariño, J., Salavedra, J., and Bonafonte, A., "Syllabic fillers for Spanish HMM keyword spotting," in *Proc. Int. Conf. Spoken Lang. Process.*, Alberta, Canada, pp. 5–8, 1992.
18. Matsunaga, S., Matsumura, T., and Singer, H., "Continuous speech recognition using non-uniform unit based acoustic and language models., in *Proc. Eurospeech '95*, Madrid, Spain, pp. 1619–1622, 1995.
19. Mirghafori, N., Fosler, E., and Morgan, N., "Fast speakers in large vocabulary continuous speech recognition:  analysis & antidotes," in *Proc. Eurospeech '95*, Madrid, Spain, pp. 491–494, 1995.
20. Morgan, N., Fosler, E., and Mirghafori, N., "Speech recognition using on-line estimation of speaking rate," in *Proc. Eurospeech '97*, Rhodes, Greece, pp. 2079–2082, 1997.

21. Plannerer, B., and Ruske, B., "Recognition of demisyllable based units using semicontinuous hidden Markov Models," in *Proc. IEEE Int. Conf. Acoust. Speech Signal Process.*, Toronto, pp. 693–696, 1991.

22. Riley, M., "A statistical model for generating pronunciation networks," in *Proc. IEEE Int. Conf. Acoust. Speech Signal Process.*, Toronto, pp. 737–740, 1991.

23. Schmid, P., Cole, R., and Fanty, M., "Automatically generated word pronunciations from phoneme classifier output," in *Proc. IEEE Int. Conf. Acoust. Speech Signal Process.*, Minneapolis, pp. 223–226, 1993.

24. Selkirk, E., *Phonology and Syntax: The Relation between Sound and Structure*, MIT Press, Cambridge, Mass., 1984.

25. Siegler, M., and Stern, R., "On the effects of speech rate in large vocabulary speech recognition systems," in *Proc. IEEE Int. Conf. Acoust. Speech Signal Process.*, Detroit, pp. 612–615, 1995.

26. Sloboda, T., and Waibel, A., "Dictionary learning for spontaneous speech recognition," in *Proc. Int. Conf. Spoken Lang. Process.*, Philadelphia, pp. 2338–2331, 1996.

27. Tajchman, G., Fosler, E., and Jurafsky, D., "Building multiple pronunciation models for novel words using exploratory computational phonology," in *Proc. Eurospeech '95*, Madrid, Spain, 1995.

28. Vicsi, K., and Vig, A., "Text independent neural network/rule based hybrid continuous speech recognition," in *Proc. Eurospeech '95*, Madrid, Spain, pp. 2201–2204, 1995.

29. Weide, R., *The Carnegie Melon Pronouncing Dictionary*, Version 0.4, Carnegie Mellon University, Pittsburgh, 1996.

# CHAPTER 24

# DETERMINISTIC SEQUENCE RECOGNITION FOR ASR

## 24.1 INTRODUCTION

In the past few chapters, we have established the basics for understanding the static pattern-classification aspect of speech recognition.

1. Signal representation: in most ASR systems, some function of the local short-term spectrum is used. Typically, this consists of cepstral parameters corresponding to a smoothed spectrum. These parameters are computed every 10 ms or so from a Hamming-windowed speech segment that is 20–30 ms in length. Each of these temporal steps is referred to as a *frame*.

2. Classes: in most current systems, the categories that are associated with the short-term signal spectra are phones or subphones,[1] as noted in Chapter 23. In some systems, though, the classes simply consist of implicit categories associated with the training data.

Given these choices, one can use any of the techniques described in Chapter 8 to train deterministic classifiers (e.g., minimum distance, linear discriminant functions, neural networks, etc.) that can classify signal segments into one of the classes. However, as noted earlier, speech recognition includes both pattern classification and sequence recognition; recognition of a string of linguistic units from the sequence of segment spectra requires finding the best match overall, not just locally. This would not be so much of a problem if the local match was always right,[2] but we are not this lucky! There are always some local errors, and our real goal is (typically) to recognize the correct sequence of words, not to find the phone identity associated with each frame. Furthermore, both timing and pronunciation can vary between our stored representation and the new speech that we wish to recognize. Therefore we must somehow trade off between different choices of frame-class identification so that the global error is minimized, that is, the error over a complete choice of a transcription for an utterance. Furthermore, even during the training phase, the boundaries of internal sound units (e.g., phones) are unknown. For these reasons, even the

---

[1] These categories are often further subdivided into classes associated with a particular phonetic context; for instance, a triphone is a phone with particular phonetic categories to its left and right.

[2] However, even if the local match always gave the correct phone, the variation in pronunciation could still generate possible confusions between words, so that higher-level information is still often necessary to decode the linguistic meaning.

simplest speech recognizer must take longer stretches of time (typically an entire utterance) into account when deciding what was said at any point. This entails doing both segmentation and classification; in many cases both are done with a single pass.

In this context, speech recognition may be seen as a particular case of sequence recognition. That is, for a sequence of frame observations (framewise signal representations or vectors) $X = (x_1, x_2, \ldots, x_n, \ldots, x_N)$, we wish to associate $X$ with a second sequence $Q = (q_1, q_2, \ldots, q_n, \ldots, q_N)$, where each $q$ represents some linguistic or quasi-linguistic unit, and where the sequence $Q$ is chosen to minimize some error criterion. For simplicity's sake, in this chapter the second sequence will be restricted to another sequence of frame observations, used as a reference. In other words, we will be matching a sequence $X^{\text{in}}$ with sequences $X_k^{\text{ref}}$ for $1 \le k \le K$, where $K$ is the number of reference sequences. We will also confine the discussion to deterministic methods, in which we establish distance metrics and methods for time normalization. In the next chapter we will move to statistical approaches, which are in some ways more general, but the intuition for sequence-integration methods may be clearer if presented first in a deterministic framework. The limitation to frame-observation references and to deterministic methods also provides some historical background, since nearly all ASR systems used these approaches before 1980.

## 24.2 ISOLATED WORD RECOGNITION

In order to make this discussion more concrete, let us consider the example of a simple template-based isolated word-recognition system. Imagine that the speech is received over the telephone, with an 8-kHz sampling rate, and with roughly a 3.4-kHz spectrum. The speech is pre-emphasized with a single-zero filter so that the spectral slope is relatively flat for voiced sounds. Once every 10 ms we apply a 25-ms Hamming window to the data, and we compute 10 mel-cepstral coefficients as explained in Chapter 22. We weight the cepstra by their index (except for $c_0$, which for simplicity's sake we ignore in this example). In principle, this weighting makes the feature variances roughly equal [5].

Thus, for each 10 ms we have 10 weighted cepstral numbers. Let us further imagine that the task is digit recognition, namely, the recognition of the words zero, one, . . . , nine.[3] Assuming that the digits are, on the average, roughly 300 ms long, we will have approximately 30 cepstral vectors for each word uttered. Each collection of cepstral vectors $X$ is called a template, illustrated in Fig. 24.1. During a training phase, we will compute and store reference templates $X_k^{\text{ref}}$. During testing, we will compare each new pattern of cepstral vectors $X^{\text{in}}$ with all of the reference templates and identify the new utterance as being the word associated with the template with the smallest distance to the new pattern. We will defer for the moment what "distance" means here.

For any computations involving the distance between two sequences of feature vectors, we must handle the issue of time normalization – comparing two sequences that are of different length.

---

[3] In real applications the word oh must also be recognized, but we ignore this detail here.

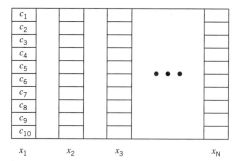

**FIGURE 24.1**  Template consisting of *N* tenth-order cepstral vectors.

### 24.2.1   Linear Time Warp

Suppose that we wish to compare two sequences of index-weighted cepstral vectors. Given a distance metric between any two vectors, if each template is the same length, we can compose the global distance (or distortion) between two templates as the sum of the local distance (still to be defined) between each corresponding frame of the templates being compared. In other words,

$$D\left(X_k^{\mathrm{ref}},\ X^{\mathrm{in}}\right) = \sum_{i=1}^{N} d\left(x_{ki}^{\mathrm{ref}},\ x_i^{\mathrm{in}}\right),\qquad(24.1)$$

where $d$ is the local distance between cepstral vectors, $D$ is the total distortion between the templates, and $N$ is the template length (the same for both templates by assumption).

In general, however, reference and input templates will have a differing number of vectors, so we cannot directly compare the sequences frame by frame. In particular, we would like two examples of the utterance "two" to have a very small overall distance, even though one of them was 300 ms long and the other was 600 ms long. The most obvious way to match the two sequences is by linear time normalization – basically, either the reference or the input vector sequence is subsampled or interpolated to yield a pair of sequences that are the same length. For the case given here, this could be done by repeating each vector of the 300-ms case, or by discarding every other vector from the 600-ms utterance. This is illustrated in Fig. 24.2 (only the first 10% on each axis is shown).

This is the approach that was used in many systems in the 1970s, and in fact it lingered in commercial systems for some time because of its simplicity. However, there are clear limitations to the idea. Generally, variations in speech duration are not spread evenly over different sounds. For instance, stop consonants vary only slightly in length, whereas diphthongs and glides (and in general vowels and vowellike sounds) vary a great deal over different utterances of the same word. For instance, in the word "two," the initial stop consonant would probably have a very similar length for both the 300-ms and 600-ms cases, whereas the following vowel would comprise most of the time normalization. Therefore, it would be preferable to warp the stop consonant less than the following vowel. In general, we would like to have a way to optimally warp the matching process. This can be done by using a nonlinear or dynamic time warp process (illustrated for a single case in Fig. 24.3).

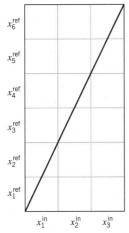

**FIGURE 24.2** Linear time normalization for the case in which the reference template has twice as many frames as the input template.

## 24.2.2 Dynamic Time Warp

***General Description*** As noted in the previous section, a linear time warping of acoustic templates does not properly compensate for the different compression–expansion factors that are observed for fast or slow instances of different phonetic classes. Therefore, it would be desirable to compensate for this variability in a *dynamic* manner, that is, finding the best possible warping of the time axis for one or both sequences that are being compared.

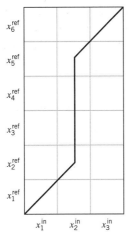

**FIGURE 24.3** Nonlinear time normalization for the case in which the reference template has twice as many frames as the input template, and the reference frame is most expanded in frames 2–5.

Vintsyuk [8] may have been the first to observe that the dynamic programming principle of Bellman [1] could be applied to this problem; however, the idea was not widely known elsewhere until the mid-1970s.

Dynamic programming is an approach to the implicit storage of all possible solutions to a problem requiring the minimization of a global error criterion. It is formulated as a sequential optimization strategy in which the current estimate of the global error function is updated for each possible step. At each step, enough information about the plausible hypotheses are retained so that at the end, when the best global error value is found, the corresponding set of choices that correspond to this value can also be discovered. A classic example of this (see, for instance, [7]), is the knapsack problem. In this problem, we wish to stuff items of $N$ different types with varying sizes and values into a knapsack that has a capacity of $M$ (in the units of size), such that we maximize the value of the take. In this case, the dynamic programming solution is an efficient approach to computing the best combination for all knapsack sizes up to $M$. Thus, the problem for $M = 10$ is solved by noting that the set of possible solutions to the $M = 9$ problem can be considered along with the possible items that could be added at that point, and so on. This can be implemented by explicitly or implicitly building up a table of cumulative costs. In addition to the costs, the local information of how each possible solution was achieved is stored; in the case of the knapsack, for each value of $M$ and each item that could have been added to complete that solution, the identity of the item is stored. Thus, once the best total value is found for $M = 10$, we can look up what was added to achieve that best number, and we can backtrack to find the sequence of items added to the knapsack to achieve the best total value.

Applied to template matching for speech recognition, this algorithm can be stated fairly simply. Imagine a matrix $D$ in which the rows correspond to frames of a reference template and the columns correspond to an input template. For each matrix element in $D$ we will define a cumulative distortion measure,

$$D(i, j) = d(i, j) + \min_{p(i,j)} \{D[p(i, j)] + T[(i, j), \; p(i, j)]\}, \qquad \textbf{(24.2)}$$

where $d$ is a local distance measure between frame $i$ of the input and frame $j$ of the reference template, and where $p(i, j)$ is the set of possible predecessors to $i, j$; in other words, the coordinates of the possible previous points on the matching trajectory between the two templates. The $T(\;)$ is a term for the cost associated with any particular transition. Thus, each matrix element is the value of the total error that arises from the best step that could lead to associating those two frames, and since this step is made after a similar optimal decision, the best cumulative distance in the final column (corresponding to the last frame in the input template) will be the distortion corresponding to the best match between reference and input. For isolated word recognition, the reference with the lowest value will be taken as the best match. The basic computational step is illustrated in Fig. 24.4.

Thus, the algorithm consists of the following steps.

1. Compute the local distance for each element in column 1 of each distortion matrix (that is, the distance between frame 1 of the input and all the frames of each reference template). Call this the cumulative distortion value for that matrix element.

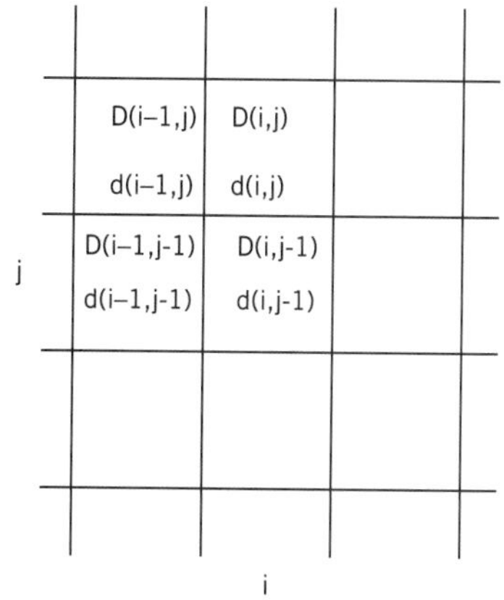

**FIGURE 24.4**  Basic DTW step for the case of simple local constraints. Each $i, j$ box is associated with a local distance $d$ and a cumulative distortion $D$; the equation at the bottom shows the basic computational step.

2. Starting with frame 2 of the input, and beginning with the bottom row (frame 1 of the first reference template), compute the local distance and add it to the best cumulative distortion value for all possible predecessors (that is, all possible matches between input and reference that could temporally precede either the current input frame or the current reference frame). Compute this value for each element in the column for each reference template.

3. Continue this operation through each of the other columns.

4. Find the best distortion number in the last column for each reference template and declare it the distortion associated with that reference.

5. Choose the word associated with the best of the reference distortions and declare it the winner.

Since this algorithm applies a dynamic programming approach to the time warp problem, it is often referred to as a dynamic time warp, or DTW.

**Global Constraints**    Although the algorithm just described gives the basic approach, there are modifications to the approach that are used in nearly all real implementations. For instance, it is a waste of computation to compare the first input frame with the last reference frame. Figure 24.5 shows a plausible set of constraints to reduce the search space.

**Local Constraints**    Similarly, the possible predecessors are limited to a few nearby matrix elements. Figure 24.6 shows a common constraint for DTW problems in which the local warping is constrained to skip no frames of either the reference or input template, but to permit repeats of either one.

Thus, for this case, the minimum of Eq. 24.2 is computed over only three predecessors.

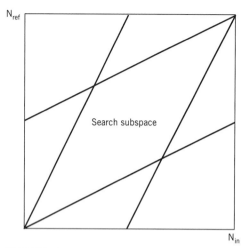

**FIGURE 24.5**  Typical global slope constraints for dynamic programming. The search-subspace section is the only region searched, since matrix elements outside of this space are considered to be unlikely matches.

***Template Clustering***    As we have described the algorithm here, the input template is compared with each reference template. However, we have left open how many reference templates will be used for each word. In general, a single template will be insufficient, given the variability inherent to speech. For speaker-dependent recognition, it is often sufficient to store a small number of reference templates for each word in the vocabulary. For speaker-independent training, many more examples are generally required. In either case, the multiple examples are often replaced with a smaller number of representative prototypes

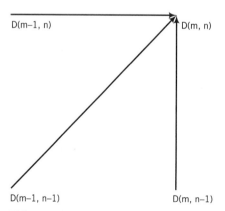

**FIGURE 24.6**  Typical local constraints for dynamic programming; *m* is the index for the frame of the speech input, and *n* is the frame for the reference template. $D(i, j)$ is the minimum distortion value for a match going through input frame *i* and reference frame *j*.

that are obtained by clustering the raw examples. Many different clustering approaches can be used, including the K-means algorithm in which cluster means are recomputed from new clusters, cluster memberships are determined for each point from the closest cluster mean, and so on. In this case, however, the distance is between two sequences (templates consisting of sequences of vectors), and the distance is taken to be the DTW distortion between the two templates. With this metric, a large number of examples for each word can be clustered into one or more prototypes that can then be used to represent the word.

### 24.2.3 Distances

In Chapters 8 and 9 we saw that a poor distance measure could lead to a bad classification performance; for instance, in the example of Chapter 8, using a simple unweighted Euclidean distance for height–weight pairs (in feet and pounds, respectively) gives too much emphasis on the weight. We saw that a reasonable strategy was to scale the features so that they covered roughly the same range; in Chapter 9 we looked at this problem from a statistical standpoint, and the Gaussian model normalized by the standard deviations for each feature. However, we noted then that even this approach did not necessarily lead to an optimal distance measure, since it did not account for the effectiveness of each feature for classification. A linear discriminant analysis was mentioned as a solution, but even in this case there is an implicit assumption about the distance between points in feature space.

Many researchers have worked on plausible approaches to measures that corresponded well to human perceptual distance. Some of these led to the kinds of feature measures we have already discussed. For instance, we mentioned in Chapter 22 the liftering that is often applied to cepstral values that result from LPC, mel-cepstral analysis, or PLP. Viewed from the perspective of perceptual distances, the design of features such as index-weighted PLP cepstra is in fact focused on generating measures that can be effectively compared with a Euclidean distance.

Once such modifications are used in the front end, however, how can we know what distance measure would be optimal? In general we must establish some mathematical framework and allow an automatic procedure to learn the best measure for distance. Using a statistical framework gives us powerful mathematical tools and a definition for optimality that corresponds well with the goals of classification; for this reason, nearly all modern systems represent distance in probabilistic terms. This will be the focus of the chapters that follow.

### 24.2.4 End-Point Detection

Because accuracy is often higher for an isolated word system,[4] pauses between words are often mandatory in commercial products (particularly for large-vocabulary dictation). For template-based systems such as the one described in this chapter, one of the main limitations to recognition accuracy has been the segmentation between the desired speech sequences and nonspeech segments; in other words, to locate the end points of the speech so that the

---

[4]Although it is generally true that the best state-of-the-art recognizer for the same vocabulary, environment, and so on is more accurate if trained for the isolated word case, it should also be noted that systems that have been designed and trained for continuous speech input can also have difficulties with isolated words.

templates can be formed. The reader may find this statement odd; after all, shouldn't it be easier to do a speech–nonspeech classification than to distinguish between different speech classes? However, it has often been observed that the former discrimination has been the source of most of the errors in such systems. One reason for this is that the vocabularies for artificial systems are typically chosen to be as orthogonal as possible – that is, most easily discriminable. In a case in which the words are necessarily similar, as in the alphabet, frequent users generally modify the vocabulary for better discrimination, as in "alpha bravo charlie." In contrast, the competing nonspeech sounds have no such characteristic and can often be confused with speech; for instance, the short-term spectrum of a breath may be confused with that of a fricative sound.

The example in Fig. 24.7 shows a click preceding the beginning of a spoken word. The click has sufficient energy that an energy threshold alone would be insufficient to screen out the click while still preserving the beginning of the speech segment. It is common to include time thresholds in the determination; for instance, in this figure the length of time

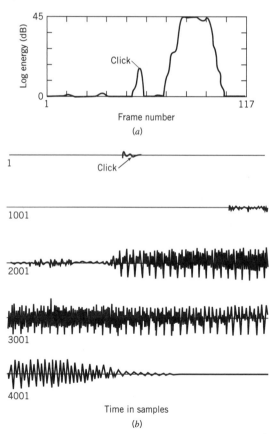

**FIGURE 24.7**    Bell Labs example of a mouth click preceding a spoken word. From [5].

between the click and the start of speech is so great that the probability of the two being part of the same word is very low. Such detection schemes have been built and often have reasonable accuracies under the conditions that the thresholds have been optimized for, but they can be troublesome as conditions change.

## 24.3  CONNECTED WORD RECOGNITION

Thus far, we have restricted the discussion to the recognition of individual words, uttered in isolation. How does the problem change if the words are uttered without pauses between them? Of course the pronunciation of each word can be dramatically altered, as we have noted earlier. However, even if we ignore this phenomenon, there are strong practical differences from the isolated word case. In principle, one could still warp the input against templates for every possible word sequence. However, in general this number is far too large – both storage requirements and the comparison time would be too great (infinite in the general case).

Therefore, in addition to the time normalization and recognition goals of isolated word DTW, connected word recognition requires segmentation into the separate words so that they can be time normalized and recognized. Here too, dynamic programming techniques can be used to search efficiently through all possible segmentations. Vintsyuk also did early work on this problem, as did Sakoe, as well as Bridle. Sakoe's algorithm [6] required two passes, one for computing distances and a second for assembling the best word sequences given these distances. Both Vintsyuk and Bridle developed one-stage approaches, and the algorithm is described in detail in [4]. Here we only briefly mention a few key features.

The basic principle of connected word DTW is (conceptually) to assemble one large matrix consisting of all the word templates, and to do dynamic programming on the whole thing a column at a time, as with the isolated word case. At word boundaries the local constraints are different (since the preceding frame can come from other words), but otherwise the basic distortion computation step is the same. However, at the end, it is necessary to backtrack from the best cumulative distortion in the last column in order to find what was said. In principle this could be done by storing a second matrix that holds the pointer to the best preceding matrix element; this can be followed back from the last column to reveal the complete warping function, even across words. Figure 24.8 shows a typical warping path that could be tracked in such a procedure.

As Ney points out in [4], the main problem with this simplified description is that it too leads to an unreasonable amount of storage requirements, particularly for large-vocabulary cases. Given a 20,000-word vocabulary, for instance, and 50 frames/word on the average, both the distortion and backtracking matrices would have $10^{12}$ elements. Fortunately, it turns out that some small complications to the algorithm provide enormous savings in storage. The distortion matrix can be reduced to storing two columns, since for most implementations the legal predecessors in the warp come from at most one frame into the past. The backtracking matrix can be replaced with two lists that are each the length of the utterance (in frames):   a "from template" list that gives the template index of the word with the lowest cost that ends

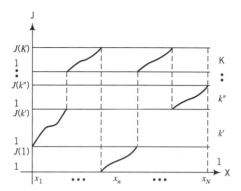

**FIGURE 24.8** Example of a DTW path for connected words. In this case, the best path was found to correspond to words $k'$, $K$, 1, $K$, and $k''$.

at a particular frame, and a "from frame" list that points to the end frame of the previous word, given that the current frame is the end frame of the current word. These lists, which we can represent as $T(n)$ and $F(n)$ respectively, can be generated on line as the DTW algorithm proceeds. To backtrack, the following procedure can be followed.

1. Let $j$ be the frame pointer, and initially let $j$ be $N$, the number of frames in the utterance. Then the word with the lowest-cost ending in the last frame would have index $T(N)$. In Fig. 24.8, this would be word $k''$.

2. Point to the last frame in the previous word using the "from frame" list $F(\ )$. In terms of the stored arrays, this value would be $F[T(j)]$; as noted above, $j$ would initially be equal to $N$. In the figure, $F[T(j)]$ would be the time index of the frame before the first frame of the best warp for word $k''$.

3. Set $j$ to the value of $F[T(j)]$. Then looking at the "from template" list $T(\ )$, one finds that the word ending in frame $j$ would be $T(j)$. In the figure, this would be word $K$.

4. Repeat the last two steps until the entire word sequence has been noted. This sequence will be associated with the optimal path through the match between the input template and all possible combinations of the reference templates.

The preceding is a simplified description of DTW. In particular, we have not discussed additive transition costs, multiplicative weightings of particular transition slopes, and word-sequence restrictions that are sometimes provided by a grammar.

## 24.4 SEGMENTAL APPROACHES

Thus far in this chapter, the only linguistic units used have been words; dynamic programming was used to find the best match to reference templates associated with candidate words. Thus, the local (per frame) distances were not associated with any linguistically defined class other than the word. However, as noted in Chapter 23, words typically are structured items, consisting of subword units that are common across many words. In principle this structure can and should be used to improve the effectiveness of any finite amount of training data.

For framewise dynamic-programming-based speech recognition, such subword units have primarily been used for statistical systems, and so we will defer their discussion to the next two chapters. However, as noted in Chapter 4, in the 1970s and 1980s a number of systems were built that made extensive use of acoustic–phonetic knowledge, some of which used statistical classification and some of which used deterministic approaches. In either case, subword units were used.

Probably the best known systems in this class were the efforts of Ron Cole's group (first at CMU [2] and more recently at OGI [3]), and related work at MIT by Victor Zue and his colleagues [9]. These systems incorporated explicit acoustic–phonetic knowledge in order to segment the speech separately into phonetic segments, and then classify them. Time normalization was thus accomplished by combining costs at a segment level rather than on a framewise basis, and phones or phonemes were the classes to be identified. By incorporation of smaller units than words, learned parameters were shared across many words. By incorporating explicit decision rules for specific phonetic examples (sometimes deterministic, sometimes statistical), the designers were not limited to defining a single distance metric for all kinds of decisions.

A prototypical example of this approach was the FEATURE system developed at CMU [2]. This system used deterministic measures for the phonetic segmentation, and then it classified the segments either with a linear discriminant function or with a Gaussian classifier.

## 24.5  DISCUSSION

The incorporation of dynamic programming into speech-recognition systems became standard by the mid 1980s, though it was used in many research systems in the 1970s. The notion of a time-flexible distance between sequences was a critical one, and essentially all systems incorporate it in some fashion now. The specific mechanism of acoustic template matching, however, is no longer commonly used, except perhaps in some voice dialing systems. It has been shown to be useful to incorporate at least some lower-level structure, either from linguistic knowledge or from some self-organized process, so that some kind of subword units are incorporated in nearly all systems.

As noted previously, a more flexible form of distance design has been developed in the form of statistical learning procedures. Distributions are estimated in a training process and then used in an optimal decision procedure during recognition. However, both training and recognition still depend on many assumptions, and the best choice of these assumptions still largely depends on designer intuition. It is often useful to consider speech recognition from the standpoint of the earlier systems described in this chapter in order to develop the insight that is necessary for these decisions.

## 24.6 EXERCISES

**24.1** As noted here and in [4], connected word recognition can be done by using only two columns from the distance–distortion matrix (in addition to the frame and template pointer lists): the current column and the previous one. Suppose that the local slope constraint prohibits entirely vertical paths; in other words, input frames cannot be repeated in the time warp. Given this restriction, how could you modify the algorithm to use only one distance–distortion column?

**24.2** Imagine an $F_1$–$F_2$ graph, and examples of formant values in frames for the phones /i/ and /u/ represented by points in the graph. In this framework, explain how discrimination from frames of nonspeech could be more difficult than discrimination between the two phone types.

## BIBLIOGRAPHY

1. Bellman, R., and Dreyfus, S., *Applied Dynamic Programming*, Princeton Univ. Press, Princeton, N.J., 1962.
2. Cole, R., Stern, R., Phillips, M., Brill, S., Specker, P., and Pilant, A., "Feature-based speaker independent recognition of english letters," in *Proc. IEEE Int. Conf. on Acoust. Speech Signal Process.*, Minneapolis, Minn., pp. 731–733, 1993.
3. Janseen, R. D. T., Fanty, M., and Cole, R. A., "Speaker independent phonetic classification in continuous English letters," in *Proc. Int. Joint Conf. Neural Net.*, Seattle, pp. II-801–808, 1991.
4. Ney, H., "The use of a one-stage dynamic programming algorithm for connected word recognition," *IEEE Trans. Acoust. Speech Signal Process.* **32**: 263–271, 1984.
5. Rabiner, L., and Juang, B.-H., *Fundamentals of Speech Recognition*, Prentice–Hall, Englewood Cliffs, N.J., 1993.
6. Sakoe, H., and Chiba, S., "Dynamic programming algorithm optimization for spoken word recognition," *IEEE Trans. Acoust. Speech Signal Process.* **26**: 43–49, 1978.
7. Sedgewick, R., *Algorithms*, Addison–Wesley, Reading, Mass., 1983.
8. Vintsyuk, T., "Speech discrimination by dynamic programming," *Kibernetika* **4**: 81–88, 1968.
9. Zue, V., "The use of speech knowledge in automatic speech recognition," *Proc. IEEE* **73**: 1602–1615, 1985.

# CHAPTER 25

# *STATISTICAL SEQUENCE RECOGNITION*

## 25.1 INTRODUCTION

In Chapter 24 we showed how temporal integration of local distances between acoustic frames could be accomplished efficiently by dynamic programming. This approach not only integrates the matches between incoming speech and representations of the speech used in training, but it also normalizes time variations for speech sounds. In the case of continuous speech, this approach also effectively segments the speech as part of the recognition search, without the need for any explicit segmentation stage. Distances can also be modified to reflect the relative significance of different signal properties for classification.

However, there are a number of limitations to the DTW-based sequence-recognition approaches described in the last chapter. As noted previously, a comparison of templates requires end-point detection, which can be quite error prone with realistic acoustic conditions. Although in principle distances can be computed to correspond to any optimization criterion, without a strong mathematical structure it is difficult to show the effect on global error for an arbitrary local distance criterion. Since continuous speech is more than just a concatenation of individual linguistic elements (e.g., words or phones), we need a mechanism to represent the dependencies of each sound or category on the neighboring context. More generally, as noted in Chapter 9, statistical distributions are a reasonable way to formally represent the variability that is observed in real speech samples.

For all of these reasons, statistical models are extremely useful for sequence recognition, particularly for speech. The use of a statistical framework provides a powerful set of tools for density estimation, training data alignment, silence detection, and in general for the training and recognition of isolated words or continuous speech. These mathematical tools have now become so widely used that they are more or less the standard methods for nearly all speech-recognition systems. Even research systems that do not strictly follow the approach given here tend to describe their algorithms in terms of the common statistical paradigm. It could also be argued that the use of statistical pattern recognition increases the generality of the methods; since the choice of any distance is equivalent to some implicit statistical assumption, one may as well directly represent the distance in terms of statistical optimality.

In statistical speech recognition, as in the deterministic case, there is a strong temporal component; in particular, during recognition a local distance is integrated over time in a way that provides some normalization for temporal variability. For most statistical approaches, speech is represented as having been generated according to some probability distributions. Since there are different speech sounds, it is necessary to assume that there is more than one

**337**

distribution; therefore, we construct models for linguistic units, each of which is associated with some statistical parameters related to the distributions. In a training process, we will estimate these parameters by choosing them to approximate a minimum probability of error (Bayes) solution. During recognition, we will search through the space of hypothesized utterances to find the hypothesis that has the maximum *a posteriori* probability. For practical reasons, we will need to incorporate some approximations, and this will make the results suboptimal. Furthermore, only estimates of the probabilities and density functions will be available, so we will never reach the true error minimum. In particular, any finite training set will be an imperfect representation of an independent test set; the latter might, for instance, have some form of acoustic noise that was not present in the training set. Thus, deterministic aspects (such as using features that are robust to spectral slope) are still important, even in a statistical system.

## 25.2 STATING THE PROBLEM

We will begin with a restatement of the Bayes rule:

$$P(M_j \mid X) = \frac{P(X \mid M_j)P(M_j)}{P(X)}, \tag{25.1}$$

where, in this case, the class $M_j$ is the $j$th statistical model for a sequence, where $0 \le j \le J$, and $X$ is the observable evidence of that sequence. As in the preceding chapter, $X$ will typically consist of cepstral vectors computed with auditory-warped spectral estimates from succeeding acoustic windows.

According to the Bayes decision rule, the minimum probability of error (of classifying $X$ into the correct category $M_c$) is attained if one always assigns $X$ to that model with the maximum $P(M_j \mid X)$. If we had many examples of each model, one could learn the parameters for some pattern classifier from the examples in the training data (as in the static pattern classification problems in Chapters 8 and 9) and apply them during recognition for each sequence. More generally, it is often necessary to break up the models into submodels that we learn to classify with some distortion so that training data is shared over the models; for instance, rather than training models for complete sentences, we tend to learn to represent words, syllables, or phones, and we put them together to make models for the utterances. Thus, when we learn the statistics of the sound [ae], we can apply them both to the representation of pat and bat.

Let's begin with the simplest case, in which the candidate models $M_j$ correspond to words, and in which we will only do isolated word recognition. In this case, it would be desirable to design a classifier that would look at a sequence of cepstral vectors (essentially the input template of Chapter 24) and categorize it as one of $J$ words. If all of the examples (both in training and test) were of exactly the same length, one could concatenate all cepstral vectors into one large vector and design a static pattern classifier by any of the approaches discussed in Chapters 8 and 9. Aside from not being generalizable to more realistic cases, though, even this idealized problem would not take advantage of any structure within the concatenated vector.

When the Bayes rule is applied to the recognition problem, we should really incorporate the dependence on the parameter set $\Theta$ that was learned during training, so that we get:

$$P(M_j \mid X, \Theta) = \frac{P(X \mid M_j, \Theta)P(M_j \mid \Theta)}{P(X \mid \Theta)}. \tag{25.2}$$

Now let us consider the speech-recognition problem per se. Ordinarily, we will have some ability to predict the probability of a word sequence even without any acoustic evidence. In the degenerate case this just consists of a uniform probability distribution over all possible utterances, but more generally there is some preference for some utterances over others. In more complex cases this preference may come from pragmatic factors (e.g., during a conversation about the weather you are less likely to discuss brain surgery), but the simplest preconceptions about the word sequence come from the statistical distributions of groups of words. In either case, we can presume that there will be some estimator available for $P(M_j)$. If we can assume that the components in $\Theta$ used for estimating $P(M_j)$ are independent of the components associated with estimating $P(X \mid M_j)$, we can separately estimate the two. Further using the fact that $P(X \mid \Theta)$ is fixed for all choices of $j$, we can state that an optimal decision rule for choosing the stochastic model $M_j$ that will lead to the minimum probability of error is

$$j_{\text{best}} = \operatorname*{argmax}_j P(M_j \mid X, \Theta) = \operatorname*{argmax}_j P(X \mid M_j, \Theta_A)P(M_j \mid \Theta_L), \tag{25.3}$$

where $\Theta_A$ is a set of parameters that have been learned for the acoustic model (representing the statistical relationship between the model and the observed sequence of feature vectors), and $\Theta_L$ is the set of parameters that have been learned for the language model (parametrizing the statistical distribution of all word sequences).

Figure 25.1 shows the general scheme, with the dependence on parameters suppressed.

If we limit our discussion to the acoustic model (and assume for the moment that the language model is perfect), then the optimum decision rule shows us that we wish to associate a large likelihood $P(X \mid M_j, \Theta_A)$ with $j_{\text{best}}$, and smaller likelihoods with the other acoustic models. More generally, there are three problems that must be addressed with the use of statistical models for speech recognition.[1]

**1.** Parametrization and probability estimation: How should $P(X \mid M_j, \Theta_A)$ be computed, and what are the necessary assumptions about the stochastic models to define the parameter set $\Theta_A$? In general it is not possible to estimate the probability density of a complete sequence without constraining assumptions; typically these assumptions involve temporal independence at some level so that the total estimate is broken up into constituents that can be effectively estimated. The rest of this chapter addresses the integration of these components into the likelihoods of acoustic sequences; the discussion of the estimation of the temporally local components will be postponed until the next chapter.

---

[1]This categorization of HMM problems was used in [1], which in turn was adapted from the discussion by Rabiner in such sources as [3] and [4].

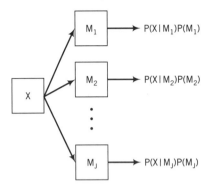

**FIGURE 25.1**   The Bayes decision rule classification. For the discussion in this chapter, the observations to be classified are a sequence of feature vectors $X$, corresponding to a speech utterance. The classes correspond to statistical models $M$, and the minimum probability of error is obtained by choosing the model with the largest product term on the right.

**2.** Training:   Given the parametrization above, and a set of training sequences associated with stochastic models, how should $\Theta_A$ be determined so that each model has the highest likelihood associated with its observed sequence of feature vectors? This will be discussed in Chapters 26 and 27.

**3.** Decoding:   Given the set of stochastic models $M_j$ with their trained parameters $\Theta_A$, how should the best sequence of these models be found to classify the input sequence $X$ according to the Bayes decision rule? We give a simplified answer to this question in this chapter, but we will give a further discussion in Chapter 28, when we discuss complete systems.

See [4] or [2] for extended discussions of all three problems.

## 25.3   PARAMETRIZATION AND PROBABILITY ESTIMATION

Even for the case of isolated words, the statistical models used must represent both temporal variability and temporal structure. A likely structure for this purpose is a stochastic finite-state automaton, which can be used as a model for a speech unit (e.g., a phone or word). The model will consist of some states, a topology of connection between them, and some associated parameters. The parameters will be learned from examples in a training phase and then will be incorporated during recognition. The automata most commonly used for speech recognition are generative models. That is, the states have outputs (rather than inputs), which are the observed feature vectors. The probability densities associated with each acoustic model are probability density functions for the generation of observed feature vectors in a sequence, conditioned on that model, as suggested by Fig. 25.2. For instance, one could imagine a density function for PLP cepstra produced by the model for the word "one." However, for the many reasons mentioned earlier, it is commonly preferable to break up the density for a complete sequence into a combination of

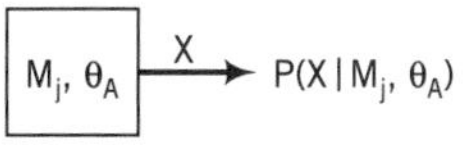

$$M_j, \theta_A \xrightarrow{\ X\ } P(X \mid M_j, \theta_A)$$

**FIGURE 25.2**  A generative model with its parameters produces an observation sequence according to some statistical distribution. To concatenate such models to generate concatenated sequences, we typically make Markov assumptions.

densities corresponding to subsequences. This is typically accomplished by using a Markov assumption.[2]

With the use of Markov assumptions, the general stochastic automata that we referred to here become Markov models. Hidden Markov models are stochastic automata that have an additional layer of indeterminacy, which we describe later in this chapter. However, we begin with an explanation of a simple Markov model (again borrowing from [4]) so that we can see the general form of such structures and how they can be used to represent the probability of sequences.

## 25.3.1  Markov Models

A Markov model is a finite-state automaton with stochastic transitions (that is, for which each transition has an associated probability) in which the sequence of states is a Markov chain. The states in each model $M$ will be designated $q_\ell [\ell = 1, \ldots, L(M)]$. Each of these states is associated with a class $\omega(q_\ell)$.

Consider the three-state Markov model shown in Fig. 25.3, where each state corresponds to one day's weather (assuming state residency is per day).

This model has a stochastic transition matrix:

$$A = \begin{bmatrix} \frac{1}{3} & \frac{1}{3} & \frac{1}{3} \\ \frac{1}{4} & \frac{1}{2} & \frac{1}{4} \\ \frac{1}{4} & \frac{1}{4} & \frac{1}{2} \end{bmatrix}.$$

By the definitions of joint and conditional probability,[3] the probability of any sequence $Q = (q^1, q^2, q^3, \ldots, q^N)$ is

$$P(Q) = P(q^1) \prod_{i=2}^{N} P(q^i \mid q^{i-1}, q^{i-2}, \ldots, q^1). \tag{25.4}$$

We typically assume that the probability of any state transition is independent of any previous transitions (that is, that the probability of moving to a particular state depends only on the identity of the previous state); this is a first-order Markov assumption. In this case,

---

[2]Recall that an $n$th-order Markov chain is a sequence of discrete random variables that depends only on the preceding $n$ variables. The value for $n$ is typically one (a first-order system) for speech recognition.

[3]See Chapter 9 for a reminder.

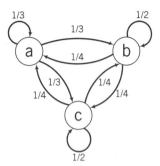

| state | | output |
|-------|-----|--------|
| a | ⟷ | sunny |
| b | ⟷ | cloudy |
| c | ⟷ | rainy |

**FIGURE 25.3**   Three-state Markov model for the weather (sunny, cloudy, or rainy). Transition probabilities are given as fractions. State residency corresponds to a day.

the previous equation reduces to

$$P(Q) = P(q^1) \prod_{i=2}^{N} P(q^i \mid q^{i-1}).$$     **(25.5)**

Thus, given this Markov property, the probability of any particular sequence can be found by multiplying the probability of the individual transitions in the state sequence.

For the weather model, we will assume (for simplicity's sake) that all initial sequences start with "a" (the first day considered is always sunny).

Then:

## EXAMPLE 25.1

The sequence "abc" occurs with a probability of $1/3 \times 1/4 = 1/12$.   ■

Further, the probability of a set of sequences (for instance, the set of all sequences of a certain length with specified start and end states) is just the sum of probabilities of all sequences within the set.

Then:

## EXAMPLE 25.2

The probability of observing the sequence "axc", where x could be anything, can be calculated as is shown in Table 25.1. If states "a" "b" and "c" corresponded to speech phones, as opposed to states of the weather, this second example could be viewed as giving the prior probability of any sequence of length three that started with phone "a" and ended with phone "c." This property will be useful later; however, it is also necessary to associate acoustic observations with the statistical models. For this, we will need to define hidden Markov models.   ■

**TABLE 25.1 Computation of the Probability of Observing Sequence "axc"**

| |
|---|
| $P(a_1 \rightarrow b_2, b_2 \rightarrow c_3) = 1/3 \times 1/4 = 1/12$ |
| $P(a_1 \rightarrow a_2, a_2 \rightarrow c_3) = 1/3 \times 1/4 = 1/12$ |
| $P(a_1 \rightarrow c_2, c_2 \rightarrow c_3) = 1/4 \times 1/2 = 1/8$ |
| $P(a_1 \rightarrow x, x \rightarrow c_3) \qquad\qquad = 7/24$ |

### 25.3.2 Hidden Markov Model

Each output of a Markov model corresponds to a deterministic event, whereas each output of a hidden Markov model (HMM) corresponds to a probabilistic density function; thus, for any observed output sequence, the generating state sequence of a HMM is hidden. As an example, imagine that there is a line of people behind a screen and the people in it are a random mix of Swedish basketball players and speech-recognition researchers. We can't see the people in the line, but each one utters a deep sigh, presumably for being in such a long line (see Fig. 25.4). Now, since (as noted in Chapter 8) the basketball players tend to be larger, their sighs are produced by a larger vocal apparatus; the associated resonances will tend to be lower in frequency. However, there will be a lot of within-occupation variability in the spectra; that is, even when one analyzes spectra only from the basketball players,

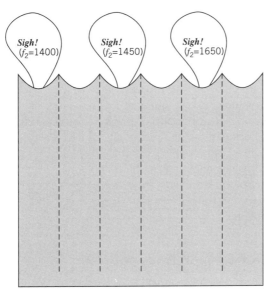

**FIGURE 25.4** The screen hides the line of basketball players and ASR researchers, but we hear their sighs.

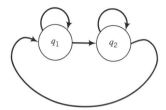

**FIGURE 25.5** Two-state HMM, with transitions permitted from each state to itself and to the other. Observations are presumed to be generated by this model, with associated density functions $P(x|q)$ for each state. Each transition is stochastic with conditional probabilities $P(q_i|q_j)$.

the spectra will still be highly variable from person to person. In the terminology of this chapter, then, the "state" refers to the job category of the hidden person (basketball player or researcher). The output associated with the state is an acoustic observation of a sigh. If we only hear the sighs and don't observe the speakers, then the job categories form a sequence of hidden variables, whereas the acoustic emissions (the sighs) are directly observed. The hidden sequence is presumed to be generated according to some distribution, and for each choice of state an acoustic signal is generated according to another distribution.

Figure 25.5 shows a two-state HMM that could represent the line-of-sighs example. State $q_1$ could correspond to a Swedish basketball player, and state $q_2$ could represent a speech researcher. Each state is associated with a density function for the emitted acoustics (where, for instance, the acoustic representation could be the average second-formant value for the high-energy portions of the sigh). In the limited (but common) case in which this density function is assumed to be independent of previous states or observations, it can be written as $P(x_n|q_i)$. The density $P(x_n|q_1)$ would then be the distribution of average formant values for a sigh from a basketball player, and $P(x_n|q_2)$ would be the distribution of average formant values for ASR researchers' sighs.

It is also commonly assumed that the sequence is a first-order Markov chain, so that the probability for a transition from state $i$ to state $j$ can be written as $P(q_j|q_i)$, where the special case of $i = j$ is referred to as a self-loop.

More formally, as with Markov models, the states in each model $M$ will be designated $q_\ell[\ell = 1, \dots, L(M)]$. Each of these states is associated with a class $\omega(q_\ell)$. Each emission probability $P(x_n|q_i)$ will actually refer to the probability of the class $\omega(q_i)$. This distinction is important since the $i$th state might not be the $i$th class; in particular, different states in the same model could be of the same class, and different incidences of the same class could occur in different models. For simplicity's sake we generally ignore the class designation in our probabilistic notation.

## 25.3.3   HMMs for Speech Recognition

As noted in Chapter 24, speech is commonly represented as a sequence of measurements that are computed over a 20-ms or 30-ms window, with a typical interframe step size of 10 ms. HMMs for speech recognition, then, are typically an interconnected group of states that are assumed to emit a new feature vector for each frame according to an emission probability density function associated with that state. Each new observation frame can in principle be associated with any state (as any frame of an input template can be associated with any frame of a reference template, as noted in Chapter 24). However, the topology of the HMM (that is, the pattern of interstate connections) and the associated transition probabilities provide temporal constraints; for instance, a high self-loop probability for a state means that it will tend to be repeated more than once, and a directed connection between [ae] and [m] in a word model means that pronunciations with [ae] followed by [m] is permitted. These models are also typically constrained to be left to right; that is, the state transitions have some temporal order.

For simplicity's sake, in most of our discussions we assume word models that associate states with subword sounds that are roughly phonelike. However, it is common that each phone is modeled by several states with their own densities, and that each subphone is modeled in context (that is, that a state represents a subphone under some contextual constraints). For instance, a state might represent the initial part of an [ae] sound for those cases in which the previous sound was a nasal. As we shall see in later chapters, however, states will often ultimately represent self-organized classes that may have something to do with linguistic categories, but that are ultimately defined implicitly by an iterative statistical procedure.

## 25.3.4 Estimation of $P(X|M)$

Given a model $M_i$ and its definition (in terms of states, transition topology, probabilities, and parameters for the estimation of probabilities), we can then consider the estimation of the likelihood of a data sequence, assuming that it was generated by this model. Initially we make first-order Markov assumptions as suggested earlier, and we ignore any details or difficulties with the estimation of the local probabilities or likelihoods. We just assume that we are able to compute the likelihoods $P(x_n | q_k)$ for each observation vector $n$ and each model state $q_k$.[4] Further, we assume that each transition probability $P(q_j | q_i)$ is known.

Given these assumptions, we can show that we can get an estimate of the data sequence likelihood from the local likelihoods, using an efficient set of recursions that can be implemented in a way that is reminiscent of the computations performed in template matching for isolated word recognition.

***Total Likelihood Estimate***   Let us assume that we can compute the likelihood of a data sequence, assuming a particular path through the model; let us call the state sequence associated with that path $Q_i$, one of $I$ paths that are of length $N$ (the data sequence length) and that are permitted in the model $M$. The joint density for $X$ and $Q_i$ could in principle be estimated by multiplying together all of the local transition probabilities and acoustic likelihoods associated with that sequence; in other words, the product would have an emission term $P(x_n | q_k)$ for each frame $n$ and another term $P(q_j | q_i)$ for each transition from $i$ to $j$. These values for each legal sequence $Q_i$ in $M$ could then be summed up to get

---

[4]Chapters 9 and 26 give a number of methods for determining these local probabilities.

the complete data likelihood:

$$P(X \mid M) = \sum_{\text{all } Q_i \text{ in } M \text{ of length } N} P(Q_i, X \mid M).$$ (25.6)

Though straightforward, this procedure requires roughly $2N \times L(M)^N$ arithmetic steps, where $L(M)$ is the number of emitting states of the model; there are $L(M)^N$ possible state sequences and each state sequence requires approximately $2N$ calculations.

Fortunately, a more efficient algorithm exists, and it is called the forward procedure.[5] For this approach, the likelihood $P(X \mid M)$ can be decomposed into the sum of joint densities with the possible final states for the length $N$ sequence, or

$$P(X \mid M) = \sum_{l=1}^{L(M)} P\left(q_l^N, X_1^N \mid M\right)$$ (25.7)

where $L(M)$ is the number of states, $N$ is the length of the observed sequence in frames, $X_a^b$ is the sequence of observations from frame $a$ through frame $b$, and $q_\ell^n$ is the state $q_\ell$ at frame $n$.

Thus, to get the complete likelihood, we need to find the joint probability of the final state and all the data leading up to it. This may not seem like a very useful expression, but in fact we can decompose it further into the product of a local term and a cumulative term in order to build up a recursive estimation scheme. Factoring the expression $P(q_l^n, X_1^n \mid M)$ into two components (again using the definitions of joint and conditional probabilities),[6] we get the following relation, called the forward recurrence:

$$P\left(q_l^n, X_1^n \mid M\right) = \sum_{k=1}^{L(M)} P\left(q_k^{n-1}, X_1^{n-1} \mid M\right) P\left(q_l^n, x_n \mid zq_k^{n-1}, X_1^{n-1}, M\right).$$ (25.8)

We then define

$$\alpha_n(l \mid M) = P\left(q_l^n, X_1^n \mid M\right),$$ (25.9)

which is the probability of the joint event that the state at time $n$ is $q_l$, and that the sequence $X_1^n$ (from the beginning to frame $n$) is observed, given the model $M$. Then we get the forward recurrence:

$$\alpha_n(l \mid M) = \sum_{k=1}^{L(M)} \alpha_{n-1}(k) P\left(q_l^n, x_n \mid q_k^{n-1}, X_1^{n-1}, M\right).$$ (25.10)

When the recurrence reaches the final frame, Eq. 25.7 can be used to obtain the complete likelihood.

---

[5]There is an analogous procedure that is called backward, which will be discussed at a later point.
[6]Recall from Chapter 9 that $P(a, b \mid c) = P(a \mid b, c) P(b \mid c)$.

The right-hand side of Eq. 25.10 is composed of two parts: the previous value for the recursion sum for each predecessor state, and a new likelihood for the state and data given all the previous values. Thus, if we could estimate the latter conditional likelihood for each time point and state, we could get the complete likelihood in an efficient number of steps. Without any simplifying assumptions, the second term can also be decomposed as follows:

$$P\left(q_l^n, x_n \mid q_k^{n-1}, X_1^{n-1}, M\right) = P\left(q_l^n \mid q_k^{n-1}, X_1^{n-1}, M\right) P\left(x_n \mid q_l^n, q_k^{n-1}, X_1^{n-1}, M\right). \qquad (25.11)$$

Unfortunately, such densities are difficult to estimate. Therefore, further assumptions are required. We typically make the Markovian assumptions that we have referred to earlier, simplifying the two terms on the right-hand side of Eq. 25.11.

First, the state chain is assumed to be first-order Markov. As with the earlier examples, this means that the probability that the Markov chain is in state $q_l$ at time $n$ depends only on the state of the Markov chain at time $n-1$, and it is conditionally independent of the past. With this assumption, we transform

$$P\left(q_l^n \mid q_k^{n-1}, X_1^{n-1}, M\right) \rightarrow P\left(q_l^n \mid q_k^{n-1}, M\right).$$

Second, observations are assumed to be independent of past features and states. This means that the probability that a particular acoustic vector will be emitted at time $n$ depends only on the state at that time, and it is conditionally independent of the past:

$$P\left(x_n \mid q_l^n, q_k^{n-1}, X_1^{n-1}, M\right) \rightarrow P\left(x_n \mid q_l^n, M\right).$$

With these simplifications, Eq. 25.11 becomes

$$P\left(q_l^n, x_n \mid q_k^{n-1}, X_1^{n-1}, M\right) = P\left(q_l^n \mid q_k^{n-1}, M\right) P\left(x_n \mid q_l^n, M\right). \qquad (25.12)$$

Thus the forward recurrence becomes

$$\alpha_n(l \mid M) = \sum_{k=1}^{L(M)} \alpha_{n-1}(k \mid M) P\left(q_l^n \mid q_k^{n-1}, M\right) P\left(x_n \mid q_l^n, M\right). \qquad (25.13)$$

Thus, given the conditional independence assumptions described here, we can compute the likelihood for a data sequence being produced by a particular model, using a recursion that is only dependent on local emission and transition probabilities. Each step consists of computing a sum over all possible predecessor states of the product of three components: the local acoustic likelihood (for that state and that frame); the transition probability from the previous state; and the value of the $\alpha$ recursion for that previous state at the previous frame. The reader may find this computation somewhat familiar – it is similar to the dynamic programming for templates discussed in the previous chapter, except that

- each of the terms is probabilistic rather than deterministic,
- the hypothesized predecessors are model states rather than actual frames,

- the local constraints do not typically permit a repeat of the current frame to match the same state (all predecessors correspond to the previous frame),
- the local and global factors are combined by multiplication rather than summation, and
- the combination of terms over different predecessors is combined by summation rather than finding the maximum (or minimum in the case of distance).

The last two differences can actually be done away with in a further simplification that we discuss in the following paragraphs.

### Viterbi Approximation:   Estimation of Best Path

Within the assumptions described here, the forward recursion yields the complete likelihood; although more efficient, it is functionally equivalent to the direct summation of the likelihoods of all possible legal paths (length $N$ sequences of state transitions) within a model. However, the procedure can be tricky to use, since it requires both multiplication and addition of likelihoods and probabilities. If only multiplication were done, then the computation could be converted to the log domain to handle the wide range of values for the products. However, since there is also addition, the log probabilities must be exponentiated.[7]

Aside from these numerical difficulties when the complete likelihood is used; it is often useful to find the single best state sequence to explain the observation sequence. This simply requires modifying the forward recurrence by replacing all the summations with the max function. Hence, the best-path (or Viterbi) approximation to Eq. 25.8 is

$$\overline{P}(q_l^n, X_1^n \mid M) = \max_k \overline{P}(q_k^{n-1}, X_1^{n-1} \mid M) P(q_l^n, x_n \mid q_k^{n-1}, X_1^{n-1}, M). \tag{25.14}$$

With the independence and the first-order assumptions, it further reduces to

$$\overline{P}(q_l^n, X_1^n \mid M) = \max_k \left[\, \overline{P}(q_k^{n-1}, X_1^{n-1} \mid M) P(q_l^n \mid q_k^{n-1}, M)\right] P(x_n \mid q_l^n, M). \tag{25.15}$$

An equivalent form can be seen for the log domain:

$$\log \overline{P}(q_l^n, X_1^n \mid M) = \max_k \left[\, \log \overline{P}(q_k^{n-1}, X_1^{n-1} \mid M) + \log P(q_l^n \mid q_k^{n-1}, M)\right] + \log P(x_n \mid q_l^n, M), \tag{25.16}$$

or

$$-\log \overline{P}(q_l^n, X_1^n \mid M) = \min_k \left[-\log \overline{P}(q_k^{n-1}, X_1^{n-1} \mid M) - \log P(q_l^n \mid q_k^{n-1}, M)\right] - \log P(x_n \mid q_l^n, M). \tag{25.17}$$

---

[7]There are, however, clever interpolation schemes for approximating $z = x + y$ (or equivalently $e^{\log z} = e^{\log x} + e^{\log y}$) while keeping all values in the log domain.

In this form, the recursive step is extremely similar to the corresponding step in the deterministic dynamic time warp. Interpreting each negative log probability as a kind of distance (since large probabilities mean a better match), Eq. 25.17 essentially says that the global distance for a particular state and frame is the sum of a local (statistical) distance $[-\log P(x_n \mid q_l^n, M)]$ and the minimum over all possible predecessors of the sum of global distances and transition costs.[8] Given such an interpretation, the approaches for recognition discussed in Chapter 24 (discrete word recognition, one-pass continuous recognition) also apply here, with the exception that we now have a formulation for how to combine the acoustic information with any prior statistics about the word sequences.

## 25.4  CONCLUSION

In this chapter, we have presented the general notion of statistical models for speech recognition, and we have then specialized them to the kind of stochastic finite-state automata that we call hidden Markov models. We have further shown how simplifying assumptions can lead us to a set of recurrences that yield acoustic sequence likelihoods. These likelihoods can, in combination with prior probabilities for word sequences, lead to word sequence hypotheses that will yield the minimum probability of error (if the assumptions are correct; alas, they aren't).

Thus far, however, we have said little about how the local probabilities used in these recursions are estimated, nor how these estimators are trained. These will be the primary topics for the next chapter.

Finally, to derive efficient and simple recursions, we found it necessary to make a number of assumptions that, as noted earlier, are almost certainly not true (but that are useful nonetheless).

1. Language model parameters and acoustic model parameters are assumed to be completely separable; that is, the language model is independent of acoustic model parameters, and the acoustic model is independent of language model parameters.

2. The state chain is assumed to be first-order Markov. This means that the probability that the Markov chain is in state $q_l$ at time $n$ depends only on the state of the Markov chain at time $n - 1$, and it is conditionally independent of the past.

3. Observations are assumed to be conditionally independent of past observations and states. This means that the probability that a particular acoustic vector will be emitted at time $n$ depends only on the transition taken at that time, and it is conditionally independent of the past.

4. Recognition is often based on the best path (Viterbi), and not on all possible state sequences (total likelihood) for a model.

---

[8]The reader should compare Eq. 25.17 with Eq. 24.2.

## 25.5  EXERCISES

**25.1**    In this chapter, we derived a forward recursion for $P(X \mid M)$ by expressing it as $\sum_l P(X_1^N, q_l^N \mid M)$. Derive a similar backward recursion. That is, derive

$$\beta_n(l \mid M) = P\left(X_{n+1}^N \mid q_l^n, X_1^n\right) = \sum_k \beta_{n+1}(k \mid M) P\left(q_k^{n+1} \mid q_l^n\right) P\left(x_{n+1} \mid q_k^{n+1}\right). \qquad \textbf{(25.18)}$$

**25.2**    The Viterbi search finds the state sequence that is the most likely match to the observed data. A search incorporating full likelihood estimates (without the Viterbi assumption) permits us to find the most likely model. How could these two yield different results? Give an example.

## BIBLIOGRAPHY

1. Bourlard, H., and Morgan, N., *Connectionist Speech Recognition:   A Hybrid Approach*, Kluwer, Boston, Mass., 1993.
2. Jelinek, F., *Statistical Methods for Speech Recognition*, MIT Press, Cambridge, Mass., 1998.
3. Rabiner, L., "A tutorial on hidden Markov models and selected applications in speech recognition," *Proc. IEEE* **37**: 257–286, 1989.
4. Rabiner, L., and Juang, B.-H., *Fundamentals of Speech Recognition,* Prentice–Hall, Englewood Cliffs, N.J., 1993.

# CHAPTER 26

# STATISTICAL MODEL TRAINING

## 26.1 INTRODUCTION

In the previous chapter, we introduced the notion of statistical models and sequence recognition; we further introduced the common assumptions of conditional independence that lead to the particular form of generative statistical model[1] called a hidden Markov model (HMM). We then showed how such models could be used to compute the likelihood of the sequence of feature vectors having been produced by each hypothetical model, given some assumptions of conditional independence. This likelihood was either a total likelihood (using the forward recursion), taking into account all possible state sequences associated with the model, or a Viterbi approximation, only taking into account the most likely state sequence. Further assuming that the language model parameters were separable from the acoustic model parameters, we showed that the Bayes rule gave us the prescription for combining the two models to indicate the model (or sequence of models) that gives the minimum probability of error.

A key component in this development was the integration of local probability values over the sequence; essentially this was a local product of state emission and transition probabilities with a cumulative value computed from legal predecessor states. In other words, we derived approaches for determining complete sequence likelihoods given all the local probabilities. However, this still left a major problem: How do we get these probabilities? Even with the simplifying assumptions we have made, this is far from a trivial problem. State densities and transition probabilities are rarely known *a priori*, so in general they must be estimated from the training data.

The methods of Chapter 9 are directly relevant here. In that chapter, we described two parametric forms (Gaussians and mixtures of Gaussians) and introduced a general methodology for training their parameters. Given a set of labels for a sequence of speech frames, then, one could associate a model with all frames that have a particular label. For example, a feature from all speech frames labeled as the [ae] sound could be represented by a mixture of Gaussians. The parameters of each such model could then be trained to maximize the likelihood of the corresponding data, for instance by using EM.

However, speech recognition requires somewhat more complex procedures, though they are qualitatively similar to this simple scenario. For one thing, EM on framewise

---

[1] A statistical model is called generative when its observations are assumed to be generated by a state occupancy or transition, according to some statistical distribution.

densities would maximize the likelihood $P(x_n \mid q_j)$ for each $j$, which might not be the same thing as maximizing the likelihood of a complete model for a word or sentence. Thus, we must optimize over the space of complete sequences; that is, we must compute expectations over the space of all possible state sequences corresponding to the models for the training data. EM in this case can be shown to increase the likelihood of a complete sequence of observations given the models. Additionally, the training data are typically labeled asynchronously to the frames; that is, we ordinarily know that the training phrase "fifty-five" corresponds to a sequence of speech frames, but we don't know where the subword units start and stop. Consequently the state identity itself must be a hidden variable in the sense that we used the term in Chapter 9.

In the next section we apply EM to HMMs.

## 26.2   HMM TRAINING

The EM derivations of Chapter 9 showed that choosing parameters that maximized the expectation of the log of the joint density for observed and hidden variables would also maximize the likelihood of the observed data. When applied to HMMs, the hidden variables are the sequence of states associated with the Markov models.[2] Starting with a general expression for an auxiliary function that could be maximized in order to ensure the data likelihood is optimized (adapted from Eq. 9.25):

$$Q = \sum_k P(k \mid x_n, \Theta_{\text{old}}) \log[P(x_n \mid k, \Theta) P(k \mid \Theta)] \tag{26.1}$$

where $k$ is a hidden variable (for which we will sum over all examples), $x_n$ is the $n$th example of the observation, and $\Theta$ are the parameters to be optimized.

For the case of a HMM, we use the state sequence as the hidden variables, and we compute the expectation conditioned on the entire sequence. As in Chapter 25, let $q_k^n$ be the state at time $n$ that is of type $k$, let $X_1^N$ refer to the complete sequence of $N$ frames, and let $Q$ be the corresponding sequence of hidden state variables. Then the corresponding expectation would be

$$Q = \sum_Q P(Q \mid X_1^N, \Theta_{\text{old}}, M) \log[P(X_1^N \mid Q, \Theta, M) P(Q \mid \Theta, M)]. \tag{26.2}$$

This is the expectation of the joint likelihood of the observed feature vectors and the unobserved HMM state sequences.[3] This expression could be maximized for each model $M$ (which has associated permissible state sequences) by adjusting $\Theta$.

---

[2]This is the minimal set of hidden variables for the HMM training problem. Other variables may also be hidden, such as the component weightings for Gaussian mixtures.

[3]In some derivations, the first probability in this equation would be the joint probability of $Q$ and $X$. Maximizing over either distribution will maximize the data likelihood, and it will also lead to estimation procedures that are essentially the same.

In Chapter 25, we noted that the probability terms in Eq. 26.2 are often simplified by a series of conditional independence assumptions. Using these, we can approximate $P(X_1^N \mid Q)$ by $\prod_{n=1}^{N} P(x_n \mid q^n)$ (a product of emission probabilities), and $P(Q)$ by $P(q^1) \prod_{n=2}^{N} P(q^n \mid q^{n-1})$ (an initial state prior multiplied by a sequence of transition probabilities). Using the usual properties of the logarithm and re-expressing the summation over all sequences $Q$ as the sum for all possible values of frame index $n$ and state type $k$ (or state pairs $k$ and $\ell$ in the case of the third term), we get

$$
\begin{aligned}
\mathcal{Q} = &\sum_{n=1}^{N} \sum_{k=1}^{L(M)} P\left(q_k^n \mid X_1^N, \Theta_{\text{old}}, M\right) \log P\left(x_n \mid q_k^n, \Theta, M\right) \\
&+ \sum_{k=1}^{L(M)} P\left(q_k^1 \mid X_1^N, \Theta_{\text{old}}, M\right) \log P\left(q_k^1 \mid \Theta, M\right) \\
&+ \sum_{n=2}^{N} \sum_{k=1}^{L(M)} \sum_{\ell=1}^{L(M)} P\left(q_\ell^n, q_k^{n-1} \mid X_1^N \Theta_{\text{old}}, M\right) \log P\left(q_k^n \mid q_\ell^{n-1}, \Theta, M\right),
\end{aligned}
\qquad (26.3)
$$

where, as in Chapter 25, $L(M)$ is the number of states in model $M$.

Although we have written this expression with a general shared notation for the parameters, typically the three terms can be optimized separately. The first term can be optimized with the same methods that were shown in Chapter 9; in particular, note the similarity to the second term of Eq. 9.26. In that chapter, we chose a reasonable parametric form for the density estimator. This permitted the computation of partial derivatives of a $\mathcal{Q}$ function with respect to the unknown parameters. Setting the resulting expression to zero produced linear equations that could be simply solved. This determined the parameter values that corresponded to the best possible increase in data likelihood for the given initial density estimates.

For simplicity's sake, here we assume that a single univariate Gaussian is associated with each HMM state.[4] Given these assumptions, we can proceed through the same steps as were taken in Chapter 9. The resulting expression for the mean associated with state (density) $j$ is then

$$
\mu_j = \frac{\sum_{n=1}^{N} P\left(q_j^n \mid X_1^N, \Theta_{\text{old}}, M\right) x_n}{\sum_{n=1}^{N} P\left(q_j^n \mid X_1^N, \Theta_{\text{old}}, M\right)},
\qquad (26.4)
$$

and the corresponding expression for the variance estimate is

$$
\sigma_j^2 = \frac{\sum_{n=1}^{N} P\left(q_j^n \mid X_1^N, \Theta_{\text{old}}, M\right)(x_n - \mu_j)^2}{\sum_{n=1}^{N} P\left(q_j^n \mid X_1^N, \Theta_{\text{old}}, M\right)}.
\qquad (26.5)
$$

(Compare these with Eqs. 9.31 and 9.32 in Chapter 9.)

---

[4]As noted in Chapter 9, the extension to vector observations is not difficult, but it complicates the mathematics. Similarly, incorporating mixture Gaussians rather than single Gaussians requires us to have hidden variables that denote both the state and the generating Gaussian, which would complicate the notation without significantly enhancing this exposition.

Continuing the analogy with the development in Chapter 9 (Section 9.8), we can use Lagrangian multipliers to optimize the second and third terms of Eq. 26.3. That is, we take partial derivatives of each term separately, summed with an additional term that incorporates constraints based on the fact that for these cases the parameters that we wish to optimize are probabilities. For the case of the second term, we want to estimate the prior probability of the first state in each model. The corresponding Lagrangian term will be

$$\lambda_\pi \left[ \sum_{k=1}^{L(M)} P\left(q_k^1 \mid \Theta\right) - 1 \right],$$
(26.6)

which expresses the constraint that the probabilities of all initial states must sum to one. Similarly, for the third term of Eq. 26.3, the Lagrangian term will be

$$\lambda_{\text{trans}} \left[ \sum_{k=1}^{L(M)} P\left(q_k^n \mid q_\ell^{n-1}\Theta\right) - 1 \right],$$
(26.7)

which expresses the constraint that the probabilities of all transitions from any particular state must sum to one. In each case we take partial derivatives of the augmented term (with respect to the variable we wish to optimize), set the result to zero, solve for the Lagrangian multiplier, and resubstitute for the final expression for the optimum value of the unknown.[5]

Taking these steps, we end up with an expression for the optimum first frame prior probabilities:

$$P\left(q_j^1 \mid M\right) = P\left(q_j^1 \mid X_1^N, \Theta_{\text{old}}, M\right).$$
(26.8)

Thus, the best estimate of this prior is just the posterior probability taken from the previous parametric representation.

Similarly, the optimum transition probabilities can be shown to be

$$P\left(q_j^n \mid q_i^{n-1}, M\right) = \frac{\sum_{n=2}^{N} P\left(q_j^n, q_i^{n-1} \mid X_1^N, \Theta_{\text{old}}, M\right)}{\sum_{n=2}^{N} P\left(q_i^{n-1} \mid X_1^N, \Theta_{\text{old}}, M\right)}.$$
(26.9)

Note in particular two points about this expression.

1. The denominator can also be expressed as an outer sum over the numerator for all states $j$, that is, $\sum_{j=1}^{L(M)} \sum_{n=2}^{N} P\left(q_j^n, q_i^{n-1} \mid X_1^N, \Theta_{\text{old}}, M\right)$.

2. The form of this equation is similar to the others we derived, in the sense that it is a kind of normalized expectation. In particular, it may be interpreted as the expected

---

[5]See Chapter 9 for the analogous computations for the probabilities that are used as the weights of a Gaussian mixture.

value of the number of transitions from state $i$ to state $j$, normalized by the expectation of the number of transitions from state $i$.

Thus, for the prior probabilities for the first state and for state transitions, we can compute optimum updates from estimates of the relevant posterior probabilities. That is, we require an estimate of posterior probability of single states and of pairs of states, conditioned on the entire observation sequence. In the case of the emission probabilities, we have just given a simple example of a parametric form (that of a single univariate Gaussian). However, it is generally true that the iterative training of HMM emission parameters is crucially dependent on the estimation of the posterior probability $P(q_j^n \mid X_1^N, \Theta_{\text{old}})$. In the following section we show how all of these posterior probabilities can be estimated. Given these estimates, though, the training procedure is (in principle) straightforward.

1. Choose a form for the local probability estimators (e.g., Gaussian) for the densities associated with each state.

2. Choose an initial set of parameters for the estimators.

3. Given the parameters, estimate the probabilities $P(q_j^n \mid X_1^N, \Theta_{\text{old}})$ for each state and time. Similarly, estimate the probabilities $P(q_j^n, q_i^{n-1} \mid X_1^N, \Theta_{\text{old}})$ for each state transition and time. These are essential terms in the estimate of the expectation $Q$ for the EM algorithm, as given by Eq. 26.3.

4. Given these probabilities, and the parametric form chosen in step 1, find the parameters that maximize $Q$. These parameters will be guaranteed to give the best possible improvement to the likelihood for each model.

5. Assess the new models according to some stopping criterion; if good enough, stop. Otherwise, return to step 3.

Although some training approaches use somewhat different criteria and probabilistic estimates, the general form of the training for all statistical sequence systems remains the same: use the current parameters to estimate posterior probabilities for the hidden variables, and then use these posteriors to determine new parameters that maximize the expectation $Q$ (and thus the data likelihood, according to the EM proof in Chapter 9). In the remainder of this chapter we describe some further specifics for some simple HMM-based approaches.

## 26.3  FORWARD–BACKWARD TRAINING

As noted in the previous section, EM-based training of HMMs requires the estimation of the probability of each type of state occurring at a frame. These probabilities can be estimated by using a combination of the forward procedure (described in Chapter 25) and a similar one called the backward recursion (see problem 2 in Chapter 25). Together these recursions will yield the required probabilities, which can then be used to generate the new set of estimator parameters. This procedure then will maximize the likelihood of the correct models (given the usual Markov assumptions).

Let's start by redefining the two recursions[6]:

$$\alpha_n(\ell \mid M) = P\big(X_1^n, q_\ell^n \mid M\big) = \left[\sum_k \alpha_{n-1}(k \mid M)P\big(q_\ell^n \mid q_k^{n-1}\big)\right] P\big(x_n \mid q_\ell^n\big), \qquad \textbf{(26.10)}$$

$$\beta_n(\ell \mid M) = P\big(X_{n+1}^N \mid q_\ell^n, X_1^n, M\big) = \sum_k \beta_{n+1}(k \mid M)P\big(q_k^{n+1} \mid q_\ell^n\big)P\big(x_{n+1} \mid q_k^{n+1}\big). \qquad \textbf{(26.11)}$$

Note that the second equation has the same form as the first, but it proceeds backward in time. The second one was also chosen to have the property that

$$\alpha_n(\ell \mid M)\beta_n(\ell \mid M) = P\big(X_1^N, q_\ell^n \mid M\big). \qquad \textbf{(26.12)}$$

In other words, the product is the joint likelihood of the complete data sequence and a particular state at a particular time. Summed up over all the possible states at that time, this will then yield $P(X_1^N \mid M)$.

Given these intermediate results, we can then compute the probability of a particular state at a particular time, given the entire data sequence:

$$P\big(q_k^n \mid X_1^N, M\big) = \frac{P\big(X_1^N, q_k^n \mid M\big)}{P\big(X_1^N \mid M\big)} = \frac{\alpha_n(k \mid M)\beta_n(k \mid M)}{\sum_\ell \alpha_n(\ell \mid M)\beta_n(\ell \mid M)} \qquad \textbf{(26.13)}$$

This can be used to update the parameters of the probability estimators for the emission density associated with each state. In particular, given a specific form for these estimators (e.g., Gaussian), the probabilities of Eq. 26.13 can be used to compute parameters that maximize the expectation $Q$ and hence the data likelihood. In section 26.4, we show this process for Gaussians and for discrete densities.

In addition to these emission densities, the state transition probabilities can also be chosen to maximize the expectation $Q$. As noted in Section 26.2, these probabilities are determined from the posterior acoustic probability estimate of the state transitions, or $P(q_\ell^n, q_k^{n-1} \mid X_1^N, M)$; see Eq. 26.9. This requires a series of approximations based on the same conditional independence assumptions that were used in Chapter 25 to develop the $\alpha$ recursion and will result in a somewhat messier expression. However, in principle, we can compute the required probability with a product of an $\alpha$ term, a $\beta$ term, and an emission and a transition probability to represent the posterior contribution of the current transition. This is illustrated in Fig. 26.1.

More formally, we wish to estimate the numerator of Eq. 26.9 (since the denominator can be obtained by summing the numerator over all possible states). Equivalently, we can estimate the quantity $P(q_j^n, q_i^{n-1}, X_1^N \mid \Theta_{\text{old}}, M)$, since both denominator and numerator

---

[6]For consistency of notation with Chapter 25, we will suppress the explicit dependence on the old parameters $\Theta_{\text{old}}$ in this section. The reader should keep in mind, however, that all of the 'local' probabilities used in these expressions come from estimators that may have been trained by a previous step (or else that use parameters that have been chosen for initialization).

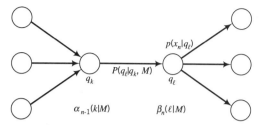

**FIGURE 26.1** A state transition and the terms that are used for training of the associated probability. The total probability of all paths terminating in state $k$ at time $n-1$ is given by $\alpha_{n-1}(k\,|\,M)$. The total probability of all backward paths (ending at time $N$) that start with state $\ell$ at time $n$ is given by $\beta_n(\ell\,|\,M)$. Local probabilities $P(q_\ell\,|\,q_k)$ and $P(x_n\,|\,q_\ell)$ are multiplied in to get an estimate of the total probability for all paths that contain the transition from $k$ to $\ell$.

of Eq. 26.9 can be multiplied by the term $P(X_1^N\,|\,\Theta_{\text{old}}, M)$. We then split up the data sequence into past, present, and future terms to express the desired quantity (suppressing $\Theta$ and $M$ for simplicity's sake), and proceed with the derivation:

$$P\big(q_\ell^n, q_k^{n-1}, X_1^{n-1}, x_n, X_{n+1}^N\big) \tag{26.14}$$

$$= P\big(q_\ell^n, x_n, X_{n+1}^N \,\big|\, q_k^{n-1}, X_1^{n-1}\big) P\big(q_k^{n-1}, X_1^{n-1}\big) \tag{26.15}$$

$$= P\big(q_\ell^n, x_n, X_{n+1}^N \,\big|\, q_k^{n-1}, X_1^{n-1}\big)\alpha_{n-1}(k\,|\,M) \tag{26.16}$$

$$= P\big(X_{n+1}^N \,\big|\, q_\ell^n, q_k^{n-1}, X_1^{n-1}, x_n\big) P\big(q_\ell^n, x_n \,\big|\, q_k^{n-1}, X_1^{n-1}\big)\alpha_{n-1}(k\,|\,M) \tag{26.17}$$

$$\approx P\big(X_{n+1}^N \,\big|\, q_\ell^n, X_1^{n-1}, x_n\big) P\big(q_\ell^n, x_n \,\big|\, q_k^{n-1}, X_1^{n-1}\big)\alpha_{n-1}(k\,|\,M) \tag{26.18}$$

$$= \beta_n(\ell\,|\,M) P\big(q_\ell^n, x_n \,\big|\, q_k^{n-1}, X_1^{n-1}\big)\alpha_{n-1}(k\,|\,M) \tag{26.19}$$

$$\approx \beta_n(\ell\,|\,M) P\big(q_\ell^n, x_n \,\big|\, q_k^{n-1}\big)\alpha_{n-1}(k\,|\,M) \tag{26.20}$$

$$= \beta_n(\ell\,|\,M) P\big(x_n \,\big|\, q_\ell^n, q_k^{n-1}\big) P\big(q_\ell^n \,\big|\, q_k^{n-1}\big)\alpha_{n-1}(k\,|\,M) \tag{26.21}$$

$$\approx \beta_n(\ell\,|\,M) P\big(x_n \,\big|\, q_\ell^n\big) P\big(q_\ell^n \,\big|\, q_k^{n-1}\big)\alpha_{n-1}(k\,|\,M) \tag{26.22}$$

Each equality in this derivation either represents a definition (for $\alpha$ or $\beta$), or else an application of the axiom $P(ab\,|\,c) = P(a\,|\,bc)P(b\,|\,c)$. Each use of the symbol $\approx$ indicates that we are making use of the conditional independence assumptions described in Chapter 25. For instance, in the expression for the joint probability of the observation and state for the $n$th frame (26.19), we can drop dependence on $X_1^{n-1}$ if there is also a dependence on $q_k^{n-1}$ (expression 26.20). This is the first-order Markov assumption. That is, given knowledge of the previous state, we assume that no other information about the past will provide any additional information about the current state.

Using the terms of Eq. 26.22 to estimate the transition probabilities of Eq. 26.9, we find that the optimum transition probabilities are

$$
\begin{aligned}
P\left(q_j^n \mid q_i^{n-1}, M\right) &= \frac{\sum_{n=2}^{N} P\left(q_j^n, q_i^{n-1} \mid X_1^N, \Theta_{\text{old}}, M\right)}{\sum_{n=2}^{N} P\left(q_i^{n-1} \mid X_1^N, \Theta_{\text{old}}, M\right)} \\
&= \frac{\sum_{n=2}^{N} P\left(q_j^n, q_i^{n-1}, X_1^N \mid \Theta_{\text{old}}, M\right)}{\sum_{n=2}^{N} P\left(q_i^{n-1}, X_1^N \mid \Theta_{\text{old}}, M\right)} \\
&= \frac{\sum_{n=2}^{N} \beta_n(\ell \mid M) P\left(x_n \mid q_\ell^n\right) P\left(q_\ell^n \mid q_k^{n-1}\right) \alpha_{n-1}(k \mid M)}{\sum_{\ell=1}^{L(M)} \sum_{n=2}^{N} \beta_n(\ell \mid M) P\left(x_n \mid q_\ell^n\right) P\left(q_\ell^n \mid q_k^{n-1}\right) \alpha_{n-1}(k \mid M)}.
\end{aligned}
\tag{26.23}
$$

Similarly, for the first frame probabilities as given by Eq. 26.8, the optimum probabilities can be computed directly from Eq. 26.13 for the case of $n = 1$, or

$$
P\left(q_k^1 \mid M\right) = \frac{\alpha_1(k \mid M)\beta_1(k \mid M)}{\sum_\ell \alpha_1(\ell \mid M)\beta_1(\ell \mid M)}.
\tag{26.24}
$$

In summary, $\alpha$ (forward) and $\beta$ (backward) recursions are used to derive estimates for the probabilities of the hidden state and state transition variables, conditioned on the sequence of acoustic feature vectors. For the unconditioned (prior) probabilities, as given by Eqs. 26.23 and 26.24, the probabilities themselves are trained parameters of the model. In the case of emission probabilities, the parameters are variables associated with some particular structure for the estimator, e.g., Gaussian.

For all of these parameters, estimation and maximization steps can be repeated until some stopping criterion is reached. The overall procedure is then referred to as forward-backward or Baum–Welch training.

# 26.4   OPTIMAL PARAMETERS FOR EMISSION PROBABILITY ESTIMATORS

For many common structures of probability estimators, update equations can be found for which the partial derivative of the expectation $Q$ (with respect to the estimator parameters) is zero, and the likelihood of the data is maximized. For any particular iteration of the Baum–Welch procedure, these update equations will be applied, incorporating probabilities that have been estimated from the previous iteration. Here we illustrate this process for the case of two simple structures:   a Gaussian density, and a discrete density.

## 26.4.1   Gaussian Density Functions

Recall that a multivariate Gaussian distribution is defined by two groups of parameters: the mean vector, and the covariance matrix. In the most general case, each of these will be unique for each state category (e.g., phone), but often the nonzero part of the covariance matrix is limited to diagonal elements (variances).

As noted previously, the expected values of each mean must be computed by weighting each feature vector by the probability that it corresponds to the particular state. This is expressed in Eq. 26.4. Note that the denominator may be interpreted as the expected number of frames associated with state $j$. Expressing both numerator and denominator in terms of the recursion results (where the $\alpha$ values come from the forward recursion and the $\beta$ values come from the backward recursion), we get

$$\mu_j = \frac{\sum_{n=1}^{N} x_n \alpha_n(j \mid M) \beta_n(j \mid M)}{\sum_{n=1}^{N} \alpha_n(j \mid M) \beta_n(j \mid M)}. \tag{26.25}$$

The variances can be computed in a similar manner.

## 26.4.2 Example: Training with Discrete Densities

For discrete densities, each $x_n$ is mapped to the nearest cluster center $y_j$ in a process called vector quantization, or VQ (to be discussed a bit more later in this chapter). A distribution is stored in a table for each $y_j$. For each feature vector $x_n$, the probability $P(x_n \mid q_\ell^n)$ is then approximated by $P(y_j \mid q_\ell^n)$, where the feature vector was closest to cluster center $y_j$.

To derive the optimum values for the emission parameters, we begin with the same term of the auxiliary function $\mathcal{Q}$ that we have been differentiating, namely the first term of Eq. 26.3. However, the parameters to be estimated for the discrete case are the probability estimates $P(y_j \mid q_\ell)$ themselves. Thus, we need a Lagrangian term in the optimization to represent the constraint that these parameters must sum to one. Putting in this term, we get

$$\mathcal{Q}^d = \sum_{n=1}^{N} \sum_{k=1}^{L(M)} P(q_k^n \mid X_1^N, \Theta_{\text{old}}, M) \log P(x_n \mid q_k^n, \Theta, M) + \lambda \left[ \sum_{i=1}^{I} P(y_i \mid q_k, \Theta) - 1 \right]. \tag{26.26}$$

Taking the partial derivative with respect to $P(y_j \mid q_\ell, \Theta)$ and setting the result to zero yields

$$\sum_{n=1}^{N} P(q_\ell^n \mid X_1^N, \Theta_{\text{old}}, M) \frac{\delta_{nj}}{P(y_j \mid q_\ell^n, \Theta, M)} + \lambda = 0, \tag{26.27}$$

where the notation $\delta_{nj}$ is used for a function that is one when $x_n$ is quantized to $y_j$ and zero when it is quantized to some other codebook entry. Here $P(y_j \mid q_\ell^n)$ is fixed to be the same for any value of $n$, so we can multiply through by this value and get

$$\sum_{n=1}^{N} P(q_\ell^n \mid X_1^N, \Theta_{\text{old}}, M) \delta_{nj} = -\lambda P(y_j \mid q_\ell^n, \Theta, M). \tag{26.28}$$

Summing over all values of $y_j$, we find that the constraint reduces the right-hand side to be $-\lambda$ and removes the $\delta$ term from the left-hand side (since the sum would now include all frames), so that we get

$$\lambda = -\sum_{n=1}^{N} P(q_\ell^n \mid X_1^N, \Theta_{\text{old}}, M). \tag{26.29}$$

Resubstituting into Eq. 26.28 and rearranging terms, we finally get

$$P(y_j \mid q_\ell, \Theta) = \frac{\sum_{n=1}^{N} P(q_\ell^n \mid X_1^N, \Theta_{\text{old}}, M) \delta_{n\ell}}{\sum_{n=1}^{N} P(q_\ell^n \mid X_1^N, \Theta_{\text{old}}, M)}. \tag{26.30}$$

This new estimate of the emission probabilities associated with a VQ value and a state can be viewed as the expected value of the number of frames associated with that VQ value and state, normalized by the expected value of the total number of frames.

Finally, we can see that the probabilities required for Eq. 26.30 are just the probabilities from Eq. 26.13, and thus they can be obtained from the normalized product of forward and backward recursions as given in that latter equation.

## 26.5  VITERBI TRAINING

In Chapter 25, we noted that the full likelihood of a model can be approximated by the likelihood associated with the most likely sequence of states. This was referred to as a Viterbi approximation. The advantage of this approximation is that sums in the $\alpha$ recursion can be replaced with a maximum, which simplifies the numerical considerations. Similarly, in Viterbi training, we will attempt to optimize the parameters to maximize the likelihood of the best path (state sequence) in the correct model. For this case, then, the posterior probabilities used in the estimation step are assumed to either be zero or one; that is, at each stage in the EM iterations, we will assume a particular state sequence for the training data. The EM steps then take the following form[7]:

1. Assume an initial set of parameters for the density estimators.
2. Determine the most likely state sequence (or assume one if initializing from segmentations rather than densities).
3. Update the parameters.
4. Assess the solution and repeat the previous two steps as necessary.

We begin by discussing the second step. Assuming that we have some probability estimators, how do we find the best segmentation of the frames into a state sequence? This question was essentially answered in a different context in the previous chapter. We know that a Viterbi decoding procedure will find the likelihood of the best path (state sequence) for each model. Further, since the most probable transition was used at each step, we will be able to backtrack and determine the corresponding state sequence. Conceptually, segmentation of training data with a known model transcription is the same as recognition, except that in the former case there are no alternate model sequences to consider.

Since we can obtain an emission probability for each frame and state category, each of these is used in a process that is often called Viterbi alignment. In this process, dynamic programming is done, essentially using the one-pass method, in which the local distances are $-\log P(y_j \mid q_\ell)$, and where there are transition costs $-\log P(q_\ell \mid q_k)$ for hypothesizing transitions from states $k$ to $\ell$. Unlike the recognition scenario, the only model sequences

---

[7]This process is conceptually much simpler than the approaches discussed previously. We have chosen to discuss the non-Viterbi case first, however, because it is more general. As noted above, Viterbi training can be considered a special case of forward–backward training in which the posterior probabilities are either one or zero.

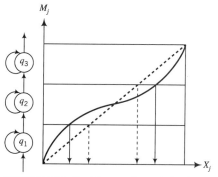

**FIGURE 26.2**  Illustration of the iterative Viterbi alignment. Each utterance $X_j$ is segmented into an initial estimate of state occupancies for the corresponding model $M_j$. For instance, this can be done with the linear match shown by the dashed line in the figure. Then the parameters are chosen to maximize the likelihood of the data. Given the new parameters, a new segmentation is found (shown here as the solid curve). This corresponds to the most likely path through the model given the new parameters. The arrows down to the X-axis show the corresponding state transition times for each of these segmentations.

that are considered are the ones associated with the known word sequence; all the word models together can be considered as a single model for the entire utterance in this case, and there is only one to be evaluated. Backtracking can be done since the best previous state can be preserved for each frame, and so the best state sequence can be found. Additionally, since only one model sequence need be evaluated, often it is not necessary to use elaborate data structures for this process – the distance and backtracking information can be held in complete matrices, since the storage is not prohibitive as it would be in the recognition case.

Figure 26.2 illustrates Viterbi alignment. In the case shown in the figure, we begin the iterative process with an assumed segmentation, but the initial segmentation can also be derived from an assumed set of densities.

The state sequence that is found through the backtracking procedure is considered to be an alignment of the states with the feature vectors. In the next step, we estimate the transition and emission probabilities, *assuming that this state sequence is correct.*

Finally, the solution must be assessed. We can do this by looking at the changes in the global likelihood and setting some threshold on the improvement. Another approach is to test for convergence of the segmentation by counting the number of frames for which the state label has changed.

Although Viterbi training can be quite effective and is simple to understand and implement, it requires an additional approximation over the Markov assumptions mentioned previously, and this is in some sense a disadvantage. However, in Viterbi training, the best path through each model is reinforced, so that during recognition the best path is more likely to correspond to the correct model.

For Viterbi training, the state transition probabilities are particularly simple to evaluate. Consider Eq. 26.9, which gave an expression for the optimum transition probabilities in the general case of probabilistic state occupation. If the posterior probabilities for a transition are assumed to be one when a state pair is present in the Viterbi alignment, and zero when it is not, then that equation becomes

$$P\left(q_j^n \mid q_i^{n-1}\right) = \frac{\text{no. trans. } i \text{ to } j}{\text{no. trans. from } i}. \tag{26.31}$$

In other words, to estimate state transition probabilities, we simply count to find the relative frequency of each particular transition.

As with the more general case of Baum–Welch training, we consider Gaussians and discrete densities in order to derive the update equations for the emission probability estimators.

## 26.5.1 Example: Training with Gaussian Density Functions

For each Viterbi iteration, the optimum means and variances are computed from all the frames labeled with each state category. The ordinary equations for these parameters are indeed optimum. This can be shown by taking the corresponding solution without any Viterbi assumptions, and substituting one and zero appropriately as the only permitted probabilities. For example, consider Eq. 26.4. Substituting a probability of one for all cases in which a segmentation yields a label of state $j$, and zero elsewhere, we get

$$\mu_j = \frac{\sum_{\text{frames labeled } j} x_n}{\text{no. frames labeled } j}. \tag{26.32}$$

Similarly, the variance is just

$$\sigma_j^2 = \frac{\sum_{\text{frames labeled } j} (x_n - \mu_j)^2}{\text{no. frames labeled } j}. \tag{26.33}$$

## 26.5.2 Example: Training with Discrete Densities

In the case of discrete density functions, since the acoustic vector has been quantized to one of a finite number of categories (typically a few hundred), we can compute relative frequencies of each of these categories for those frames that have a particular state label. Starting from Eq. 26.30, we again substitute probabilities of one for frames within a segmentation for state $\ell$ and zero elsewhere. Given the $\delta$ function in that equation, which further eliminates frames from the numerator sum for VQ values other than $j$, we end up with

$$P(y_j \mid q_\ell) = \frac{\text{no. frames labeled } \ell \text{ and } y_j}{\text{no. frames labeled } \ell}. \tag{26.34}$$

Given these new probabilities, another Viterbi alignment can be run, and so on, until some stopping criterion is met.

# 26.6 LOCAL ACOUSTIC PROBABILITY ESTIMATORS FOR ASR

Here we give a little more detail about the estimator structures used in the previous examples, as well as a few more complex types.

## 26.6.1 Discrete Probabilities

As noted previously, discrete probabilities for speech frames (given a particular state category) are estimated by counting co-occurrences between state labels and the VQ index for each frame. Prior to the training methods described above, a VQ training phase is implemented. In this phase, the training data are clustered by using one of a number of common methods, such as the K-means method. In a variant on this approach, sometimes hierarchical clustering is done – first the feature vectors are clustered into two clusters, then into four, and so forth, with each step consisting of a complete K-means clustering, and with the next step initiated by choosing a split that satisfies some reasonable criterion. A common number of cluster centers to end up with is somewhere between 128 and 512. The smaller the number the more robust the quantization will be to small variations, but the larger the number is the better match each cluster center value is to the unquantized version of each feature vector.

Thus, in this training phase, a list of prototypical feature vectors is computed; this list is commonly called a codebook. When it is used (either during training or recognition), each incoming feature vector is mapped to the closest prototype vector in the codebook and the resulting index is used to represent the frame; as noted earlier, this can be used as an index into a table of probabilities.

In practice, it is better to use separate codebooks for static features (e.g., mel cepstra), dynamic features (e.g., delta mel cepstra), and energy-related features. This leads to multiple probabilities that must be combined. Typically, they are combined by multiplication, which is tantamount to assuming conditional independence between these features. Thus even though the use of discrete distributions seems to be free of strong distributional assumptions (e.g., Gaussian), their practical use requires quite strong statistical assumptions. If a single codebook is used, then the use of Euclidean distances for the quantization is also equivalent to a strong assumption about the statistical distribution of the features. Finally, recent practical experience for many researchers has shown that continuous densities can often be used to achieve better performance.

Despite these seeming difficulties, the simplicity and low computation requirements for these methods make discrete (VQ-based) HMMs a popular methodology for local probability estimation.

## 26.6.2 Gaussian Densities

As noted earlier, feature vectors associated (deterministically or probabilistically) with each state can be assumed to be generated by a multivariate Gaussian distribution. This is a fairly strong assumption, but it is not so bad if a full covariance matrix is used (including nondiagonal elements). Unfortunately, when enough state categories are used to make this assumption as reasonable one, the number of parameters for full covariance matrices can be prohibitive.

For this reason, it is now more common that simpler Gaussians are used (variance only) and are combined in mixtures, as described below. This is often found to be a more effective use of the parameters.

### 26.6.3  Tied Mixtures of Gaussians

Each emission probability density can also be assumed to be the weighted sum of a common pool of Gaussian densities. We can express this as

$$P\left(x_n \mid q_k^n\right) = \sum_{j=1}^{J} c_j P\left(x_n \mid q_k^n, j\right), \qquad (26.35)$$

which is axiomatically correct if $c_j = P(j \mid q_k^n)$.

Another way of looking at such a system is as a discrete system for which variances are also considered, and for which the probability is not just associated with a single cluster center, but with several (or all in the unconstrained case). Thus, this method is often referred to as soft VQ. The HMMs that result are sometimes referred to as semicontinuous for similar reasons. Practically, these methods have become very popular, as they share some of the simplicity of the discrete methods and yet permit continuous representations of the densities.

### 26.6.4  Independent Mixtures of Gaussians

In the previous case, the density for each state was estimated by a different weighting on the same pool of base densities. In the more general case of mixture Gaussians, each state has its own set of mixtures densities (not just the weights, but separate Gaussians with their own means and variances). This provides a more detailed estimate of the densities, but as such generally requires more training data.

Many systems now actually use something in between tied and independent Gaussians; there is generally some kind of automatic procedure to determine to what extent the data will support an independent estimator for each state, and to what extent the data are sparse enough that tying is required.

### 26.6.5  Neural Networks

As noted in Chapter 9, neural networks can also be used to estimate probabilities. However, they are sufficiently different (different assumptions, optimization criteria, etc.) that they will be treated separately in the next chapter.

## 26.7  INITIALIZATION

Thus far, we have largely skipped over the initialization of parameters. This is required for any form of EM. There are a number of common ways in which this is done. Sometimes the models are initially estimated from a manually transcribed data base such as TIMIT (briefly described in Chapter 23). For instance, if three states are used for each phone, the phone segmentations can be used as marked in TIMIT, and three segments for each phone can initially be assumed to be of equal length. Models can be trained from this, iterated on TIMIT, and then used as the initial models for the new data base. Alternatively, TIMIT phone models could be used to provide an initial segmentation of the new data base, and that could be iterated upon in a similar manner. Sometimes systems use a signal-processing approach to come up with preliminary segmentations, which are used to generate the initial models. One can also just divide up the sequences according to the average relative lengths of phones. Even simpler approaches have been used, but particularly for Viterbi training, the initialization can have a significant effect on the results.

## 26.8  SMOOTHING

For all of these methods, there is a fundamental difficulty that is essentially always present to some degree:   we wish to capture the variability inherent in the data, which pushes us toward ever finer representations (for instance, modeling triphones, or phones with specific left and right context, rather than phones); however, these finer categories have fewer examples. Indeed, poor estimates for categories that occur relatively infrequently may also hurt recognition for other categories significantly, since recognition requires integration of the probabilistic estimates over complete utterances.

Therefore, it is common that we will require the combination of good estimates of coarse categories with noisy estimates of fine categories, in order to stabilize the estimate of the latter:   this process is often called smoothing. A number of techniques have been developed for this purpose; here we describe two.

**1.** Backoff smoothing:   this is a simple but often effective method. Thresholds are set for a minimum number of examples for each level of granularity, and when the minimum is not present, the estimator backs off to a coarser level. For instance, if there are enough examples of a particular triphone, its emission probability may be used directly; if there are not enough, a biphone might be used; if there are not enough examples of the biphone, then the phone probabilities are used. The setting of proper thresholds is obviously a tricky point, but for many purposes simple heuristics work for this.

**2.** Deleted interpolation:   this is a much more sophisticated process, and it appears to work quite well. Instead of choosing which estimator to use, all of them are used with weighting factors that are learned from testing on data that are disjoint from the training data. Often, this is done by partitioning the training data into $N$ pieces (e.g., 10) and then training up $N$ different sets of parameters by using all the different choices for $(N - 1)/N$ of the data. For the case of two estimators, the goal is to choose $\epsilon$ optimally to combine the estimator parameters, that is,

$$\Theta_T = \epsilon\Theta_1 + (1 - \epsilon)\Theta_2. \tag{26.36}$$

In one approach, $\epsilon$ is set to the fraction of utterances for which $\Theta_1$ was better. Another solution is to treat $\epsilon$ as missing information (a hidden variable) and to use EM training.

## 26.9   CONCLUSIONS

In this chapter we have shown the basic approaches used for the training of HMMs for speech recognition. As usual we have just scratched the surface; consult [1]–[5] for more of the theory (convergence proofs, etc.). Conference papers from the International Conference on Acoustics, Speech, and Signal Processing (ICASSP), the International Conference on Spoken Language Processing (ICSLP), and Eurospeech all have many more recent examples of practical approaches that have been taken in complete systems.

In all of these approaches, however, we have made a set of assumptions that we actually know to be unrealistic; in particular, trying to represent the longer-term dependence between observations is a recurring theme in much current research. Further, the basic training approaches are entirely based on improving the likelihood scores of the correct models, and they do nothing to assure us that the likelihoods of the incorrect models will be low. This issue will lead us to a study of discriminant models and estimators.

## 26.10   EXERCISES

**26.1**   Show all the steps that lead to the expression for optimum first frame prior probabilities as given in Eq. 26.8.

**26.2**   Show all the steps leading to Eq. 26.25, starting from Eq. 26.4 and using the definitions for the forward and backward recursions.

**26.3**   Show all the steps leading to Eq. 26.31 from Eq. 26.9.

## BIBLIOGRAPHY

1. Baum, L. E., and Petrie, T., "Statistical inference for probabilistic functions of finite state Markov chains," *Ann. Mathemat. Stat.* **37**: 1554–1563, 1966.
2. Bourlard, H., and Morgan, N., *Connectionist Speech Recognition: A Hybrid Approach*, Kluwer, Boston, Mass., 1993.
3. Dempster, A. P., Laird, N. M., and Rubin, D. B., "Maximum likelihood from incomplete data via the EM algorithm," *J. R. Stat. Soc.* **39**: 1–38, 1977.
4. Jelinek, F., *Statistical Methods for Speech Recognition*, MIT Press, Cambridge, Mass., 1998.
5. Rabiner, L., and Juang, B.-H., *Fundamentals of Speech Recognition,* Prentice–Hall, Englewood Cliffs, N.J., 1993.

# CHAPTER *27*

# *DISCRIMINANT ACOUSTIC PROBABILITY ESTIMATION*

## 27.1  INTRODUCTION

In the previous chapters we introduced the notion of trainable statistical models for speech recognition, in particular focusing on the set of methods and constraints associated with hidden Markov models (HMMs). In both training and recognition phases, the key values that must be estimated from the acoustics are the emission probabilities, also referred to as the acoustic likelihoods. These values are used to derive likelihoods for each model of a complete utterance, in combination with statistical information about the *a priori* probability of word sequences. In other words, probabilities of the local acoustic measurements' having been generated by each hypothesized state are ultimately integrated into a global probability for a complete utterance's having been generated by a complete HMM (either by considering all possible state sequences associated with a model, or by considering only the most likely).

In Chapter 26 we provided examples of two common approaches to the estimation of these acoustic probabilities: codebook tables associated with vector quantized features, giving probabilities for each feature value conditioned on the state; and Gaussians or mixtures of Gaussians associated with one or more states. For both of these examples, EM training is used to maximize the likelihood of the acoustic feature sequence's having been generated by the correct model. However, when the parameters are modified in this way, there is no guarantee that they will also reduce the likelihoods of the incorrect models. Training that explicitly guarantees the relative improvement of the likelihood for the correct versus incorrect models is referred to as being discriminant. Discriminant training for sequence-recognition systems thus has the same goal as discriminant training for static tasks; the parameters for the classifier are trained to distinguish between examples of different classes. In the limit of infinite training data and convergence to optimal parameters, maximum likelihood training is also discriminant, in that it converges to the Bayes solution that guarantees a minimum probability of error. However, given practical limitations, it is often helpful to focus more directly on discrimination during training, which is the topic discussed in this chapter. There are a number of approaches to discriminant training in ASR, including the use of neural networks for state probability estimation. We briefly survey a range of approaches, and then we focus on the use of neural networks for discriminant ASR systems.

## 27.2  DISCRIMINANT TRAINING

Recall that in Chapter 25, we used the Bayes rule to describe the fundamental equation for statistical speech recognition. For convenience, we repeat this here:

$$P(M_j \mid X) = \frac{P(X \mid M_j)P(M_j)}{P(X)}, \tag{27.1}$$

where, as before, the class $M_j$ is the $j$th statistical model for a sequence, $0 \le j \le J$, and $X$ is the observable evidence of that sequence.

In real systems, the actual probabilities are unknown, and instead we have estimates that depend on parameters that we will learn during training. Again repeating from Chapter 25, we can explicitly incorporate dependence on a parameter set $\Theta$ to get

$$P(M_j \mid X, \Theta) = \frac{P(X \mid M_j, \Theta)P(M_j \mid \Theta)}{P(X \mid \Theta)}. \tag{27.2}$$

Recall that, for example, $\Theta$ could include means and variances for Gaussian components of the density associated with each state.

During training, $\Theta$ is changing. Typically $\Theta$ is changed to maximize the likelihood $P(X \mid M_j, \Theta)$. However, this will also change the denominator $P(X \mid \Theta)$, and we cannot be assured that the quotient will increase. To illustrate this potential difficulty further, if we expand the latter probability to a sum of joint probabilities

$$\sum_{k=1}^{K} P(X, M_k \mid \Theta) = \sum_{k=1}^{K} P(X \mid M_k, \Theta)P(M_k \mid \Theta), \tag{27.3}$$

then the Bayes rule expression becomes

$$P(M_j \mid X, \Theta) = \frac{P(X \mid M_j, \Theta)P(M_j \mid \Theta)}{\sum_{k=1}^{K} P(X \mid M_k, \Theta)P(M_k \mid \Theta)}, \tag{27.4}$$

or

$$P(M_j \mid X, \Theta) = \frac{1}{1 + \left\{ \left[ \sum_{k \ne j} P(X \mid M_k, \Theta)P(M_k \mid \Theta) \right] / \left[ P(X \mid M_j, \Theta)P(M_j \mid \Theta) \right] \right\}}. \tag{27.5}$$

Thus, given some speech acoustics and some model parameters, the probability of a particular model is simply related to the ratio of the likelihood of that model weighted by its prior probability, divided into the sum of this product for all of the other models. Thus, to be sure that this probability estimate increased for a change in the parameters, we would aim to reduce the likelihood for incorrect models as well as to increase the likelihood for correct models. Training procedures that attempt to do this (or at least increase the likelihood ratio between correct and incorrect models) will be discriminant (between models).

There are several major categories of discriminant training that have been developed. We briefly mention three of them, and then we proceed in somewhat more detail on another discriminant approach that is based on neural networks.

## 27.2.1 Maximum Mutual Information

A quantity that is closely related to the fraction of Eq. 27.4 is the mutual information between the models and the acoustics, or

$$I(M, X \mid \Theta) = E\left[\log \frac{P(M_j, X \mid \Theta)}{P(M_j \mid \Theta)P(X \mid \Theta)}\right], \tag{27.6}$$

where $E$ is the expectation operator over the joint probability space for the models and the acoustics.

For a particular choice of model and acoustic pair,

$$\begin{aligned} I(M_j, X \mid \Theta) &= \log \frac{P(M_j, X \mid \Theta)}{P(M_j \mid \Theta)P(X \mid \Theta)} \\ &= \log \frac{P(X \mid M_j, \Theta)}{\sum_{k=1}^{K} P(X \mid M_k, \Theta)P(M_k \mid \Theta)}, \end{aligned} \tag{27.7}$$

where the last transformation is obtained by dividing the numerator and denominator by $P(M_j \mid \Theta)$ and then expanding out the denominator as in the previous section.

This differs from Eq. 27.4 only in that it lacks a prior probability term in the numerator, and in that there is a log function. It is clear that this too is a discriminant formulation, in that alterations to $\Theta$ that increase the mutual information will increase the earlier quantity [ignoring the prior probability term $P(M_j \mid \Theta)$ in the numerator of Eq. 27.4].

In work at IBM (see, e.g., [1]), methods were developed to train parameters $\Theta$ in order to increase this criterion. These approaches have been referred to as maximum mutual information (MMI) methods. Training is done with a gradient learning approach, in which the parameters are modified in the direction that most increases the mutual information.

MMI approaches have been incorporated in a number of speech-recognition research systems. One practical problem with using MMI for continuous speech recognition is the need for probability estimates for each of the terms in the denominator of Eq. 27.7. In the general case, there are an infinite number of such terms, since there are an infinite number of possible word sequences. One solution is to approximate the denominator by estimating probabilities for a model that permits any phoneme sequence. Another approach has been to approximate the sum over all models (i.e., all possible word sequences) by just using the sum over the $N$-most probable models (word sequences).

## 27.2.2 Corrective Training

Corrective training was a term applied by the IBM group to an MMI-like approach in which the parameters were modified only for those utterances in which the correct models had a lower likelihood than the best models. This can be seen as an approximation to MMI in which the fraction in Eq. 27.7 is only modified for a reduced set of examples (only the cases in which the most likely hypothesis was incorrect). For these cases, the acoustic probabilities are adapted upward for the correct model and downward for the incorrect models. In other words, if

$$P(X \mid M_r, \Theta) \geq P(X \mid M_c, \Theta) + \Delta,$$

then

$$\Theta \Longrightarrow \Theta^*$$

such that

$$P(X \mid M_c, \Theta^*) \geq P(X \mid M_c, \Theta); \quad P(X \mid M_r, \Theta^*) \leq P(X \mid M_r, \Theta).$$

Here $\Delta$ is a margin that must be exceeded before an utterance is considered to be recognized so poorly as to suggest correction of the models, $M_c$ is the correct model, and $M_r$ is the recognized model. The method was described in greater detail in [2].

### 27.2.3   Generalized Probabilistic Descent

A generalization of corrective training and MMI approaches was developed by Katagiri et al. [12]. Given parameters $\Theta$, they define a discriminant function associated with each model $M_i$ as $g_i(X; \Theta)$. This discriminant function can be any differentiable distance function or probability distribution. Often the discriminant function is defined as

$$g_i(X; \Theta) = -\log P(X \mid M_i, \Theta). \tag{27.8}$$

Another solution could be to define $g_i(X; \Theta)$ as the MMI in Eq. 27.7.

Classification will then be based on this discriminant function according to the rule

$$X \in M_j \text{ if } j = \operatorname*{argmax}_i g_i(X; \Theta). \tag{27.9}$$

Given this discriminant function, we can define a misclassification measure that will measure the distance between one specific class and all the others. Here again, several measures can be used, each of them leading to different interpretations. However, one of the most general ones given in [12] is

$$d_j(X; \Theta) = g_j(X, \Theta) - \log \left\{ \frac{1}{K-1} \sum_{k \neq j} \exp[\eta g_k(X; \Theta)] \right\}^{1/\eta}, \tag{27.10}$$

in which $K$ represents the total number of possible reference models. It is easy to see that if $\eta = 1$, Eq. 27.10 is then equivalent to Eq. 27.7, in which all the priors are assumed equal to $1/K$.

The error measure (Eq. 27.10) could be used as the criterion for optimization by a gradientlike procedure, which would result in something very similar to MMI training. However, the goal of generalized probabilistic descent is to minimize the actual misclassification rate, which can be achieved by passing $d_j(X; \Theta)$ through a nonlinear, nondecreasing, differentiable function $F$ (such as the sigmoidal function given in Chapter 8, Eq. 8.8) and then by minimizing

$$E(\Theta) = \sum_j \sum_{X \in M_j} F[d_j(X; \Theta)]. \qquad (27.11)$$

Other functions can be used to approximate the error rate. For example, we can also assign zero cost when an input is correctly classified and a unit cost when it is not properly classified, which is then another formulation of the minimum Bayes risk.

This approach is very general and includes several discriminant approaches as particular cases. In the case of continuous speech recognition (for which all incorrect models cannot typically be enumerated), this approach requires an approximation of the incorrect model scores. Approximations such as those used with MMI (e.g., $N$-best hypotheses) can also be used for this case.

## 27.2.4  Direct Estimation of Posteriors

Model probabilities are usually estimated from likelihoods using the Bayes rule. However, it is also possible to estimate the posterior probabilities directly and to incorporate the maximization of these estimates (for the correct model) directly in the training procedure. The most common structure for such an approach is a neural network, typically a multilayer perceptron such as that described in Chapter 8; sometimes recurrent (feedback) connections are also used. Neural networks can, under some very general conditions, estimate HMM state posterior probabilities. These can then be used to estimate model probabilities.

It has been shown by a number of authors ([4], [9], [18]) that the outputs of gradient-trained classification systems can be interpreted as posterior probabilities of output classes conditioned on the input. See the Appendix at the end of this chapter for a version of the proof originally given in [18].

These proofs are valid for any neural network (or other gradient-trained system), given four conditions.

1. The system must be trained in the classification mode; that is, for $K$ classes (e.g., state categories), the target is one for the correct class and zero for all the others.

2. The error criterion for gradient training is either the mean-squared difference between outputs and targets, or else the relative entropy between the outputs and targets.

3. The system must be sufficiently complex (e.g., contain enough parameters) to be trained to a good approximation of the mapping function between input and the output class.

4. The system must be trained to a global error minimum. This is not really achieved in practice, so the question is whether the local minimum that might actually be obtained will be good enough for our purpose.

It has been experimentally observed that, for systems trained on a large amount of speech, the outputs of a properly trained MLP or recurrent network do in fact approximate posterior probabilities, even for error values that are not precisely the global minimum. When sigmoidal functions (e.g., the function given in Eq. 8.8) are used as output nonlinearities, for instance, it is often found that the outputs roughly sum to one, at least on the average. However, since individual examples do not identically sum to one, many researchers use some form of normalized output. One of the most common approaches to this is to use a *softmax* function rather than a sigmoid. A common form for this nonlinearity is

$$f(y_i) = \frac{e^{y_i}}{\sum_{k=1}^{K} e^{y_k}}, \tag{27.12}$$

where $y_i$ is the weighted sum of inputs to the $i$th output neuron. In other words, each such neuron value is exponentiated, and then the results are normalized so that they sum to one.

Thus, neural networks can be trained to produce state posteriors for a HMM, assuming that each output is trained to correspond to a state category (e.g., a phone). HMM emission probabilities can then be estimated by applying the Bayes rule to the ANN outputs, which estimate state posterior probabilities $P(q_k \mid x_n)$. In practical systems, we most often actually compute

$$\frac{P(x_n \mid q_k)}{P(x_n)} = \frac{P(q_k \mid x_n)}{P(q_k)}. \tag{27.13}$$

That is, we divide the posterior estimates from the ANN outputs by estimates of class priors. The scaled likelihood of the left-hand side can be used as an emission probability for the HMM, since, during recognition, the scaling factor $P(x_n)$ is a constant for all classes and will not change the classification.

Figure 27.1 shows the basic hybrid scheme, in which the ANN generates posterior estimates that can be transformed into emission probabilities as described here, and then can be used in dynamic programming for recognition.

Since posterior probabilities for an exhaustive set of state categories sum to one, the network training is discriminant at the state level; that is, changing the parameters to boost the correct state will also move the system farther away from choosing the incorrect states. In fact, the backpropagation process explicitly includes the effects of negative training from the targets associated with the incorrect states.

However, the goal of discriminant training is not to improve discrimination at the level of states, but rather at the level of complete models (i.e., words or utterances). Is there any reason to believe that this training is discriminant between models? Although there is no real proof of this, there are several types of observations that can be made.

**1.** Intuitive: a system that is better at distinguishing between submodels should be better at distinguishing models. Although mismatches at higher levels (e.g., poor pronunciation models) can interfere with this, it nonetheless would be hard to argue that it should be better to have poor local discrimination.

**2.** Empirical: when a neural network is trained, performance at the state (frame) level and at the word level can be tested after each epoch. The number of epochs that yields the best performance is often not exactly the same for the different levels; however, the general tendencies are very similar, and the optimum stopping point is typically very close

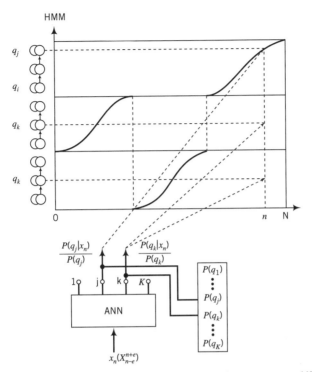

**FIGURE 27.1** Use of a neural network to generate HMM emission probabilities for speech recognition. At every time step $n$, acoustic vector $x_n$ with right and left context is presented to the net (Fig. 27.2). This generates local probabilities $P(q_k \mid x_n)$ that are used, after division by priors $P(q_k)$, as local scaled likelihoods in a Viterbi dynamic programming algorithm. Here, the arrows coming up from each ANN output symbolize the use of these scaled likelihoods (after taking the negative logarithm) as distances from the acoustic input to their corresponding state at time $n$. The solid curves show the best path at each time point.

for the two criteria. Therefore, to the extent that the word error rate is a measure of model discrimination, the state-based discriminant training of neural networks tends to lead to systems that are more discriminant at the model level.

**3.** Theoretical:   although there is no proof per se that the training described here is discriminant for words or utterances, it can actually be proved that an idealized system that incorporates dependencies on all previous states is discriminant at the complete model level. The simpler system here can be viewed as the same training regimen with some strong simplifying assumptions.

On the last point, work over the past few years has shown that relaxing these simplifying assumptions somewhat (for instance, including a dependency on the previous state explicitly in the network training) can demonstrate some improvement [5]. However, since the improvement is small, it is still likely that the simpler system is typically improving model-level discrimination.

Aside from improved discrimination, there are other reasons for researchers' interest in the use of neural networks for probability estimation in statistical ASR. The structure of the network permits flexible inclusion of a range of features, such as acoustic inputs from long temporal contexts. This can facilitate a wide range of experiments that might otherwise be quite tricky with Gaussian mixtures or discrete densities. These approaches also tend to be unconstrained by implicit or explicit assumptions about the feature distributions (e.g., conditional independence within a state). Finally, in practice it has often been observed that fewer parameters are required for posterior-based systems, in comparison with equivalently performing systems that use likelihood estimators. This may be due to the tendency of the former systems to incorporate more parameter sharing.

The next section elaborates on some basic characteristics of hybrid HMM–ANN systems that are used for ASR.

## 27.3 HMM–ANN BASED ASR

### 27.3.1 MLP Architecture

As described earlier, scaled HMM state emission probabilities can be estimated by applying the Bayes rule to the outputs of neural networks that have been trained to classify HMM state categories. Such estimates have been used in a significant number of ASR systems, including large-vocabulary speaker-independent continuous speech-recognition systems. Sometimes these systems have used MLPs, such as those described in Chapter 8. The MLPs could either consist of one single large trained network (systems have been trained with millions of parameters)[16], or of a group of separately trained smaller networks [8]. Sometimes the systems use recurrent networks, typically with connections from a hidden layer back to the input [19], [7]. These can be tricker to train than the fully feedforward systems, but they typically can get very good results with fewer parameters than the feedforward systems.

A typical single-net feedforward implementation is illustrated in Fig. 27.2. Acoustic vectors usually incorporate features such as those discussed in Chapter 22 (e.g., mel cepstra or PLP). A temporal context of such vectors (e.g., 9) are input to the network. For simple implementations, the output categories correspond to context-independent acoustic classes such as phones, and each phone uses a single density. More complex implementations can use multiple states per phone, context-dependent phones, or both (see Chapter 23 for a discussion of triphones, for instance). These more complex designs typically use multiple networks (see, for instance, [8]).

### 27.3.2 MLP Training

A number of techniques have been developed to improve the performance of these networks. Some of these are as follows.

**1.** On-line training:  neural network theory is somewhat more complete for so-called batch training than for on-line training. In the former, modification of the weights is only done once per pass through the data, whereas weights are modified for every new pattern in

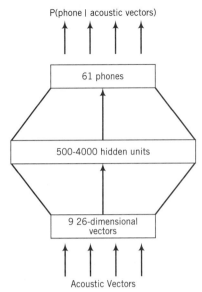

P(phone | acoustic vectors)

61 phones

500-4000 hidden units

9 26-dimensional vectors

Acoustic Vectors

**FIGURE 27.2** Single large MLP used for probability estimation in speech recognition. The acoustic input consists of feature vectors for the current frame, four previous, and four following frames. The output corresponds to phonetic categories for HMM states.

on-line training. However, the latter tends to be much faster for realistic data sets, since the data tend to be quite redundant, so that one on-line pass is in practice equivalent to many batch passes.

**2.** Cross-validation: the goodness of the network must be evaluated during training in order to determine whether sufficient learning has occurred, or sometimes even to optimize training parameters (such as the learning rate). In practice it is often useful to test the system on an independent data set after each training pass; depending on the learning algorithm, the resulting test set performance can be used either for comparison with a stopping criterion or to assess how to change the learning rate. For many cases, failure to do this testing can result in networks that are overtrained. That is, they can overfit the training data, which can lead to poor generalization on independent test sets.

**3.** Training criterion: the relative entropy criterion referred to here tends to perform better (certainly in terms of convergence) than the mean-squared error.

See [16] for further discussion on these and other practical points.

### 27.3.3 Embedded Training

In Chapter 26 we described two principle approaches for iteratively improving the model likelihoods, both based on EM. In Viterbi training, we alternately segmented word-labeled data and updated parameters for the models; these steps were repeated until some stopping criterion was reached. In forward–backward training, explicit segments were were not computed; rather, recursions were used to generate state probability estimates for each frame, and model parameters were estimated from these. These iterative or embedded procedures were necessary because of the lack of an analytical solution to the optimization of the model parameters. In particular, segment boundaries or probabilities are not typically known; rather, training is usually done with speech utterances for which the phone sequence (or sometimes only the word sequence) is known, but not the exact timing of phonetic segments.

Similarly, ANN training can be embedded in an EM-like[1] process. In the Viterbi case, dynamic programming is used to segment the training data (using scaled likelihoods computed from the network outputs). The resegmented data are then used to retrain the network. There is also an approach that is quite analogous to the typical Baum–Welch procedure of Chapter 26; scaled likelihoods are used with the usual forward–backward equations to estimate posterior probabilities for each state and frame. The network is then retrained, using these probabilities as targets [10]. In some cases the recursions can be modified to accommodate dependencies on previous states for more complex models [5].

This section has focused on the use of feedforward neural networks in HMM-based speech recognition. For a brief description of a recent large-vocabulary system based on recurrent networks, see [7]; the system itself is described more completely in [19].

## 27.4  OTHER APPLICATIONS OF ANNs TO ASR

For brevity's sake, we have only discussed neural networks in the context of their application to discriminant training for HMM probability estimation. However, there are a range of other applications of neural networks to speech recognition; some examples follow.

1. Predictive networks:   training networks are used to estimate each frame's acoustic vector given some number of previous vectors, assuming some state category [14], [21].

2. ANN models of HMMs:   networks can be designed to implement the Viterbi algorithm [15] or the forward recursion of the forward–backward algorithm [6].

3. Nonlinear transformation:   networks may also be used to transform the observation features for HMMs [3].

4. Clustering:   networks have been used to cluster the data prior to classification stages, for instance, using the Kohonen feature maps [13].

5. Adaptation:   networks have been used to adapt to the acoustic vectors for a particular speaker [11], or to map from noisy to clean acoustic conditions [20].

6. Postprocessing:   networks trained for more complex models (e.g., statistics of a complete phonetic segment) can be used to rescore word-sequence hypotheses that

---

[1]The procedures that are actually used correspond more to what is sometimes called generalized EM. In generalized EM, the M step doesn't actually maximize the likelihood, but simply increases it, typically by a gradient procedure.

are generated by a first-pass system that may use more conventional approaches [22].

We also refer the interested reader to [16] and [17] for more information on many of the topics discussed in this chapter.

## 27.5 EXERCISES

**27.1** Consider Eq. 27.4. Suppose that all K models have entirely disjoint parameters; that is, changing the parameters for one model has no effect on the likelihoods for another model. Does an increase in $P(X \mid M_k, \Theta)$ (as a result of a change in $\Theta$) imply an increase in $P(M_k \mid X, \Theta)$? What if each parameter change affects the likelihood estimates for all models?

**27.2** Briefly describe the principles that underly the use of maximum mutual information for HMM parameters. Explain how these methods differ from maximum likelihood estimation. Under what circumstances might these alternative training techniques be of benefit?

**27.3** It was stated in the chapter that HMM-based systems that use neural networks for posterior probability estimation often require fewer parameters than systems that are trained to estimate state likelihoods by using Gaussians or Gaussian mixtures. State some reasons why this might be true.

**27.4** Systems that are trained to be more discriminant should, in principle, make fewer errors than systems than have not been so trained. Describe some testing condition in which the discriminant training could hurt performance.

**27.5** It is sometimes said that neural networks do not require the selection of an intermediate representation (i.e., features) but rather can automatically extract the optimum representation. However, in practice, researchers have found that predetermined features are essential for good speech-recognition performance, even when neural networks are used. Why might someone think that the neural networks were sufficient, and why might they not be in practice?

## 27.6 APPENDIX: POSTERIOR PROBABILITY PROOF

Here we briefly repeat the proof originally given in [18], which gives a clear explanation of how networks trained for one-of-K classification can be used as estimators of probabilities for the K classes.

We assume that the network training criterion will be the mean-squared error (MSE) between the desired outputs of the network, which we represent as $d_i(x)$ for the $i$th output and an input of $x$, and the actual outputs $g_i(x)$. In practice, other common error criteria will lead to the same result.

For continuous-valued acoustic input vectors, the MSE can be expressed as follows:

$$E = \int P(x) \sum_{k=1}^{K} \sum_{\ell=1}^{K} P(q_k \mid x) \left[ g_\ell(x) - d_\ell(x) \right]^2 dx. \qquad (27.14)$$

Since $P(x) = \sum_{i=1}^{K} P(q_i, x)$, we have

$$E = \int \sum_{i=1}^{K} \left\{ \sum_{k=1}^{K} \sum_{\ell=1}^{K} [g_\ell(x) - d_\ell(x)]^2 \, P(q_k \,|\, x) \right\} P(q_i, x) \, dx.$$

After a little more algebra, using the assumption that $d_\ell(x) = \delta_{k\ell}$ if $x \in q_k$, we find that adding and subtracting $p^2(q_\ell \,|\, x)$ in the previous equation leads to

$$\begin{aligned}
E &= \int \sum_{i=1}^{K} \left\{ \sum_{\ell=1}^{K} \left[ g_\ell^2(x) - 2g_\ell(x)P(q_\ell \,|\, x) + P^2(q_\ell \,|\, x) \right] \right\} P(q_i, x) \, dx \\
&\quad + \int \sum_{i=1}^{K} \left\{ \sum_{\ell=1}^{K} [P(q_\ell \,|\, x) - P^2(q_\ell \,|\, x)] \right\} P(q_i, x) \, dx \\
&= \int \sum_{i=1}^{K} \left\{ \sum_{\ell=1}^{K} [g_\ell(x) - P(q_\ell \,|\, x)]^2 \right\} P(q_i, x) \, dx \\
&\quad + \int \sum_{i=1}^{K} \left( \sum_{\ell=1}^{K} \{ P(q_\ell \,|\, x)[1 - P(q_\ell \,|\, x)] \} \right) P(q_i, x) \, dx.
\end{aligned} \tag{27.15}$$

Since the second term in this final expression 27.15 is independent of the network outputs, minimization of the squared-error cost function is achieved by choosing network parameters to minimize the first expectation term. However, the first expectation term is simply the MSE between the network output $g_\ell(x)$ and the posterior probability $P(q_\ell \,|\, x)$. Minimization of Eq. 27.14 is thus equivalent to minimization of the first term of Eq. 27.15, that is estimation of $P(q_\ell \,|\, x)$ at the output of the MLP. This shows that a discriminant function obtained by minimizing the MSE retains the essential property of being the best approximation to the Bayes probabilities *in the sense of mean-squared error*. A similar proof was also given in [18] for the relative entropy cost function (computing the relative entropy between target and output distributions).

# BIBLIOGRAPHY

1. Bahl, L., Brown, P., de Souza, P., and Mercer, R., "Maximum mutual information of hidden Markov model parameters," in *Proc. IEEE Int. Conf. Acoust. Speech Signal Process.*, Tokyo, pp. 49–52, 1986.
2. Bahl, L., Brown, P., de Souza, P., and Mercer, R., "A new algorithm for the estimation of hidden Markov model parameters," in *Proc. IEEE Int. Conf. Acoust. Speech Signal Process.*, New York, pp. 493–496, 1988.
3. Bengio, Y., De Mori, R., Flammia, G., and Kompe, R., "Global optimization of a neural network-Hidden Markov Model hybrid," *IEEE Trans. Neural Net.* **3**: 252–259, 1992.
4. Bourlard, H., and Wellekens, C., "Links between Markov models and multilayer perceptrons," *IEEE Trans. Pattern Anal. Machine Intell.* **12**: 1167–1178, 1990.

5. Bourlard, H., Konig, Y., and Morgan, N., "A new training algorithm for statistical sequence recognition with applications to transition-based speech recognition," *IEEE Signal Process. Lett.* **3**: 203–205, 1996.

6. Bridle, J., "Alpha-Nets: a recurrent neural network architecture with a hidden Markov model interpretation," *Speech Commun.* **9**: 83–92, 1990.

7. Cook, G., and Robinson, A., "Transcribing broadcast news with the 1997 Abbot system," in *Proc. IEEE Int. Conf. Acoust. Speech Signal Process.* Seattle, pp. 917–920, 1998.

8. Fritsch, J., and Finke, M., "ACID/HNN: clustering hierarchies of neural networks for context-dependent connectionist acoustic modeling," in *Proc. IEEE Int. Conf. Acoust. Speech Signal Process.*, Seattle, pp. 505–508, 1998.

9. Gish, H., "A probabilistic approach to the understanding and training of neural network classifiers," in *Proc. IEEE Int. Conf. Acoust. Speech Signal Process.*, Albuquerque, N.M., pp. 1361–1364, 1990.

10. Hennebert, J., Ris, C., Bourlard, H., Renals, S., and Morgan, N., "Estimation of global posteriors and forward-backward training of hybrid HMM/ANN systems," in *Proc. Eurospeech '97*, Greece, pp. 1951–1954, 1997.

11. Huang, X., Lee, K., and Waibel, A., "Connectionist speaker normalization and its application to speech recognition," in *Proc. IEEE Workshop Neural Net. Signal Process.* IEEE, New York, pp. 357–366, 1991.

12. Katagiri, S., Lee, C., and Juang, B., "New discriminative training algorithm based on the generalized probabilistic descent method," in B. H. Juang, S. Y. Kung, and C. A. Kamm, eds., *IEEE Proc. NNSP*, IEEE, New York, pp. 299–308, 1991.

13. Kohonen, T., "The 'neural' phonetic typewriter," *IEEE Comput.* **21**: 11–22, 1988.

14. Levin, E., "Speech recognition using hidden control neural network architecture," in *Proc. IEEE Int. Conf. Acoust. Speech Signal Process.*, Albuquerque, N.M., pp. 433–436, 1990.

15. Lippmann, R., "Review of neural networks for speech recognition," *Neural Comput.* **1**: 1–38, 1989.

16. Morgan, N., and Bourlard, H., "Continuous speech recognition: an introduction to the hybrid HMM/connectionist approach," *Signal Process. Mag.* **12**: 25–42, 1995.

17. Morgan, N., and Bourlard, H., "Neural networks for statistical recognition of continuous speech," *Proc. IEEE* **83**: 742–770, 1995.

18. Richard, M., and Lippmann, R., "Neural network classifiers estimate Bayesian a posteriori probabilities," *Neural Comput.* **3**: 461–483, 1991.

19. Robinson, A., "An application of recurrent nets to phone probability estimation," *IEEE Trans. Neural Net.* **5**: 298–305, 1994.

20. Sorenson, H., "A cepstral noise reduction multi-layer network," in *Proc. IEEE Int. Conf. Acoust. Speech Signal Process.*, Toronto, pp. 933–936, 1991.

21. Tebelskis, J., and Waibel, A., "Large vocabulary recognition using linked predictive neural networks," in *Proc. IEEE Int. Conf. Acoust. Speech Signal Process.*, Albuquerque, N.M., pp. 437–440, 1990.

22. Zavaliagkos, G., Zhao, Y., Schwartz, R., and Makhoul, J., "A hybrid segmental neural net/hidden markov model system for continuous speech recognition," *IEEE Trans. Speech Audio Process.* **2**: 151–160, 1994.

# CHAPTER 28

# SPEECH RECOGNITION AND UNDERSTANDING

## 28.1  INTRODUCTION

The incorporation of linguistic constraints into speech recognition will be the major focus of this chapter.[1] Thus far, we have described ASR as a pattern-recognition problem, requiring signal representations, distance or probability estimators, and temporal integration. We have largely ignored linguistic structure, except where it was required to describe the classification units (Chapter 23). However, the Bayes rule formulation of ASR requires an estimate of the prior probability of a hypothesized sequence of words. Since we often do not have enough examples of any given complete utterance to estimate its likelihood accurately, we must be concerned with strategies for training word-sequence probability estimators with insufficient data.[2] Finally, it is also necessary to represent the pronunciation of words as a succession of smaller linguistic units such as phones.

We show how these aspects are incorporated in the decoding process for recognition. We also discuss a number of aspects of complete system integration, including one recent example of a speech-understanding system, that is, a system that includes a functional interpretation of the recognized word sequences (for a limited task).

For the purposes of this chapter, we establish some simple definitions:

- Decoding:  given local probabilities or distances between inputs and references, search for the best model sequence.[3]
- Recognition:  given the input speech, give the best word sequence.
- Understanding:  given the input speech, give the best meaning (or action).

Figure 28.1 shows a typical architecture for a simple speech-understanding system.

---

[1] This chapter owes much to earlier documents and talks by Dan Jurafsky, Gary Tajchman, and Eric Fosler-Lussier.

[2] Note that this problem is quite analogous to acoustic probability estimation, in which estimators for the probability of speech sounds must be trained despite typically having an insufficient representation of all possible acoustic variations.

[3] Because of the limitations of space we do not provide any serious description of decoding strategies here. In practical systems the design of decoders can be critical, providing the best possible trade-off between resource requirements and search errors (where the latter are errors arising from the decoding process that would not have occurred if all possible word sequences had been considered) and resource requirements. The heuristics of good decoders can be quite varied, depending strongly on the details of each individual system. For more on this topic, see [1], [5], [6], and [7].

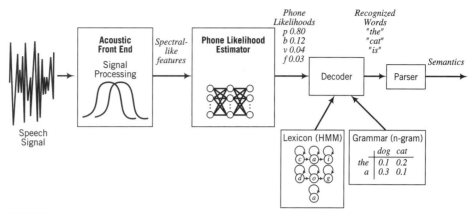

**FIGURE 28.1**  Generic speech-understanding system.

## 28.2  PHONOLOGICAL MODELS

In Chapter 23, we described some of the basic elements of word pronunciation, particularly phones and phonemes. In principle, one could simply permit any pronunciation for any word in the training set, and let the iterative training techniques of the last two chapters be used to discover the pronunciations. To some extent, this kind of self-organized pronunciation model has been found useful; for instance, IBM has developed systems in which the fundamental linguistic unit has been the *fenone* rather than the phoneme, where the former is a self-organized unit (that is, one that is learned in an unsupervised manner from the data).

Still, there do appear to be regularities in pronunciation that can be exploited, and since these regularities are tied to other phenomena (such as speaking rate, accent, stress, etc.), it appears that there can be some value in explicit models of pronunciation. In any event, the majority of ASR systems initialize their statistical training based on a lexicon including one or more pronunciations for each word. A training procedure is then used to further refine the pronunciations.

Figure 28.2 shows sample pronunciations for two short words. Each of the states in these models can represent a phonetic HMM, which might comprise several states with tied or separate acoustic distributions. Note also that the models incorporate phonetic transition probabilities, which are generally learned from examples but which can be initialized from good guesses. An initial lexicon may actually consist of many more pronunciations than these, arising from a combination of dictionaries, application of phonological rules, text-to-speech systems, examples from manually transcribed data bases, and even hand editing. Given all of these potential pronunciations, a working set of pronunciation models can be derived from the following procedure.

1. Run the local acoustic probability estimator (e.g., discrete density estimator) and incorporate in a Viterbi alignment of the training data; in other words, use dynamic

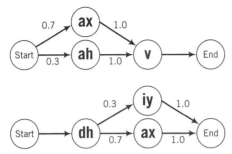

**FIGURE 28.2** Pronunciation models for "of" and "the." Each state in these models could consist of several HMM states, including possible self-loops.

programming with backtracking to find the most likely state sequence for the training observations.

2. Repeat the alignment procedure as necessary (including retraining of the acoustic estimators).

3. With the final alignments, count the occurrences of each pronunciation and normalize to get pronunciations probabilities.

Figure 28.3 shows a set of paths that could be considered for the different pronunciations of the words "of the."

At the surface, this is no different than any other Viterbi training procedure; however, here we specifically isolate the phonological model aspects of specific groups of states. This

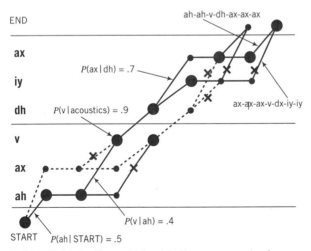

**FIGURE 28.3** Forced Viterbi alignment paths for two short words, each with two possible pronunciations. In this case we assume that a phone corresponds to a single HMM state with a possible self-loop.

**TABLE 28.1  Some Phonological Rules for Casual Speech**

| Name Reductions | Rule | Pr[a] |
|---|---|---|
| Mid vowels | -stress [aa ae ah ao eh er ey ow uh]→ ax | 0.60 |
| High vowels | -stress [iy ih uw] → ix | 0.57 |
| R-vowel | -stress er → axr | 0.74 |
| Syllabic n | [ax ix] n → en | 0.35 |
| Syllabic m | [ax ix] m → em | 0.35 |
| Syllabic l | [ax ix] l → el | 0.72 |
| Syllabic r | [ax ix] r →axr | 0.77 |
| Flapping | [t d]→ dx /V ___ [ax ix axr] | 0.87 |
| Flapping-r | [t d]→ dx /V r ___ [ax ix axr] | 0.92 |
| H-voicing | hh → hv / [+voice] ___ [+ voice] | 0.92 |

[a]This column shows the probability of the rule's application, as taken from transcriptions of natural speech input to a speech-understanding system, using forced Viterbi alignment to determine the pronunciations.

can sometimes lead to estimates of the probability of broader categories of pronunciation, which can be generalized to words that were not seen during training. For instance, we can observe a systematic variation of pronunciations based on such factors as rate of speech, dialect, accent, and words in context.

Table 28.1 gives phonological rule probabilities that have been obtained through such a procedure; these can be used to generate reasonable pronunciations for words that have never been observed in the training set.

## 28.3  LANGUAGE MODELS

In Chapter 9 we showed that the Bayes rule provides a simple transformation to reverse the arguments in a conditional probability or density. In earlier chapters we applied this rule for the specific example of speech-observation frame and word sequences, in which the parameters of the acoustic and language models are assumed disjoint. For convenience, we repeat this formulation here:

$$P(M \mid X, \Theta) = \frac{P(X \mid M, \Theta_A)P(M \mid \Theta_L)}{P(X \mid \Theta_A)}. \tag{28.1}$$

In the previous chapters, we focused on the acoustic probabilities $P(X \mid M, \Theta_A)$, and we assumed that the language models would provide the prior model probabilities. The reader should note that the simple expression $P(M \mid \Theta_L)$ hides a host of theoretical and practical difficulties. At the core of this difficulty is the fact that, in general, we don't have enough data to estimate the probability of word strings of unrestricted length. Even if we were to limit ourselves to word strings with three elements, for a vocabulary of 10,000 words there would be a trillion possible word strings. Of course many of these would never occur, but with that many possible strings, we would never know how to interpret the nonoccurrence of a word string in the training set. In other words, if a word triple never occurs in training, it still could occur during recognition.

For this reason, the estimation of $P(M \mid \Theta_L)$ typically requires the use of multiple levels of approximation, in which poor estimates of multiple word probabilities are combined with better estimates of shorter-length sequences. This situation is not unique to language modeling, of course, and we have already spoken of the need for smoothing in estimates of local acoustic probabilities. In the case of language models, though, estimates also commonly incorporate a large number of training sequences that may come from written text, which differs in fundamental ways from word sequences in natural speech; the relative scarcity of word-transcribed data for spoken language often necessitates the additional use of the text-based materials.

Although iterative statistical training methods such as EM learn directly from the data, they can nearly always benefit from the prior imposition of some structure. In speech-recognition systems that are focused on a limited domain (such as data-base query for restaurants information), the syntactic and semantic structure can be used to further constrain the word-sequence hypotheses for recognition. In the more general domain of English, our limited knowledge of the structure of spoken language has not yet really been shown to provide strong advantages over unstructured statistical approaches.

Some of the fundamental questions about language modeling for speech recognition bear a strong resemblance to our questions about statistical modeling in general, as described in Chapter 26:

**1.** Parametrization and probability estimation: How should $P(M \mid \Theta_L)$ be computed, and what are the necessary assumptions about the stochastic models to define the parameter set $\Theta_L$? In general it is not possible to estimate the probability density of a complete sequence without constraining assumptions; as with the acoustic parameters, these assumptions typically involve conditional independence so that the total estimate is broken up into constituents that can be effectively estimated.

**2.** Training: Given the parametrization here, how should $\Theta_L$ be determined so that the known sequence of words has the highest possible likelihood?

**3.** Decoding: How can the trained parameters $\Theta_L$ be incorporated along with the acoustic models to optimally classify the input sequence $X$ according to the Bayes decision rule?

As noted in Chapter 5, the uncertainty about each word given the the predictive power of a language model is measured by its *perplexity*, which is just $2^H$, where $H$ is the average entropy for a word given the language model. In the simple case in which each word has exactly $N$ equally likely successors, the perplexity would then be $N$. In general, adding linguistic predictive power will reduce the perplexity, and this tends to improve recognition performance (though this is not guaranteed; many researchers have been disappointed by recognition results using techniques that reliably reduce perplexity). Explicit representations of syntactic and semantic structure can be useful for perplexity reduction, but estimators for $P(M)$ have been dominated by simple statistical approaches, most commonly the *n*-gram approach [2].

### 28.3.1  *n*-Gram Statistics

In many recognition systems, the language model is a simple one:  for each word (including a placeholder "START" word) there is a list of possible words that can follow. This is sometimes called a word-pair grammar (for obvious reasons). A generalization of this approach is to associate a probability with each word in the follow set, and this is called a *bigram* model. Similarly, a word list with associated probabilities that can follow any pair of words is called a *trigram* model. In general, such models are called *n*-grams, where *n* is a small integer.

How can such probabilities be used to estimate $P(M)$? First, let $M = w_1^{N_w} = w_1, w_2, \ldots, w_{N_w}$ for $N_w$ words in the hypothesized HMM $M$. Then, for each sequence $w_1^N$,

$$P\left(w_1^N\right) = P(w_1)P(w_2 \mid w_1)P\left(w_3 \mid w_1^2\right) \ldots P\left(w_N \mid w_1^{N-1}\right) = \prod_{k=1}^N P\left(w_k \mid w_1^{k-1}\right). \tag{28.2}$$

We then approximate

$$P\left(w_N \mid w_1^{N-1}\right) \approx P\left(w_N \mid w_{N-n+1}^{N-1}\right) \tag{28.3}$$

so that

$$P\left(w_1^N\right) \approx \prod_{k=1}^N P\left(w_k \mid w_{k-n+1}^{k-1}\right), \tag{28.4}$$

where $P(w_1)$ and $P(w_1 \mid w_{2-n}^0)$ are typically taken to be equivalent to $P(w_1 \mid \text{START})$.

This approximation should look familiar:  it uses an $(n-1)$-order Markovian assumption. Thus, for the trigram case, $n$ is equal to 3 and we assume that we only need two predecessor words to compute the linguistic priors for the next word.

Let's take the specific example of the word sequence "I eat dinner." For a bigram grammar, the probability for this sequence is given as

$$P(\text{I eat dinner}) = P(\text{I} \mid \text{start})P(\text{eat} \mid \text{I})P(\text{dinner} \mid \text{eat}).$$

Since the word transcription is presumed to be known for the training set, in principle the *n*-gram probabilities can simply be estimated by counting co-occurrences of the words. For instance, for the case above, a simple estimate of $P(\text{eat} \mid \text{I})$ would be

$$P(\text{eat} \mid \text{I}) = \frac{P(\text{I eat})}{P(\text{I})} = \frac{\text{no. occurrences of "I eat"}}{\text{no. occurrences of "I"}}. \tag{28.5}$$

When there are are many occurrences of these words, good estimates can be obtained in this simple way. However, as noted previously, it is difficult to estimate the likelihoods of infrequently occurring *n*-grams. For large-vocabulary recognition, the number of possible bigrams or trigrams becomes huge, and many of the probabilities are poorly estimated. Furthermore, the predictive power is often stronger for words that are not the immediately preceding words. For instance, in the phrase "eating a large number of hamburgers," the word "eating" would most likely be a better predictor of "hamburgers" than "number of." Modeling these longer-distance relationships with simple $n$-grams would typically require a fairly large value for $n$, which would not be feasible with typical training set sizes. Some word types tend to be more important than others in general (e.g., verbs as opposed to conjunctions), and simple $n$-grams do not have a notion of structure that would explicitly handle syntactic and semantic relations. Some of these problems are ameliorated by using written text to train language models, since huge amounts of on-line text is now available. This also, however, raises a new problem: spoken language is quite different from written language, and the differences may limit the utility of the written corpora for spoken language model training.

For these reasons, a major concern in speech-recognition research has been the improvement of statistical language models. Some of the most successful of these techniques have been similar to those discussed in previous chapters in the context of acoustic likelihood estimation – for instance, backoff or deleted interpolation techniques for the incorporation of trigram, bigram, and unigram probabilities. Experiments continue with hybrids between more structured grammars and word-based ones [e.g., class-based $n$-grams, such as $P(\text{Foodclass} \mid \text{eat})$].

As of this writing, formal grammars such as context-free grammars are most commonly used for the interpretation of recognizer output in a speech-understanding system. One example of such an approach is given later in this chapter.

## 28.3.2 Smoothing

As noted earlier, there is never enough training data to learn all the statistical relationships between words in speech communication. This fact has been one of the motivations for the interest in more structured language models. However, even for $n$-grams, it can be difficult to make good probability estimates for infrequent combinations of words. This requirement is similar to the one discussed in Chapter 26 for acoustic density estimation. In that chapter, we noted that the need for capturing acoustic variability motivates the use of finer state categories such as triphones (as opposed to phones), but that such finer categories have fewer examples. A similar observation can be made for the language model. Trigrams are stronger models (in the sense of potentially being more predictive) than bigrams, but many more sequences of three words will occur infrequently than sequences of two words.

With language modeling as with acoustic modeling, it is common that we will require the combination of good estimates of coarse categories with noisy estimates of fine categories, in order to stabilize the estimate of the latter: as noted in Chapter 26, this process is often called *smoothing*. The same techniques that are used for acoustic model smoothing are commonly used for language modeling as well, namely the following.

**1.** Backoff smoothing: this is a simple but often effective method. Thresholds are set for a minimum number of examples for each level of granularity, and when the minimum is not present, the estimator backs off to a coarser level. For instance, if there are enough examples of a particular trigram, the probability derived from relative frequencies may be used directly; if there are not enough, a bigram might be used; if there are not enough examples of the bigram, then the unigram probability (relative frequency of the word) is used. Backoff grammars are extremely common in ASR systems.

**2.** Deleted interpolation:  this is the same process that is described in Chapter 26, applied to the combination of $n$-gram probabilities for different values of $n$. As in the acoustic case, two sets of parameters are combined with weighting factors that are learned from testing on data that are disjoint from the training data. Often, this is done by partitioning the data into $N$ pieces (e.g., 10) and then training $N$ different sets of parameters using all the different choices for $N - 1/N$ of the data. For the case of two estimators, the goal is to optimally choose $\epsilon$ to combine the estimator parameters, that is

$$\Theta_T = \epsilon\Theta_1 + (1 - \epsilon)\Theta_2. \qquad (28.6)$$

In one approach, $\epsilon$ is set to the fraction of utterances for which $\Theta_1$ was better. Another solution is to treat $\epsilon$ as missing information and to use EM training. Finally, in practice this value is often set by trial and error.

## 28.4  DECODING WITH ACOUSTIC AND LANGUAGE MODELS

As noted earlier in this chapter, the Bayes rule provides a mathematical structure that permits the easy integration of prior word-sequence probabilities with the acoustic probabilities. However, there are some important practical considerations. For instance, when first-order HMMs are used, the only word-sequence probability incorporated in the search is (by definition) a bigram (which may back off to a unigram). In other words, since the probability of exit from a word-final state is only dependent on the word transition probability (that is, probability of the new word given the old), there is no obvious way to include an $n$-gram with an $n$ of 3 or more in an on-line Viterbi search with the usual first-order Markov assumptions.

However, there are three ways in which such longer-range models are sometimes incorporated in the search.

**1.** Use a higher-order model. If the model explicitly includes higher-order dependencies, and if the search is built to handle them, then even a Viterbi search can use trigrams. This does, however, definitely complicate the search.

**2.** Use a depth-first search. Some recognition decoders incorporate a strategy commonly called stack decoding, in which a prioritized queue (not actually a stack) holds the best current hypotheses, which are extended and reshuffled for each new frame. Since a complete hypothesis is evaluated for each new extension, long-range models ($n$ greater than 2) can be used.

**3.** Postprocess the word sequence. Some number of proposed hypotheses for the complete utterance can be evaluated by a separate module that can do the scoring based on long-range criteria. This can be done explicitly in what is called an $N$-best list, in which the first module submits a list of the most likely utterance hypotheses, along with an estimate of their probabilities. These probabilities are rescored by the second module, using new sources of information (such as matching scores for syntactic or semantic rules). It can also be done implicitly by generating a lattice (i.e., graph) of possible word sequences, with associated probabilities from the on-line acoustic and language models. Here too the lattice can be rescored, using the second set of information sources.

## 28.5   A COMPLETE SYSTEM

In the early-mid-1990s, Berkeley speech researchers designed a speech-understanding system called the Berkeley Restaurant Project (BERP), a medium-vocabulary (1200 word) speaker-independent spontaneous speech-understanding system [3]. BERP functions as a knowledge consultant whose domain is restaurants in Berkeley. A brief description of this system (which drew inspiration from earlier systems such as VOYAGER from MIT [8]) may serve to illustrate some of the points discussed earlier.

Figure 28.4 shows the complete speech-understanding system, in which the decoder uses bigrams in addition to a stochastic context-free grammar (SCFG)[4] for language model probabilities. In the current implementation of BERP, a SCFG is used to generate bigram probabilities that are smoothed with the bigrams computed from the training set transcriptions. The SCFG consists of 1389 hand-written rules, which were trained on a 4786-sentence corpus using EM. A sample rule would be

$$\text{WANTING} - S \rightarrow i \text{ WANT} - V \text{ WANT} - OBJ \text{ (Prob} = 0.35).$$

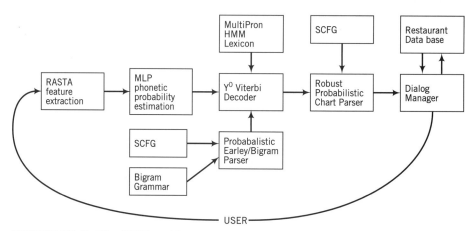

**FIGURE 28.4**   The BERP architecture.

[4]An SCFG is a four-tuple (W, N, A, R) in which W is a set of terminal symbols (in our case words), N is a set of nonterminal symbols (e.g., VERB), A is a start symbol, and R is a set of rules of the form $N^i \rightarrow X^j$, where $X^j$ is a string of terminals and nonterminals and where there is an associated probability $P(N^i \rightarrow X^j)$.

**TABLE 28.2  A Berkeley Restaurant Project Interaction**

| Name | Address | Cost | Distance |
|------|---------|------|----------|
| Thai Bar-B-Q | 1958 Shattuck | $$ | 5 |
| Plearn Thai Cuisine | 2050 University | $$ | 5 |
| Berkeley Thai House | 2511 Channing | $$ | 20 |
| Siam Cuisine | 1181 University | $$$ | 18 |

BERP: *Welcome to BERP. How may I help you?*
User: I'm looking for a Thai restaurant.
BERP: *When would you like to go out?*
User: I'd like to go for lunch on Saturday.
BERP: *How far from ICSI are you willing to travel?*
User: Anywhere within walking distance.
BERP: *How much money do you want to spend?*
User: Doesn't matter.
BERP: *Here are some restaurants that meet your* requirements...

The decoder is a first-order Viterbi decoder, and the result of the recognition is interpreted by a parsing routine (using the SCFG) so that fields in a data-base query can be filled in. A simple dialog manager handles the interaction between the user and the data base, making use of the filled-in field. The natural language component knows a little bit about the world; for instance, it knows what day it is so that "tonight" gets reinterpreted as the specific day so that the system can check to see if a restaurant is open.

Table 28.2 gives a sample interaction with the program.

An interesting aspect of this research project was analyzing the different rule probabilities that were observed for different groups of speakers; this was studied as part of a larger project to try to understand the nature of accent. For instance, initiating a sentence with the word "PLEASE" occurred with probability 0.00057 for American speakers of English, while for German speakers of English the corresponding probability was 0.1.

## 28.6  ACCEPTING REALISTIC INPUT

We conclude this chapter with a brief discussion of some practical concerns for complete speech-recognition–speech-understanding systems.

Early in this book we mentioned a kind of recognition paradigm called word spotting, in which the system is intended to recognize some limited vocabulary but which can accept an unlimited vocabulary input. A typical form for such a system is to look for something like digits but permit the user to say "Uh, I guess the number is five." In such systems, we are not only interested in a low error rate, but also in rejection of irrelevant words. Often we are willing to have more word errors (in the sense of not getting them right) in exchange for having better confidence about the ones that we have not rejected. Formally, such systems take the form of a signal detection scheme, in which the designer trades off false acceptances against missed detections.

However, in practice, rejection and measures of confidence are significant for *all* speech-recognition–speech-understanding systems. All such systems will make errors, and it is important to have a good strategy for those cases in which the system believes it may be making an error. This means that several other facilities must be available besides the recognition algorithm we have emphasized, including:

- A representation for words and sounds other than the ones in our lexicon
- A measure of confidence in the recognition–understanding output
- A rejection policy given that a measure of confidence

The first requirement is often called a garbage model (sometimes also called a filler model). This is sometimes implemented by using a fully connected model with all phones; another approach is to use an average log emission probability for the top $N$ states other than the best one (for each frame). In any case, the difference between the log likelihoods of the most probable model and this filler model (a log likelihood ratio) is often used as a measure of confidence. Other measures can be derived from the range between log likelihood for the most likely and second most likely utterance, or from estimates of the average entropy of the phonetic probabilities. Measures have also been developed based on the behavior of the decoder itself, such as the number of alternate hypotheses that remain within some pruning threshold.

During development tests a signal detection curve (sometimes called a receiver-operating characteristic, or ROC) is used to associate confidence levels with trade-offs between false acceptances and false rejections for the utterance recognition. Having such measures can also be important for assessing a fielded technology; although the recognition error rate is often not easily determined for systems that are being used commercially, statistics about confidence measures can be unobtrusively gathered.

Given these models and measures, any particular application will have a rejection policy. For instance, if the end use is a telephone application for which the cost is closely related to time use (time is money), the best strategy may simply be to fall back to a human operator whenever any confidence is low. For other cases in which a few retries are tolerable, the system may just request a repeat by the user. Of course, users will often just talk louder in this case.

As noted earlier, speech-understanding systems incorporate natural language processing for the recognized utterances. However, for fluent speech, utterances often do not follow the expected word patterns. Furthermore, users will insert disfluencies (filled pauses such as "um" and false starts such as "I'm going to Lon – Paris") and nonspeech sounds (such as coughs). Although filler models will handle some of these problems, the resulting hypothesized transcriptions will still often not be parsed by using a simple grammar. Consequently, it is essential that the natural language processing component be able to handle speech fragments; that is, it has to be able to pick out useful groups of words out of a word sequence that otherwise does not make sense to the parser. In the case of BERP, for instance, it can pick out "Chinese food" out of the phrase "I um wanna have let's see Chinese food". Other speech-understanding systems have similar capabilities.

## 28.7 CONCLUDING COMMENTS

As with most other chapters in this text, we have just touched the surface on many topics, and we have had to leave out many others entirely. As an example of the former, we have just briefly discussed search (decoding) techniques, and we only mentioned the use of formal grammars. We have ignored speaker adaptation, in which model parameters are adjusted during use to better match a particular speaker; this is a critical feature of recognition systems used on personal computers, e.g., for dictation. We also have not described real-time constraints for practical systems and the impact that these have on algorithmic choices. For many applications, the integration with a natural language processing component is not merely a static one as we have reported here; the generation and interpretation of dialog is quite important. Complete systems are not merely recognizers or even understanding engines, and such components must be integrated with synthesis or playback, speaker verification, and so on. Nonetheless, this chapter is an introduction to the rest of the recognizer.

We recommend that the reader consult [4] for a range of topics related to language modeling and computational linguistics that are essential to the design of complete recognition (and synthesis) systems.

## BIBLIOGRAPHY

1. Gopalakrishnan, P., and Bahl, L., "Fast matching techniques," in C. H. Lee, F. Soong, and K. Paliwal, eds., *Automatic Speech and Speaker Recognition*, Kluwer, Norwell, Mass., 1996.
2. Jelinek, F., "Self-organized language modeling for speech recognition," in A. Waibel and K.-F. Lee, eds., *Readings in Speech Recognition*, Morgan Kaufmann, San Mateo, Calif., 1990.
3. Jurafsky, D., Wooters, C., Tajchman, G., Segal, J., Stolcke, A., Fosler, E., and Morgan, N., "The Berkeley restaurant project," in *Proc. ICSLP-94*, Yokohama, Japan, pp. 2139–2142, 1994.
4. Jurafsky, D., and Martin, J., *Speech and Language Processing: An Introduction to Natural Language Processing, Speech Recognition, and Computational Linguistics*, Prentice–Hall, Englewood Cliffs, N.J., 1999.
5. Ney, F., and Aubert, X., "Dynamic programming search: from digit strings to large vocabulary word graphs," in C. H. Lee, F. Soong, and K. Paliwal, eds., *Automatic Speech and Speaker Recognition*, Kluwer, Norwell, Mass., 1996.
6. Renals, S., and Hochberg, M., "Efficient search using posterior phone probability estimates," in *Proc. IEEE Int. Conf. Acoust. Speech Signal Process.*, Detroit, pp. 596–599, 1995.
7. Schwartz, R., Nguyen, L., and Makhoul, J., "Multiple-pass search strategies," in C. H. Lee, F. Soong, and K. Paliwal, eds., *Automatic Speech and Speaker Recognition*, Kluwer, Norwell, Mass., 1996.
8. Zue, V., Glass, J., Goodine, D., Leung, H., Phillips, M., Polifroni, J., and Seneff, S., "Integration of speech recognition and natural language processing in the MIT VOYAGER system," in *Proc. IEEE Int. Conf. Acoust. Speech Signal Process.*, Toronto, pp. I.713–716, 1991.

# PART VII

# SYNTHESIS AND CODING

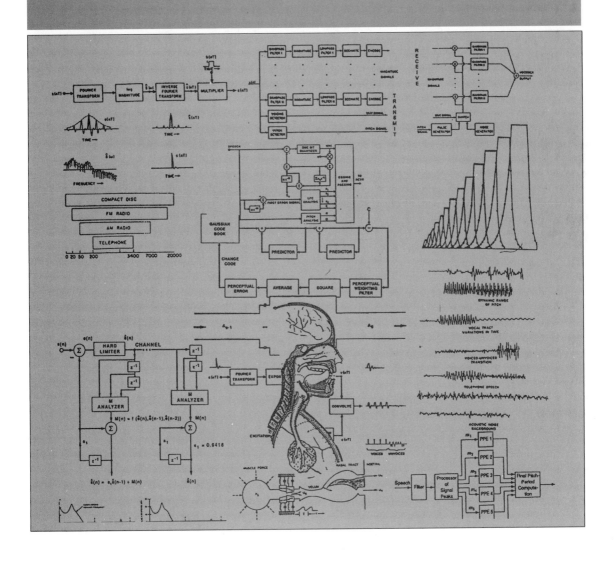

And computers are getting smarter all the time; scientists tell us that soon they will be able to talk with us. (By "they," I mean computers; I doubt scientists will ever be able to talk to us.)

—Dave Barry

I**N PART** VII we proceed to applications for which a speech signal is generated for human listeners. As with recognition, we begin with classical methods and then extend the discussion to the more refined approaches that are current.

Chapter 29 describes the basics of speech-synthesis systems. Although many modern systems are concatenative, the initial emphasis in this chapter is on analysis–synthesis, since many of the ideas are also fundamental to speech-coding systems. The chapter concludes with a discussion of overlap-add-based concatenation. Chapter 30 proceeds to describe basic approaches to pitch detection, which is still a difficult problem. Many modern coding systems avoid this difficulty by not making an explicit pitch determination per se, but this information is still important for some applications. Chapter 31 provides a general background for the different classes of vocoders; as with Chapter 24's description of deterministic recognition systems, this chapter does not so much describe a method in current practice as provide background material for more complex methods that are still related to the older approaches. Chapter 32 then proceeds to discuss vocoders that are designed for a low bit rate (typically 2.4 kbits/s or less), starting with simpler approaches and proceeding to more modern refinements. Finally, Chapter 33 describes higher-quality vocoders, leading up to the class of systems that is currently used in cellular telephony.

# CHAPTER 29

## SPEECH SYNTHESIS

## 29.1 INTRODUCTION

The goal of this chapter is to introduce engineering approaches for "talking" machines that use sequences of subword units. The use of such elements provides flexibility for extended or even arbitrary vocabularies, as required for applications such as the unlimited translation from written text to speech. Text-to-speech synthesis involves many steps, most of which are beyond the scope of this book. The basic components are as follows.

**1.** Text preprocessing:    This step translates the text string of characters into a new string with ambiguities resolved. An example of this is the translation of "Dr." into either "Doctor" or "Drive," depending on the linguistic context.

**2.** Text to phonetic–prosodic translation:    The processed text is parsed to determine syntactic (and, to some extent, semantic) structure. The sequence of words and their derived structure are then used to generate prosodic information and preliminary strings of sound units (e.g., phonemes or diphones). The sound unit generation is generally more complex than merely looking up the words in a dictionary, as the pronunciation of words is highly dependent on context. A sequence of modifications to canonical pronunciations is typically encoded in a series of rewrite rules, such as the ones described in Chapter 23 (Subsection 23.4.4). Prosodic rules determine quantities such as pitch, duration, and amplitude for each of these sound units.

**3.** Given the sequence of these quantities (sound unit label, pitch, duration, amplitude), the signal-processing component generates speech.

The first component is mundane but critical to high-quality synthesis. The second component is a key one, and by some researchers' estimation may be the source of the most disturbing errors in current speech-synthesis systems. The third component is the topic of this chapter, and it is still critical to the intelligibility and quality of the resulting synthetic speech.

The sound unit to speech translation is done with one of three basic approaches:

1. Articulatory synthesis
2. Source–filter synthesis (synthesis by rule)
3. Concatenative synthesis

The first category consists of physical models for articulators and their movements. In a sense the early mechanical systems from von Kempelen and Wheatstone were articulatory synthesizers, but in our own age a more typical approach would be that described in a number of publications by Coker and colleagues; see, for example, [8], [10], or [47]. In these approaches computational models for the physical system are used directly to estimate the resulting speech signal. The method is appealing because of its directness, but the difficulty of deriving the physical parameters by analysis and the large computational resources required for synthesis have made this approach more interesting from a scientific standpoint than a practical one.

Historically, what we have called source–filter synthesis here has been formant synthesis, in the sense that the classic synthesizers based on a spectral shaping of a driving excitation have most often used formants to characterize the spectral shape. Formants have a straightforward acoustic–phonetic interpretation, while being computationally simple in comparison to full articulatory models; a formant is usually represented by a second-order filter. This characterization was particularly conducive to developing rules that handled the parameter modification for a particular context. However, sometimes the spectral characteristics have been specified in terms of the short-time cepstrum, and sometimes linear prediction has been used. Much of this chapter describes the basic approaches of this type.

Finally, as integrated circuit density and speed have continued to increase exponentially, concatenative approaches to speech-signal generation have become increasingly important. Both direct-time waveform storage and parametric storage for speech segments have become more common in recent years, although the currently dominant approach is to use diphone segments, as defined in Chapter 23. A number of other units have also been tried, such as demisyllables.

## 29.2 PARAMETRIC SOURCE–FILTER SYNTHESIS

As noted earlier, von Kempelen's synthesizer could be viewed as an early articulatory system. However, even the 1939 Voder of Chapter 2 already had abstracted the model of speech production from that of a direct analog of articulation to one that focused on key features that could be easily parametrized. As noted previously, this simple model was based on separately specifying the excitation from the spectral shaping that modeled the filtering action of the vocal tract. Although this characteristic of the Voder foreshadowed later developments, the Voder still required hand control of the parameters, as did von Kempelen's synthesizer. In both cases, virtuosic performances by the human controller were required. By the 1950s, this requirement was softened by the introduction of an intermediate step. In the OVE [19] and PAT [37] synthesizers,[1] parameters were entered by hand after careful spectrographic study of the utterance to be synthesized. (Thus, these were not real-time devices like the Voder or von Kempelen's machine.) OVE was a serial combination of formant resonators, whereas PAT was a parallel network. In each case, though, intense human effort was involved, but not to play the synthesizer. Rather, researchers inspected spectrograms to determine formant positions for different sounds, which were catalogued and recalled.

This human labor was eliminated in the Pattern Playback of Haskins Laboratory, [12] which synthesized speech directly from printed spectrograms or hand-drawn spectrographic cartoons. The device is shown in Fig. 29.1. Parameters were applied by means of an optical system; the tone wheel generated optical pulses at a fixed rate (120 Hz). Thus, harmonics of roughly equal amplitude were generated. These were modulated by the moving transparent

[1]OVE is the acronym for the Latin *orator verbis electris*, and PAT stands for parametric artificial talker.

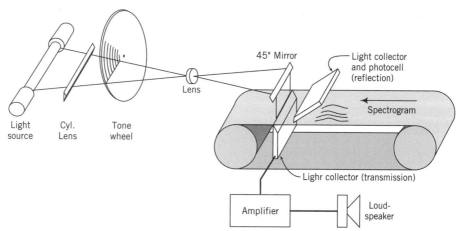

**FIGURE 29.1** The Haskins Pattern Playback, consisting of an optical system for modulating the amplitudes of a set of harmonics of 120 Hz over time, depending on patterns painted on a moving transparent belt. From [35].

belt that contained the spectrographic information. Photocells transcribed the modulated light into electrical signals, which were then amplified and played out through the loudspeaker. Despite the constant pitch of the system and the rather artificial-sounding results, this device played an important role in early speech-perception research [38].

The 1960s witnessed the development of synthesis-by-rule programs. The researcher specified the sequence of phonemes for the utterance to be synthesized. The system had to figure out how to convert this phoneme sequence into the parameters needed to drive the synthesizer.[2] For a pure synthesis-by-rule system, the system also had to derive prosodic information such as pitch, intensity, and duration. Early programs were compromises; prosodic features were still generated by the research worker while spectral parameters were produced automatically.

As noted earlier, complete text-to-speech translation necessitated an additional major step: grapheme- (text symbols) to-phoneme transcription. When a trained linguist looks at printed or written text, he or she can easily write a sequence of phonemes for a spoken version. To do this automatically was a major research undertaking. Umeda et al. [49] demonstrated the first complete text-to-speech system, and research proceeded through the 1970s and into the 1980s, culminating in commercially available systems.

## 29.2.1　Formant Synthesizers

The early OVE of Fant was capable of little more than vowel-like utterances ("I love you"); but by 1962 at the Stockholm Speech Communication Conference, Fant et al. [20] had developed OVE II, a much more sophisticated synthesizer; it is shown in Fig. 29.2.

---

[2]By this time, especially toward the end of the decade, speech synthesizers were implemented as computer programs.

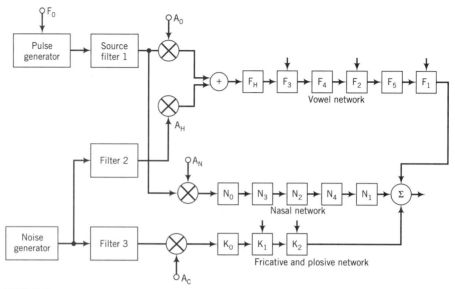

**FIGURE 29.2** The OVE II Speech Synthesizer of Gunnar Fant. From [20].

OVE II had three parallel channels; the top branch could synthesize vowels, semivowels, and glides; the middle branch was devoted to nasals; and the bottom branch was for fricatives and plosives. The excitation structure was also quite advanced for its time. It had the ability to control the balance of periodic and noise excitation for most of the sounds. Thus, for example, the top branch could produce whispered vowels and the bottom branch could produce mixed buzz and hiss excitation for voiced fricative sounds.

The Holmes synthesizer [30] is shown in Fig. 29.3. Note the 6 dB/octave block at the output, which is a high-pass filter that represents the radiation characteristic of the mouth-to-air junction. Also, the 12 dB/octave block following the glottal wave generator is a low-pass filter to give an approximation to the spectrum of the glottal waveform.

An important distinction between the Holmes and Fant synthesizers is the way in which amplitude information is used. In the Fant device, the different branches are controlled separately by $A_H$, $A_N$, $A_C$, and $A_O$. For Holmes' synthesizer, the five formant filters, $F_N$, $F_1$, $F_2$, $F_3$, and $F_4$, are individually controlled to produce *all* sounds. Intuitively, the Fant structure more closely resembles the human vocal tract, yet Holmes claimed that his configuration yielded a closer approximation to human speech [31].

Since most synthesizers (and indeed most vocoders) produce speech that is inferior to the same utterance spoken by a person, a question arises as to the most important reasons why this is so. Some claim that an imprecise simulation of the excitation function is the primary limitation [22]. Others seem to believe that small spectral differences are the root of the problem [21]. In any event, it is certainly true that in the case of full text-to-speech translation, the prosodic contours make synthetic speech sound unnatural for many utterances.

The synthesizer developed by Dennis Klatt [34] was a compromise between the designs of Fant and Holmes. A block diagram is shown in Fig. 29.4 (also see Fig. 29.12 in later text). Klatt incorporated Fant's generation of voiced sound by using cascade formant

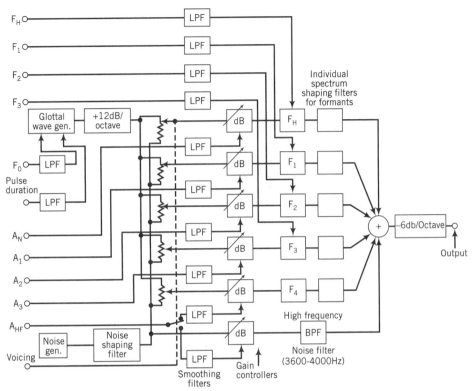

**FIGURE 29.3** The Holmes Parallel-Formant Synthesizer. From [30].

resonators. For the sounds created by frication, he used a channel vocoder structure, with fixed filters of variable gain. A study of Fig. 29.4 reveals that there are a total of 19 parameters to vary; five of these are for the excitation sources. This basic structure has served as the synthesizer for many text-to-speech systems, as exemplified by Klattalk and DecTalk (Klattalk's later commercial form).

## 29.2.2 Other Source–Filter Synthesizer Structures

The systems of Fant, Lawrence, Holmes, Klatt, and others are representative of key structures in all source–filter synthesizers. In this section we categorize and discuss the various configurations. We begin by listing the major variants:

1. All-pole synthesizers, derived from formant analysis.
2. All-pole synthesizers, derived from linear predictive coding analysis.
3. All-zero synthesizers, derived from cepstral analysis.
4. Fixed poles and variable zeros, derived from channel vocoder analysis.
5. Variable poles and variable zeros, derived for a parallel-formant synthesizer.

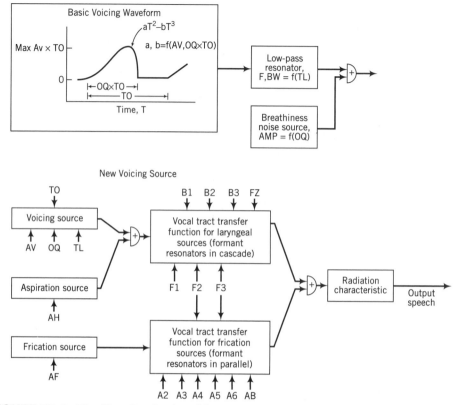

**FIGURE 29.4** The Klatt Synthesizer. Nineteen variable control parameters are identified, including the new voicing source parameters OQ (open quotient) and TL (spectral tilt). From [35].

All-pole synthesizer derived from formant analysis:    the structure is that of the vowel portion of Fig. 29.2, Fant's OVE II. The digital network is a cascade of second order poles that have been derived either by inspection of a spectrogram, or automatically.

All-pole synthesizer derived from linear predictive coding analysis:    linear predictive coding (LPC) analysis was discussed in Chapter 21. Here we simply want to show the structure derived from such an analysis; Fig. 29.5 shows two possible configurations resulting from LPC analysis.

Figure 29.5(a) is the digital network corresponding to an $n$th-order difference equation. Since the present output $y(n)$ is a function of only the past outputs, the result is an all-pole network. The derived parameters $a_1$ through $a_p$ can be obtained, for example, by using the method of Levinson, described in [29]. A modification of this method leads to Fig. 29.5(c). In either case, the analysis is done automatically. It is important to realize that the two networks have identical transfer functions. However, the bottom network is the preferred one because it is less sensitive to quantization noise. The difference equation could also be realized with a cascade of second-order polynomials, which generally can be implemented without much sensitivity to coefficient inaccuracy.

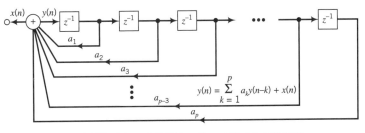

$$y(n) = \sum_{k=1}^{p} a_k y(n-k) + x(n)$$

(a) Direct-form digital filter with variable "$a$" coefficients

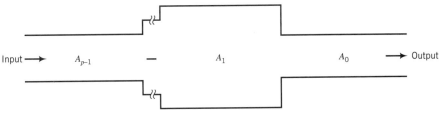

(b) Acoustic tube with variable area functions

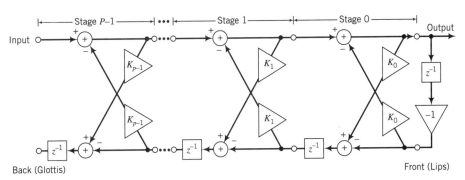

(c) All-pole lattice network with variables "$k$" parameters

**FIGURE 29.5** Two configurations for all-pole synthesizers based on LPC analysis. (a) Direct-form implementation of the difference equation giving a synthesizer output as a weighted sum of its past values plus the excitation input. (b) Model of the acoustic tube with a variable cross-sectional area that could give rise to such a characteristic. (c) Interpretation of this model that suggests a lattice form for the filter.

All-zero synthesizer derived from cepstral analysis: taking the cepstrum of the speech signal results in a reasonably good separation of the vocal tract function and the excitation function in the cepstral domain. The speech can now be synthesized by convolving the time versions of the two functions; the temporal vocal tract function is an all-zero function. Figure 29.6 shows the corresponding digital network.

Fixed poles and variable zeros, derived from channel vocoder analysis: the channel vocoder consists of a relatively large number of bandpass filters (the number varies from 14 to 30). These filters are fixed but their gains are variable. If we assume that the filters are all-pole resonators, then taking the weighted, summed output does not affect the pole

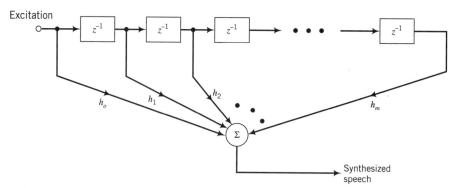

**FIGURE 29.6**   All-zero synthesizer based on cepstral analysis.

positions but the positions of the zeros vary with the weights, which, in turn, are a function of the signal. Figure 29.7 shows the channel vocoder synthesizer.

Variable poles and variable zeros synthesizer:   a version of such a system is shown in Fig. 29.3; a simplified version is shown in Fig. 29.8 below. This looks quite similar to the channel vocoder synthesizer of Fig. 29.7. However, since the filters themselves as well as their gains are allowed to vary, fewer filters are needed.

### 29.2.3   Talking Chips

As previously noted, early devices such as the Voder were manually operated, and later devices still required major investments of equipment. The clearest sign that technology

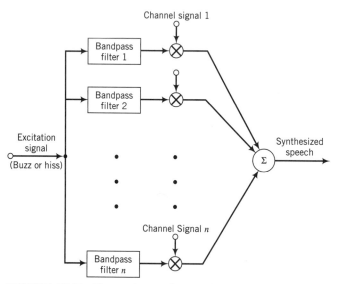

**FIGURE 29.7**   Channel vocoder synthesizer.

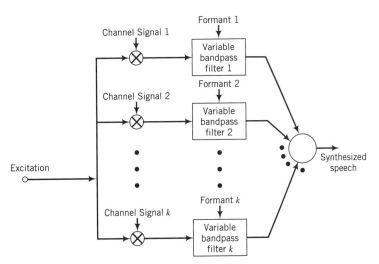

**FIGURE 29.8** Parallel-formant synthesizer.

had brought synthesis to the masses was the 1978 release of a three-chip system for a child's toy called Speak and Spell, in which the synthesizer itself was implemented on a single chip. The chip used LPC parameters that had previously been stored, so in some sense it was doing storage and playback with a limited vocabulary; but by use of the source–filter model, the amount of storage was greatly reduced (at the cost of some additional computation). The chip implemented a digital lattice filter, as in Fig. 29.5(c), and it is described in [51]. Later other chips were introduced that also used a source–filter approach, though some used cascades of second-order filters rather than a single lattice. The range of chips introduced during this period are discussed in some detail in [43] and included a concatenative system that will be described here.

## 29.3  CONCATENATIVE METHODS

With the exception of a brief mention of articulatory synthesis, this chapter has focused on speech-synthesis methods based on short-term spectral estimates that are derived from a source–filter model and then used during synthesis. However, many systems use what are called concatenative approaches; in a nutshell, speech waveforms (or compressed representations) are stored and then concatenated during synthesis. For instance, in the approach developed by Forrest Mozer and used by some semiconductor manufacturers in the 1980s, simplified waveforms were stored and concatenated during playback [43]. Voiced sounds were compressed by manipulating a pitch period waveform to reduce the number of signal samples that were required to have a power spectrum that was sufficiently close to the original.

For each pitch period, setting the phase of each harmonic to 0 or $\pi$ reduced the Fourier representation to a sum of positive or negative cosines, while leaving the total magnitude spectrum unchanged. Since the resulting waveform was symmetric, half of the samples of polarity for each cosine satisfy two criteria: (1) minimizing the energy in the first and last quarters of the symmetric waveform, and (2) minimizing the quantization error for a 4-bit representation of each sample. Finally, the quantized samples from the third quarter of the symmetric waveform were stored whenever their spectrum deviated significantly from the spectrum for the previous stored pitch period; otherwise a repeat count was incremented. For unvoiced sounds, the procedure was simpler, relying on a table of prototypical waveforms.

Mozer's synthesis was targeted at very low bit rate, stored speech synthesis. The motivating factor was the reduction of cost, since memory was still expensive in the early 1980s. However, as memory densities continued their exponential climb, the significance of concatenative approaches began to increase for more general applications of speech synthesis, such as text-to-speech systems. A technique that has been used in many systems is pitch synchronous overlap and add (PSOLA) [6]. In this approach, diphones are concatenated pitch synchronously. The alignment to pitch periods permits a variation of pitch frequency by varying the timing of repeats for each waveform type. We will elaborate on this transformational aspect of PSOLA in Chapter 34.

Through the late 1980s to 1990s, a wide range of alternatives to PSOLA have been developed; in fact, the original PSOLA is now often called TD-PSOLA (time-domain PSOLA) to distinguish it from its many variants. Although PSOLA's simplicity is appealing, operation in the time domain often results in discontinuities at the overlap points, and it tends to have a high data rate. Some of the alternative approaches for concatenation-based systems are as follows.

**1.** LP-PSOLA:   rather than storing diphone waveforms per se, linear predictive coefficients (or transformed values, such as reflection coefficients or log-area ratios) can be stored to represent a segment. The PSOLA algorithm is then used to provide prosodic modifications by operating on the excitation signal, and linear smoothing is applied to the LP coefficients [44]. In its simplest form, the input is just a voiced-unvoiced choice of the classical variety. In multipulse linear predictive coding-PSOLA (MPLPC-PSOLA) the input is a multipulse sequence of pulses. Other approaches include codebook excited linear predictive PSOLA (CELP-PSOLA) and residual-excited linear predictive PSOLA (RELP-PSOLA) [17]. All of these are advanced LP techniques that are described in Chapter 33. In each case the more complex excitation model results in more storage than simple LP-PSOLA (but significantly less than in TD-PSOLA), and more computation than either LP-PSOLA or TD-PSOLA.

**2.** Sinusoidal (harmonic) modeling:   an alternate parametrization of speech, which can also be applied to the storage of variables representing segments, is that of a combination of sinusoidal signals. In general this would be accomplished by taking a DFT of the segment in question, but this would not result in any compression. Certainly in the case of voiced signals this representation can easily be made more compact, as the only magnitudes and phases of interest correspond to harmonics of the fundamental frequency. In the case of unvoiced sounds, many sounds can be represented by a group of sinusoids (not necessarily harmonic) that are chosen in an optimization procedure [39], [3].

**3.** ABSOLA:   a recent modification of the sinusoidal approach was introduced by George [24], [25] and further developed by Macon and Clements [40]. The acronym refers to analysis-by-synthesis overlap add. Particularly in its more recent form, this modification permits smoothing between concatenated segments using representations of the spectral envelope. This reduces discontinuities at the connecting points between diphone segments. As suggested by the name, the analysis is done by an iterative optimization to minimize an error criterion between the original speech and the synthesizer output.

**4.** MBR-PSOLA:  in multiband resynthesis PSOLA, a TD-PSOLA data base is analyzed and resynthesized for voiced frames, using a multiband analysis. For voiced frames, the analysis is done with a mixed periodic–noise model. The resynthesized frames are stored and then used during synthesis, much as the original frames were in TD-PSOLA, but with improved concatenation properties. A variant of this scheme is multiband resynthesis overlap add (MBROLA) [17].

## 29.4  SPECULATION

As noted by Dutoit in [17], speech synthesis is not a solved problem, at least in the sense that its quality is still not as good as that of natural speech (particularly for general text-to-speech synthesis). Although machine storage capabilities have enabled particularly simple concatenative techniques, these approaches make very little use of phonetic knowledge of any sort, and their interpolation techniques seem *ad hoc*. Additionally, the use of segment instances rather than derived properties of these instances seems suboptimal. In some ways the current condition seems comparable to the state of ASR when we merely stored a large number of prototypes and used DTW to find the closest one. In the case of synthesis, couldn't we derive parameters from a large speech corpus so that speech waveforms could be generated based on statistical properties, rather than just examples? Dutoit refers to such an approach as *corpus-based models of speech segments*. Some work has already been done in this direction, such as [16].

Dutoit further notes that natural speech has a kind of variability that is not observed in synthetic speech, but that adding randomness to the parameter streams in current systems just makes them sound worse. Further work in this area seems warranted.

Although this chapter has focused on the phonetic and spectral aspects of synthesis, generating good prosodic contours (pitch, duration, and amplitude) is a critical aspect of the process. This area clearly requires a great deal of work, judging by the intonation of random-input text for current systems. Developing better statistical models for these prosodic quantities seems to be a fruitful area for future work. d'Alessandro notes that there is in fact a move in this direction in prosody generation [14]. It is likely that many of the statistical methods described in Chapters 25–28 in this book will begin to see greater application to speech synthesis in general, and to prosody analysis and generation in particular.

Finally, Dutoit [17] has noted that a critical component to further progress in speech synthesis is the availability of a synthesis system with full control of its parameters for research purposes. He has made such a system available that uses MBROLA; interested readers can download the system from http://tcts.fpms.ac.be/synthesis/mbrola.html. Similarly, the Edinburgh speech center has made their Festival system publically available at http://www.cstr/ed/ac/uk/projects/festival.html. It currently supports both residual LPC and PSOLA algorithms.[3]

## 29.5   EXERCISES

**29.1**   When a speech transformation (changing male-sounding speech to female-sounding speech) is performed, both pitch and formants are raised. Give a physiological explanation for why this strategy seems to work.

**29.2**   Show mathematically that a channel vocoder synthesizer transfer function has fixed poles and variable zeros.

**29.3**   Sketch the parameter sequences needed to synthesize the following utterances: "pie," "map," and "sky," on OVE II.

**29.4**   The Flanagan–Ishizaki model (Fig. 29.10 below) assumes interdependence of excitation and vocal tract filter. Give arguments both for and against the need to complicate the model in this manner.

**29.5**   Which type of error is more damaging to intelligibility in a given synthesizer – excitation function errors or vocal tract filter errors? Explain.

**29.6**   Briefly describe the different approaches to speech synthesis for the following systems (include such topics as motivation, implementation, and synthesis strategy):

**(a)** The Voder of Homer Dudley.

**(b)** John Holmes and his work on OVE II.

**(c)** The Pattern Playback of Haskins laboratory.

**(d)** The articulatory synthesizer of George Rosen.

**(e)** The DecTalk of Dennis Klatt.

**(f)** Synthesis by rule.

**29.7**   Describe at least three important steps in a text-to-speech synthesizer.

**29.8**   Give four examples (for English) where grapheme-to-phoneme conversion could get into trouble. Can you design a system to avoid or alleviate these errors?

## 29.6   APPENDIX:   SYNTHESIZER EXAMPLES

### 29.6.1   The Klatt Recordings

Beginning with the Voder, much research (interrupted by World War II) has been dedicated to the development of speech synthesizers. Much of this work was summarized in 1987 in a monumental article by Dennis Klatt [35]. Included with Klatt's paper was a plastic record

---

[3]Unfortunately, this printed page is fairly static, and the Web changes rather frequently. If the Web pages cited here do not lead to the desired systems, we apologize. Check with the our text's home page at the Wiley site for hints and suggestions.

with a large number of synthesizer recordings.[4] We list a number of these recordings here with a few comments.

### 29.6.2 Development of Speech Synthesizers

*Voder:* This is a recording of the Voder that was described in Chapter 2. A block diagram of the Voder was shown in Figs. 2.4 and 2.5.

*Pattern Playback:* See Fig. 29.1 and the accompanying discussion.

*PAT, the Parametric Artificial Talker:* This device, consisting of three resonators in parallel with three frequency controls, was introduced by Walter Lawrence to American audiences at an MIT conference in 1956. Three additional parameters were used for excitation control. A moving glass slide converted painted patterns to control parameters.

*OVE or OVE I:* OVE was also introduced at the 1956 MIT conference as a cascade connection of three formants, with formants 1 and 2 controlled by the movement of a mechanical arm. Hand-held potentiometers controlled the frequency and amplitude of the voicing source. Since OVE spoke only vowel-like sounds, no noise source was needed.

*PAT in 1962:* Updated version of PAT. Amplitude controls and a separate fricative were added. At the 1962 Stockholm Speech Communication Seminar, the synthesizer attempted to match a naturally spoken sentence.

*OVE II in 1962:* At the same seminar, the updated OVE II also matched, quite successfully, the same spoken utterance. See Fig. 29.2.

*Holmes' Sentence on OVE II:* John Holmes worked very hard to try to match the sentence "I enjoy the simple life" to make it indistinguishable from the original.

*The Same Sentence by Holmes:* Using his own parallel synthesizer, Holmes tried again. See Fig. 29.3.

*Male-to-Female Transformation with DecTalk:* The fundamental frequency was multiplied by 1.7, and the formants and glottal wave shape were varied.

*The DAVO Articulatory Synthesizer:* George Rosen built the original model in 1958 [48]. It was modified by Hecker [28] to include a nasal tract. The original Rosen model is shown in Fig. 29.9.

*Vocal Cord–Vocal Tract Synthesis:* Flanagan and Ishizaki [22] simulated a speech synthesizer in which the vocal cord and vocal tract operations were interdependent. A block diagram is shown in Fig. 29.10, and the details of an individual T section are shown in Fig. 29.11.

---

[4]See the Wiley Web site associated with this book for information about access to these recordings.

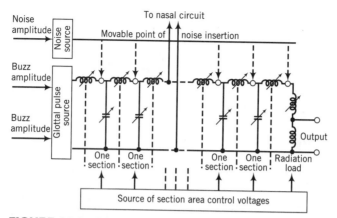

**FIGURE 29.9**     DAVO (dynamic analog of the vocal tract). From [48].

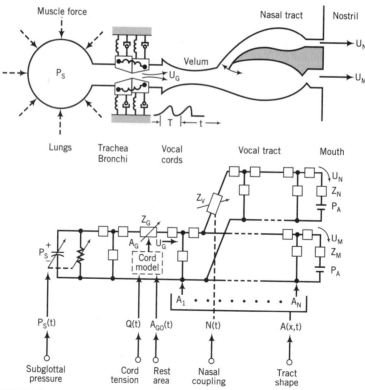

**FIGURE 29.10**     Schematic of the vocal cord–vocal tract system. From [22].

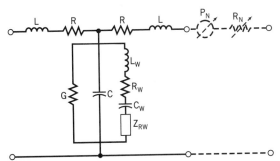

**FIGURE 29.11**  Circuit of an individual T section. From [22].

The independence of source and filter in nearly all synthesizer models neglects the possible effect that the dynamics of the vocal tract can alter the dynamic properties of the excitation function. By overtly including cord–tract interaction, Flanagan et al. have edged one step closer to a true physiological model, but it is still an open question as to the perceptual effect of this step. Notice the turbulent noise source in Fig. 29.11; this allows generation of noise anywhere in the model for fricative sounds. Could this system be improved by using digital waveguides?

### 29.6.3  Segmental Synthesis by Rule

Early synthesis-by-rule programs began with a phoneme string; prosodic features were entered by hand to match the original utterance.

***Kelly–Gerstman [32] Synthesis by Rule:***   They used a basic element that resembles a digital waveguide. A section was shown in Fig. 11.3. This system was demonstrated at the Copenhagen 1961 International Conference on Acoustics.

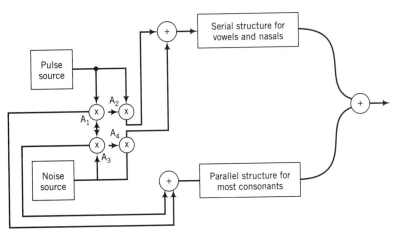

**FIGURE 29.12**  Structure of the Klatt Synthesizer.

***British Synthesis by Rule, 1964:***   Holmes et al. demonstrated this program at the fall meeting of the Acoustical Society in Ann Arbor, Michigan, 1964.

***Diphone Concatenation Synthesis by Rule:***   Dixon and Maxey [15] of IBM demonstrated this program at the 1967 MIT Conference on Speech Communication and Processing.

***Synthesis by Rule with an Articulatory Model:***   Coker [8] also demonstrated this device at the same 1967 MIT Conference.

## 29.7   SYNTHESIS BY RULE OF SEGMENTS AND SENTENCE PROSODY

The previous examples did not include prosodic features. These examples, in addition to using phoneme strings as inputs, also include stress marks and some syntactic information.

***First Prosodic Synthesis by Rule:***   Mattingly discussed this as part of his report on speech research [42].

***Sentence-Level Phonology Incorporated in Rules:***   Klatt [33] used phonology to generate segmental durations and a fundamental frequency contour, as well as sentence-level allophonic variations. The inputs were phoneme strings plus stress and syntactic symbols.

***Rules from Linear Prediction Diphones and Prosody:***   Olive [45] demonstrated this system at ICASSP '77.

***Rules from Linear Prediction Demisyllables:***   Browman [4] also used prosodic rules.

## 29.8   FULLY AUTOMATIC TEXT-TO-SPEECH CONVERSION

***First Text-to-Speech System:***   Umeda et al. [49] designed this system, based on an articulatory model.

***Bell Laboratories Text-to-Speech System:***   Coker et al. [9] demonstrated this system at the 1972 International Conference of Speech Communication and Processing in Boston.

***Haskins Text-to-Speech System:***   Cooper et al. [13] used the Mattingly phoneme-to-speech rules, coupled with a large dictionary [42].

***Reading Machine for the Blind:***    Kurzweil [36] demonstrated this commercially available machine, which included an optical scanner, on the CBS Evening News.

***Votrax Type-n-Talk System:***    Gagnon [23] demonstrated this cheap device at the 1978 ICASSP. He implemented the research by Elovitz et al. [18] that converted letters to sounds.

***Echo Low-Cost Diphone Concatenation:***    This was demonstrated in 1982.

***MITalk System:***    Allen et al. [1], [2] demonstrated a full-blown text-to-speech system, using complicated heuristics to translate graphemes to phonemes. The synthesizer was created by Klatt.

***Multi-Language Infovox System:***    This was a commercial system [41] that was based on the research of Carlson et al. [5]. It was developed at the Royal Institute of Technology in Stockholm and demonstrated at ICASSP '76 and ICASSP '82.

***The Prose-2000 Commercial System:***    Original research was done at Telesensory Systems by James Bliss and associates [26], [27].

***Klattalk and DECtalk:***    This was Dennis Klatt's final system, which was licensed to Digital Equipment Corporation.

***Bell Laboratories Text-to-Speech System:***    Olive and Liberman [46] used the Olive diphone synthesis strategy [45] in combination with a large morpheme dictionary [11] and letter-to-sound rules [7]. The system was demonstrated at the 1985 ASA meeting.

## 29.8.1  The van Santen Recordings

A recent book [50] also included a set of synthesizer demonstrations; given the move toward concatenative synthesis in the past 10 years, most of the examples are of this variety. Here we briefly summarize the audio demos on the CD accompanying this reference as well as being described in the text.

***DECtalk:***    This demo is from an updated version of the system described above.

***Bell Laboratories Text-to-Speech System:***    Similarly, this is an updated version of the diphone-based system described above. The system has migrated to a commercial product called TrueTalk.

***Orator:***    Bellcore has a commercial system that is based on demisyllable concatenation.

***Eurovocs:***    This is a diphone concatenation system.
    We also recommend [50] for its treatment of a broad set of synthesis-related issues, including prosodic analysis and synthesis.

# BIBLIOGRAPHY

1. Allen, J., Hunnicut, S., Carlson, R., and Granstrom, B., "MITalk-79: the MIT Text-to-Speech system," *J. Acoust. Soc. Am. Suppl.* 1 **65**: S130, 1979.

2. Allen, J., Hunnicut, S., and Klatt, D. H., *From Text to Speech; The MITalk System*, Cambridge Univ. Press, London/New York, 1987.

3. Almeida, L. B., and Silva, F. M., "Variable frequency synthesis: an improved harmonic coding scheme," in *Proc. IEEE Int. Conf. Acoust. Speech Signal Process.*, San Diego, pp. 27.5.1–27.5.4, 1984.

4. Browman, C. P., "Rules for demisyllable synthesis using lingua, a language interpreter," in *Proc. IEEE Int. Conf. Acoust. Speech Signal Process.*, Denver, 1980.

5. Carlson, R., Granstrom, B., and Hunnicut, S., "A multi-language text-to-speech module," in *Proc. IEEE Int. Conf. Acoust. Speech Signal Process.*, Paris, pp. 1604–1607, 1982.

6. Charpentier, F., and Stella, M., "Diphone synthesis using an overlap-add technique for speech waveform concatenation," in *Proc. IEEE Int. Conf. Acoust. Speech Signal Process.*, Tokyo, pp. 2015–2018, 1986.

7. Church, K. W., "Stress assignment in letter-to-sound rules for speech synthesis," in *Proc. 23rd Meet. Assoc. Comp. Ling.* Chicago, pp. 246–253, 1985.

8. Coker, C. H., "Speech synthesis with a parametric articulatory model," reprinted in J. L. Flanagan and L. R. Rabiner, eds., *Speech Synthesis*, Dowden, Hutchinson and Ross, Stroudsburg, Pa., 1968.

9. Coker, C. H., Umeda, N., and Browman, C. P., "Automatic synthesis from ordinary english text," *IEEE Trans. Audio Electroacoust.* **AU-21**: 293–297, 1973.

10. Coker, C. H., "A model of articulatory dynamics and control," *Proc. IEEE* **64**: 452–460, 1976.

11. Coker, C. H., "A dictionary-intensive letter-to-sound program," *J. Acoust. Soc. Am. Suppl.* 1 **78**: S7, 1985.

12. Cooper, F. S., Gaitenby, J. H., Liberman, A. M., Borst, J. M., and Gerstman, L. J., "Some experiments on the perception of synthetic speech sounds," *J. Acoust. Soc. Am.* **24**: 597–606, 1952.

13. Cooper, F. S., Gaitenby, J. H., Mattingly, I. G., Nye, P. W., and Sholes, G. N., "Audible outputs of reading machines for the blind," *Stat. Rep. Speech Res.* SR-35/36, Haskins Laboratory, New Haven, Conn., pp. 117–120, 1973.

14. d'Alessandro, C., and Liénard, J-S., "Synthetic speech generation," in R. Cole, J. Mariani, H. Uszkoreit, G. Battista, A. Zaenen, A. Zampolli, and V. Zue, eds., *Survey of the State of the Art in Human Language Technology*, Cambridge Univ. Press, London/New York, 1997.

15. Dixon, N. R., and Maxey, H. D., "Terminal analog synthesis of continuous speech using the diphone method of segment assembly," *IEEE Trans. Audio Electroacoust.* **AU-16**: 40–50, 1968.

16. Donovan, R. E., and Woodland, P. C., "Automatic speech synthesizer parameter estimation using HMMs," in *Proc. IEEE Int. Conf. Acoust. Speech Signal Process.*, Detroit, pp. 640–643, 1995.

17. Dutoit, D., *An Introduction to Text-to-Speech Synthesis*, Kluwer, Dordrecht, the Netherlands, 1997.

18. Elovitz, H., Johnson, R., McHugh, A., and Shore, J., "Letter-to-sound rules for automatic translation of English text to phonemes," *IEEE Trans. Acoust. Speech Signal Process.* **ASSP-24**: 446–459, 1976.

19. Fant, G., "Speech communication research," *Ing. Vetenskaps Akad.* **24**: 331–337, 1953.

20. Fant, G., Martony, J., Rengman, U., and Risberg, A., "OVE II synthesis strategy," presented at the Stockholm Speech Communications Seminar, Stockholm, 1962.

21. Fant, G., "Non-uniform vowel normalization," Speech Trans. Lab., R. Inst. Technol., Stockholm QPSR 2–3, pp. 1–19, 1975.

22. Flanagan, J. L., Ishizaka, K., and Shipley, K. L., "Synthesis of speech from a dynamic model of the vocal cords and vocal tract," *Bell Syst. Tech. J.* **54**: 485–506, 1975.

23. Gagnon, R. T., "Votrax real time hardware for phoneme synthesis of speech," in *Proc. IEEE Int. Conf. Acoust. Speech Signal Process.*, Tulsa, Oklahoma, pp. 175–178, 1978.

24. George, E. B., "*An analysis-by-synthesis approach to sinusoidal modeling applied to speech and music signal processing*," Ph.D. Thesis, Georgia Institute of Technology, 1991.

25. George, E. B., and Smith, J. T., "Speech analysis/synthesis and modification using an analysis-by-synthesis/overlap-add sinusoidal model," *Trans. Speech Audio Process.* **5**: 389–406, 1997.

26. Goldhor, R. S., and Lund, R. T., "University-to-industry technology transfer: a case study," *Res. Policy* **12**: 121–152, 1983.

27. Groner, G. F., Bernstein, J., Ingber, E., Pearlman, J., and Toal, T., "A real-time text to speech converter," *Speech Technol.* **1**: 73–76, 1982.

28. Hecker, M. H. L., "Studies of nasal consonants with an articulatory speech synthesizer," *J. Acoust. Soc. Am.* **34**: 179–188, 1962.

29. Hofstetter, E. M. "An introduction to the mathematics of linear predictive filtering as applied to speech analysis and synthesis," Tech. Note **36**, rev. 1, MIT Lincoln Laboratory, Lexington, Mass., 1973.

30. Holmes, J. N., "The influence of the glottal waveform on the naturalness of speech from a parallel formant synthesizer," *IEEE Trans. Audio Electroacoust.* **AU-21**: 298–305, 1973.

31. Holmes, J. N., "Formant synthesizers; cascade or parallel," *Speech Commun.* **2**: 251–273, 1983.

32. Kelly, J., and Gerstman, L., "Digital computer synthesizes human speech," *Bell Labs. Rec.* **40**: 216–217, 1962.

33. Klatt, D. H., "Structure of a phonological rule component for a speech synthesis by rule program," *IEEE Trans. Acoust. Speech Signal Process.* **ASSP-24**: 391–398, 1976.

34. Klatt, D. H., "Software for a cascade/parallel formant synthesizer," *J. Acoust. Soc. Am.* **67**: 971–995, 1980.

35. Klatt, D. H., "Review of text-to-speech conversion for English," *J. Acoust. Soc. Am.* **82**: 737–792, 1987.

36. Kurzweil, R., "The Kurzweil reading machine: a technical overview," in M. R. Redden, and W. Schwandt, eds., *Science, Technology and the Handicapped*, AAAS Rep. 76-R-11, Washington, D.C., pp. 3–11, 1976.

37. Lawrence, W., "The synthesis of speech from signals which have a low information rate," in W. Jackson, ed., *Communication Theory*, Butterworths, London, pp. 460–469, 1953.

38. Liberman, A. M., Cooper, F. S., Shankweiler, D. P., and Studdert-Kennedy, M., "Perception of the speech code," *Psychol. Rev.* **74**: 431–461, 1967.

39. MacAulay, R. J., and Quatieri, T. F., "Magnitude only reconstruction using a sinusoidal speech model," in *Proc. IEEE Int. Conf. Acoust. Speech Signal Process.*, San Diego, pp. 27.6.1–27.6.4, 1984.

40. Macon, M., and Clements, M., "Speech concatenation and synthesis using an overlap-add sinusoidal model," in *Proc. IEEE Int. Conf. Acoust. Speech Signal Process.*, Atlanta, pp. 361–364, 1996.

41. Magnusson, L., Blomberg, M., Carlson, R., Elenius, K., and Granstrom, B., "Swedish researchers team up with electronic venture capitalists," *Speech Technol.* **2**: 15–24, 1984.

42. Mattingly, I. G., "Synthesis by rule of general american English," *Suppl. Stat. Rep. Speech Res.*, Haskins Laboratory, New Haven, Conn., pp. 1–223, 1968.

43. Morgan, N., *Talking Chips*, McGraw–Hill, New York, 1984.

44. Moulines, E., and Charpentier, F., "Diphone synthesis using multipulse LPC technique," in *Proc. FASE Int. Conf.*, Edinburgh, pp. 47–51, 1988.

45. Olive, J. P., "Rule synthesis of speech from diadic units," in *Proc. IEEE Int. Conf. Acoust. Speech Signal Process.*, Hartford, Connecticut, pp. 568–570, 1977.

46. Olive, J. P. and Liberman, M. Y., "text-to-speech-an overview," *J. Acoust. Soc. Am. Suppl.* 1 **78**: S6, 1985.

47. Parthsarathy, S., and Coker, C. H., "Automatic estimation of articulatory parameters," *Comput. Speech Lang.* **6**: 37–75, 1992.

48. Rosen, G., "A dynamic analog speech synthesizer," *J. Acoust. Soc. Am.* **30**: 201–209, 1958.

49. Umeda, N., Matsui, E., Suzuki, T., and Omura, H., "Synthesis of fairy tales using an analog vocal tract," in *Proc. 6th Int. Cong. Acoust.*, Tokyo, pp. B159–162, 1968.

50. van Santen, J. P. H., Sproat, R., Olive, J., and Hirschberg, J., *Progress in Speech Synthesis*, Springer Pub., New York, 1997.

51. Wiggins, R., "An integrated circuit for speech synthesis," in *Proc. IEEE Int. Conf. Acoust. Speech Signal Process.*, Denvet, Colorado, pp. 398–401, 1980.

# CHAPTER 30

# *PITCH DETECTION*

## 30.1  INTRODUCTION

In some of the previous chapters, we have stressed the model of speech and music production as consisting of one or more excitations that drive a time-variable filter. In this chapter we focus on the excitation model, and in particular on the extraction of pitch frequency. The time-variable filter that results in the spectral envelope can be estimated in different ways, including filter banks, cepstra, and linear prediction (see Chapters 19, 20, and 21), as well as combinations of these approaches (see Chapter 22).

Modeling of the excitation function of speech requires paying particular attention to the following components:  (a) the periodic or nearly periodic opening and closing of the glottis during voicing; (b) the shape of the glottal pressure pulse; (c) the position in the vocal system of the constriction that creates turbulent flow during unvoiced sound; (d) the nature of the excitation function during stop consonant articulation; (e) how voicing and turbulence combine during articulation of the voiced fricative sounds; and (f) possible nonlinear interactions between excitation and acoustic tube response.

It is probably fair to state that accurate modeling of the excitation parameters is more complex than modeling of the time-varying linear filter that we use to represent the vocal tract. Channel vocoder researchers in the 1950s must have been somewhat aware of this when they stated that vocoders of that era lacked good pitch detectors. During this period and later (into the 1960s and 1970s), many novel methods of tracking the voice fundamental frequency were invented; these included algorithms for distinguishing buzz (quasi-periodic voicing) from hiss (turbulent air flow; therefore, noiselike excitation). Yet, it is still true that a completely accurate model of the excitation function of human speech does not exist. However, much progress has been made; this chapter outlines some of the basic results.

## 30.2  A NOTE ON NOMENCLATURE

As discussed in Chapter 16, the word pitch (in the context of speech processing), as defined operationally by psychoacousticians, is the frequency of a pure tone that is matched by the listener to a more complex (usually periodic) signal. This is a subjective definition. When engineers speak of a "pitch detector," they usually refer to a device that measures the fundamental frequency of an incoming signal; this is an objective definition. In this chapter, pitch perception refers to the subjective result and pitch detection refers to an objective result. Pitch detection and fundamental frequency estimation are often used interchangeably.

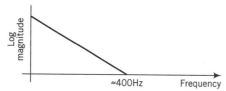

**FIGURE 30.1**  Dudley's slope filter to enhance the fundamental frequency component.

## 30.3  PITCH DETECTION, PERCEPTION AND ARTICULATION

In Chapter 16, several models of pitch perception were treated. Many ideas in pitch detection (but not all of them) are reminiscent of these models, which yielded insights that help us invent better pitch detectors. However, we gain additional insight by considering how the excitation function in speech or music is produced.

Homer Dudley's design of the original channel vocoder included a pitch detector. At that time, many psychoacousticians believed Helmholtz's assertion that the fundamental frequency component must exist at some level in order to perceive the pitch. It is interesting to speculate whether Dudley was influenced by this belief. The Dudley pitch detector was based on the articulatory premise that the voiced speech signal always included the fundamental frequency component; his design consisted of a slope filter designed to enhance this component to make it easier for the hardware to correctly measure its frequency. The slope filter is shown in Fig. 30.1.

It can be seen from the sketches of Fig. 30.2 that passage through the slope filter of a signal with almost equal first and second harmonics greatly reduces the second harmonic relative to the fundamental.

A simple way to extract the fundamental period is shown in Fig. 30.3. First, the positive peaks of the signal are found; this is followed by the detection algorithm shown in the figure. Dudley made use of the knowledge that unadulterated speech almost always (perhaps always) contains a significant fundamental component. Thus, for example, Charles Vaderson successfully demonstrated a channel vocoder with Dudley's pitch detector in 1950. However, many practical communication systems (e.g., telephones) are band limited, and the fundamental component of the speech may be completely missing.[1] Furthermore, environmental noise may completely mask the fundamental. This leads us to describe more complex signal-processing algorithms for conditioning the speech prior to detection.

## 30.4  THE VOICING DECISION

Certain speech sounds, such as the voiceless fricatives /s/, /sh/, /th/, and /f/, can be modeled as the output of an acoustic tube complex when a portion of the tube has very narrow cross

---

[1]In many telephone systems the components below 300 Hz are strongly attenuated.

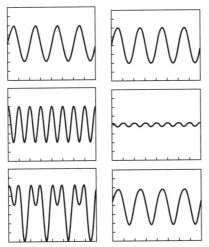

**FIGURE 30.2** Effect of the slope filter on a two-component signal. The three panels on the left show the first and second harmonics of a complex tone and the sum of the two. The three panels on the right show that the second harmonic has been greatly attenuated by the slope filter, leading to a sum that closely resembles the first harmonic of the tone.

section (a constriction), causing the airflow to become turbulent. For our present discussion, it is sufficient to equate turbulence to the presence of a random noise source. For example, in the production of /s/, the narrow, turbulent cross section is located between the tongue tip and upper teeth. Thus, the source is close to the mouth opening and the excitation is shaped by various reflections in the vocal tract. A reasonable model of the excitation for these sounds is that of white noise, which is then shaped by the vocal tract for that sound.

The voiced fricatives /z/, /th/ (as in the), /zh/ (as in azure), and /v/ are controlled by the same vocal tract shape as their voiceless counterparts, but, in addition, the vocal

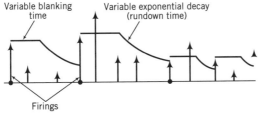

**FIGURE 30.3** Extraction of the pitch period. Detections are only permitted when a peak value (indicated by the vertical lines with arrows) exceeds a time-varying threshold determined by the previous detection's amplitude and time, as shown by the solid horizontal and exponentially decreasing curves.

cords are simultaneously vibrating. Thus, there are two sources of excitation in this case; furthermore, since the periodic excitation is formed at the glottis and the noise is formed near the lips, the two sources excite the vocal tract quite differently.

The voiceless plosives, /p/, /t/, and /k/, involve a transient burst followed by noiselike aspiration. As discussed in Chapter 17, the formant transitions at the start of voicing are auditory cues for distinction among these three sounds, so this has to be part of an articulator model. In Chapter 31, we relate how these modeling issues have been dealt with in vocoder design.

## 30.5   SOME DIFFICULTIES IN PITCH DETECTION

Figure 30.4 illustrates some of the problems encountered in pitch detection. Figure 30.4(a) shows two speech waveforms; the bottom signal has a period approximately one-fourth of the top signal. This illustrates the large dynamic range of the voice fundamental frequency. The pitch of some male voices can be as low as 60 Hz, whereas the pitch of children's voices can be as high as 800 Hz. Figure 30.4(b) shows how the period can fluctuate drastically and almost instantaneously. The leftmost period is quite short, but the next five periods are more than twice as long before snapping back to shorter periods. This kind of behavior makes pitch tracking difficult. Figure 30.4(c) shows a rapid change in the spectrum caused, for example, by sudden closure as in a vowel-to-nasal transition. Although the fundamental frequency has not changed drastically, pitch detection based on waveform analysis can suffer. Figure 30.4(d) shows a transition region from aperiodic (hiss) excitation to quasi-periodic (buzz) excitation. For the precise transition instant to be caught, a fast-acting time-domain detector would be best. Finally, Figs. 30.4(e) and 30.4(f) show the effect of speech degradation that is caused by telephone transmission and added acoustic noise, causing extra problems in pitch extraction.

## 30.6   SIGNAL PROCESSING TO IMPROVE PITCH DETECTION

Several methods of conditioning the speech signal to improve pitch detection have proved useful. Among these, we include low-pass filtering, spectral flattening and correlation, inverse filtering, comb filtering, cepstral processing, and high-resolution spectral analysis.

Low-pass filtering:   We know from Chapter 16 that human pitch perception pays more attention to the lower frequencies. Interestingly, estimating the pitch period by eye is typically easier with low-passed waveforms such as those shown in Fig. 30.6 than with full-band waveforms such as those shown in Fig. 30.5. It thus seems plausible that a pitch-detection device would have less trouble finding the correct period by analyzing the signal of Fig. 30.6 than that of Fig. 30.5. This has proved true in practice.

Spectral flattening and correlation:   A more sophisticated concept was proposed by Sondhi [16]. It is based on the observation that a Fourier series representation of harmonics of equal amplitude and zero phase results in a signal that is very much like a pulse train. Sondhi proposed that the original signal first be spectrally flattened. An approximation to

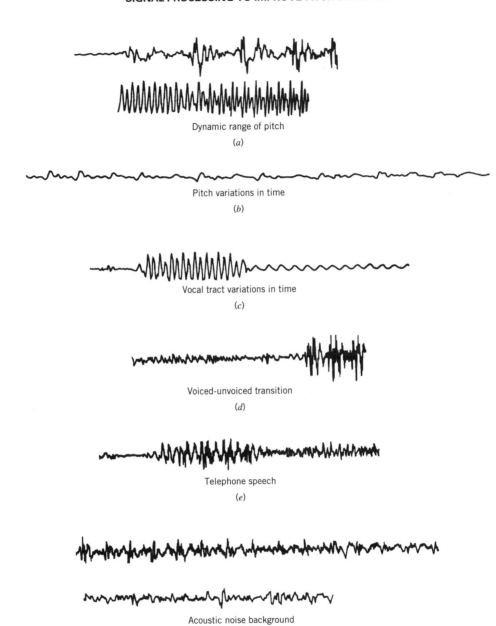

Dynamic range of pitch

(a)

Pitch variations in time

(b)

Vocal tract variations in time

(c)

Voiced-unvoiced transition

(d)

Telephone speech

(e)

Acoustic noise background

(f)

**FIGURE 30.4** Six examples of difficulties in pitch detection.

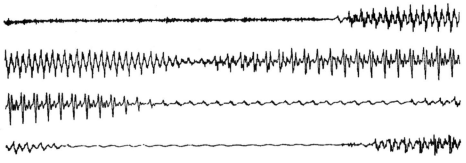

**FIGURE 30.5**  Full-band speech signal.

this operation is shown in Fig. 30.7, where the outputs of a bank of bandpass filters (BPFs) are divided by their own energy and the components added.

The sum is now sent through an autocorrelator, which creates a zero-phase time function, thus approximating the equal harmonic–zero-phase criterion proposed by Sondhi. Figure 30.8 shows the effect of autocorrelation.

Inverse filtering:   This concept begins with the hypothesis that the speech signal is the convolution of an excitation and a vocal tract filter. If one were able, in some manner, to specify the time-varying vocal tract at all times, then the speech signal could be passed through a filter with a spectrum *inverse* to that of the vocal tract filter; the output, ideally, should be the glottal waveform, again simplifying pitch tracking. Chapters 19–21 describe methods for estimating the spectral envelope; inverse filtering, at least for the relatively simple vowel sounds, consists of building a linear system having zeros where the original spectral envelope has poles. Markel [9] has implemented inverse filtering as part of his SIFT algorithm for fundamental frequency estimation.[2]

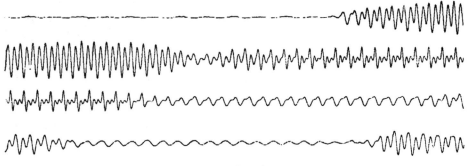

**FIGURE 30.6**  Low-pass filtered speech signal.

[2]SIFT stands for simplified inverse filter tracking.

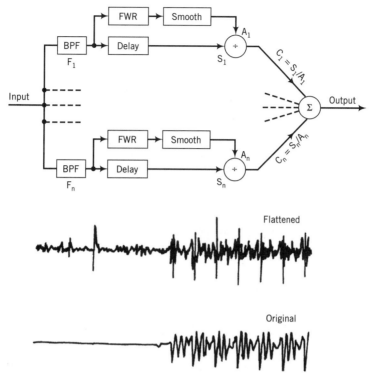

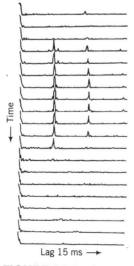

**FIGURE 30.7** Spectral flattening and its effect on the speech signal; FWR, full-wave rectification. From [16].

**FIGURE 30.8** Autocorrelation function of spectrally flattened speech, with successive 30-ms sections with 15-ms overlap. From [16].

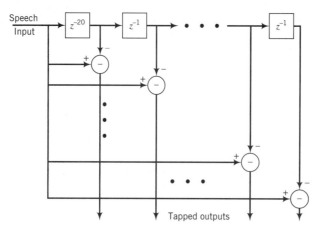

**FIGURE 30.9**   Comb filtering of the speech wave.

Comb filtering:    The speech signal is sent through a multitude of delays, corresponding to all possible (discrete) periods of the input. The system is shown in Fig. 30.9.

For 10-kHz sampling and a fundamental frequency range of 50–500 Hz, the number of possible periods (in samples) ranges from 20 to 200. Thus the comb filter of Fig. 30.9 must contain at least 181 taps; at each tap the signal and its delayed version are subtracted. If the signal is periodic, one of the tapped outputs should be zero, so an estimate can be made by examining the tap outputs.

Ross et al. [12] have implemented a comb-filter pitch detector.

Cepstral pitch detection:    As elucidated in Chapter 20, cepstral analysis performs deconvolution of the source and filter. In Chapter 20 we stressed the application to the spectrum envelope but, as shown by Noll, [11] the high-time portion of the cepstrum contains a very clear hint about the fundamental frequency. Figure 30.10 shows sequences of log spectrum cross sections and the resulting cepstra for a male (two left columns) and a female (two right columns). Note the large peak corresponding to the pitch period in the second and fourth columns.

Finally, we should mention that simply measuring the spectrum with a high resolution illuminates the positions of the harmonics. In the next section, we show how this straightforward operation can lead to a powerful pitch-detection algorithm.

## 30.7   PATTERN-RECOGNITION METHODS FOR PITCH DETECTION

It has often been found that multiple sources of information (or multiple estimators of a variable) provide a more reliable estimator. For instance, it can be easily shown that the estimate of a variable's mean formed by averaging $N$ independent measurements has a variance that is $1/N$ times the variance of a single measurement. More generally, the use of multiple estimators permits a secondary decision process to consider agreement among estimators. In practice, it is often difficult to determine whether measurements are independent, but even if there is *some* degree of dependence, improvement is often made.

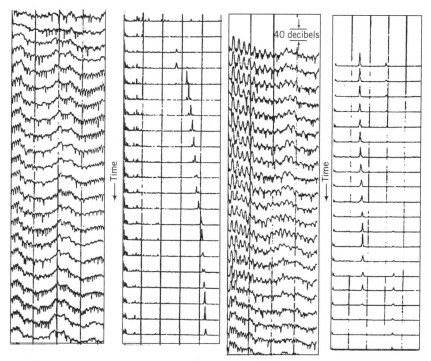

**FIGURE 30.10** Cepstral analysis for pitch detection. Panels 1 and 3 show successive spectral cross-sections and panels 2 and 4 show the corresponding cepstra. From [11].

An example of the use of parallelism for pitch detection [4] was the development of such a program that consisted of four major steps:

1. A low-pass filter to smooth the speech wave.
2. A processor that generated six functions of the peaks of the filtered speech.
3. Six identical elementary pitch-period estimators (PPE), each working on one of the functions.
4. A global, statistically oriented computation based on the results of step 3.

Figure 30.11 depicts the six measurements. Each is input to a PPE; the task of the PPE is to eliminate spurious peaks and save those that are separated by the correct period. Each PPE performs the function described by Fig. 30.3.

The box labeled final pitch-period computation in Fig. 30.11 compiles a histogram of all measured intervals between peaks as outlined in Fig. 30.12. To avoid delays, the only candidates for most probable period are chosen from among the most recent, and one of these six is selected, based on the histogram of all periods. This set of measurements can be repeated as often as desired; typically, a new selection is made every 5–15 ms.

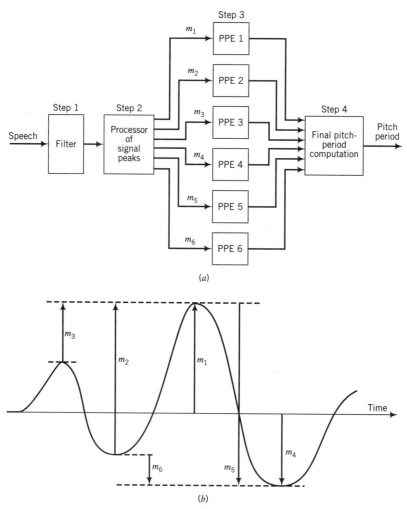

**FIGURE 30.11**   Estimation of periods by elementary pitch detectors.

The histogram obtained from this algorithm can also be used to produce a buzz–hiss decision [5].

As computer processing has increased in speed, there has been an evolution toward new algorithms that require such speeds to operate in real time. One set of algorithms extends the histogram idea discussed here by including more measurements. For example, instead of preprocessing with just a single low-pass filter, the speech is passed through a bank of 19 bandpass filters covering the range 200–2000 Hz. The output of each filter is passed through an elementary pitch detector similar to the PPE described earlier. There now exists a total of 38 outputs (since both positive and negative peaks are represented). Also, the histogram of times between pitch peaks is generalized somewhat to include times between

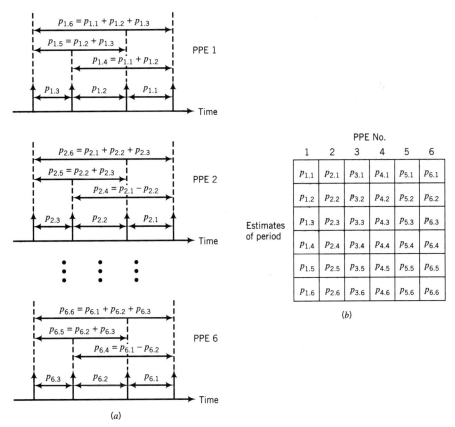

**FIGURE 30.12** Final estimate of the pitch period: (a) outputs of the six PPEs, (b) matrix of the six outputs; each of the entries in the first row of the matrix is a candidate for the final estimate. Adapted from [4].

peaks that are separated by several other peaks; this is equivalent to the original proposal by Licklider [7] to compute the correlation function of the spike train from simulated neurons.

Licklider's concept, plus later ideas that were similar [1], [10], [8], [2], based its measurements on the speech wave directly or on a low-pass filtered version of the speech. The same notion of computing a histogram, but one based on high-resolution spectral analysis, was carried out by Schroeder [13] and Seneff [15]; the latter will be briefly summarized.

Figure 30.13 shows a spectral magnitude cross section containing seven peaks. (This spectrum is based on a 20-ms windowed section of the speech.) The peaks are ordered, as shown. Then the frequencies of peaks 1 and 2 are marked. Then peak frequencies 1, 2, and 3 are marked; then 1, 2, 3, and 4 are marked, and so on, until all seven peak frequencies have been marked in this manner. A histogram is then computed (bottom right) of the intervals shown in Fig. 30.13, and the winner is picked to be the interval that occurs with the greatest probability.

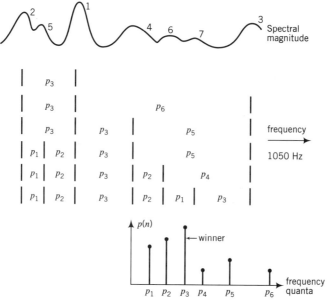

**FIGURE 30.13** Harmonic pitch-detection algorithm. From [15].

The above two algorithms have in common the concept that performing a collection of procedures on the conditioned speech can lead to improvement. A somewhat different statistical approach was developed by Goldstein [6] and implemented by Duifhuis [3]. In their method, a single, powerful algorithm is employed but the parameters are adjusted to be successively tuned to the specific fundamental frequency. In other words, the hypothesis is advanced that the result is, for example, $f_1$. This hypothesis is then tested by comparing the spectrum of the signal with the spectrum of the hypothetical signal, and a score is obtained. The procedure is now repeated for $f_2$, $f_3$, and so on, and the best score determines the winner. The crucial point is that *all permissible* hypotheses go through the test. An implementation of this algorithm is shown in Fig. 30.14. This procedure exemplifies the maximum likelihood approach of testing all reasonable hypotheses and choosing the one having the greatest probability.

## 30.8  MEDIAN SMOOTHING TO FIX ERRORS IN PITCH ESTIMATION

Physiological constraints on fundamental frequency variations can often be advantageous in correcting errors made by pattern-recognition techniques. A popular technique invented by Tukey [18] looks at a sequence of final decisions and treats this sequence as a collection of points on a histogram. Thus, for example, the sequence 5, 6, 12, 7, 8 is plotted as a discrete probability density in the lower part of Fig. 30.15 and as a cumulative probability distribution in the upper part.

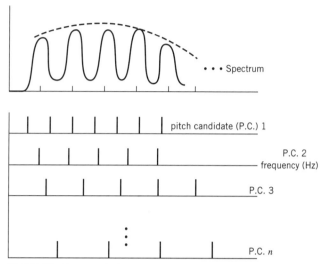

**FIGURE 30.14**   Goldstein–Duifhuis optimum processor algorithm.

The median is the $x$ position for which $P$ is one-half. In Fig. 30.15, the median is 7; thus, the center of the sequence is replaced by 7, so the new sequence becomes 5, 6, 7, 7, 8. In this example, the outlier 12 was replaced. In this case, as in many others, median smoothing is preferable to a linear filter, for which the effect of an outlier would spread to other samples. In the case in which a one represents a buzz (voiced) excitation and a zero represents a hiss (unvoiced), a sequence of 1, 1, 0, 1, 1, the zero gets changed to a one and the modified sequence is 1, 1, 1, 1, 1. Thus, median smoothing can also be used to fix buzz–hiss errors.

In practice, median smoothing is applied to sequences in much the same way as a symmetric FIR filter. That is, for each window of $N$ points, where $N$ is an odd integer, the

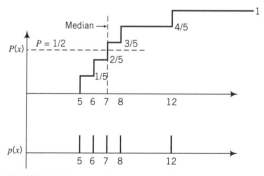

**FIGURE 30.15**   Illustration of median computation for a sequence of numbers. The lower graph shows a discrete density; the upper shows the cumulative distribution; median corresponds to the point where cumulative distribution is 0.5.

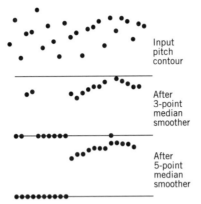

Input
pitch
contour

After
3-point
median
smoother

After
5-point
median
smoother

**FIGURE 30.16** Example of median smoothing for a sequence derived from a speech signal. Each point is computed every 10 ms. Zero values correspond to the decision that speech for that frame is unvoiced. From [15].

value of point $(N + 1)/2$ in the window (for a new, smoothed sequence) is set equal to the median of the points in the window. The window then is stepped along by one sample point, and the function is recomputed. This repeated sequence of operations is then referred to as an $N$-point smoothing.

   By cascading a three-point median smoother with a five-point median smoother (Fig. 30.16), one can transform the presmoothing pitch contour (top figure) into the result shown on the bottom. In this modification of the basic median smoother, two additional constraints are imposed:    (a) if the low-pass signal energy is below a threshold, the result is set to hiss (zero in the figure), and (b) if the variance of three successive results is too large, the median smoother output is also set to hiss.

   Another approach to smoothing pitch estimates is dynamic programming. This was incorporated in systems such as the one reported in [14]. More recently this was applied to normalized cross-correlation coefficients in the "robust algorithm for pitch tracking" algorithm as reported in [17].

## 30.9  EXERCISES

**30.1**   Using Klatt's synthesizer as a model (Chapter 29), determine the sequence of excitation functions for:
   **(a)** Voiced fricatives,
   **(b)** Voiced plosives,
   **(c)** Affricates (ch, dj),
   **(d)** Voiceless plosives.

**30.2**   Figure 30.7 shows a way of spectrally flattening a speech signal. Another method is to pass the speech through a bank of bandpass filters and hard limit each output, which is then bandpass filtered by an identical filter. Can you compare the two methods? How are they the same? How do they differ?

**30.3** How might the measurements described in connection with Figs. 30.11 and 30.12 be used to create a buzz–hiss decision?

**30.4** Given a sequence of detected periods of 90, 90, 94, 73, 85, 40, 78, 95, 97, 50, 100, 105, 110:

(a) Find the new sequence after three-point median smoothing.

(b) Find the resulting sequence after processing the result of (a) with a five-point median smoother.

**30.5** Build a circuit or write a program to generate a pulse train. Include the ability to vary the repetition frequency. For reference frequencies of 50, 100, 200, 400, and 800 Hz, measure the just noticeable deviation from that reference. Design a convenient way to plot results. Discuss.

**30.6** The pitch detection circuit used by Dudley consisted of a slope filter followed by a zero-crossing meter. The filter had a log magnitude versus frequency shape that approximated a straight line (see Fig. 30.1).

(a) Design a digital filter to approximate the magnitude response of the slope filter.

(b) Consider a signal defined by the equation

$$y(t) = \cos \omega t + \cos 2\omega t + \cos 4\omega t \tag{30.1}$$

as input to the slope filter.

Write the equation for the output $q(t)$. Measure the zero crossings of both the input and output and discuss how they compare as pitch detectors.

**30.7** Describe the perceptual effects of the following types of errors in modeling the excitation function:

(a) Mistakenly changing buzz to hiss.

(b) Mistakenly changing hiss to buzz.

(c) Doubling the detected pitch.

(d) Halving the detected pitch.

**30.8** Write a brief essay (less than 1000 words), giving your views of the following pitch detection algorithms:

(a) Comb filtering.

(b) Gold–Rabiner parallel-processing time-domain algorithm.

(c) Cepstral analysis.

(d) Spectral flattening and autocorrelation as described by Sondhi.

(e) Implementation of Goldstein's model as described by Duifhuis.

(f) Seneff's Harmonic pitch detector.

# BIBLIOGRAPHY

1. Aarset, T. C., and Gold, B., "Models of pitch perception," Tech. Rep. 964, MIT Lincoln Laboratory, Lexington, Mass., 1992.
2. Delgutte, B., and Cariani, P. A., "Coding of the pitch harmonic and inharmonic complex tones in the interspike intervals of auditory-nerve fibers," in M. E. H. Schouten, ed., *The Auditory Processing of Speech*, Mouton–De Gruyter, Berlin, pp. 37–45, 1992.
3. Duifhuis, H., Willems, L. F., and Sluyter, R. J., "An implementation of Goldstein's theory of pitch perception," *J. Acoust. Soc. Am.* **71**: 1568–1580, 1982.

4.  Gold, B., and Rabiner, L. R., "Parallel processing techniques for estimating pitch periods of speech in the time domain," *J. Acoust. Soc. Am.* **46**: 442–448, 1969.

5.  Gold, B., "A note on buzz-hiss detection," *J. Acoust. Soc. Am.* **36**: 1659, 1964.

6.  Goldstein, J. L., "An optimum processor for the central formation of pitch of complex tones," *J. Acoust. Soc. Am.* **54**: 1496–1516, 1973.

7.  Licklider, J. C. R., "A duplex theory of pitch perception," *Experientia* **7**: 128–138, 1951.

8.  Lyon, R. F., "Computational models of neural auditory processing," in *Proc. ICASSP'84*, San Diego, 1984.

9.  Markel, J. D., "The SIFT algorithm for fundamental frequency estimation," *IEEE Trans. Audio Electroacoust.* **AU-20**: 367, 1972.

10. Meddis, R., and Hewitt, M. J., "Virtual pitch and phase sensitivity of a computer model of the auditory periphery I:  pitch identification," *J. Acoust. Soc. Am.* **89**: 6, 1991.

11. Noll, A. M., "Cepstrum pitch determination," *J. Acoust. Soc. Am.* **41**: 293, 1967.

12. Ross, M. J., Schaffer, H. L., Cohen, A., Freudberg, R., and Manley, H., "Average magnitude difference function pitch extractor," *IEEE Trans. Acoust. Speech Signal Process.* **ASSP-22**: 353–362, 1974.

13. Schroeder, M. R., "Period histogram and product spectrum:  new methods for fundamental frequency measurement," *J. Acoust. Soc. Am.* **43**: 829–834, 1968.

14. Secrest, B., and Doddington, G., "An integrated pitch tracking algorithm for speech systems," in *Proc. ICASSP'82*, Boston, pp. 1352–1355, 1983.

15. Seneff, S., "Real-time harmonic pitch detector," *IEEE Trans. Acoust. Speech Signal Process.* **ASSP-26**: 358–364, 1978.

16. Sondhi, M. M., "New methods of pitch extraction," *IEEE Trans. Audio Electroacoust.* **AU-16**: 262–266, 1968.

17. Talkin, D., "A robust algorithm for pitch tracking," W. B. Kleijn and K. K. Paliwal, eds., in *Speech Coding and Synthesis*, Elsevier, Amsterdam/New York, pp. 495–518, 1995.

18. Tukey, J. W., "Nonlinear (nonsuperposable) methods for smoothing data," in *Proc. Eascon '74*, Washington, D.C., pp. 673, 1974.

# CHAPTER 31

# VOCODERS

## 31.1 INTRODUCTION

We are now in a position to define complete vocoder systems, often called analysis–synthesis systems. Detailed descriptions of spectral envelope estimations employing three analysis methods were outlined in Chapters 19, 20, and 21. These methods are filter banks, cepstral analysis, and linear prediction. In addition, speech synthesis was treated in Chapter 29 and pitch detection was treated in Chapter 30. In this chapter, we combine this material into complete analysis–synthesis structures. The following chapters will elaborate on specific forms of vocoders.

The primary application of vocoder systems is *source coding* to reduce the required rate of transmission. Source coding is also useful for efficient storage of speech and music, for example, in voice answer-back systems or in compact disks. As Dudley pointed out (see the preface to Chapter 3), source coding is useful in secrecy systems; it is also useful as an aide to help us understand how speech is produced, which in turn may be helpful for exploring diagnostic tools in speech and hearing pathologies. Another application is speech transformations, which we discuss in Chapter 34.

## 31.2 STANDARDS FOR DIGITAL SPEECH CODING

Figure 31.1 shows the standardized bandwidths and coding rates for four speech and music applications. The purpose of source coding research is to devise methods of lowering the required coding rates while maintaining the quality and robustness of the transmitted or stored speech. (Notice that the standard for telephony decreased fourfold within the 1972–1991 period, as a result of research).

The range of coding rates for these systems varies over nearly an order of magnitude. Intuition tells us that different applications will require different system designs. Furthermore, given certain robustness requirements, it may often be necessary to employ systems with rates that can be time variable to help combat environmental obstacles such as noise or (for long-distance transmission) atmospheric uncertainties.

## 31.3 DESIGN CONSIDERATIONS IN CHANNEL VOCODER FILTER BANKS

A version of a channel vocoder system is displayed in Fig. 31.2. These days, systems such as these are usually composed entirely of digital hardware once the speech is sampled and quantized.

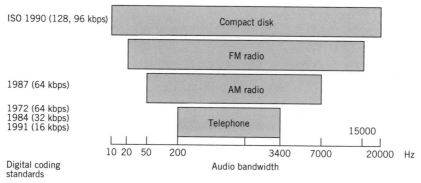

**FIGURE 31.1**   Digital coding standards.

The first set of design issues centers around the bandpass filters. We should begin by assuming that the analyzer and synthesizer filters covering the same section of bandwidth are identical. (This is not an absolute requirement, and some vocoder designs do not obey this dictum.) Given this, we now list some of the design parameters that have to be studied:

1. What should $N$ be; that is, how many filters in the bandpass filter bank?
2. What should be the filter bandwidth as a function of its center frequency?
3. Which of the many known design methods works best for channel vocoders?
4. How can the FFT algorithm be adapted to meet criteria?

Number of bandpass filters:    To look at this issue, we first need to specify the overall system bandwidth. According to the standards of Fig. 31.1, a compact disk uses an approximately 20-kHz bandwidth compared to a 3-kHz one for a telephone channel. Does a vocoder with a CD bandwidth then require seven filters for every one of the telephone? Certainly

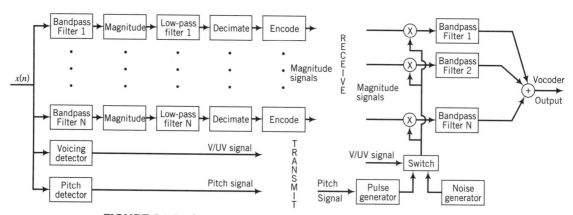

**FIGURE 31.2**   Channel vocoder analyzer and synthesizer.

not! Recall from Chapter 15 on Psychoacoustics that the frequency resolution of the ear diminishes with increasing frequency. Figure 15.4 shows that the auditory bandwidth at 10 kHz is approximately 2000 Hz!

Consider for the moment the telephone example. In addition to the signal bandwidth, one must specify the available transmission bandwidth, which in turn controls the overall bit rate. Assume that the vocoder must transmit information at 2400 bps (bits per second). Further assume that 400 bps will be allotted to transmission of the excitation parameters. Finally, also assume that each of the magnitude signals of Fig. 31.2 has to be updated 50 times per second and that 4 bits are needed to represent each of these signals; thus 200 bps per channel are needed. The result is that $N = 10$. However, we know from Dudley's early results that 10 bandpass filters yield speech of substandard quality (see Chapter 3).

The lesson to be learned is this: any one parameter in the channel vocoder cannot be designed in isolation! Design parameters must be a compromise among the number of channels, the update rate per channel, the bits allocated per channel, and the division of rate between the spectral envelope and excitation function. Much experience has established that satisfactory vocoded speech for telephony may require from 15 to 25 channels; later we will show how such numbers fit into the overall design.

Filter bandwidth specification: We begin with the knowledge that filters can have an increased width with center frequency without degrading the output speech. Many early channel vocoders were designed with the *same* bandwidth; the reason was the increased ease of filter design and implementation.

One possible design criterion is again to refer to Fig. 15.4. The auditory bandwidths are approximately 100 Hz for center frequencies below approximately 800 Hz and go up to approximately 250 Hz for 3-kHz center frequencies. One can design filter passbands that more or less follow this criterion. Again, a compromise for ease of implementation is usually invoked. The critical bandwidth curve is approximated in a stepwise fashion; for example, six filters of 100 Hz width from 200 Hz to 800 Hz; six filters of 150 Hz width at center frequencies 950, 1100, 1250, 1400, 1550, and 1700; five filters of 200 Hz width at 1800, 2000, 2200, 2400, and 2600; and three filters of 300 Hz width at 2800, 3100, and 3400. There is a total of 20 filters.

Another possible design criterion is to specify that, for the great majority of speakers, each filter should encompass a single harmonic of the voiced speech. If this could be done, it might be argued that the data rate of the following low-pass filters could be lowered, since the detected envelope when a single harmonic is detected (rather than two harmonics) within a given filter may require less updating and thus a lower transmission rate. If this notion were followed for voice fundamental frequencies as low as 100 Hz, a total of 32 filters would be needed to encompass the 200- to 3400-Hz telephone band. (For CDs, for which music storage is the primary requirement, this argument no longer holds.) Furthermore, equal bandwidth filters would be used.

An advantage of equal bandwidth design is based on the role of the synthesizer filter bank. A reasonable criterion for good spectral envelope representation is to examine the spectrum of the synthesizer output when the excitation consists of a single sharp impulse and all filter magnitude signals are constant; the resultant spectrum should be constant over all frequencies. Another way to state this is that the *sum* of all synthesizer filter impulse responses should be very much like a perfect impulse. Designing such a filter bank is significantly simpler when all filters are of equal bandwidth and shape.

Filter designs:     Chapter 7 has a discussion of the various criteria for digital filters based on previous analog designs. One point worth reiterating is that sharper skirts don't improve the quality of the vocoded speech because reverberation is introduced. Filters with linear phase characteristics can be incorporated into the design to minimize reverberations. These can be implemented in at least four ways. Bessel filters [23] are minimum phase filters that produce a very linear phase; Lerner filters [15] that are not at minimum phase (they have zeros outside the unit circle) also produce an excellent phase response. Both of these designs are of the IIR variety, but linear phase can also be attained with FIR designs, either by use of the direct nonrecursive implementation, or with frequency-sampling filters that are implemented by using IIR elements.

## 31.4  ENERGY MEASUREMENTS IN A CHANNEL VOCODER

The next step in spectrum analysis of the incoming speech is to process the outputs of the bandpass filters to produce an estimate of the spectrum. These estimates, called magnitude signals in Fig. 31.2, are derived as shown in the figure. A simple example in Fig. 31.3 illustrates the progression from the output of a single bandpass filter that generates a 1000-Hz harmonic of the input signal. The box marked "magnitude" (Fig. 31.2) immediately following the bandpass filter can be, for example, a full-wave or half-wave rectifier (or it can be a system involving Hilbert transforms, as discussed later). A half-wave rectified signal is shown in Fig. 31.3(b). The spectrum of this signal is shown in Fig. 31.3(c).

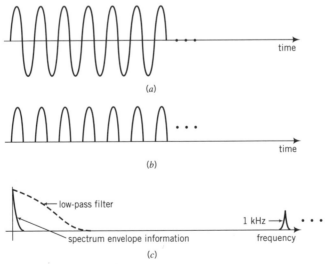

**FIGURE 31.3**  Example of energy measurement with a half-wave rectifier.

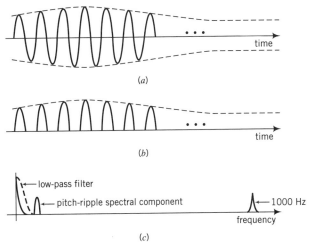

**FIGURE 31.4**　Effect of pitch ripple in a spectral estimate.

We have assumed a sampling rate at the input and at the bandpass filter levels of 10 kHz. The low-pass filter that follows has, in this case, the simple task of eliminating frequencies of 1 kHz and above while preserving the spectrum near dc that reflects the slow variation caused by spectral changes. Typically, this variation is of the order of 5–15 Hz. Therefore, decimation (downsampling) can allow for a new sample every 10–20 ms. Each discrete sample must now be further quantized, using as few bits as possible to keep transmission rates as low as possible. At this point, the magnitude signals of Fig. 31.2 are ready to be multiplexed and transmitted to the synthesizer.

What happens if a particular bandpass filter is wide enough to accept two successive harmonics of the speech? Figure 31.4 shows the bandpass filter output consisting of harmonics 12 and 13 of a speech wave with a fundamental frequency of 80 Hz. The half-wave rectified signal is shown in part (b) and its spectrum is shown in part (c).

Notice that there is now a substantial component at the fundamental frequency; this component is unwanted pitch ripple. Now the low-pass filter has an appreciably more difficult task to remove this spurious component while faithfully passing the legitimate spectral variations. Again, we are faced with the fact that good design (in this case, of the low-pass filters) is a *compromise* depending on a number of factors: spectral variations, bandpass filter widths, pitch-ripple attenuation, and so on. The final choice of low-pass filter bandwidth determines the permissible degree of decimation, and this in turn affects the distribution of coding bits for transmission.

Spectral magnitude may also be estimated by Hilbert transform techniques [21]. In the simplest case, if the bandpass filter output is a sinusoid, then the Hilbert transform is a pure tone exactly 90° out of phase but of the same magnitude; therefore, a diagram such as Fig. 31.5 will generate the magnitude of the original bandpass filter output. In this idealized case, no low-pass filter is required.

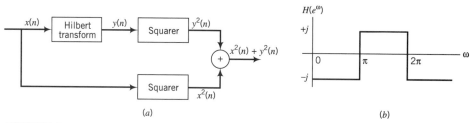

**FIGURE 31.5**    Hilbert transform method for finding the squared magnitude of a tone.

The block marked "Hilbert Transform" in Fig. 31.5(a) has the idealized frequency response shown in part (b). The magnitude is constant and the phase oscillates between $+j$ and $-j$.

## 31.5   A VOCODER DESIGN FOR SPECTRAL ENVELOPE ESTIMATION

For many years, requirements for many digital vocoders included operation at 2400 bps. If we assume, tentatively, that 400 bps is enough to encode excitation, that leaves us 2000 bps to play with. If 20 channels are initially chosen, this comes to 100 bps per channel. Now, subjective measurements indicate that channel quantizing paradigms are efficiently implemented by means of differential pulse code modulation (DPCM) across bands; this means that the differences between adjacent bands are transmitted, instead of the actual band values. Intuitively, this makes sense, since we expect a high degree of correlation between adjacent filters [5]. The first (lowest frequency) channel serves as a three-bit reference, and successive channels use two-bit DPCM coding. For 20 channels, the total is $3 + 38 = 41$ bits. For a frame rate of 50 samples/s, this comes to 2050 bps, which is just a bit more than desired. Obviously, the numbers can be varied in many different ways. For example, if the designer believes that he or she can get away with only 15 channels, each channel can be coded more accurately with three-bit DPCM plus four bits for the reference, a total of 45 bits per frame, but now the frame rate must be reduced to 44.4 frames/s.

## 31.6   BIT SAVING IN CHANNEL VOCODERS

Several useful tricks for bit-rate reduction are available; in this section, we discuss three such items. These are (a) efficient quantization, (b) linear transformations, and (c) frame fill.

   Efficient quantization:    The human ear and brain judge relative sound intensities more or less logarithmically (see Chapter 15). Thus, it makes sense to quantize the channel energy in a nonuniform manner. Naturally, the idea of some sort of logarithmic quantization comes to mind, but care is needed. The log of zero is $-\infty$; this is not a useful value in a quantizer. Smith [22] analyzed the $\mu$-law quantizer that results in a useful logarithmic type

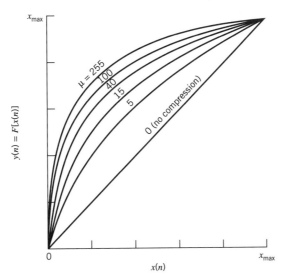

**FIGURE 31.6** $\mu$-law characteristic. From [22].

of quantizer. The signal is first passed through a nonlinearity with no memory given by

$$y = X\left\{\frac{\log[1 + \mu(x/X)]}{\log(1 + \mu)}\right\}, \tag{31.1}$$

where it is assumed that the input $x$ is always positive, $X$ is the maximum value of $x$, and $\mu$ is a parameter.

Figure 31.6 shows input–output relations of a $\mu$-law characteristic for several values of $\mu$, and Fig. 31.7 is an example of a two bit $\mu$-law quantizer.

A more general approach to the quantization problem is given by Max [16]. Under the hypothesis that the probability density function of the signal is known, one constructs a set of quantizing steps that correspond to equal probabilities of signal magnitudes occurring within that quantization interval; thus if the signal magnitude is very unlikely to fall within a given range, the resulting quantization interval can be large, causing a large distortion; this, however, happens very rarely. An example of a three-bit quantizer using Max's concept is seen in Fig. 31.8.

Linear transformations of the spectral data: Since the magnitude signals (see Fig. 31.2) from a channel vocoder are, to some degree, correlated, it seems natural to search for a way that takes advantage of this correlation to reduce the data rate. These magnitude signals are correlated over both time and frequency. Taking advantage of these two-dimensional correlations has generally been considered to be too difficult; the work cited below separates the two dimensions. First, we deal with the frequency-domain correlation; thus, for example, in a channel vocoder, each frame (consisting of perhaps 20 spectral values) is analyzed separately and a specific linear transformation is invoked that tends to *order* the results proportional to their significance.

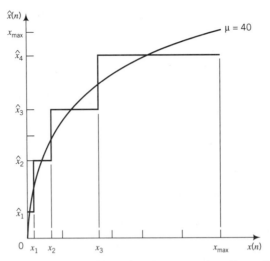

**FIGURE 31.7** Two-bit $\mu$-law quantizer. From [22].

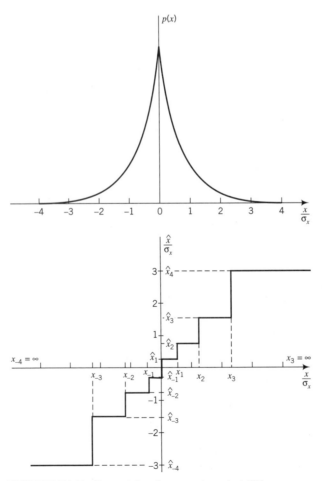

**FIGURE 31.8** Quantizing for equal probabilities.

**TABLE 31.1 Authors and Methods of Linear Transformations to Reduce Bit Rate**

| Author | Method |
|---|---|
| Crowther | Walsh–Hadamard transform |
| Kramer | PCA |
| Pols | PCA |
| Andrews | Karhunen–Loev transform |
| Zoharian | PCA |
| Ahmed | DCT |

A very simple example helps explain the concept. Assume a frame consisting of only two spectral samples, $x_1$ and $x_2$. Now, apply the transformation

$$y_1 = x_1 + x_2, \quad y_2 = x_1 - x_2. \tag{31.2}$$

If $x_1$ and $x_2$ are both equal to some number $x$, then $y_1 = 2x$ and $y_2 = 0$. Thus, only $y_1$ need be transmitted and both $x_1$ and $x_2$ can be reconstructed from the inverse transformations

$$x_1 = \frac{y_1 + y_2}{2}, \quad x_2 = \frac{y_1 - y_2}{2}. \tag{31.3}$$

More generally, we need to find a linear transformation

$$y_j = \sum_{i=1}^{N} a_{ij} x_i, \quad j = 1, 2, 3, \ldots, N, \tag{31.4}$$

such that the transformed set $y_j$ is arranged in *size place*. It is then up to the intuition of the designer to estimate how many bits can be saved by transmitting the $y_j$ set. Since Eq. 31.4 is a linear transformation, the full set of $x_i$ can be reconstructed. These ideas have been tried by numerous researchers, among them Crowther and Rader [7], Kramer and Mathews [14], Pols [20], Andrews [2], Zoharian and Rothenberg [26], and Ahmed et al. [1]. Table 31.1 lists these authors and the particular transformations they used. Note that in several cases the authors used principal components analysis (PCA), which we briefly mentioned in Chapter 8 as a general approach to dimensionality reduction.

This transformation scheme, which is based on the correlation properties of signals, was derived in 1933 by Hotelling [13]. In this approach, the linear transformation is chosen to rotate the $x$ vector so that the new components correspond to eigenvectors of the sample covariance matrix. The components with the largest eigenvalues (or, equivalently, variances) are viewed as the *principal* ones. Thus, for example, if one begins with 20 variables and successively finds the first four or five transformed variables, these may be sufficient to preserve nearly all the information of the original variables. Clearly, bit savings can result.

A simpler approach that does not require the determination of the optimal transformation is to compute the discrete cosine transform (DCT). For many speech applications, the

DCT leads to similar results as a procedure such as PCA that decorrelates the transformed variables.

The DCT of a sequence $x(n)$ of $N$ points is defined by the equation

$$\text{DCT}(k) = \frac{2}{N} \sum_{n=0}^{N-1} x(n) \cos \frac{(2n+1)k\pi}{2N}, \quad k = 1, 2, \dots, (N-1). \tag{31.5}$$

For $k = 0$, the term preceding the sum is $\sqrt{2}/N$.

The reader is referred to the cited references for more details. It should also be mentioned that the straightforward method of DPCM across adjacent channels takes advantage of correlations in frequency.

## 31.7 DESIGN OF THE EXCITATION PARAMETERS FOR A CHANNEL VOCODER

The original Dudley channel vocoder employed a pulse generator, a noise generator, and a buzz–hiss switch. The underlying assumption was that speech is composed of sounds that are periodic or noisy (aperiodic). The purpose of a voicing detector at the analyzer was to make a one-bit voicing decision to control the synthesizer buzz–hiss switch. It is an interesting fact that although this model includes some fallacious assumptions, it produced speech that was quite intelligible.

What are these fallacies? First, for the voiced fricative sounds, the excitation is really a combination of buzz and hiss. This immediately makes the job of the analyzer more difficult. It must first identify the sound as a voiced fricative and then determine the relative values of the two sources.

However, this is not the only difficult aspect of the problem. The voiceless plosives, typically, follow a silence, then begin with a transient burst, followed by noiselike aspiration, before finally beginning pure voicing if followed by a vowel (the situation is even more complex if the plosive is followed by another consonant such as a plosive or fricative.) If the plosive is followed by a transition sound such as "r" or "l", both buzz and hiss may be simultaneously present.

The synthesizer configurations of Fant and Klatt (see Chapter 29) confront this issue by including buzz–hiss combinations in their structures. Their synthesizers are controlled either by humans who know what sounds they want or by text, where again the desired sound output is preordained. This fine degree of control is a formidable task for an automatic analyzer system.

Some systems have been built which do estimate periodic and aperiodic components, however, such as the hybrid harmonic/stochastic analysis described in [9]. In this approach, least squares optimization is used to find the best set of harmonics (frequency, amplitude, and phase) to match the voiced spectrum. The error spectrum, which consists of the original minus the harmonics, is approximated by a filtered noise source. The parameters of the noise filter are estimated in a second optimization procedure.

Both the voiced and unvoiced plosives feature transient bursts, which are often important perceptual cues for recognition by people (see Chapter 17). These bursts can be very short, lasting perhaps 5–15 ms. For low-rate vocoder systems to transmit this information

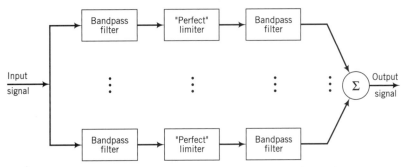

**FIGURE 31.9** Bank of filter–limiter–filter for spectral flattening.

adequately requires too great a bit rate for both excitation and spectral parameters. (Very rapid spectral changes also occur in transitions to nasal murmurs.) Again, the major difficulty is for the analyzer to detect when these modifications are needed.

At this writing, there is little to report on efforts to rectify this situation. The fact remains that present-day vocoders operating at 2400 bps are not able to synthesize speech that is indistinguishable from the original.

Spectral flattening: Spectrum analysis for vocoders is almost always carried out on a frame-by-frame basis; for each frame, a windowed speech segment of 20–40 ms is analyzed. Within this window, many periods of the speech may be present. If the signal is not quite periodic (if, for example, there is pitch jitter), the excitation function can contribute to the overall spectral shape. Thus, if this excitation function is faithfully produced at the synthesizer and since it is also measured by the spectrum analyzer, a degree of pitch-induced spectral distortion is introduced into the system. As a way to combat this problem, spectral flattening can be introduced to reduce this distortion; an example of such a system is shown in Fig. 30.7. In the present application, the input signal will be the excitation function generated by the chosen source functions.

The notion of spectral flattening was first introduced by David et al. [8] in a different context. They assumed that all of the relevant excitation information resided in the so-called baseband part of the speech; this is the speech bandwidth from approximately 300 to 900 Hz. By passing this signal through a simple nonlinearity such as a rectifier to generate higher frequencies and then passing this new signal through the spectrum flattener, either like the one in Fig. 30.7 or the one in Fig. 31.9, the resulting excitation spectrum would be very close to constant with frequency. Figure 31.10 shows a complete voice-excited vocoder.

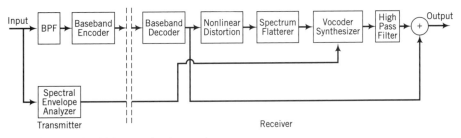

**FIGURE 31.10** Voice-excited vocoder.

## 31.8  LPC VOCODERS

LPC (linear predictive coding) vocoders were briefly discussed in Chapter 3. In Chapter 21, a more detailed discussion of the LPC analyzer was undertaken. In this section we describe several synthesis structures.

The relative roles of the excitation parameters in LPC and channel vocoders is certainly an interesting and useful topic for discussion. In a low-rate system, for example, 2400 bps, both types of vocoders usually rely on algorithms that generate these parameters by a direct analysis of the speech wave. However, a notable difference between the two systems is the presence, in LPC, of the *error signal*. This can be a powerful tool that allows for fascinating LPC vocoder variations. The use of the unadulterated error signal as excitation results in synthetic speech that is a replica of the original. No such parameter is available for channel vocoders. Of course, simply transmitting the full error signal does not result in bit saving; however, if we assume that LPC spectral analysis has captured much of the spectral information, this means that the error signal will be primarily a function of the excitation parameters and ought to be codable at a lower rate. Many schemes have tried to take advantage of this fact, and in Chapter 33 we explore several. In this section, we focus on spectral estimation for LPC vocoders of low rate and assume that excitation analysis and synthesis take place in the same manner as for channel vocoders.

We know from previous chapters that the LPC synthesizer is an all-pole digital filter. Figure 31.11 is a block diagram of a complete system.

One possible LPC synthesizer configuration is shown in Fig. 21.2. This is the standard direct-form digital filter, in which the coefficients are obtained by one of the matrix inversion methods described in Chapter 21. Another possible structure is that of a cascade

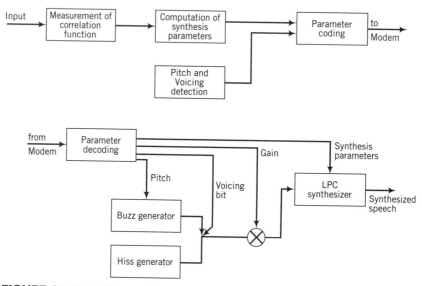

**FIGURE 31.11**  LPC vocoder.

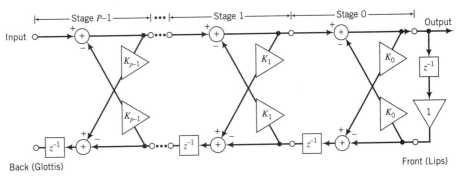

**FIGURE 31.12**   Lattice synthesizer for LPC.

of digital pole pairs; for this, the analyzer must compute the roots of the polynomial. A third, commonly used structure is the lattice of Fig. 31.12.

An observed advantage of the lattice is its decreased vulnerability to quantization error.

## 31.9   CEPSTRAL VOCODERS

Chapter 20 describes cepstral analysis. The first vocoder employing cepstral methods was by Oppenheim [19]. Figure 31.13 shows the analyzer configuration, and Fig. 31.14 shows the synthesizer structure. By liftering the signal obtained after the inverse Fourier transform with short-time window $w(n)$ to eliminate the high-time components, the cepstrum $c(n)w(n)$ emphasizes vocal tract filter information. As discussed in Chapter 30 on Pitch Detection, the high-time components can be used to obtain pitch and voicing information. In this way, deconvolution of the excitation and vocal tract filter components is performed.

Synthesis involves the reverse process of convolution, as seen in Fig. 31.14.

The excitation generator produces pulses at the measured fundamental period or, for voiceless sounds, closely spaced pulses of random polarity. These are convolved with the estimated impulse response of the vocal tract $h(n)$ to produce the synthesized output speech.

## 31.10   DESIGN COMPARISONS

In the sections on channel vocoders, we detailed several design issues. In this section we briefly review these issues on a comparative basis for the three basic algorithms:   channel, LPC, and cepstral vocoders.

Discreteness of analysis in the three systems:   we discussed the issue of how many filters were appropriate for channel vocoder analysis in Section 31.3. By performing an energy measurement for each channel, we get a discrete representation of the spectral magnitude. In LPC analysis, the least-squares criterion leads to direct computation of the parameters of the synthesis filter. The computation involves the solution of a set of $n$ linear equations, so the specification of $n$ corresponds to the choice of the number of filters

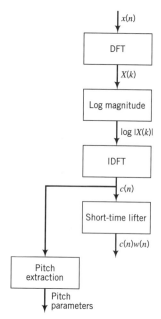

**FIGURE 31.13** Cepstral vocoder analysis. The input speech is a discrete-time sequence $x(n)$. Its Discrete fourier transform (DFT) is $X(k)$. The next two operations convert the spectrum to cepstrum. The cepstrum is used to determine voicing and pitch, and a window that is nonzero for cepstral indices less than some value is multiplied by $c(n)$, the cepstrum, yielding a truncated cepstrum $c(n)w(n)$.

in the channel vocoder. It is interesting to note that $n = 10$ has been considered to be a reasonable specification for a 2400-bps LPC vocoder; 10 filters were not adequate for Dudley's original channel vocoder design. (This issue is explored in the problem set of this chapter.)

In cepstral analysis, the low-time part of the cepstrum yields information on the vocal tract filter. Since the cepstrum is computed from a discrete Fourier transform, the low-time portion consists of a finite number of values. It is much like a Fourier series representation of the slowly varying spectral envelope shown as the output of the log magnitude block in Fig. 31.13. In Oppenheim's system, he chose the first 32 points of the cepstrum as his spectral representation.[1] In later work [25], Weinstein and Oppenheim represented the homomorphically computed vocal tract impulse response by linear prediction; they were able to reduce the overall bit rate to approximately 4000 bps by sending the resulting LPC coefficients.

Bandwidth specification:  in Section 31.2, issues of choosing a set of bandwidths for the channel vocoder filter bank were discussed. In traditional LPC systems, once the value of

---

[1] Since Oppenheim's motivation was to prove the feasibility of a brand new algorithm, he did not design a 2400-bps system; it would be difficult to devise a 2400-bps cepstral vocoder with so many parameters without some clever coding of the cepstral points.

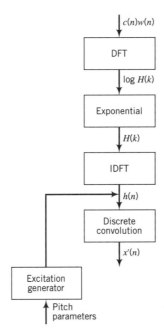

**FIGURE 31.14** Cepstral vocoder synthesis. The truncated cepstrum given by $c(n)$ $w(n)$ is converted to a log spectrum and then to a corresponding impulse response. This is convolved with a periodic or noiselike excitation determined by the pitch and voicing to yield $x'(n)$, an approximation to the original speech.

$n$ is chosen, the resultant synthesizer and its spectrum are determined, so there is no need for any additional specification. In cepstral analysis, the designer has the choice of the DFT size. Both the LPC and cepstrum can be modified to be more perceptually oriented. A popular technique (in speech-recognition research) to create a perceptually oriented cepstrum starts with direct smoothing of the log spectrum, as shown in Fig. 31.15.

Mel scale approximation to smoothing is accomplished by multiplying the log spectrum by each of the $N$ weighting functions of Fig. 31.15(b) and summing to produce an $N$-point smoothed log spectrum. The resultant cepstrum is thus primarily a function of the vocal tract filter; pitch ripples have been mostly eliminated [3].

In traditional LPC, the required correlation matrix is computed in the time domain. However, the correlation function can also be computed as the Fourier transform of the power spectrum, and the power spectrum can be computed with variable bandwidth to resemble the responses of auditory filters, as in mel scale cepstral analysis. As noted in Chapter 22, Hermansky [12] has proposed a perceptually linear predictive technique that employs this trick, plus an equal loudness approach.

Pitch ripple: in the channel vocoder, pitch ripple was caused by channel filters that were too wide. In the cepstrum, a sufficiently low pitch will cause cepstral components of the excitation function to appear in the *liftering* window designed to try to isolate the vocal tract filter component. In LPC, if the pitch is too low, the fit of the resultant synthesis

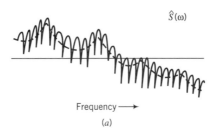

$\hat{S}(\omega)$

Frequency ⟶

(a)

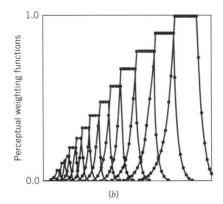

(b)

**FIGURE 31.15**  Smoothing of the log spectrum by multiplying (a) by the set of windows (b).

filter will be a function of the transition from one harmonic to the next as well as from one formant to the next.

Quantization of the parameters:    in all systems, design comparisons depend on the perceptual effects of quantizing the particular parameters. With direct spectral measurements, as in the channel vocoder, more bits are needed to code the lower-frequency components.

## 31.11  VOCODER STANDARDIZATION

The research and development efforts directed toward the deployment of different vocoder systems has resulted in a veritable Tower of Babel. Vocoder transmitters and receivers must be compatible; a channel vocoder transmission would not be understood by a LPC or homomorphic receiver without specialized interfaces. A transmission system operating at 8 kb/s is not automatically compatible with a 2.4 kb/s receiver.

At present, there are many vocoder systems of different designs, different bit rates, and different error-correcting methods. As a result, there are various organizations to set standards. The International Telecommunications Union sets global telecommunications standards. This group formulates speech-coding standards, evaluates and tests potential standards, and requests new standards to meet new applications. The European Telecommunications Standards Institute sets standards for digital cellular communications in Europe. In North America, these standards are set by the Telecommunication Industry Association, and in Japan they are set by the Research and Development Center for Radio Systems. The International Maritime Satellite Corporation sets standards for satellite-based telephony applications. Finally, the U.S. Government and NATO have the responsibility to standardize secure telephony. A detailed study of these standards is given by Cox [6].

Almost all standards in the bit rate range of 2.4 kb/s through 8 kb/s are variations on the CELP algorithms described in Chapter 33; these algorithms are extensions of the basic LPC approach with a more elaborate excitation model. The notable exception is the secure 2.4 kb/s standard [24]; thus far, the CELP algorithm has not led to a sufficiently intelligible system at this low rate.

A 4.15 kb/s coder was standardized by the International Maritime Satellite Corporation for worldwide ship to shore via satellite telecommunications. This is the Improved Multi-Band Excitation (IMBE) coder that transmits voice (including error correction) at 6.4 kb/s [10], [11]. It is a variation on the Sine Tranform Coder[2] [17], with voicing decisions made independently in different frequency bands.

Finally, we should make a few remarks about compatibility even given standards. Two standards operating at different bit rates are not compatible without a carefully designed interface. A simple example of a two-rate system is described in [4], in which a channel vocoder operating at 2.4 kb/s is augmented by subband signals derived from the low-frequency channel filters. When transmitting to a compatible terminal, the subband spectrum is given precedence over the same spectral portion generated by the 2.4 kb/s vocoder; but when transmitting to a 2.4 kb/s compatible terminal, only the channel vocoder portion of the analyzed speech is sent.

## 31.12 EXERCISES

**31.1** Consider the following.

(a) Consider two adjacent filters of the same bandwidth. The designer has specified that the amplitude responses of these filters cross at the 3-dB point. Now apply a sinusoidal input to both filters at the frequency at which they cross. What should be the relative phases at this crossover point so that the summed output is exactly of the same magnitude as that of a sinusoid at either of the filter center frequencies? (Assume that the adjacent filter has no effect at the center frequency of each filter.)

(b) If the relative phase at crossover is between $135°$ and $225°$, this will result in a large dip in magnitude at the crossover. Can you think of a simple way to improve the situation?

**31.2** If the input to the Hilbert transform scheme of Fig. 31.5 is the sum $A \cos(\omega_1 t) + B \cos(\omega_2 t)$, find the output at point $P$. Is there a need for low-pass filtering at this point? Explain.

**31.3** How can the $\mu$-law equation (Eq. 31.1) be modified to behave as a prequantizer for a signal that goes through negative as well as positive excursions?

---

[2]This is a complete vocoding technique based on the analysis technique briefly described in Chapter 29. Chapter 34 will describe the use of this approach for speech transformations.

### TABLE 31.2   Parameters of Four Speech Files to be Synthesized

| $F_1$ | $F_1$(BW) | $F_2$ | $F_2$(BW) | $F_3$ | $F_3$(BW) | $F_0$(Pitch) |
|-------|-----------|-------|-----------|-------|-----------|--------------|
| 270 | 40 | 2290 | 70 | 3010 | 170 | 100 |
| 270 | 40 | 2290 | 70 | 3010 | 170 | 300 |
| 300 | 50 | 870 | 60 | 2240 | 220 | 100 |
| 300 | 50 | 870 | 60 | 2240 | 110 | 300 |

**31.4**   This problem is a small research project. It was pointed out in the text (Section 31.10) that a tenth-order LPC performed better than a 10-channel vocoder. Perhaps a filter bank that resembled the auditory system (instead of an equal bandwidth filter bank) might give better results. The project, therefore, is to program a tenth-order LPC and to compare it with *your* choice of a filter bank for a 10-channel vocoder. Keep in mind that auditory filters increase in bandwidth as the frequency rises. For this exercise, do not attempt to quantize the parameters to achieve a specified bit rate.

**31.5**   (The following four problems were donated by Professor Don Johnson, who teaches a speech-processing course at Rice University.) The purpose of this laboratory exercise is twofold. The first is to obtain experience in the generating of synthetic speech. The second is to apply the theory of short-time spectral analysis to speech.

(a) Generate four files containing synthetic speech. This speech is to be generated on the computer by applying a train of unit samples to a cascade of second-order IIR filters. Let the effective sampling rate be 10 kHz.

The parameters of these speech files are in Table 31.2 (all values are in hertz).

Use impulse invariance to design the digital filters. In your response, demonstrate the design procedures and tabulate the filter coefficients.

(b) Compute the short-time spectrum of each of the synthetic speech files. Use rectangular and Hanning windows with durations of 51.2 ms and 12.8 ms. Compute and display the log magnitudes of the DFT of the windowed speech segments. Note if the phase of the window relative to the location of the excitation made any difference in the results.

(c) Extract from each of the synthetic speech files one pitch period. Compute the log magnitude of the DFT of these periods and compare to the results of (b).

(d) Compare measurements of formant location and bandwidths to the known values.

**31.6**   The purpose of this laboratory exercise is to determine how well homomorphic deconvolution works on speechlike signal.

(a) Compute and display the cepstrum for each of the synthetic speech files you generated in the previous laboratory assignment. Note the various aspects of the cepstrum (e.g., do different pitch values change the spectral envelope much, how well does the pitch line stand out, etc.). Use a 51.2-ms Hanning window in comparing the required short-time spectra.

(b) Compute the smoothed spectral envelope for these cepstra in two ways:   one in which the pitch is known exactly and one with a fixed window duration of 3.0 ms.

(c) Compare the homomorphically smoothed spectra with their corresponding short-time (un-smoothed) spectra.

(d) Measure formant frequencies and bandwidths and compare them with their actual values and also with the values obtained in the previous laboratory assignment.

**31.7** In this laboratory exercise you are to study the smoothed spectra obtained with linear prediction.

(a) Use the correlation method of linear prediction to compute the coefficients of the predictor polynomial for the four speech files. Apply a Hanning window with a duration of 51.2 ms to the data prior to the computation of the correlation function. Evaluate these coefficients for orders 4, 6, and 12.

(b) Compute the log magnitude spectral envelopes for each case. Compare the effects of pitch and order on the spectral envelope.

(c) Measure formant frequencies and bandwidths from the linear predictive spectra. Compute the roots of the predictor polynomial. Do these roots correspond to the location of the poles you used to generate the synthetic speech files? Compare.

**31.8** Spectral analysis in the presence of noise: it is claimed that the spectral analysis schemes that are based on an explicit model of speech production (e.g., homomorphic or linear predictive analysis) are sensitive to parametric errors in the model. In this laboratory exercise, changes in the spectral envelope caused by the presence of additive white noise will be explored.

(a) Use a Gaussian random-noise generator to produce a white-noise sequence. Produce cases in which the signal-to-noise ratio is 0 dB, 10 dB, and 20 dB for each of the low-pitch speech files.

(b) Use short-time DFT's (12.8-ms Hanning window) to determine the spectra of these records.

(c) Apply cepstral and LPC analysis on these data records to obtain spectral envelope information. Note how these methods are sensitive to the presence of noise. Note how the spectra change with assumed model order.

**31.9** In Section 31.10 it is stated that more bits are needed to code the lower-frequency parameters. Propose some perceptually based arguments to verify this statement.

# BIBLIOGRAPHY

1. Ahmed, N., Natarajan, T., and Rao, K. R., "Discrete cosine transform," *IEEE Trans. Comput.* **L-23**: 90–93, 1974.
2. Andrews, A. C., "Multidimensional rotations in feature selections," *IEEE Trans. Comput.* **C-20**: 1045–1051, 1971.
3. Applebaum, T. H., Hanson, B. A., and Wakita, H., "Weighted cepstral distance measures in vector quantization based speech recognizers," in *Proc. ICASSP '87*, Dallas, pp. 1155–1158, 1987.
4. Bially, T., Gold, B., and Seneff, S., "A technique for adaptive voice flow control in integrated packet networks," *IEEE Trans. Commun.* **28**: 325–333, 1980.
5. Blankenship, P. E., and Malpass, M. L., "Frame-fill techniques for reducing vocoder data rates," Tech. Rep. 556, MIT Lincoln Laboratory, Lexington, MA, 1981.
6. Cox, R. V., "Speech coding standards," in W. B. Kleijn and K. K. Paliwal, eds., *Speech Coding and Synthesis*, Elsevier, Amsterdam/New York, Chap. 2, 1995.
7. Crowther, W. R. and Rader, C. M., "Efficient coding of vocoder channel signals using linear transformations," *Proc. IEEE* **54**: 1594–1595, 1966.
8. David, E. E., Schroeder, M. R., Logan, B. F., and Prestigiacomo, A. J., "Voice-excited vocoders for practical speech bandwidth reduction," presented at the Speech Communications Seminar, Stockholm, 1963.
9. Dutiot, D., *An Introduction to Text-to-Speech Synthesis*, Kluwer, Dordrecht, the Netherlands, 1997.
10. Griffin, D. W., and Lim, J. S., "Multiband excitation vocoder," *IEEE Trans. Acoust. Speech Signal Process.* **ASSP-36**: 1223–1235, 1988.

11. Hardwick, J. C., and Lim, J. S., "The application of the IMBE speech coder to mobile communications," in *Proc. ICASSP '91*, Toronto, pp. 249–252, 1991.

12. Hermansky, H., "Perceptual linear predictive (PLP) analysis of speech," *J. Acoust. Soc. Am.* **87**: 1738–1752, 1990.

13. Hotelling, H., "Analysis of a complex of statistical variables into principle components," *J. Educ. Psychol.* **24**: 417–441, 498–520, 1933.

14. Kramer, H. P., and Mathews, M. V., "A linear coding for transmitting a set of correlated signals," *IRE Trans. Inform. Theory* **IT-2**: 41–46, 1956.

15. Lerner, R. M., "Bandpass filters with linear phase," *Proc. IEEE* **52**: 249–268, 1964.

16. Max, J., "Quantizing for minimum distortion," *IRE Trans. Inform. Theory* **IT-6**: 7–12, 1960.

17. McAulay, R. J., and Quatieri, T. F., "Speech analysis/synthesis based on a sinusoidal representation," *IEEE Trans. Acoust. Speech Signal Process.* **ASSP-34**: 744, 1986.

18. McLarnon, E., "A method for reducing the frame rate of a channel vocoder by using frame interpolation," in *Proc. ICASSP '78*, Washington, D.C., pp. 458–461, 1978.

19. Oppenheim, A. V., "Speech analysis-synthesis based on homomorphic filtering," *J. Acoust. Soc. Am.* **45-2**: 458–465, 1969.

20. Pols, L. C. W., "Real-time recognition of spoken words," *IEEE Trans. Comput.* **C-20**: 972–978, 1971.

21. Rabiner, L. R., and Gold, B., *Theory and Applications of Digital Signal Processing*, Prentice–Hall, Englewood, Cliffs, N.J., 1975.

22. Smith, B., "Instantaneous companding of quantized signals," *Bell Syst. Tech. J.* **36**: 653–709, 1957.

23. Storch, L., "Synthesis of constant time-delay ladder networks using Bessel polynomials," *Proc. IRE* **42**: 1666–1675, 1954.

24. Tremain, T. E., "The government standard linear predictive algorithm:   LPC-10," *Speech Technol.* **1**: 40–49, 1982.

25. Weinstein, C. J., and Oppenheim, A. V., "Predictive coding in a homomorphic vocoder," *IEEE Trans. Audio Electroacoust.* **AU-19**: 243–248, 1971.

26. Zoharian, A. S., and Rothenberg, M., "Principal-components analysis for low redundancy encoding of speech spectra," *J. Acoust. Soc. Am.* **69**: 832–845, 1981.

# CHAPTER *32*

# *LOW-RATE VOCODERS*

## 32.1 INTRODUCTION

To a first approximation, a teletype system transmitting at 75 bps can transmit textual information at almost the same rate as a person speaking the same text. Of course, the speech has much more information than the text. The speaker's identity, emotional state, and prosodic nuances are all information, though not all of this information may be necessary for speech communication per se. For this chapter, it is assumed that a good 2400-bps vocoder contains all of the relevant information. Given this assumption, we will consider low-rate vocoders to encompass bit rates between 75 and 2400 bps.

In Chapter 31 we examined two methods of bit-rate reduction: efficient quantization schemes and linear transformations. In this chapter we extend this discussion to report on several other bit-saving methods. First, we describe the benefits obtainable by taking advantage of the time correlation of the spectral and excitation components. When the sampling rates of these components are lowered, bit saving automatically takes place; the trick will be to find interpolation algorithms that do not inordinately degrade the output speech.

A different approach to bit-rate reduction comes from the original work by C. P. Smith [18], [19] on channel vocoders, which was later applied to LPC vocoders by Buzo et al. [3]. This approach was called pattern matching by Smith and vector quantization (VQ) by Buzo et al. (We shall, for the most part, stick with the more popular latter description.) The idea was (as discussed more generally in Chapters 9 and 26) this: the number of perceptually distinguishable spectra is far smaller than the number that is typically generated by a speech device such as a channel vocoder. Therefore, (a) if the spectrum were treated as a multi-dimensional vector and a way could be found to store all the perceptually distinguishable spectra, (b) if each of the stored spectra was given a label, and (c) when a new spectrum arrived, it would be matched against the stored vectors and the label of the best match would be transmitted and decoded at the receiver using the same codebook.

Another traditional way of reducing bit rate is to reduce the number of parameters that have to be sent. For example, in a typical channel vocoder, 15–20 channel signals are used; in LPC, 8–12 predictor coefficients are generally needed. Formant vocoders, in contrast, require as few as four spectral parameters; the problem here is the difficulty of tracking formants accurately.

Finally, technological advances in the field of ASR have led to the concept of recognition synthesis, in which longer speech segments (e.g., phonemes, diphones, and syllables [8], [21]) are identified at the analyzer and regenerated at the synthesizer. This is an interesting research topic, but as explained further later, it is unlikely to achieve general use (over languages, speakers, and acoustic conditions) in the foreseeable future.

In this chapter our goal is to study several systems that make use of one, or some combination, of the above concepts.

## 32.2   THE FRAME-FILL CONCEPT

The basic idea of frame fill, proposed by McLarnon [10], is to transmit from analyzer to synthesizer every $M$th frame, thereby achieving an $M$:1 reduction in bit rate. The savings is not quite that great, since some control information must be sent, instructing the receiver how to reconstruct (or fill in) missing information. Thus with the choice of $M = 2$, close to a 1200-bps rate is feasible if a 2400-bps system is started with.

Frame fill for a channel vocoder:   Let's define frame $(N - 1)$ and frame $(N + 1)$ as the two frames that are sent, and frame $(N)$ as the frame computed by the analyzer but not sent. Then:

1. Compare frame $(N)$ of data to be omitted with frame $(N - 1)$ and frame $(N + 1)$.
2. In accordance with some reasonable distance measure, decide which neighbor matches the omitted frame.
3. Also, consider as a match candidate some weighted combination of the information contained in the two neighboring frames.
4. Select the option (three choices) representing the best match and append its I.D. code (two bits) to the frame that is to be transmitted.

McLarnon chose a distance metric to be

$$d = \sum_{k=1}^{K} |\log[S_c(k)] - \log[S_r(k)]|, \qquad (32.1)$$

where $K$ is the number of channels and where $S_c(k)$ refers to the magnitude of the $k$th spectral component for the candidate $N$th frame and $S_r(k)$ refers to the equivalent magnitude for either frame $(N - 1)$ or frame $(N + 1)$. Table 32.1 (after [2]) summarizes the coding

**TABLE 32.1   Coding Conventions for a Low-Rate Channel Vocoder[a]**

| System (bps) | Voicing | Pitch | Ref. Channel | DCPM Two-Bit | DCPM One-Bit | Ctrl | Total Bits | Frame Rate (Hz) |
|---|---|---|---|---|---|---|---|---|
| 2400 | 1 | 6 | 3 | 18 | 0 | 0 | 48[b] | 50 |
| 1200 | 2 | 6 | 3 | 17 | 1 | 2 | 48 | 25 |
| 800(V) | 2 | 6 | 3 | 1 | 17 | 4 | 33 | 24,24 |
| 800(UV) | 2 | 0 | 3 | 6 | 12 | 4 | 33 | 24,24 |

[a]From [2].
[b]Two bits are unused.

**TABLE 32.2  Coding Conventions for a Low-Rate LPC Vocoder[a]**

| Parameter | 1200 bps | 800 bps |
|---|---|---|
| Sync | 1 | 0 |
| V/UV this frame | 1 | 0 |
| V/UV next frame | 1 | 1 |
| Strategy bits | 4 | 3 |
| Pitch pointer | 6 | 6 |
| Energy | 5 | 4 |
| $k_0$ | 5 | 4 |
| $k_1$ | 5 | 4 |
| $k_2$ | 4 | 4 |
| $k_3$ | 4 | 3 |
| $k_4$ | 4 | 2 |
| $k_5$ | 4 | 1 |
| $k_6$ | 3 | 1 |
| $k_7$ | 2 | 1 |
| $k_8$ | 2 | 1 |
| $k_9$ | 2 | 1 |

[a]Note that $k_5$–$k_9$ are omitted in unvoiced frames. From [2].

strategy to reduce the bit rate from an original 2400-bps channel vocoder to 1200 bps and 800 bps.

Frame fill for a LPC vocoder:  Bit-rate savings in LPC vocoders is more subtle than for channel vocoders and depends greatly on the synthesizer structure. For example, if the predictor coefficients were transmitted, assigning bit rates to each would be an arduous empirical task because of the obtuse relationship between the coefficients and the more physically intuitive spectrum. Both psychoacoustic experiments [1] and engineering considerations favor the use of the reflection coefficients. Blankenship and Malpass [2] performed frame-fill experiments on a 2400-bps LPC vocoder and arrived at the breakdown of Table 32.2 for both 1200 bps and 800 bps.

Notice that the reflection coefficients $k_4$ through $k_9$ are not sent when the excitation is hiss. This knowledge allows the analyzer to perform a fourth-order rather than a tenth-order analysis during hiss. The strategic control of frame fill is somewhat more complex for LPC and, as shown in the table, requires four bits for the 1200-bps version and three bits for the 800-bps version. Table 32.3 gives comparative DRT (diagnostic rhyme test) results for all three versions of both systems (recall that Chapter 17 briefly describes the DRT).

**TABLE 32.3 Summary
of Three-Speaker DRT
Results[a]**

| | Vocoder | |
| --- | --- | --- |
| Rate (bps) | Channel | LPC |
| 2400 | 89.6 | 91.6 |
| 1200 | 87.6 | 85.1 |
| 800 | 84.0 | 82.0 |

[a]From [2].

# 32.3  PATTERN MATCHING OR VECTOR QUANTIZATION

Let's assume that a listener can tell any two spectral patterns apart from a total population of $2^{20}(= 1,048,576)$ patterns. Given a strategy that permits identification of the  storage location of any of these million or so patterns, we see that the transmitter needs to transmit *only* the storage location, with confidence that the receiver (possessing the same set of stored patterns) can generate the correct spectrum.

So far, so good; but there are difficulties. How do we determine the particular subset of $2^{20}$ out of a much larger set? How do we implement an efficient search procedure at the receiver to find the pattern?

Two conditions for some success are clear. First, a large amount of data must be collected and analyzed, and second, a distance metric must be formulated so that most entries are perceptually distinct. An attractive geometric way of looking at these issues was proposed by Buzo et al. [3]. They defined a single frame of information as a  vector. If we think of each frame as a single point in a multidimensional space, we try to fill the space with points that are as far apart from each other as possible. This line of reasoning leads straightforwardly to the term vector quantization.[1]

When C. P. Smith began his work in the 1950s, bulk memory was best obtained by rotating machines, such as drums. Progress was slow, although by 1962, Smith was able to perform a feasibility experiment. Buzo, Markel, and Gray reintroduced the concept using LPC vocoders. At the Lincoln Laboratory, the same concepts were applied to a channel vocoder by Gold [6] and to the SEEVOC system of Paul [11]. Each used a different vocoder configuration, a different distance metric, and a different strategy for building a system.

Gold's pattern-matching channel vocoder:   listening and visual observation were the primary vehicles; this made the procedure interactive but time consuming. The system begins with an empty table of stored patterns. The first sentence is processed by the channel vocoder analyzer, and *all* spectral cross sections are entered into memory. By visual inspection of these cross sections, the experimenter enters the nonredundant spectra into the pattern table.

---

[1]VQ was also discussed in Chapters 9 and 26 in the context of statistical pattern recognition.

The next sentence is matched against this embryonic table. The process continues, using both visual and auditory feedback, until the experimenter decides that a sufficient number of patterns has been stored.

Paul's adaptive vector quantization:   adaptation is accomplished by continuing alteration of the pattern table to match the current speaker and environment. An incoming spectrum is matched against all existing reference patterns. If the best match fails to satisfy a fixed criterion, the new pattern is incorporated into the pattern table, replacing that pattern that has not been transmitted for the longest time.

The spectral envelope estimation (SEE) vocoder algorithm was used as a base in these experiments [12]. (SEE uses cepstral processing to obtain spectral patterns.) Performance was quite impressive. When a new speaker began, a brief period of quasi-intelligible speech was followed by adaptation; the system quickly tuned in on the new speech. It should be noted that the system requires that updated reference sets be periodically transmitted. To maintain the low bit rate requires the detection of silent intervals during which new pattern sets can be sent.

## 32.4  THE KANG–COULTER 600-bps VOCODER

Kang and Coulter [9] developed a 600-bps vocoder that employs LPC methods followed by formant tracking based on the LPC parameters, as well as vector quantization of the formants. Their device is a useful example of how the different data-reduction techniques can be combined to achieve significant bit-rate reduction. A block diagram of their system is shown in Fig. 32.1.

LPC vocoders have already been discussed in Chapters 21 and 31. Here, we want to refer back to the predictor coefficients and show how to manipulate them to improve formant tracking. Let's consider, as an example, an $n$th-order predictor. We know that the $z$-transform of the synthesizer can be expressed in terms of an $n$th-order polynomial in $z$.

$$H(z) = \frac{1}{1 - a_1 z^{-1} - a_2 z^{-2} \cdots - a_n z^{-n}}. \tag{32.2}$$

For a tenth-order system, as $a_{10}$ approaches unity, the 10 poles of the system gravitate toward the unit circle, as shown in Fig. 32.2. As the poles progress, the resultant spectrum changes as indicated in Fig. 32.3. (In the figure the parameter $k_{10}$ corresponds to $a_{10}$ in our notation.) As $k_{10}$ gets very close to the unit circle, in ($f$), the peaks of the spectrum are very obvious and relatively easy to identify.

However, these values of peak frequencies are only *approximations* to the formants. As seen in Fig. 32.2, the pole trajectories are not radial. Thus, it is necessary to backtrack by gradually returning $k_n$ to its original value and iteratively recomputing the formants in steps.

Vector quantization of the formants:   just as perceptually distinct speech spectra are a relatively small subset of all possible speech spectra, so are the perceptually significant formant patterns a small subset of all such patterns. Kang and Coulter worked with a stored table of a mere 128 vector formant patterns. They also vector quantized six partial correlation coefficients for use with unvoiced sounds. Their parameter coding is shown in Table 32.4.

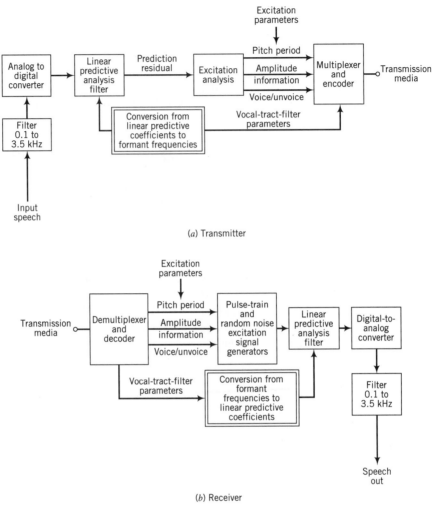

**FIGURE 32.1**   Kang–Coulter 600-bps voice digitizer. From [9].

Finally, Kang and Coulter present a summary of DRT results, using the Voiers version of feature comparison (see Chapter 17); these results are shown in Table 32.5.

## 32.5   SEGMENTATION METHODS FOR BANDWIDTH REDUCTION

In Section 32.1 it was noted that the lowest reasonable data rate for a speech-transmission system is of the order of 75 bps, equivalent to a teletype rate. Let us try to imagine how one would go about building such an ideal system. First, a very good automatic speech recognizer would be the front end of the system. This would mean that the transmission

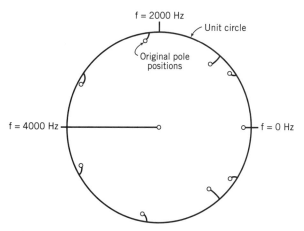

**FIGURE 32.2** Loci of the poles as $k_{10}$ approaches unity. From [9].

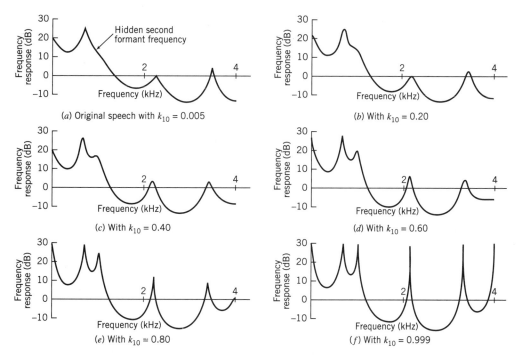

**FIGURE 32.3** Spectra of an all-pole system as poles migrate toward the unit circle. From [9]. As the poles get closer to the unit circle, the peaks associated with the poles become more distinct and easier to identify. (See text for further explanation.)

**TABLE  32.4   Parameter Coding for a 600-bps Voice Digitizer[a]**

| Parameter | Coding | |
|---|---|---|
| | Typ. 2400-bps Linear Predictive Encoder | 600-bps Voice Digitizer |
| Frame Rate | 44.444 Hz | 40 Hz |
| Vocal tract filter parameters | 40 bits/frame | 7 bits/frame |
| Excitation parameters | | |
| Voiced–unvoiced decision | 1 bit/frame | 1 bit/frame |
| Amplitude | 6 bits/frame | 4 bits/frame |
| Pitch | 6 bits/frame | 5 bits/ double frame |
| Synchronization | 1 bit/frame | 1 bit/double frame |
| Tot. no. of bits | 54 bits/frame | 30 bits/double frame |

[a]From [9].

**TABLE  32.5   DRT Summary:   Comparison of a 600-bps System with the Mother 2400-bps System[a]**

| Feature | Perception | 600 bps Voice Digitizer | 2400 bps LPE |
|---|---|---|---|
| Voicing | Distinguishes /b/ from /p/, /d/ from /t/, /v/ from /f/, etc. | 99.9 | 89.6 |
| Nasality | Distinguishes /n/ from /d/, /m/ from /b/, etc. | 84.4 | 93.6 |
| Sustention | Distinguishes /f/ from /p/, /b/ from /v/, /t/ from /Θ/, etc. | 78.1 | 77.0 |
| Sibilation | Distinguishes /s/ from /Θ/, /ʃ/ from /d/, etc. | 60.2 | 93.2 |
| Graveness | Distinguishes /p/ from /t/, /b/ from /d/, /w/ from /r/, /m/ from /n/, etc. | 68.0 | 81.5 |
| Compactness | Distinguishes /y/ from /w/, /g/ from /d/, /k/ from /t/, /ʃ/ from /s/, etc. | 88.3 | 93.0 |
| Average | | 79.9 | 88.0 |

[a]For the most part, the lower bit-rate system is not as good at providing cues for feature perception; we don't know why the voicing result appears to have been better for the lower rate system. From [9].

system has available the equivalence of printed text material that can now be sent by teletype. At the receiver, a text-to-speech (system) would be needed to reproduce the spoken version of the text.

There are two strong limitations to this scenario. First, even a very good text-to-speech system does not reproduce the characteristics of the speaker. This means that we really need more bits with which to (hopefully) reproduce the style of the speaker and convince the listener of the speaker's identity. To determine how many more bits might be needed for this task, we would have to invent methods and perform psychoacoustic testing on a fairly large scale, in order to arrive at some estimation of the extra bandwidth needed, and how to do the job of approximating the speaker's voice. To our knowledge, such research has not been done. However, as long ago as 1980, speech researchers were beginning to grapple with such problems, and work centered around these issues is currently quite active.

Even if we are willing to sacrifice speaker identity and style, however, one would still need to accurately recognize phonemes or some other linguistically sufficient speech unit. This has been demonstrated for Japanese speech in [8] and [21]. In the latter reference, for instance, phoneme HMMs were used to recognize phoneme strings. Japanese phonotactics permit relatively strong constraints on phoneme sequences (in comparison, say, with English), and the resulting recognition was good enough to provide 150-bps synthetic speech with comparable subjective evaluations as with a VQ system that required 400 bps (where neither of these figures included pitch information). This was very impressive; however, the experiments were done for a single speaker, under pristine acoustic conditions, and in a language that is particularly conducive to such an approach. We are not aware of any current application of this approach to the general case of unconstrained acoustic conditions, speaker style, and without speaker-specific training.

Nonetheless, let's look more closely at an idealized speech-transmission system, one that recognizes and transmits individual phonemes. If we assume an average speaking rate of 10 phonemes/s and a total of 64 phonemes, this idealized system would transmit the spectral components (coded as phonemes) at 60 bps. The synthesized speech quality would be determined exclusively by the phonemes stored at the receiver, independent of the specific speaker. Further, since the acoustic properties of a phoneme are strongly dependent on the surrounding sounds, the resultant synthetic speech intelligibility would almost surely be severely compromised.

A somewhat better (but far from perfect) facsimile of the speech could be obtained by recognizing and storing *allophones* at both ends. Assuming 1000 allophonic variations of all the phonemes, and again choosing a speaking rate of 10 phonemes/s, our new system now requires 100 bps for the transmission of vocal tract information.

Examining Table 32.4, we see that the frame rate for the Kang–Coulter 600-bps vocoder is 40 Hz. We can speculate that the application of frame fill would reduce the bit rate to approximately 300 bps. With this modification we are beginning to approach the limits as defined by the idealized analysis–synthesis system. Such an extension has been tried on different basic vocoders by various researchers and is called segmental vocoding. Instead of simply omitting the transmission of alternate frames (as in frame fill), strongly correlated contiguous frames are merged into segments, and these segments are then vector quantized. If there are, on average, $N$ frames per segment, the resultant equivalent bit rate is reduced by almost a factor of $N$. Note that this approach does not require explicit ASR for the segments.

It should be noted that merging of contiguous frames is an extension of the frame-fill concept discussed briefly in Section 32.2. Rather than halving the frame rate, the segmentation methods divide the frame rate by a signal-dependent integer. Since much of the speech signal consists of quasi-stationary voiced segments that change very slightly from frame to frame, segmentation algorithms can lead to greater savings in the average bit rate.

Segmentation algorithms lead to variable length segments, with resultant complications. For example, the Kang–Coulter vocoder would search for a stored segment with formant tracks that most closely resembled the formant tracks of the new segment.

Representation of the speech spectrum can take many forms, depending on the specific processing system. For the Kang–Coulter vocoder, the derived spectral parameters are formants, and it seems reasonable to compare adjacent segments by inspecting first-order formant differences. Roucos et al. [15] worked with the LPC-derived parameters and, for the segmentation determination, they proposed inspection of the log-area ratio differences.[2]

A key issue is the comparison of the derived segments with stored versions of the codebook segments. In Section 32.3 there was a brief discussion of adaptive vector quantization, in which the system went through a transient period, entering new vectors into the codebook that were based on a new speaker's speech properties. Alternate strategies are the  multispeaker approach, in which a fixed codebook is created, based on the data from a variety of speakers, with the hope that any new speaker's segments will resemble stored segments from this variety. Still another strategy is the single-speaker approach, wherein a fixed collection of segments is computed for just one speaker – the sole user. This idea can be extended to a population of users, each one having his or her private codebook.

Obviously, the single-speaker approach requires a smaller codebook and thus uses a lower bit rate than the multispeaker system. The adaptive VQ should compare in bit rate with the single-speaker VQ at the price of more computation and greater complexity, since it has to include an algorithm that adds new code words to the codebook and an algorithm to determine which code words to eliminate.

Many strategies exist for deciding on the appropriate code to transmit, given an analyzed segment. A standard criterion is the $L_2$ measure; that is, the mean-squared difference between the components of the new vector and those of each of the stored vectors. These components differ from system to system, whether it be LPC, a channel system, or a homomorphic system. In addition, as discussed in [7], different distance measures are available, such as log spectral distance, cepstral distance, and various criteria based on likelihood ratios.

A variety of methods for very low bit-rate coding have been the object of research during the past two decades; see, for example, [4], [5], [14]–[17], [20]–[22].

---

[2] See Rabiner and Schafer [13] for definitions of various LPC parametric representations, including log-area ratios.

## 32.6 EXERCISES

**32.1** Prove that for the direct-form $n$th order LPC synthesizer, the poles migrate toward the unit circle as the $n$th PARCOR coefficient $k_n$ approaches unity.

**32.2** Devise an algorithm to obtain good approximations to the formant frequencies after correctly identifying the spectral peaks when $k_n$ is very close to unity.

**32.3** Prove that the poles corresponding to the formants remain inside the unit circle as $k_n$ approaches unity and then backtracks.

**32.4** Given the formant frequencies obtained by the above methods, find the corresponding predictor coefficients for a sixth-order system.

## BIBLIOGRAPHY

1. Barnwell, T. P., and Voiers, W. D., "An analysis of objective measures for user acceptability of voice communications systems," Final Rep. DCA100-78-C-0003, DCA was the issuer, Wash., D.C. 1979.
2. Blankenship, P. E., and Malpass, M. L., "Frame-fill techniques for reducing vocoder data rates," Tech. Rep. 556, MIT Lincoln Laboratory, Lexington, MA, 1981.
3. Buzo, A., Gray, A. H., Gray, R., and Markel, J., "Speech coding based on vector quantization," *IEEE Trans. Acoust. Speech Signal Process.* **ASSP-28**: 562–574, 1980.
4. Cernocky, J., Baudoin, G., and Chollet, G., "Segmental vocoding – going beyond the phonetic approach," in *Proc. ICASSP '98*, Seattle, pp. 605–608, 1998.
5. Ghaemmaghami, S., and Deriche, M., "A new approach to modelling excitation in very low-rate speech coding," in *Proc. ICASSP '98*, Seattle, pp. 597–600, 1998.
6. Gold, B., "Experiments with a pattern-matching vocoder," in *Proc. ICASSP '81*, Atlanta, pp. 32–34, 1981.
7. Gray, A. H., Jr., and Markel, J. D., "Distance measures for speech processing," *IEEE Trans. Acoust. Speech Signal Process.* **ASSP-24**: 380–391, 1976.
8. Hirata, Y., and Nakagawa, S., "A 100 bits/s speech coding using a speech recognition technique," in *Proc.* Eurospeech '89, Paris, pp. 290–293, 1989.
9. Kang, G. S., and Coulter, D. C, "600-bit-per-second voice digitizer (linear predictive formant vocoder)," NRL Rep. 8043, Naval Research Laboratory, Wash., D.C. 1976.
10. McLarnon, E., "A method for reducing the frame rate of a channel vocoder by using frame interpolation," in *Proc. ICASSP '78*, Washington, D.C., pp. 458–461, 1978.
11. Paul, D. B., "An 800 bps adaptive vector quantization vocoder using a perceptual distance measure," in *Proc. ICASSP '83*, Boston, 1983.
12. Paul, D. B., "The spectral envelope estimation vocoder," *IEEE Trans. Acoust. Speech Signal Process.* **ASSP-29**: 562–574, 1981.
13. Rabiner, L. R., and Schafer, R. W., *Digital processing of speech signals*, Prentice–Hall, Englewood Cliffs, N.J., 1978.
14. Roucos, S., Schwartz, R. M., and Makhoul, J., "A segment vocoder at 150 b/s," in *Proc. ICASSP '83*, Boston, pp. 61–64, 1983.
15. Roucos, S., Schwartz, R. M., and Makhoul, J., "Segment quantization for very-low-rate speech coding," in *Proc. ICASSP '82*, Paris, France, pp. 1565–1568, 1982.

16. Schwartz, R. M., and Roucos, S. E., "A comparison of methods for 300–400 b/s vocoders," in *Proc. ICASSP '83*, Boston, pp. 69–72, 1983.

17. Shiraki, Y., and Honda, M., "LPC speech coding based on variable-length segment quantization," *IEEE Trans. Acoust. Speech Signal Process.* **36**: 1437–1444, 1988.

18. Smith, C. P., "An approach to speech bandwidth compression," Tech. Rep. AFCRC-TR-59-198, U.S. Air Force Cambridge Research Center, Cambridge, 1959.

19. Smith, C. P., "Perception of vocoder speech processed by pattern matching," *J. Acoust. Soc. Am.* **46**: 1562–1571, 1969.

20. Soong, F. K., "A phonetically labeled acoustic segment (PLAS) approach to speech analysis-synthesis," in *Proc. ICASSP '89*, Glasgow, paper S11.7, pp. 584–587, 1989.

21. Tokuda, K., Masuko, T., Hiroi, J., Kobayashi, T., and Kitamura, T., "A very low bite rate coder using HMM-based speech recognition/synthesis techniques," in *Proc. ICASSP '98*, Seattle, pp. 609–612, 1998.

22. Wong, D. Y., Juang, B. H., and Cheng, D. Y., "Very low data rate compression with LPC vector and matrix quantization," in *Proc. ICASSP '83*, Boston, pp. 65–68, 1983.

# CHAPTER 33

# MEDIUM-RATE AND HIGH-RATE VOCODERS

## 33.1 INTRODUCTION

In Chapter 31 we were introduced to vocoders and the technique of analysis–synthesis systems. The channel vocoder was discussed in some detail. In Chapter 32, low-rate vocoders were presented that coded speech at rates below 2400 bps, with an expected sacrifice in sound quality. In this chapter[1] we discuss speech coders that operate at bit rates greater than 2400 bps (typically 4800–16000 bps) but that can deliver more robust and higher-quality speech. This always involves some form of waveform coding *in addition* to the customary modeling of the vocal tract parameters. Also, we briefly discuss higher rate, strictly waveform coding systems such as subband coding and differential pulse code modulation.

A variety of medium-rate systems have been proposed and implemented, including voice-excited systems, LPC residual-excited systems, split-band systems and, more recently, code-excited systems, employing analysis-by-synthesis methods. It should be noted that analysis by synthesis requires much more computational power than previous systems so that its use in real-time environments is a result of advances in high-speed computing.

## 33.2 VOICE EXCITATION AND SPECTRAL FLATTENING

Much of the thinking that has gone into the efforts to improve vocoder quality derives from the understanding that poor representation of the excitation function plays an important role in quality deterioration. An early attempt to confront this issue was the development of the voice-excited vocoder [35]. The idea behind it is illustrated in Fig. 33.1.

Figure 33.1 shows the short-time spectrum of a voiced sound. The trick is to generate the spectrum of Fig. 33.1c, a spectrally flattened version of the original sound, which can serve as a suitable excitation signal. In their original work, Schroeder and David passed the low-pass signal (Fig. 33.1b) through a zig-zag network (Fig. 33.2) to produce many distortion products.

A more effective scheme that was later developed by David et al. [13] is shown in Fig. 33.3.

The speech is low-pass filtered to create a baseband signal and then passed through a simple distortion network (such as a half-wave rectifier) to generate upper harmonics. This

---

[1]This chapter was jointly written by Jeff Gilbert and the authors.

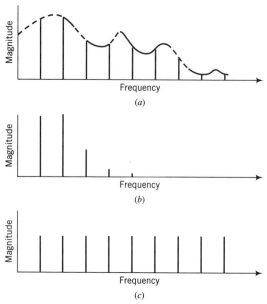

**FIGURE 33.1**   Spectral flattening of the baseband signal to produce the vocoder excitation signal.

new signal passes through a filter bank extending over the entire spectrum of interest, and the bandpass filter outputs are hard limited.

## 33.3   VOICE-EXCITED CHANNEL VOCODER

This method of spectrum flattening is not ideal. If, for example, two harmonics appear in the same filter, the weaker harmonic tends to be suppressed by the stronger one. Nevertheless, this concept has been implemented as part of the voice-excited channel vocoder shown in Fig. 33.4.

Typical bandwidths of the baseband signal vary from 600 to 900 Hz; the wider bandwidth signals lead to higher-quality results, but also to greater transmission rates. The

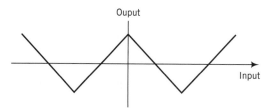

**FIGURE 33.2**   Zig-zag network.

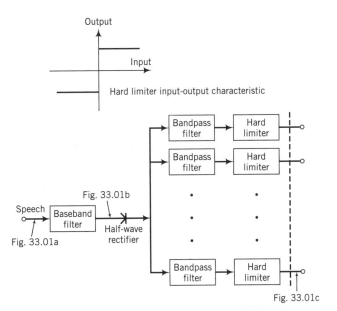

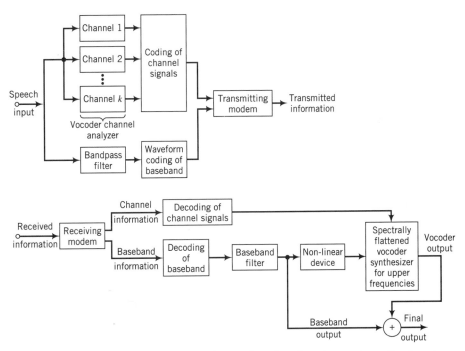

**FIGURE 33.3** Implementation of spectral flattening.

**FIGURE 33.4** Voice-excited channel vocoder, showing the analyzer (top) and synthesizer (bottom).

remaining speech bandwidth is analyzed in the conventional channel vocoder manner, as described in Chapter 31. The baseband signal is usually transmitted by some standard waveform coding technique (discussed in Section 33.5). At the receiver, the baseband signal is decoded, spectrally flattened, applied as excitation to vocode for frequencies higher than the baseband, and also added to the vocoded speech to produce the output.

Figure 33.3 shows the receiver baseband filter and part of the synthesizer filter bank. The nonlinear device is the hard limiter shown in the figure. In order to eliminate harmonics produced by the limiters, an extra filter bank (not shown in the figure) follows the limiters. The received vocoder channel signals produced by the analysis system at the top of Fig. 33.4 are applied to modulators that normally are inserted between the limiters and this final filter bank.

## 33.4 VOICE-EXCITED AND ERROR-SIGNAL-EXCITED LPC VOCODERS

The principles behind voice-excited LPC vocoders are the same as for channel vocoders; a spectrally flattened excitation signal must be generated and then integrated into an LPC vocoder configuration.

There is an additional component in LPC systems: the residual, or error signal. We have pointed out that exciting the LPC synthesizer with the complete error signal reproduces the input speech, and so the problem is to find a compressed version of the error signal that can be conditioned to act as an appropriate excitation. Magill and Un [26] invented a residual-excited LPC vocoder that was able to operate at a 9600-bps rate. Soon after, Weinstein [39] developed a voice baseband-excited LPC system. Whether one uses the error signal or the original speech, the problem is the same; at the analyzer, the signal must be low-pass filtered to conserve the transmission bandwidth. At the synthesizer, nonlinear distortion plus spectral flattening must take place. In their original work, Magill and Un used the old trick of David et al. [13] – passing the residual through a nonlinear distorting network to bring back the high frequencies. This is not a satisfactory method, since the degree of deviation from the desired spectrum flattening depends to a great extent on the time-varying spectrum of the speech. Makhoul and Berouti [27] proposed an alternative method, illustrated in Fig. 33.5. Here the baseband is replicated by appropriate downsampling.

Such a scheme has two deviations from the ideal; first the replications contain the same spectral envelope as the baseband, and second the harmonic structures of the replicated spectra are not (except by accident) at multiples of the fundamental frequency.[2]

Another scheme, given in [39], uses the LPC error-signal concept to perform spectral flattening. In this scheme, known as VELP (for voice-excited linear prediction, as opposed to RELP, for residual-excited linear prediction), an LPC analysis is performed on the received, distorted, baseband signal to create a new error signal. If we assume that this analysis

---

[2]It's worth noting the results of Ritsma (Chapter 16) that the human auditory system is less sensitive to pitch distortions at frequencies above 1500 Hz.

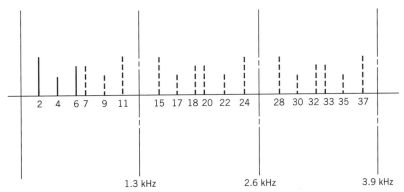

**FIGURE 33.5** Alternate method of spectral flattening, using downsampling. The solid spectral lines are those of the original baseband speech. By reducing the original sampling rate by a factor of 6 (from 7.8 kHz to 1.3 kHz), the spectrum is aliased, producing the additional spectral lines shown as dashed.

accurately represents the spectrum, it follows that the resultant error signal has a flat spectrum and is thus a useful excitation.

Figure 33.6 shows simplified block diagrams of VELP and RELP. In VELP, since the excitation is derived directly from the received, low-pass filtered speech, this baseband can be added to the high-pass vocoded speech. In RELP, the processed error signal is used only as excitation.

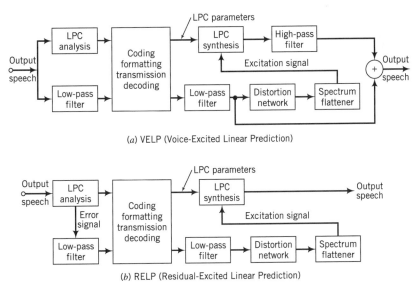

**FIGURE 33.6** Block diagrams of VELP and RELP.

Split-band systems versus voice or error-signal-excited systems: in split-band systems, the baseband signal is transmitted; but instead of distorting this signal at the receiver to regenerate higher-frequency components, the system makes explicit use of a pitch detector. Thus, the baseband frequencies are waveform coded while the higher frequencies are vocoded by one of several possible methods.

How do split-band systems compare with voice-excited or residual-excited systems? Possible advantages for the latter are that (a) pitch and voicing detection are not needed, thus saving hardware and channel capacity, and (b) quality in a noisy environment is improved since there are no pitch parameters to be estimated or bad voicing decisions to be made.

Although it is true that split-band systems require the extra hardware (or software), it can be argued that with the present state of technology, the difference in implementation cost is insignificant. A more important issue is that of robustness. Psychoacoustic tests demonstrate that for frequencies higher than 1500 Hz, pitch errors are not very intrusive. For example, a channel vocoder can be totally noise excited for frequencies above 1800 Hz with very little effect on the overall quality [18].

## 33.5 WAVEFORM CODING WITH PREDICTIVE METHODS

Since the speech spectrum is not, in general, flat, and this implies correlation of waveform samples, bit-rate savings are possible by means of delta modulation techniques. We assume that fewer bits are required to code the *difference* between two adjacent samples than are needed to quantize the samples directly. A simple mathematical statement of the problem is this: let $s(n)$ be the signal, and compute the difference $e(n) = s(n) - s(n-1)$. Next, quantize $e(n)$ so that $\hat{e}(n)$ is the quantized version of $e(n)$, and transmit this quantized error signal. The system, shown in Fig. 33.7, is called differential pulse code modulation (DPCM) and shows the coding and decoding of the signal.

From this simple notion have grown concepts and schemes of varying degrees of sophistication. First, we note that speech intensity varies widely with phonemes and speakers; this suggests that an adaptive quantization technique could save additional bits. A possible implementation of such a concept is shown in Fig. 33.8.

The derived volume signal, $V$, is presumed to be slowly varying at roughly syllabic rates. It controls the compression and expansion of the quantizer. At the receiver, true speech volume is approximately restored by multiplying the quantized error signal by $V$. An important feature of such a system is the need for frame synchronization.

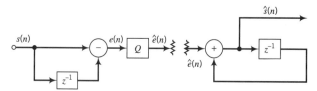

**FIGURE 33.7** Differential pulse code modulation.

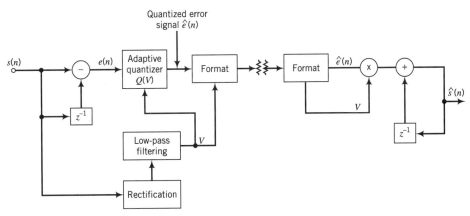

**FIGURE 33.8** Adaptive differential pulse code modulation (ADPCM).

A difficulty with the concept corresponding to Fig. 33.8 is the slope-overload problem. If the integrator in Fig. 33.8 has a fixed step size, the response to a rapidly rising signal would cause the decoded signal to fall behind. To overcome this problem, we introduce a quantizer with a variable step size; such a system is shown in Fig. 33.9 and is called CVSD (continuously variable slope delta) modulation.

First, the speech synthesizer is implemented at the transmitter so that its output can be monitored continuously and compared to the input signal. Then the error signal is taken to be the difference between the input signal and this synthesized signal. In addition, the slope analyzer, $M$, changes the quantization steps according to the following rule:   if three

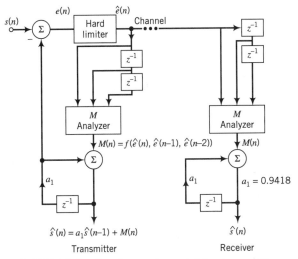

**FIGURE 33.9** Continuously variable slope delta modulation.

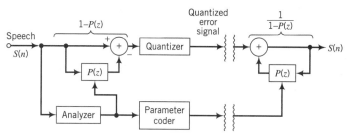

**FIGURE 33.10** Rudimentary linear prediction concept.

successive error signal samples are the same sign (indicating an inability to follow rapid changes), the step size is increased; in all other cases the step size decreases. Increases and decreases are not equal. Also, both increases and decreases are clamped.

## 33.6 ADAPTIVE PREDICTIVE CODING OF SPEECH

The ideas of adaptive PCM can be generalized and to some extent incorporated into speech-modeling concepts. Figure 33.10 shows a rudimentary linear prediction model. Here $1 - P(z)$ represents a linear filtering of speech samples, and the analyzer alters the parameters of $1 - P(z)$ according to some chosen algorithm (e.g., a least-squares minimization). Both the error signal and parameters are quantized and transmitted. At the synthesizer, the inverse filter is excited by the received error signal to produce the output speech.

The error signal of the system of Fig. 33.10 grows proportionally to the speech volume. To prevent this, we introduce Fig. 33.11.

The synthesizer filter is placed in the analyzer feedback loop, and the quantized error signal magnitude is adjusted to try to minimize the error signal. This gain control in the

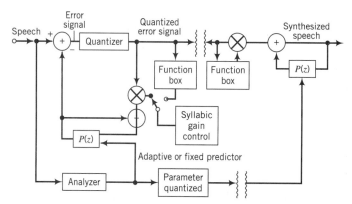

**FIGURE 33.11** Predictor and quantizer in a feedback loop.

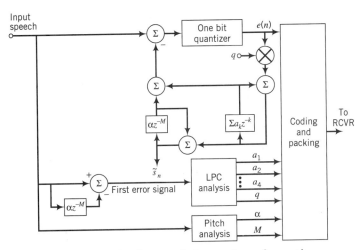

**FIGURE 33.12** Adaptive predictive coding of speech.

feedback loop can be adjusted for each new speech sample (as in CVSD), or it can be a slowly varying adjustment operating, for example, at a syllabic rate.

A sophisticated realization of this concept, shown in Fig. 33.12, is called adaptive predictive coding (APC) [6].

In this system, prediction of the speech wave is based on (a) the measured value of the speech wave one fundamental period back and (b) an LPC analysis of the error signal resulting from (a). The fundamental period is computed by a pitch detector and $\alpha$ is computed to minimize the mean-squared error of the first error signal. The parameters $a_1$, $a_2$, $a_3$, and $a_4$ are computed by performing a LPC analysis of the first error signal. The transmitted error signal $e(n)$ is computed as shown in the figure. The parameter $q$ is controlled at syllabic voice rates. The transfer function between the signals $e(n)$ and $\hat{s}_n$ is seen to be the appropriate synthesis filter.

APC embodies the concepts of both modeling and prediction. By incorporating a degree of both vocal tract and excitation modeling, the APC algorithm has managed to reduce the error-signal information rate by quite a bit; thus APC systems work well at bit rates of approximately 8 kbps.

## 33.7 SUBBAND CODING

Waveform coding is different from vocoding because it does not make explicit use of a speech-production model. Nevertheless, certain properties of the speech signal and the human auditory system can be put to good advantage in designing a waveform coded speech-transmission system. For example, it is known that the speech signals in an average telephone conversation are present less than 50% of the time. We also know that a narrow-band signal can more effectively mask quantization noise if the noise spectrum is in the

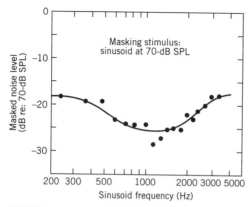

**FIGURE 33.13**   Masking threshold of noise to a sinusoidal masker. From [22].

neighborhood of the signal frequency. This leads to the notion of subband coding [30], [12] in which the speech is first passed through a filter bank and then each band is individually quantized. Krasner [22] ran an experiment to study the masking properties of a tone on noise within the critical band of the noise. The results are shown in Fig. 33.13.

We see that the noise should be from 18 to 28 dB below the signal before it is completely masked. With a 16-channel subband coder, Krasner was able to show that approximately 35 kbps is required to make 4-kHz speech sound as good as the raw speech.

Estaban and Galand [15] proposed the use of quadrature mirror filters (QMFs). These filters have rather magical properties:   under certain conditions, aliasing caused by overlapping bands can be shown to cancel. A QMF splits the total bandwidth into two symmetrical subbands. The upper filter can be obtained precisely from the lower one by replacing all odd coefficients by their negatives. Figure 33.14 shows a four-channel QMF decomposition and reconstitution system.

## 33.8   MULTIPULSE LPC VOCODERS

Another way to transmit the excitation signal is used in the multipulse LPC vocoder, which could be considered a direct predecessor of CELP.[3] In a multipulse vocoder, the residual is modeled by the summation of a number of scaled impulses that can be characterized by their amplitude and location. Thus in addition to transmitting the LPC coefficients, the amplitude and location of the excitation impulses are also transmitted. The question then remains how to determine the location and height of the impulses. This is done by an analysis-by-synthesis

---

[3]It is interesting to note that one of the inventors of this method, Bishnu Atal, also co-authored the first paper on code-excited linear prediction (CELP). Atal was one of the original inventors of LPC itself.

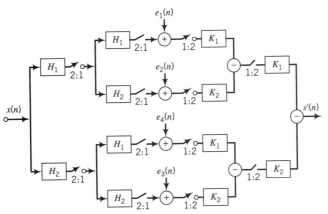

**FIGURE 33.14**  Four-channel subband coder with QMFs.

method of choosing the impulses such that they minimize the difference between the actual speech signal and the speech signal that the receiver will reconstruct. (See Fig. 33.15.)

The impulses are filtered through the LPC synthesis filter, which will then reconstruct the coded version of the signal. This is then subtracted from the actual signal and passed through a perceptual weighting filter. The perceptual weighting filter is of the form $W(z) = A(z)/A(z/c)$, where $c$ is a constant controlling the amount of weighting, with 1 denoting no weighting and typical values ranging from 0.7 to 0.9. $A(z)$ is the prediction filter. The perceptual weighting filter accounts for the effect on human hearing of the interaction of the excitation sequence with the LPC parameters [36]. The perceptually weighted error is then minimized by an optimization procedure for choosing the optimum impulse locations and amplitudes. For simplicity, the impulses are determined sequentially, although an increased

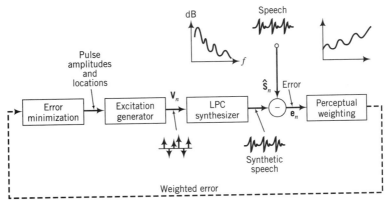

**FIGURE 33.15**  Multipulse LPC. From [4].

performance could be obtained at the expense of increased computational requirements if the impulses were determined simultaneously. The time complexity of the sequential optimization is linear in the number of possible impulse locations and linear in the total number of impulses to be determined.

Multipulse LPC vocoder with pitch synthesis:   although multipulse LPC is fairly compact and can yield high-quality synthetic speech, several modifications can further improve it. One of these modifications is particularly relevant to CELP and is described here. The modification is based on the observation that the excitation sequences are often highly correlated from one pitch period to the next.[4] By adding a pitch synthesis filter, one can exploit this redundancy. The pitch synthesis filter is of the following form, where $c$ is a scale factor, $b$ controls the amount of prediction ($0 < b < 1$), and $T$ is the pitch period:

$$\theta_p(z) = \frac{c}{1 - bz^{-T}}. \tag{33.1}$$

Thus the addition of the pitch synthesis filter means that only the difference between the excitation signal and a scaled version of what it previously was has to be transmitted.

## 33.9  CODE-EXCITED LINEAR PREDICTIVE CODING

Although multipulse LPC with pitch synthesis yields an improved performance over previous versions of LPC, by the application of vector quantization techniques to the problem of coding the excitation signal, the performance can be increased even further, leading to a lower level of audible distortion as well as less required bandwidth. The resultant technique is called Code-excited linear prediction.

Basic CELP:   as pointed out by Schroeder and Atal [37], "the speech synthesizer in a code-excited linear predictive coder is identical to the one used in adaptive predictive coders (APC)."

In CELP, the difference in excitation from frame to frame is not characterized by a few impulses, but rather as one of a fixed number of sequences in a codebook. Thus, to resynthesize the signal, the CELP synthesizer shown in Fig. 33.16 can be used. In the basic CELP implementation, the sequences in the codebook are predetermined, zero mean, unity variance, Gaussian random numbers.

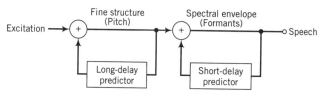

**FIGURE 33.16**  CELP synthesizer.

[4]This concept was introduced in the discussion of the APC system.

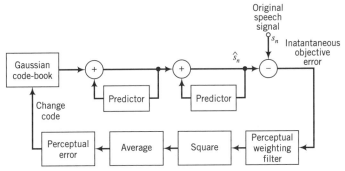

**FIGURE 33.17** Basic CELP coder. From [4].

A codebook of Gaussians was chosen because an investigation of a large corpus of speech data revealed that the excitation sequence needed in an LPC vocoder with a pitch synthesis filter was indeed Gaussian by nature [5]. Non-Gaussian codebook designs are described later.

The CELP coder then has the burden of choosing the codebook sequence[5] that will produce the least perceived distortion. This is performed by an analysis-by-synthesis method as shown in Fig. 33.17. The address of the code word (for each frame) that minimizes the perceptual error is then sent to the receiver, along with the standard LPC parameters.

In the most basic version of CELP, an exhaustive search over the whole codebook is performed. This, however, can be a quite costly operation. The codebooks are typically 512–1024 entries; a typical entry consists of approximately 40 samples (which would be a 5-ms interval for an 8-kHz sampling rate). Thus, each excitation sequence in the codebook must be filtered with the pitch synthesis and LP synthesis filters, subtracted from actual speech signal, and then filtered with the perceptual weighting filter. This represents a large computational burden. Optimizations to reduce the coding time are discussed in the next section.

The gain factor for each excitation sequence can be adjusted in a manner similar to the way that the impulse amplitudes are adjusted in multipulse LPC. The LP spectral parameters are coded by using vector quantization as previously described.

Since the pitch period and excitation sequence change more rapidly than the spectral characteristics of the speech signal, the codebook index, gain, and pitch estimate are transmitted more often than the LP spectral coefficients.

The U.S. Federal Standard 1016 specifies a 4800-bps CELP coder as follows:

1. The 10 LP spectral parameters are updated every 30 ms, consuming 34 bits or 1133.3 bps.

2. The pitch estimate (both period and long-term feedback gain) is sent four times as often (every 7.5 ms). The feedback gain is coded as a five-bit number, whereas every

---

[5]Also sometimes called the excitation sequence, since it is what excites the LP synthesis filter, or the innovations sequence, since it represents what is new in the signal (i.e., what cannot be predicted by the LP synthesis filter).

other pitch period is coded as an eight-bit and a six-bit number. When it is coded as an eight-bit number, the actual pitch period is sent. In between the eight-bit values, six-bit difference values are sent. Since the pitch does not often change extremely rapidly, this is sufficient. The pitch estimate totals 1600 bps.

3. Finally the excitation codebook indices are sent every 7.5 ms, using a nine-bit index to specify which of the 512 codebook vectors to use and a five-bit gain to specify its amplitude. This requires 1866.67 bps. When added to 200 bps of synchronization and forward error correction, this yields the specified 4800 bps.

### 33.9.1 Modifications to CELP

As with most new techniques, several modifications to the basic CELP vocoder have been proposed and implemented. They are divided into two categories. The modifications that are aimed particularly at reducing the computation involved in the codebook search are presented in Section 33.10. Other modifications targeted at higher signal fidelity or other improvements are discussed as follows.

### 33.9.2 Non-Gaussian Codebook Sequences

Although Gaussian noise can model much of the variation of the excitation, it does not model well the onset of voicing from silence or unvoiced segments. Where multipulse LPC was able to target particular places in the excitation sequence and quickly add the periodic impulses required for voiced speech, a Gaussian-only codebook is unable to be so specific in its alterations. This, however, can be remedied by including some deterministic excitation sequences in the codebook [42].

The improved codebook is divided into four sections representing four different excitation classes. The first class, occupying half of the 512-entry codebook, is the standard stochastic excitation sequences. Another class consisting of single pulses is included to allow rapid voicing onset. A third class representing periodic pulses at the current pitch period (as estimated in the LP Analysis section and used in the pitch synthesis filter) is included for more gradual voicing onset. A fourth class is that of a glottal pulse for more detailed excitation at lower pitch periods.

Another implementation of the CELP algorithm uses sinc and cosc functions as part of the codebook [25]. Using deterministic codebook sequences can not only improve performance but can also lower the computational requirements.

### 33.9.3 Low-Delay CELP

As described here, CELP vocoders buffer of the order of 20 ms of speech and can thus delay speech 20–40 ms. This can be a problem for use over a telephone line if echo cancellation is not used [29]. For applications in which the delay is critical, a low-delay implementation of CELP is possible [8], [9]. One-way delays of the order of 2 ms have been achieved. The low-delay CELP does have the disadvantage of a higher resultant bit rate – typically 16,000 bits/s are required for toll-quality speech.

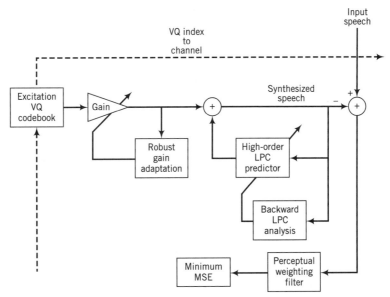

**FIGURE 33.18**  A low-delay CELP coder. Adapted from [8].

The reduced delay is achieved by using much smaller blocks (five samples or 0.625 ms) and additional adaptation mechanisms to predict signal characteristic trends (See Fig. 33.18). Prediction and gain adaptation entail having the synthesis section look at the recent past history of the excitation sequence and filtered excitation sequence, and from that update auxiliary LP generators that produce new estimates of the gain coefficient and spectral filter coefficients.

The coder models this by computing the synthesizer estimates of the gain and spectral filter coefficients, and it compensates by adjusting the excitation sequence so as to produce the desired output. Thus, the only information that is sent over the channel is the excitation codebook indices. The advantage is that only very recent information is used. All adaptation occurs based on the excitation sequences, which are decoded by the synthesizer very shortly after they are present at the coder. Therefore, no long-term (20 ms) framing of the speech occurs.

Since the pitch predictor would require a history at least the length of a pitch period, it would introduce delay and thus cannot be used. Because of this, the LP filter order must be increased to approximately 50.

As previously mentioned, the excitation sequences are also drastically reduced in length to five samples. The codebook index is still 10 bits, but three bits are designated for the gain control and seven for the excitation waveform shape. Since only 128 shapes are now possible, standard Gaussian random sequences do not provide an adequate selection. Instead, a closed-loop optimization procedure is performed that chooses and optimizes the vectors in the codebook by using real test data to reduce anticipated distortion.

Channel errors are often an important issue in speech coding and have been dealt with in an interesting manner in this implementation. The codebook indices are coded with a binary gray code, such that codes that differ by one bit are the most similar. (The gray code is a binary code in which sequential bits differ by only one bit. For instance, a three bit gray code would be: $0 = 000, 1 = 001, 2 = 011, 3 = 010, 4 = 110, 5 = 111, 6 = 101, 7 = 100$. Thus, bit errors can be handled more robustly.)

Other implementations of low-delay CELP coding have been investigated. In particular, [29] discusses an implementation that uses an adaptive codebook (see section 33.10.6 for definition).

## 33.10   REDUCING CODEBOOK SEARCH TIME IN CELP

In its initial incarnation, CELP was far from operating in real time. In fact, the original implementation by Schroeder and Atal required 125 s of Cray-1 CPU time to process 1 s of speech [37]. Since then, aside from the general improvements in CPU speed,[6] many optimizations have been proposed and many real-time solutions have been engineered.

The rate-limiting step in the CELP system is the search through the code-word dictionary to find the excitation sequence that minimizes the perceived distortion. As the algorithm has been stated, this requires passing each sequence in the codebook through three all-pole infinite impulse response (IIR) filters, namely, the pitch synthesis filter, the LP synthesis filter, and the perceptual weighting filter (as described in Section 33.9). It is then necessary to compute an associated error for the sequence. This exhaustive search can be quite demanding, even for the fastest digital signal processors. A number of solutions to this problem have been proposed, some of which will be discussed here.

### 33.10.1   Filter Simplification

By careful investigation of the arrangement of the filters in the CELP coder, it is possible to modify the basic structure so as to reduce the three IIR filtering operations to one filtering operation with no memory. A brief explanation of the modifications and results will be presented here; the reader is referred to [2] for further details.

The first modification is to move the perceptual weighting filter from the output of the summer ($\Sigma$) to its two inputs. This is equivalent as the filter is linear. The perceptual weighting filter is of the form $A(z)/A(z/c)$, and thus when placed after the pitch synthesis filter [which is of the form $1/A(z)$], the two will combine to form a single filter of the form $1/A(z/c)$. (See Figs. 33.19 and 33.20.)

Next, the pitch synthesis filter can be removed from the codebook search loop. The pitch synthesis filter is of the form $B(z) = 1 - bz^{-T}$, where $b$ is a gain factor and $T$ is akin to the predicted pitch period. Thus, the output of the pitch synthesis filter is equal to the sum of its input and what its output was $T$ samples ago. It can be assumed that $T$ is longer than the

---

[6]The Cray 1, considered a supercomputer at the time referred to here, had a clock rate of 80 MHz, which would be considered extremely slow by the standards of today's home PCs. It did have great memory bandwidth, however.

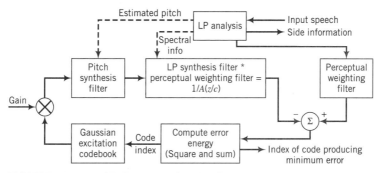

**FIGURE 33.19** CELP coder after the first reduction step. Adapted from [2].

excitation sequence length. This means that the output of the pitch synthesis filter $T$ ago was due to a previous excitation sequence and hence *the current choice of excitation sequence will not affect it*. For this reason, it is constant with respect to the codebook search for the current block and can just be subtracted from the input speech once and then forgotten. A simplified picture of the reduced CELP coder is shown in Fig. 33.21.

Thus the CELP coding has been reduced to some processing of the original speech segment (including removing the effects of the memory of the pitch synthesis filter), and then a filtering operation with no memory on each of the codebook sequences, followed by an error computation of the sum squared difference between the filtered code word and modified original speech.

## 33.10.2  Speeding Up the Search

One advantage of the previously described filter simplification is that it allows the search procedure to be simplified to the maximization of a scaled inner dot product of each codebook sequence with a modified version of the original signal [2]. For the first step, it is shown that the search procedure can be reduced to the scaled inner dot product of

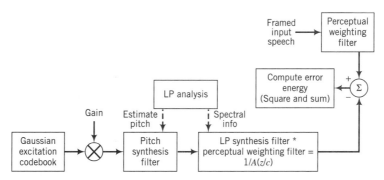

**FIGURE 33.20**  Another view of Fig. 33.19.

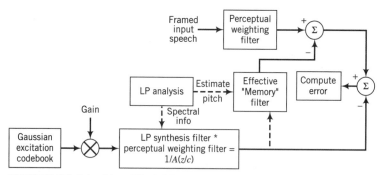

**FIGURE 33.21** Simplified CELP coder with memory removed. Modified and simplified version from [2].

a filtered version of each codebook sequence with the original signal. This is shown as follows.

The goal is to minimize the error $E$ over the current input block $x_n$ and find the best gain $G$, with $g_n$ representing the response of $1/[A(z/c)]$.

The error metric $E$ used is the mean-squared error:

$$E = \sum_n (x_n - Gg_n)^2. \tag{33.2}$$

Find $G$ by setting the derivative of $E$ with respect to $G$ to 0:

$$0 = \frac{\partial E}{\partial G} = -2\sum_n x_n g_n + 2G\sum_n g_n^2, \tag{33.3}$$

$$G = \frac{\sum x_n g_n}{\sum g_n^2}, \tag{33.4}$$

so

$$E = \sum_n x_n^2 - 2\frac{\sum x_n g_n}{\sum g_n^2}\sum x_n g_n + \left(\frac{\sum x_n g_n}{\sum g_n^2}\right)^2 \sum g_n^2, \tag{33.5}$$

$$E = \sum_n x_n^2 - \frac{\left(\sum x_n g_n\right)^2}{\sum g_n^2}. \tag{33.6}$$

However, $\sum x_n^2$ only depends on the input and $\sum g_n^2$ is a scale factor depending only on the LPC spectral filter and code word, so we just precompute $\alpha_k = \sqrt{\sum g_n^2}$ and choose the best $g_n$ to maximize $(\sum x_n g_n)/\alpha_k$.

Thus the inner product of the filtered code words and the input sequence, scaled by a factor inversely proportional to the energy of the codebook sequence, is to be maximized. This scale factor does not, however, have to be computed in real time. Since the LPC spectral coefficients are vector quantized, and the codebook sequences are quantized, a simple look-up table can be implemented to obtain the correct value for a given codebook index and set of LPC spectral coefficients.

This still does appear to require that each code word be filtered by $1/[A(z/c)]$ before its inner product with the input sequence is computed. However, this can be reduced to computing an inner product of the unfiltered code word with a filtered version of the input sequence as follows:

The goal is to maximize $P = \sum_n x_n g_n$, where $x_n$ is the input sequence and $g_n$ is the response of the inverse filter with no memory, $1/[A(z/c)]$, to the current code word, $c_n$. Here $g_n$ can be written as the convolution product of $c$ and the impulse response $f_n$ of $1/[A(z/c)]$:

$$g_n = \sum_i c_i f_{n-i}, \tag{33.7}$$

so

$$P = \sum_{n=0}^{N-1} \left( x_n \sum_{i=0}^{N-1} c_i f_{n-i} \right) = \sum_{i=0}^{N-1} \left( c_i \sum_{n=0}^{N-1} x_n f_{n-i} \right), \tag{33.8}$$

or

$$P = \sum_{i=0}^{N-1} c_i d_i \ \text{ with } \ d_i = \sum_{n=0}^{N-1} x_n f_{n-i}. \tag{33.9}$$

Thus we compute only the inner product of the backward-filtered input $d_i$ with un-filtered code-word vectors $c_i$. Hence the codebook search has effectively been reduced to taking an inner dot product of each code word with another vector, which is only computed once per frame. This is much more computationally feasible than an entire filtering operation per code-word comparison.

### 33.10.3 Multiresolution Codebook Search

Given a particular distance measure for computing which codebook entry matches best, there are several ways to search that are more effective than the exhaustive full-search method. A multistage codebook search entails performing preliminary, less computationally intensive and thus less accurate searches on the data base to narrow the number of possible candidate sequences down from the entire size of the data base to some more reasonable subset, which is then searched in the slow, accurate manner. This type of optimization also is used frequently in other fields such as image processing.

In one version of the multistage codebook search, preliminary comparison stages are performed at lower temporal resolution by low-pass filtering and decimating by five the 1024 codebook entries and desired excitation sequence [28]. The scores from these reduced-resolution comparisons are used to determine the best 70 candidates, which are then compared at full resolution.

The authors report that a speedup by a factor of 9 is achieved with those parameters. The decimation factor and subset search size were chosen by simulation so that 99% of the time the reduced resolution search would not eliminate the correct sequence. Using two preliminary stages at decimations of five and two allows a speedup by a factor of 13, again with 99% accuracy.

The reduced-resolution codebook entries can be computed off line since they are independent of the speech signal. This results in a slight increase in required storage, but it does not impose a computation time penalty.

### 33.10.4  Partial Sequence Elimination

Another approach to multistage searching is not to compare the full codebook sequences at reduced resolution, but rather to compare only part of the sequences at full resolution [10]. In this scheme, the first $k$ samples of the desired excitation sequence are compared with the first $k$ samples of each of the codebook vectors. This is used to narrow the second stage, which is a full-length search, to some fixed number of the most promising codebook entries. This was reported to reduce the number of multiply and add operations by 75%.

### 33.10.5  Tree-Structured Delta Codebooks

Trees-structured data bases often lend themselves to an efficient search. One requirement, however, is that the parents and children must be related in some meaningful way. Thus, if the values in a conventional linear codebook were assigned to the nodes in a tree, there would be no computational benefits. Gaussian random sequences may not appear to lend themselves well to ordering, but by use of a limited set of random sequences to form a basis for a larger set, the larger set can be encoded in a tree whose height is proportional to the number of basis vectors. This can be done with tree-structured delta codebooks [38].

A tree-structured delta codebook is constructed hierarchically from a limited number, $L$, of precomputed basis vectors, $\Delta C_0$ through $\Delta C_{L-1}$. (See Fig. 33.22). The codebook entries can be found by starting with the code vector at the root and then adding or subtracting

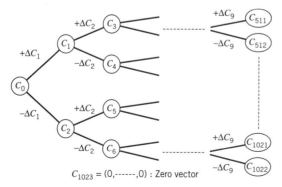

**FIGURE 33.22**   Tree-structured delta codebook. Taken from [38].

the delta code vectors encountered while traversing the path to the desired node. (Note that $C_0 = \Delta C_0$.) For example, the value of $C_5$ is $\Delta C_0 - \Delta C_1 + \Delta C_2$. In this way, the $2^L - 1$ vectors in the codebook can be generated from the $L$ delta code vectors, $\Delta C_0 \ldots \Delta C_{L-1}$.

Tree-structured delta codebooks not only reduce memory requirements but also reduce computational requirements. (The former is reduced by a factor of 100 whereas the latter is reduced by a factor of 70 as compared to conventional codebooks.) Because of the incremental nature of the tree structure, the best-match codebook searching can be performed on the codebook deltas rather than on the entire codebook. This reduces the amount of filtering and correlation (or vector dot products in reverse terminology) to be linear with $L$. As before, the energies of the filtered codebook sequences have to be computed, and used, to inversely scale the inner product, which is computationally proportional to the square of $L$. This, however, could be eliminated if a look-up table was used as seen previously in section 33.10.2.

The choice of delta code vectors is computed off line by analyzing large amounts of speech data and choosing principal orthogonal vectors that lead to the least distortion when used to code the test data. Iterative methods allow the delta code vectors to adapt.

It should be noted that because of the tree nature of the codebook, and the fact that valid codebook indices lie both at the leaves of the tree and also in interior nodes, some of the delta code vectors are used more often than others. In particular, the delta code vectors with low indices are used more often than those with higher indices. The extreme cases are $\Delta C_0$, which is a component of all vectors in the codebook, and $\Delta C_{L-1}$, which is only a component of half of the codebook vectors. Thus the ordering of the delta codebook vectors is important. Delta vectors that are more tuned to the current excitation sequence spectral distribution should be placed near the root of the tree, whereas the delta vectors that are tangential to the excitation sequence should be placed in lower levels. This motivates a technique called delta vector sorting, in which the order of the delta vectors is keyed to the spectral content of the desired excitation sequence. The delta vectors are all filtered through the LPC spectral shaping filter and sorted by weighted energy. The delta vectors with the greatest amount of energy can be used most effectively to build the desired excitation sequence, and they are thus placed nearest to the root of the codebook tree. Since the coefficients of the LPC spectral shaping filter are known to both the coder and synthesizer, no additional information has to be transmitted.

## 33.10.6  Adaptive Codebooks

In the previous implementations of CELP, the codebooks used were precomputed and remained constant through the coding and decoding operation. This, however, does not take advantage of the fact that the speech signal's characteristics vary over time. Thus, it might be advantageous to have a codebook that *adapts* to the speech signal. This approach has been used in many of the more recent implementations of CELP; see, for example, [16], [17], and [29]. (The concept is also described for use in a self-excited vocoder in [34]).

The most common form of an adaptive codebook has the codebook contain entries taken from the previously synthesized excitation signal. Thus think of the codebook not as a set of $K$ discrete $N$-sample code vector sequences that are selected from, but rather a recorded linear history of the previous $KN$ excitation samples (actually $KN + N - 1$). Then choosing a code word involves determining where in the past $KN$ samples to select $N$ consecutive samples for reuse as the current code vector.[7] Since the code index is now in the range of zero to $KN - 1$, it would require $\log_2(K) + \log_2(N)$ bits, so $KN$ is chosen to be usually approximately 128 with $N$ remaining at approximately 40. Thus the total codebook is quite small. ($K$ no longer has any real significance, as the code words are all part of one continuous sequence.) The code index is sometimes called a lag or represented by an $L$, since it represents the amount of time to go back to retrieve the excitation sequence.

An adaptive codebook also has the computational advantage of not requiring the perceptual weighting filter. Since the excitation sequence is taken from recent excitation sequences, it will be of the correct perceptual weighting since the previous excitations sequences were. Although this argument may seem circular, a proof and simulations are given in [16]. Some implementations filter or scale the excitation sequence before adding it to the codebook [29].

Adaptive codebooks are good for steady-state voiced excitation sequences, but they are not well suited for the voicing onset or for unvoiced sounds. To remedy this, some predetermined static codes are included in another section of the codebook. It should be noted that these sequences can bypass the perceptual weighting filter as well, since they are more random in nature and it is hoped that the noise shaping is not necessary.

### 33.10.7   Linear Combination Codebooks

Although forming codebooks from a linear combination of basis vectors was described as it applied to the tree encoding of code words in section 33.10.5, an alternative implementation is described here as it is sufficiently different to warrant individual treatment. The codebook arrangement described here is used in the system described in the next section.

Instead of using a tree structure to combine the basis vectors, it is possible to use the actual binary representation to control how the basis vectors are combined [17]. Consider $K$ basis vectors used to form a $2^K$ entry codebook. In this $2^K$ entry codebook, each codebook entry is a linear combination of all basis vectors, where the basis vectors can be scaled only by $-1$ or 1. Thus it is easy to see how all $2^K$ entries are formed from a $K$-bit codebook index.

This structure is very useful for a number of reasons. First, it is quite memory efficient, since again only the basis vectors need to be stored as the combinations are inherent in the binary code. This coding also makes the system resistant to channel errors since changing one bit of the code will only change the sign of one of the basis vectors and the others will remain the same.

The structure also facilitates an efficient codebook search. If the codebook entries are arranged by gray code order, then neighboring vectors differ only by one basis vector. This means that transformed versions of all of the codebook vectors can be obtained sequentially and incrementally by simply adding or subtracting twice the value of one of the transformed basis vectors.

---

[7]Thus the term adaptive codebook makes it sound more complex than it actually is, but this is the accepted terminology for a fairly simple concept.

### 33.10.8 Vector Sum Excited Linear Prediction

Finally, another method to reduce the codebook search time is called vector sum excited linear prediction (VSELP) [17]. VSELP entails using not just one codebook, but rather a sum of the outputs of multiple (typically two or three) codebooks to form an excitation sequence. Each of the codebooks is smaller than in a single codebook system, typically 128 vectors in size, and has its own associated gain. The codebook search then can be performed by sequentially searching through the individual codebooks and optimizing the gain parameters individually. Since the sum of the sizes of the codebooks is less than the size of a large codebook, the search can be performed more rapidly. In the system described, one of the codebooks used was of the adaptive variety, whereas the two others were standard stochastic codebooks so that the next excitation could be a combination of the two types of excitation. Vector quantization is employed to efficiently code the three gain parameters.

## 33.11 ADAPTIVE TRANSFORM CODING

It was pointed out in Section 31.6 that bit savings were possible if the vocoder channel signals were linearly transformed prior to quantization. An appropriate transformation would cause the resultant variables to be uncorrelated and lead to a lower transmission rate. Transform coding is a method of applying these ideas to the original signal samples, rather than the channel signals.

If the transmitter and receiver agree in advance on the bit representation of the transformed variables, the receiver can parse the received bit sequence to identify the variables and then proceed to perform an inverse transformation to generate the coded speech. However, if the best bit assignments are time varying, parsing can only be done by means of the transmission of low-rate side information. Such a scheme is called adaptive transform coding (ATC).

In Chapter 31, we described principal component analysis (PCA), a transformation and dimensionality reduction scheme that is based on finding linear transformations that result in components with the largest variances [21]. We also noted that similar results can often be achieved with a discrete cosine transform (DCT). The DCT is often used in ATC schemes [3] and yields results that are similar to those found with the use of schemes such as PCA. Unlike PCA, the DCT computation does not require the explicit measuring of a (time-varying) correlation matrix. However, the optimum bit assignment can still vary with time, so that side information transmission is necessary, or at least desirable. References [11] and [41] treat the problem of ATC for speech, whereas [40] is an exhaustive treatment of image coding using transformations but also contains useful results that can be applied to speech signals.

## 33.12 CONCLUSIONS

In this chapter, speech coders covering bit rates that were higher than the traditional 2400-bps systems were studied. A relatively large number of such systems have been proposed and implemented, but it's worth noting that these systems are all variations of the basic notions developed decades earlier. The primary such notion was that advantages can be gained by modeling human speech production as that of a time-variable linear filter with appropriate excitation. Additionally, though, the speech wave can be digitized with resultant improvements in transmission despite the apparent need for greater bandwidth.

The search for the best excitation function has punctuated a good deal of the work reported in this chapter. It began with the ideas that led to the voice-excited vocoder and continued through the many inventions that characterized code-excited LPC vocoders. Several developments were crucial to these advances:   one was the work of C. P. Smith that led to the many uses of vector quantization (see Chapter 32); another was the work of Atal and Schroeder in creating the LPC concepts. In particular, the LPC error signal played a vital role. Of the more recent systems, the CELP coder has received much attention because it strikes a requisite balance between coding efficiency and speech quality. It considers both technological and biological constraints to achieve this goal. The CELP algorithm can be implemented readily in real time by using standard low-cost digital signal-processing chips, allowing easy entrance into the commercial sector. The VSELP vocoder described in Section 33.10.8 has been adopted by the Telecommunications Industry Association as the standard speech coder for use in North American digital cellular telephone systems.

In addition to the approaches described here, there has been a significant amount of work over the past 15 years in the general area of sinusoidal (sometimes called harmonic) coders, as briefly mentioned in Chapters 29 and 31. Since such coders are particularly amenable to prosodic transformations, they will also be discussed in the next chapter.

# 33.13  EXERCISES

**33.1**   Consider the use of voice-excited vocoders in the public telephone network. In a typical network, frequencies below 300 Hz are not transmitted. Design a voice-excited vocoder that operates at 9600 bps. First, determine the required band of the baseband filter and estimate the bit rate to transmit this signal. Then, choose a reasonable number of filters for spectral analysis and design the coding rates for each channel so as to fulfill the overall specified rate.

**33.2**   Design one or more split-band systems using channel vocoders. Make a block diagram sketch and estimate the bit rate needed for each of the parameters you choose.

**33.3**   Repeat the split-band design problem for an LPC-based system.

**33.4**   Explain the primary differences between the multipulse and CELP vocoders.

**33.5**   The perceptual weighting filter is described in the text as a frequency-domain weighting of the LP spectrum. Describe some other properties of human hearing that might potentially be of some importance for making the error minimization relevant to speech perception.

**33.6**   Explain why ADPCM requires frame synchronization.

**33.7**   Find the inverse to the discrete cosine transform.

**33.8**   Show how the DCT can be computed by using fast transform methods.

# BIBLIOGRAPHY

1. Adoul, J-P., and Lamblin, C., "A comparison of some algebraic structures for CELP coding of speech," in *Proc. ICASSP '87*, Dallas, pp. 1953–1956, 1987.
2. Adoul, J-P., Mabilleau, P., Delprat, M., and Morissette, S., "Fast CELP coding based on algebraic codes," in *Proc. ICASSP '87*, Dallas, pp. 1957–1960, 1987.
3. Ahmed, N., Natarajan, T., and RAO, K. R., "Discrete cosine transform," *IEEE Trans. Comput.* **C-23**: 90–93, 1974.
4. Atal, B. S., and Remde, J. R., "A new model of LPC excitation for producing natural-sounding speech at low bit rates," in *ICASSP '82*, Paris, pp. 614–617, 1982.
5. Atal, B. S., "Predictive coding of speech at low bit rates," *IEEE Trans. Commun.* **COM-30**: 600–614, 1982.
6. Atal, B. S., and Schroeder, M. R., "Predictive coding of speech signals," in *Proc. Int. Cong. Acoust.*, Tokyo, 1968.
7. Benyassine, A., and Abut, H., "Mixture excitations and finite-state CELP speech coders," in *Proc. ICASSP '92*, San Francisco, pp. I-345–I-348, 1992.
8. Chen, J., "High-quality 16 kb/s speech coding with a one-way delay less than 2 ms," in *Proc. ICASSP '90*, Albuquerque, pp. 453–456, 1990.
9. Chen, J., Melchner, M., Cox, R., and Bowker, D., "Real-time implementation and performance of a 16 kb/s low-delay CELP speech coder," in *Proc. ICASSP '90*, Albuquerque, pp. 181–184, 1990.
10. Copperi, M., and Sereno, D., "CELP coding for high-quality speech at 8 kbits/sec," in *Proc. ICASSP '86*, Tokyo, pp. 1685–1688, 1986.
11. Cox, R. V., and Crochiere, R. E., "Real-time simulation of adaptive transform coding," *IEEE Trans. Acoust. Speech Signal Process.* **ASSP-29**: 147–154, 1981.
12. Crochiere, R. E., Webber, S. A., and Flanagan, J. L., "Digital coding of speech in subbands," *Bell Syst. Tech. J.* **55**: 1069–1085, 1976.
13. David, E. E., Schroeder, M. R., Logan, B. F., and Prestigiacomo, A. J., "Voice-excited vocoders for practical speech bandwidth reduction," Presented at the Speech Communications Seminar, Stockholm, 1963.
14. Dunham, M., and Gray, R., "An algorithm for the design of labeled-transition finite-state vector quantizers," *IEEE Trans. Commun.* **COM-33**: 83–89, 1985.
15. Estaban, D., and Galand, C., "Application of quadrature mirror filters to split-band voice coding schemes," in *Proc. ICASSP '77*, Hartford, Conn., pp. 191–195, 1977.
16. Galand, C., Menez, J., and Rosso, M., "Adaptive code excited predictive coding," *IEEE Trans. Signal Process.* **40**: 1317–1326, 1992.
17. Gerson, I., and Jasiuk, M., "Vector sum excited linear prediction (VSELP) speech coding at 8 kbps," in *Proc. ICASSP '90*, Albuquerque, 461–464, 1990.
18. Gold, B., Informal experiment, Lexington, Mass., 1972.
19. Hernández-Gómez, L., Casajús-Quirós, F., Figueira-Vidal, A., and Garcia-Gomez, R., "On the behavior of reduced complexity code excited linear prediction (CELP)," in *Proc. ICASSP '86*, Tokyo, pp. 469–472, 1986.
20. Johnson, M., and Tanginuchi, T., "On-line and off-line computational reduction techniques using backward filtering in CELP speech coders," *IEEE Trans. Signal Process.* **40**: 2090–2093, 1992.
21. Hotelling, H., "Analysis of a complex of statistical variables into principle components," *J. Educ. Psychol.* **24**: 417–441, 498–520, 1933.
22. Krasner, M. A., "Digital encoding of speech and audio signals based on the perceptual requirements of the auditory system," Tech. Rep. TR-535, MIT Lincoln Laboratory, Lexington, 1979.

23. Kroon, P., and Deprettere, E., "A class of analysis-by-synthesis predictive coders for high quality speech coding at rates between 4.8 and 16 kbits/s," *IEEE J. Sel. Areas Commun.* **6**: 353–363, 1988.

24. Kroon, P., and Atal, B., "Quantization procedures for the excitation in CELP coders," in *Proc. ICASSP '87*, Dallas, pp. 1649–1652, 1987.

25. Lin, D., "Ultra-fast CELP coding using deterministic multi-codebook innovations," in *Proc. ICASSP '92*, San Francisco, pp. 317–320, 1992.

26. Magill, D. T., and Un, C. K., "Residual-excited linear predictive coder," JASA Vol. 55, Supplement abstract NN3 in the April 24–26 87th meeting of the Acoustical Society, p. 581, New York.

27. Makhoul, J., and Berouti, M., "High frequency regeneration in speech coding systems," in *Proc. ICASSP '79*, Washington, D.C., pp. 428–431, 1979.

28. Mauc, M., and Baudoin, G., "Reduced complexity CELP coder," in *Proc. ICASSP '92*, San Francisco, pp. I-53–I-56, 1992.

29. Menez, J., Galand, C., and Rosso, M., "A 2 ms-delay adaptive code excited linear predictive coder," in *Proc. ICASSP '90*, Albuquerque, pp. 457–460, 1990.

30. Morrow, W. E., "Transmission of high quality speech signals by means of PCM at low digital rates," Intern. Memo, MIT Lincoln Laboratory, Lexington, MA, 1965.

31. Parsons, T., *Voice and Speech Processing*, McGraw–Hill, New York, 1986.

32. Rabiner, L. R., and Schafer, R. W., "Digital processing of speech signals," Prentice–Hall, Englewood Cliffs, N.J., 1978.

33. Rose, R., and Barnwell, T., "Quality comparison of low complexity 4800 bps self excited and code excited vocoders," in *Proc. ICASSP '87*, Dallas, pp. 1637–1640, 1987.

34. Rose, R., and Barnwell, T., "The self excited vocoder – an alternate approach to toll quality at 4800 bps," in *Proc. ICASSP '86,* Tokyo, pp. 453–456, 1986.

35. Schroeder, M. R., and David, E. E., Jr., "A vocoder for transmitting 10 kHz speech over a 3.5 kHz channel," *Acoustica* **10**: 35–43, 1960.

36. Schroeder, M., and Atal, B., "Speech coding using efficient block codes," in *Proc. ICASSP '82*, Paris, pp. 1668–1671, 1982.

37. Schroeder, M., and Atal, B., "Code-excited linear prediction:  high-quality speech coding at very low bit rates," in *Proc. ICASSP '85*, Tampa, pp. 937–940, 1985.

38. Taniguchi, T., Tanaka, Y., and Ohta, Y., "Tree-structured delta codebook for an efficient implementation of CELP," in *Proc. ICASSP '92*, San Francisco, pp. I-325–I-328, 1992.

39. Weinstein, C. J., "A linear prediction vocoder with voice excitation," in *Proc. Eascon '75*, Washington, D.C., pp. 30A–30G, 1975.

40. Wintz, P. A., "Transform-picture coding," in *Proc. IEEE* **60**: 880–920, 1972.

41. Zelinsky, R., and Noll, P., "Adaptive transform coding of speech signals," *IEEE Trans. Acoust. Speech Signal Process.* **ASSP-25**: 299–309, 1977.

42. Zinser, R., and Koch, S., "CELP coding at 4.0 kb/sec and below:  improvements to FS-1016," in *Proc. ICASSP '92*, San Francisco, pp. 313–316, 1992.

# OTHER APPLICATIONS

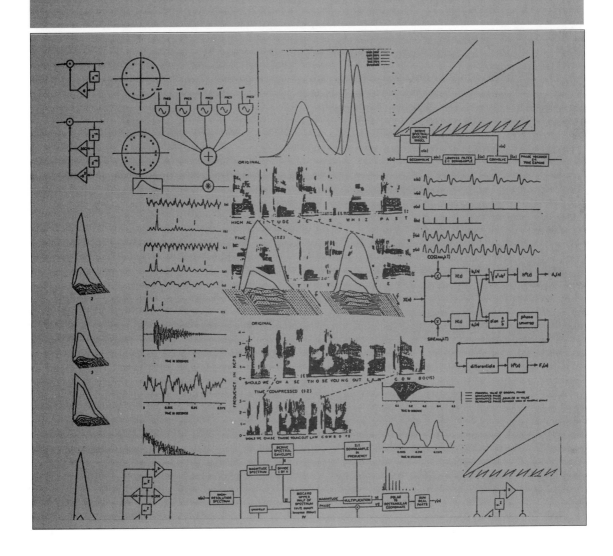

The Universe is full of magical things patiently waiting for our wits to grow sharper.
—Bertrand Russell

**P**ARTS VI and VII focused on speech recognition, synthesis, and coding as the primary applications of the engineering methods introduced in this book. These tasks have been the object of study for decades, and they can also be seen as archetypes for a wide range of potential applications. However, there are certainly many more ways to use the technologies discussed here. Part VIII provides introductory information about a few other applications to give a sense of the possibilities. The techniques developed for the analysis and synthesis of speech can be adapted to provide a range of transformations, as described in Chapter 34. Similar techniques can also be applied to music synthesis, as described in Chapter 35. And finally, the approaches to statistical pattern recognition that were described in Part VI can be adapted to the task of processing speech to confirm the identity of the talker. This application, called speaker verification, is described in Chapter 36. Other applications are left as an exercise for the reader to discover.

*34*

# *SPEECH TRANSFORMATIONS*

## 34.1 INTRODUCTION

There are a variety of techniques to modify the speed, pitch, and spectrum of a speech signal. Some methods work directly on the speech wave to modify the time scale or pitch. Other methods are based on analysis–synthesis systems, in which the derived parameters can be controlled to modify the synthetic output. Various medium- and high-rate vocoder systems do not explicitly compute the fundamental frequency, and this complicates pitch modification.

Speech modification techniques have many applications. For instance, as noted in Chapter 29, pitch and duration must often be modified for concatenative synthesis. Speeding up a voice response system can save time for a busy, impatient user. It may also be a useful addition in speech communication channels subject to fading. Compressing the spectrum could potentially be of help to people with hearing disabilities.

The following three sections explain some of the fundamental issues in speech transformations. This is followed by a study of speech modification in analysis–synthesis systems, that is, channel vocoders, LPC vocoders, and homomorphic vocoders. The chapter concludes with a review of three specific systems:    the phase vocoder [4], the Seneff system [19], and the sine-transform coder of Quatieri and McAulay [16].

## 34.2 TIME-SCALE MODIFICATION

A popular application of speech processing is time-scale modification. In this section, several systems are presented that perform this function while preserving the original excitation and spectrum.

Schemes for time-scale compression and expansion include work by Lee [7], Garvey [5], and Fairbanks et al. [3]. These early works utilized the sampling method. Time was divided into segments; for example, for a 2:1 increase in speedup, all segments were 30 ms; for a 1.5:1 speedup, segment intervals alternated between 30 ms and 10 ms. Thus, to time compress the speech by a factor of 2 [as shown in Fig. 34.1(b)], alternate segments were deleted and the remaining segments abutted. To time expand by a factor of 2, each segment was repeated, as shown in Fig. 34.1(c). Previously, Miller and Licklider [10] had demonstrated that speech could be chopped (segments merely zeroed out), with no appreciable intelligibility loss for small chops.

Figure 34.1 shows that the sampling method introduces artifacts, such as probable discontinuities at the segment boundaries.[1]

---

[1] Sampling here refers to the selection of the speech segments. It should not be confused with the usual use of the word to denote analog-to-digital (A/D) and digital-to-analog (D/A) conversion rates.

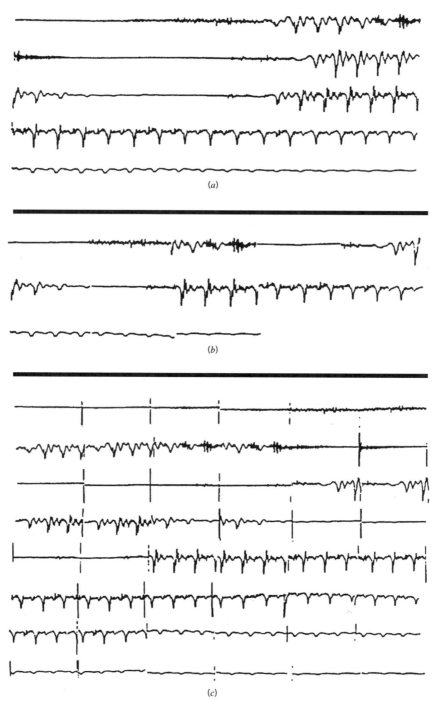

**FIGURE 34.1** Sampling method of speech compression and expansion: (a) original utterance of the histogram, (b) speedup, and (c) slowdown.

Scott and Gerber [18] performed a pitch synchronous time-scale modification and reported an increase in word intelligibility from 88.1% for the sampling method to 92.1% for their method. Their experiment was restricted to words that were completely voiced, as in the example of Fig. 34.2.

A more recent method of time-scale modification described in [20] makes use of the pitch synchronous overlap and add (PSOLA) algorithm that was mentioned in Chapter 29.

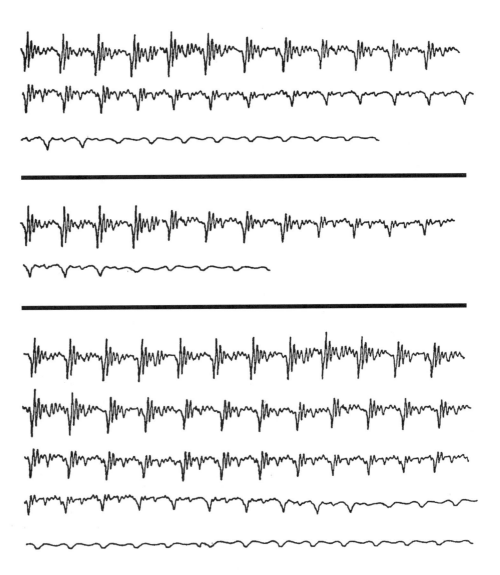

**FIGURE 34.2** Pitch synchronous time compression and expansion of "on."

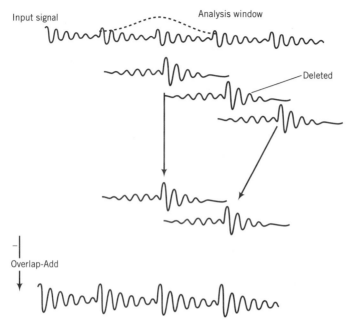

**FIGURE 34.3**   PSOLA for time compression. From [20].

The overlap–add procedure can be performed directly on the speech signal or on a wave-form representation of a derived excitation signal. The first step in this procedure is to compute the pitch marks. Then, a window (e.g., a Hanning window) is erected around each pitch mark, of an integer number of periods for voiced speech. The resultant win-dowed speech fragments (let's call them frags), if properly added, produce the original signal. If, now, some frags are selectively removed and the remaining frags are added in an overlap–add way, the result is a sped-up version of the original signal with the same pitch and spectrum as the original. The important provision of this technique is that a frag following a deleted frag assumes the time position of the latter. An example is shown in Fig. 34.3.

## 34.3   TRANSFORMATION WITHOUT EXPLICIT PITCH DETECTION

There are several ways to generate an excitation function that is useful for exciting a vocal tract model to produce synthetic speech.

Homomorphic analysis:   in Fig. 20.1, the high-time filtered version of the cepstrum represents the excitation function. Therefore, when an inverse FFT is performed followed by exponentiation followed by an FFT, the appropriate excitation function is obtained.

Inverse filtering:   by performing inverse filtering on the incoming speech, the output represents the excitation function. For the all-pole model hypothesized for LPC, the error

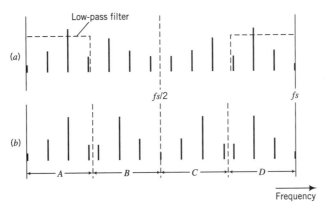

**FIGURE 34.4**   Effect of low-pass filtering followed by downsampling. The original spectrum, in (a), was obtained by sampling an analog signal at rate $f_s$. When (a) is low-pass filtered and downsampled by 2:1, (b) results. The segments marked B and C in the top spectrum (a) have disappeared, and B, below has been replaced by D, above. C, below, has become A, above.

signal thus serves as the excitation function. Inverse filters can be derived by computing the mathematical inverse of a derived spectral envelope.

Spectral flattening:   this method removes the spectral envelope component of the speech, leaving the excitation function. Different techniques are available to perform this function.

An approximation to spectral flattening can be realized by low-pass filtering of the speech followed by downsampling. An example is shown in Fig. 34.4. Notice that the resulting excitation spectrum is not flat. Also, the frequency position of the spectral lines are probably not harmonics of the voice fundamental frequency. Nevertheless, if the baseband signal (i.e., the low-pass filter output) is of sufficiently high bandwidth (e.g., 1500 Hz), the perceptual effect of these distortions is minimal (see Chapter 16 on Pitch Perception).

## 34.4   TRANSFORMATIONS IN ANALYSIS–SYNTHESIS SYSTEMS

Vocoders are analysis–synthesis systems. Thus, once the parameters of a given speech model are analyzed, it is possible to *intervene* before synthesis to produce some *transformed* version of the speech. For example, we can change the fundamental frequency from its measured value to some function of that value. The spectrum and the timing may also be altered. We will first show how such transformations can be handled in a channel vocoder. Analogous results are obtainable with LPC and cepstral vocoders.

Speeding up the speech:   we are familiar with the result of playing a tape back at a higher speed than was used during recording. The pitch increases and the formants get higher, thus distorting the spectrum. In a digital channel vocoder, the same effect can be obtained by raising the D/A clock rate relative to the A/D clock rate. However, it is usually desirable to speed up the speech without changing the pitch or distorting the spectrum. How can this be done?

In a channel vocoder, analysis is performed on a frame basis. In each frame (typically 10–20 ms long), the energy in a frequency band is estimated (see Chapter 31). During synthesis, the number of samples synthesized is made equal to the number of samples analyzed; analysis and synthesis frames are of the same duration. Now, let's imagine that for every 100 input samples to the analyzer, only 50 samples are synthesized. This effectively shortens the duration of the output speech relative to the input speech; the result is a speedup. The fundamental frequency and the spectrum have been parametrized so they are unchanged.

It is clear that speeding up speech cannot work in real time. However, this use of the channel vocoder can be applied to a practical real-time situation [2]. Consider a long-distance speech-communication link, in which atmospheric conditions result in occasional fading of the signal. A two-way signaling path can be set up, in which the receiver notifies the transmitter that a fade has occurred. When the transmitter gets this message, it stores the analysis frames rather than transmitting; meanwhile, it continues to send a probe signal that presumably will not be received until the fade passes. When this happens, the receiver sends an all-clear signal and now the transmitted speech is sped up until the buffer is cleared, at which time normal transmission resumes.

Pitch change:   In the early days of channel vocoders, it was demonstrated that pitch could easily be varied in real time by turning a dial. In a frame-oriented digital vocoder, there is no difficulty in combining pitch modification with slowdown (see Fig. 34.5) or speedup.

Spectral modifications:   There are several reasons for interest in this type of transformation. Deep-sea divers speak in a helium-rich environment, and this gives the speech a Mickey Mouse effect that is due to the spectral changes caused by changes in the velocity of sound. These spectral distortions can be reversed in a channel vocoder.

Another possible application is spectral modification to cram the frequency space into the hearing portion of a partially deaf person's ear. A popular game is to change a male voice to female and vice versa; such a game could conceivably be part of a psychological gender experiment.

Voice-modification methods can also be applied to various automatic speech-recognition tasks. One such application is to modify the reference (input) voice into a target voice on which the recognizer has been trained [11], [13], [20]. Another application is, for a

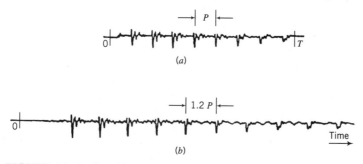

**FIGURE 34.5**  Combined pitch change and slowdown.

multispeaker environment, to generate many modified versions of one or several speakers to be used as data for training the recognizer.

To summarize:  a channel vocoder, because it parametrizes excitation and spectrum separately and because the number of output samples need not be equal to the number of input samples, is capable of modifying speed, pitch, and spectrum in any combination.

The classical LPC and homomorphic vocoders (see Chapters 20 and 21) transmit the voice fundamental frequency and the voiced–unvoiced decision in ways very similar to that of a channel vocoder. Thus, pitch modifications in these systems can be the same as in the channel vocoder.

However, the parameters encoding vocal tract information are very different for these three classical algorithms. The channel vocoder estimates spectral envelope with a filter bank, and it transmits an encoded version of these estimates. In LPC, the transmitted spectral parameters are closely associated with the synthesizer; for example, the reflection coefficients might be transmitted. The homomorphic vocoder typically sends an encoded version of the low-time liftered cepstrum.

In an LPC vocoder, spectral modifications can be implemented in various ways. For example, once the analyzer has determined the synthesizer parameters, the spectral envelope can be computed, either directly or by computing the DFT of the synthesizer impulse response. A new set of autocorrelation values are then computed from the modified spectrum and the reflection coefficients recomputed.

Alternately, a DFT of the computed correlation values yields the square of the spectral magnitude, which can now be modified and an inverse DFT computed to create the modified correlation function, which can then be used to compute the modified parameters for transmission.

In a homomorphic vocoder the cepstrum can be  low-time liftered to preserve the part that pertains to the spectral envelope. The liftered cepstrum is encoded and transmitted. As seen in Fig. 34.6, a DFT will produce the log spectral envelope, and this spectrum is modified and exponentiated, and then another DFT generates the impulse response corresponding to the modified log spectrum.

One method of performing time-scale modification in a channel vocoder is mentioned above, where speedup or slowdown is obtained by synthesizing a different number of samples that were analyzed. Another technique is to alter both the fundamental frequency parameters and the spectral parameters and then modify the ratio of the input to output sampling rates. As an example, we wish to double the vocoded speech rate. First, we transmit one-half of the fundamental frequency parameter and a version of the spectrum that has been scrunched (compressed in frequency) by a factor of 2. Then the D/A clock at the receiver is set to double the A/D clock at the transmitter. Thus, the vocoded speech is perceived to

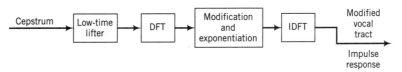

**FIGURE 34.6**  Spectral modification in a homomorphic vocoder.

have the original pitch and spectrum, but twice the rate. Comparable manipulations allow for time-scale modifications in LPC and homomorphic vocoders.

## 34.5   HYBRID SYSTEMS

Improved performance can be realized (at the expense of an increased bit rate) with hybrid schemes, in which the excitation is obtained by a bandwidth reducing, nonlinear operation on the original speech (or, in the case of LPC, on the error signal). In these situations, how to manipulate excitation becomes less obvious.

The accurate analysis and parametrization of the excitation can be a tricky task. In Chapter 33, we discussed the voice-excited vocoder (VEV), which produced an excitation signal by direct processing of a band-limited version of the speech. We also presented a variety of similar systems. In all these cases, excitation is no longer modeled in terms of a few simple parameters, and no longer fits into the analysis–synthesis transformation scheme. Examples include split-band systems, VELP, RELP, APC, ADPCM, and so on. In the following sections, methods are described for performing useful transformations despite these constraints.

## 34.6   SPEECH MODIFICATIONS IN PHASE VOCODERS

The phase vocoder [4] begins by performing a spectral analysis of the incoming signal; typically, if performed by means of FFT, the result is available as a real and imaginary component at each frequency position. A rectangular-to-polar coordinate transformation is then implemented, and the magnitude and phase derivative may now be modified independently. As an example, consider speeding up the speech by a factor of 2. First, we multiply the phase derivative of each channel by one-half. This means that the frequencies present in the modified spectrum are half of the previous values. The spectrum then undergoes a polar-to-rectangular transformation, another FFT is taken, and the resulting waveform is played out at double the input rate. The final result, therefore, is a sped-up version of the original speech but with no noticeable change in spectrum or pitch.

Modification of the phase derivatives while playing the synthetic speech at the same rate as the input rate modifies the frequencies of the speech harmonics and thus has the effect of changing the pitch.

A phase vocoder implemented with a filter bank for spectrum analysis and synthesis is most accurate if each filter contains at most a single component during voicing.

Examples of both speedup and slowdown are shown in Fig. 34.7.

The original phase vocoder [4] was implemented by using a filter bank. Portnoff [15] worked out the rate modification details with a short-time Fourier transform analysis (STFT) to effectively emulate the phase vocoder. This work served as a model for future work [12], [17] employing the STFT. Nawab et al. [14] went a step farther; showing how the modified signal could be reconstructed from the *magnitude* of the STFT. Griffin and Lim [6] developed the LSEE-MSTFTM (least-squares error estimation of the modified short-time Fourier transform magnitude).

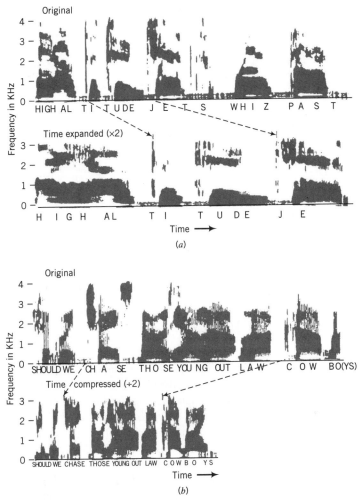

**FIGURE 34.7** Spectrographic illustrations of slowdown and speedup. From [4].

# 34.7  SPEECH TRANSFORMATIONS WITHOUT PITCH EXTRACTION

Having established that there are a large number of systems that bypass vocal source modeling, we now describe how such systems might be used to implement transformations. Seneff's approach [19] is shown in Fig. 34.8. The figure shows the steps leading to a doubling of the fundamental frequency without changing the spectrum and without parametrizing the fundamental frequency estimation.

First, the spectral envelope is measured; we know from Chapter 31 that there are several available approaches. Spectral envelope estimation is used to create a time-domain inverse filter that has the effect of deconvolving the excitation and the spectral envelope.

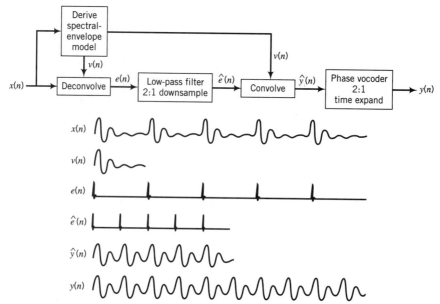

**FIGURE 34.8** Example of transformation without pitch extraction:   doubling of fundamental frequency. From [19].

Passage of the original signal through the inverse filter generates an approximation to the excitation (shown for voiced speech in the figure). By low-pass filtering and downsampling by 2:1, excitation pulses are generated having half the period. Meanwhile, the impulse response of the vocal tract is obtained by the inverse transform of the spectral envelope, and this function is convolved with the downsampled excitation function to produce the transformed signal, time compressed by 2:1. Finally, this signal is sent through a phase vocoder; we see that the output has twice the pitch of the input.

Figure 34.9 shows how these ideas can be incorporated into a complete system, in which phase vocoding is integrated into the overall system structure.

A high-resolution spectrum is obtained; the magnitude and phase are treated separately. Items I, II, and III illustrate how a spectrally flattened version of the spectrum can be obtained. The phase spectrum is  unwrapped. Our next job is to reduce the extent of the spectrum as shown in the figure; this is done by 2:1 downsampling of the spectrum envelope to produce V. Multiplying the downsampled spectral envelope by the flattened spectrum produces VI. It is then a simple matter to combine item VI with the phase-multiplied spectrum (VII) to finally (by means of polar-to-rectangular transformation) get back the original spectrum (I) but with twice the pitch.

Phase unwrapping and multiplying are illustrated in Fig. 34.10.

The original phase is essentially linear for any harmonic component of the spectrum, as shown by the solid lines that traverse zero to $2\pi$. Unwrapping the phase yields the lower straight line, and doubling the phase yields the dotted straight line above; when it is rewrapped, we get the result shown in the figure.

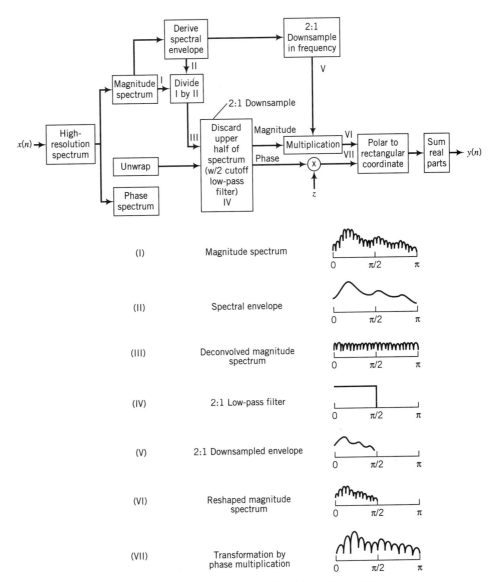

**FIGURE 34.9** Seneff speech-modification algorithm. From [19].

## 34.7.1 Frequency Compression and Gender Transformation

Hearing loss is a term that covers a wide range of symptoms. Here we speculate on the possibility that severe loss of high-frequency hearing (e.g., above 1 kHz) can to some degree be compensated by employing one of the transformation tricks described above. We speculate that some improvement may occur if, for instance, a 4-kHz speech spectrum is

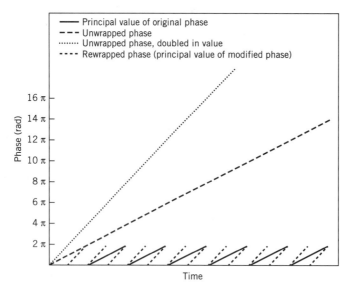

**FIGURE 34.10** Phase unwrapping and multiplication. From [19].

scrunched into the 0- to 1-kHz band. In a channel vocoder, this can be done by keeping the pitch undisturbed and designing the synthesizer with the *same* number of filters as the analyzer, but covering only the low band. Thus, for example, the 4-kHz analyzer filter magnitude signal modulates the 1-kHz synthesis filter, the 2-kHz analyzer filter modulates the 500-Hz synthesis filter, and so on.

Some experiments along these lines have been tried; unfortunately, we know of no successful results thus far [8]. An example of spectral scrunching is shown in Fig. 34.11 parts (e) and (f). Also shown in Fig. 34.11 [parts (c) and (d)] is the male-to female transformation, in which both fundamental frequency and effective formant frequencies have been increased.

## 34.8 THE SINE TRANSFORM CODER AS A TRANSFORMATION ALGORITHM

In Chapter 29 we briefly discussed the use of sinusoidal analysis for the representation of segments for concatenative synthesis. These approaches are applied to vocoding in the sine transform coder (STC) of Quatieri and McAulay [16]. In this method, the synthesizer is excited by a collection of sinusoidal signals. The frequencies and magnitudes of these signals are derived with an analysis procedure based on a high resolution, short-time DFT.[2]

---

[2]The phase is also derived in the STC analysis; the reader is referred to their papers for more details on their sophisticated analysis procedure.

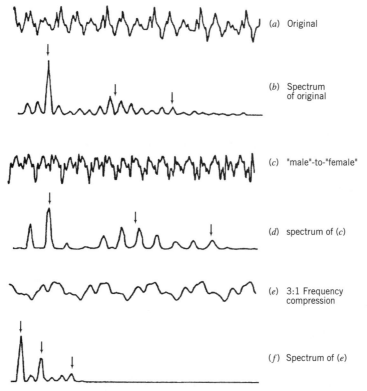

**FIGURE 34.11** Several transformation examples: (a) 31-ms section of waveform of the vowel in "great," (b) spectrum of waveform in (a), (c) male-to- female conversion, (d) spectrum of waveform in (c), (e) 3:1 spectral scrunch, and (f) spectrum of waveform in (e). From [19].

The sum of these sinusoids represents the resultant synthesis. Given this model, the STC is capable of time-scale modification, pitch modification, and spectrum modification.

Time-scale modification in the STC: the analysis procedure computes the frequencies and magnitudes of the sinusoids at a rate corresponding to the rate at which successive DFTs are performed. When the rate of presentation of these parameters to the synthesizer is changed, the rate of the resultant synthetic speech is also changed.

Spectral modification in the STC: given the high-resolution DFT, a number of options are available for finding the spectral envelope (e.g., cepstral analysis, spline interpolation, and LPC analysis). The spectral envelope is now scrunched, as illustrated in Fig. 34.12, and new magnitudes are assigned to the sinusoids based on sampling the scrunched spectrum.

Pitch modifications in the STC: given the spectral envelope, pitch modification can be done by changing the derived frequencies and then sampling the spectral envelope at the new frequencies to generate new magnitudes for the shifted frequencies.

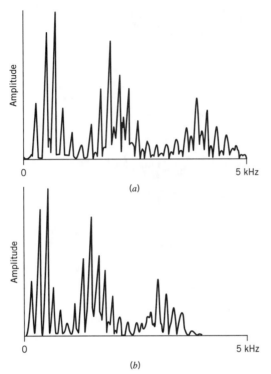

**FIGURE 34.12**    Illustration of spectral modification in STC. From [16].

The above methods can be combined with changes in the A/D to D/A clock-rate ratio to produce a great variety of modifications.

## 34.9   VOICE MODIFICATION TO EMULATE A TARGET VOICE

In Section 34.4 it was stated that in certain applications it is desirable to transform a voice to match that of a specific target voice. Valbret et al. [20] discuss this problem. To implement their scheme, it is first necessary to have sufficient data on the reference voice (the input) and the target voice. To train the system, words from each speaker are time aligned, using DTW (dynamic time warping) methods (see Chapter 24), followed by a training algorithm, using vector quantization, to set up a correspondence between target and reference vectors. When this training is complete, it is straightforward to map the input speaker into the target speaker.

The results of Valbret et al. indicate that the average value of the fundamental frequency is a more important cue than the spectrum to identify a given speaker.

Childers [1] describes a method of modeling the glottal source for voice conversion. He uses a polynomial model and enters 32 versions of the glottal source function into a VQ table.

## 34.10  EXERCISES

**34.1**  Show how spectral scrunching can be realized using the system of Fig. 34.9.

**34.2**  In Sec. 34.6 it is stated that a phase vocoder implemented as a filter bank works best if there is a single harmonic in each filter during voicing. Explain why this is so.

**34.3**  How would you modify the spectrum in a phase vocoder without affecting speed or pitch?

**34.4**  Consider a high-frequency speech-communication system in which it is desired to maintain speech continuity despite fades. Imagine that during the first 10 s there are no fades; at 10 s there is a 5-s fade. Make a sketch of the resulting timing of the transmitted, buffered, and received speech, indicating the beginning and end of the fade. The end result should be that the receiver gets all of the speech, but occasionally it will receive a time-scaled version of the transmitted speech.

**34.5**  It is desired to slow the output of the speech synthesized by the STC algorithm by 70%. Describe the steps needed to do this. Assume that the analyzer computes a high-resolution DFT every 10 ms.

**34.6**  Using pitch synchronous speedup by 2, design an algorithm to speed up the utterance histogram of Fig. 34.1. Present your research as a computer program with audio results, if possible, or as a block diagram or flow chart.

**34.7**  Given a cepstrum, how would you modify the pitch of the utterance without explicitly estimating the pitch?

**34.8**  In Fig. 34.4(a), assume that the frequencies present are 300, 600, 900, and 1200 Hz, and so on. What frequencies appear in Fig. 34.4(b)? If the corresponding signal is now upsampled by 2:1 to restore the original sampling rate, what frequencies appear in the new signal?

**34.9**  Using the PSOLA method, sketch a design for modifying pitch, leaving other parameters intact.

## BIBLIOGRAPHY

1. Childers, D. G., "Glottal source modeling for voice conversion," *Speech Commun.* **16**: 127–138, 1995.
2. Gold, B., Lynch, J., and Tierney, J., "Vocoded speech through fading channels," in *Proc. ICASSP '83*, Boston, p. 101, 1983.
3. Fairbanks, G., Everitt, W. L., and Jaeger, R. P., "Method for time or frequency compression-expansion of speech," *IRE Trans. Audio Electroacoust.* **AU-2**: 7–12, 1954.
4. Flanagan, J. L., and Golden, R. M., "Phase vocoder," *Bell Syst. Tech. J.* **45**: 1493–1509, 1966.
5. Garvey, W. D., "The intelligibility of abbreviated speech patterns," *Quart. J. Speech* **39**: 296–306, 1953.
6. Griffin, D. W., and Lim, J. S., "Signal estimation from modified short-time Fourier transform," *IEEE Trans. Acoust. Speech Signal Process.* **ASSP-32**: 236–242, 1984.
7. Lee, F. F., "Time compression and expansion of speech by the sampling method," *J. Audio Eng. Soc.* **20**: 738–742, 1972.
8. Lippmann, R. P., "Experiments with frequency compression hearing aids," personal communication, Lincoln, Laboratory, Lexington, 1980.

9. McAulay, R. J., and Quatieri, T. F., "Speech analysis/synthesis based on a sinusoidal representation," *IEEE Trans. Acoust. Speech Signal Process.* **40**: 744–754, 1986.

10. Miller, G. A., and Licklider, J. C. R., "The intelligibility of interrupted speech," *J. Acoust. Soc. Am.* **22**: 167–173, 1950.

11. Mizuno, H., and Abe, M., "Voice conversion algorithm based on piecewise linear conversion rules of formant frequency and spectrum tilt," *Speech Commun.* **16**: 153–164, 1995.

12. Moulines, E., and Laroche, J., "Non-parametric techniques for pitch-scale and time-scale modification of speech," *Speech Commun.* **16**: 175–205, 1995.

13. Narendranath, M., Murthy, H. A., Rajendran, S., and Yegnanarayana, B., "Transformation of formants for voice conversion using artificial neural networks," *Speech Commun.* **16**: 207–216, 1995.

14. Nawab, S. H., Quatieri, T. F., and Lim, J. S., "Signal reconstruction from short-time Fourier transform magnitude," *IEEE Trans. Acoust. Speech Signal Process.* **ASSP-31**: 986–998, 1983.

15. Portnoff, M. R., "Time-scale modification of speech based on short-time Fourier analysis," *IEEE Trans. Acoust. Speech Signal Process.* **ASSP-29**: 374–390, 1981.

16. Quatieri, T. F., and McAulay, R. J., "Speech transformations based on a sinusoidal representation," *IEEE Trans. Acoust. Speech Signal Process.* **ASSP-34**: 1449–1464, 1986.

17. Roucos, S., and Wilgus, A. M., "High quality time-scale modification for speech," in *Proc. ICASSP '85*, Tampa, pp. 493–496, 1985.

18. Scott, R. J., and Gerber, S. E., "Pitch synchronous time compression of speech," in *Proc. Conf. Speech Commun. Process.*, Newton, Mass., 63–65, 1972.

19. Seneff, S., "System to independently modify excitation and/or spectrum of speech waveform without explicit pitch extraction," *IEEE Trans. Acoust. Speech Signal Process.* **ASSP-30**: 566–578, 1982.

20. Valbret, H., Moulines, E., and Tubach, J. P., "Voice transformation using PSOLA technique," *Speech Commun.* **11**: 175–187, 1992.

CHAPTER **35**

# SOME ASPECTS OF COMPUTER MUSIC SYNTHESIS

## 35.1  INTRODUCTION

In this chapter, we first briefly discuss the possible reasons for the use of a computer to generate musical sounds. Then we try to categorize the different approaches to computer music synthesis, that is, the kinds of processing algorithms that are commonly used. The reader will undoubtedly realize that many of the methods of speech processing are very closely related to those of music processing; both depend strongly on the computer and on the ideas of digital signal processing. However, whereas speech processing is tied to the physiology of speaking and listening to speech, music processing is closely associated with the acoustic production of music by traditional instruments and the physiology of listening to music. For some researchers in the field, the computer frees them from the classical sounds of strings, horns, and the like and allows them to create new "instruments."

Much of our discussion will be based on the survey paper by Moorer [4].

Why computer music? In the late 1950s, several speech researchers began to realize the potential power of the computer to help in their research. At the time, signal processing, such as audio bandpass filtering, was far too expensive to do on computers. However, with the development of digital signal processing (DSP) techniques and the availability of high speed, small, and cheap integrated circuits, it became apparent that digital methods were the answer to future growth. Research workers in music quickly adopted this new technology and used it for their own ends – freedom of expression, precision, and greatly enhanced implementation of new ideas. An early visionary was Mathews [2], who realized that computer-based block-diagram compilers could be applied to music synthesis.

## 35.2  SOME EXAMPLES OF ACOUSTICALLY GENERATED MUSICAL SOUNDS

Figures 35.1, 35.2, and 35.3 show the waveforms and spectra for the cello, clarinet, and trumpet, respectively. From these figures, it can be seen that a single note can be divided into three segments:  the attack, the steady state, and the decay. Notice, for example, that the attack time of the cello is appreciably longer than that of the clarinet or trumpet. These three

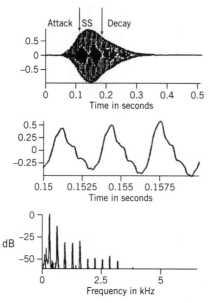

**FIGURE 35.1**    Waveforms and spectrum of the cello. From [4].

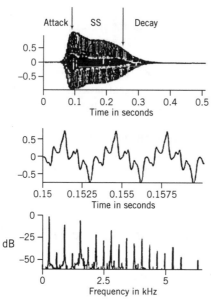

**FIGURE 35.2**    Waveforms and spectrum of the clarinet. From [4].

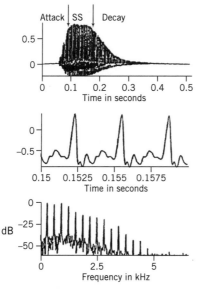

**FIGURE 35.3** Waveforms and spectrum of the trumpet. From [4].

instruments have in common their periodicity. There is, however, a noticeable lowering of the even harmonics of the clarinet at low frequencies. During the steady-state portion, the note may be altered by *tremolo* (an amplitude modulation) or *vibrato* (a frequency modulation).

Figure 35.4 is a bass drum note; it is not periodic, more or less resembling noise. But the sound is distinct to our ears, so there is certainly an underlying structure.

## 35.3 MUSIC SYNTHESIS CONCEPTS

The computer can be used to help the composer develop his or her compositional ideas. It can also be used as a real-time performer. An interesting example of performance was demonstrated by Vercoe [9]. A piano–violin sonata of Caeser Franck was performed; the violin part was played by a young woman. The computer performed the keyboard part from a score in its memory but also by detecting the pitch of the violin and adjusting its tempo to that of the live performer.

The three main elements in computer music are:

**1.** The development of software to emulate the acoustics of a real or imaginary instrument.

**2.** The editing capability to deal with a musical score.

**3.** Software to integrate the above elements into a performance.

The use of block-diagram compilers and specialized computer languages has been a standard line of development. The programmer (often the composer) defines "instruments"

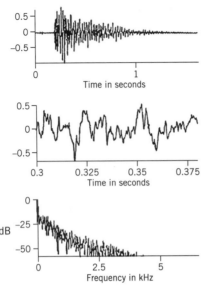

**FIGURE 35.4**  Example of an isolated bass drum note. From [4].

and then enters a score, which consists of durations, pitch, and loudness. These methods have been named *direct synthesis* [4].

Direct synthesis may be implemented in a variety of ways. The approach pioneered by Mathews [2] was to devise a programming language to simulate various instruments. A specific software module is called a unit generator. Examples of unit generators are oscillators, noise generators, frequency shifters, envelope generators, and other blocks that the programmer may wish to add to this collection. A crude simulation of the cello note of Fig. 35.1 is shown in Fig. 35.5.

The first five harmonics of the cello spectrum are implemented by oscillators, each with its own amplitude and frequency control, and the sum is then multiplied by the envelope

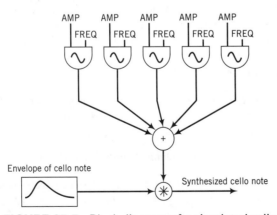

**FIGURE 35.5**  Block diagram of a simulated cello note.

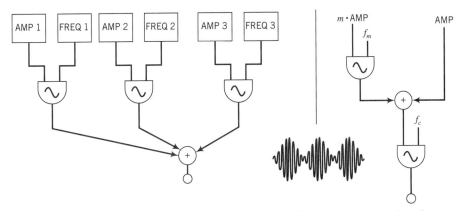

**FIGURE 35.6** Two flow charts to generate an amplitude-modulated signal.

generator, obtained by indexing a prestored version of the envelope of Fig. 35.1. Thus, for a single note, the programmer must specify five amplitudes and five frequencies as well as the envelope. If this note, with all its parameters, is repeated often, the programmer can install the complete algorithm for that note as a subroutine. Different notes, with different tempos, may be treated as separate subroutines. We can see that creating a program for even a relatively simple musical piece is quite complicated.

The top part of Fig. 35.5 is often called additive synthesis, referring to the addition of a collection of sinusoidal oscillators. Musical waveforms can be created in other ways. Figure 35.6 shows two block diagrams (or flow charts) that generate the same waveform.

Amplitude modulation is a simple way of generating several harmonics without resorting to additive synthesis. More complex forms of modulation produce richer signals. Chowning [1] employed a form of frequency modulation (FM).

We know of broadcast FM as the modulation of a high-frequency carrier by a relatively low-frequency modulation. The mathematical basis of FM resides in the following formula:

$$\sin[\theta + \alpha \sin(\beta)] = J_0(\alpha)\sin(\theta) + \sum_{k=1}^{\infty} J_k(\alpha)[\sin(\theta + k\beta) + (-1)^k \sin(\theta - k\beta)], \qquad \textbf{(35.1)}$$

where $\theta = 2\pi f_c nT$, $f_c$ is the carrier frequency, $\beta = 2\pi f_m nT$, $f_m$ is the modulation frequency, $J_k$ is the $k$th-order Bessel function, and $\alpha$ is the modulation index. By manipulation of these parameters and the modulation index, a variety of musical sounds can be synthesized without the sine generators needed for each component of additive synthesis.

## 35.4 ANALYSIS-BASED SYNTHESIS

Analysis-based synthesis closely resembles the methods of vocoder research. A given musical piece is first analyzed and parameters are extracted; these parameters are used to drive a synthesizer. Of course, given the ability to perform transformations, such systems are more than just musical vocoders.

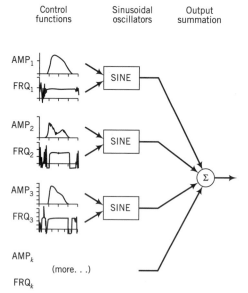

**FIGURE 35.7**   System for additive synthesis. From [4].

The most common ways of employing analysis-based synthesis are known as additive synthesis and subtractive synthesis. Additive-synthesis concepts dates back a full century, to the Telharmonium (see Chapter 2). Recently, additive synthesis has proved to be a useful speech-processing algorithm [3].

Figure 35.7 shows the concept of additive synthesis.

The system consists of an array of sinusoidal generators, each controlled by two time-varying parameters, the amplitude and frequency.

Figure 35.8 shows how to perform analysis to generate the parameters for additive synthesis.

The right-hand side of Fig. 35.8 is a harmonic extractor for a single harmonic of a periodic signal. It is important to realize that the fundamental frequency must be known in advance. The design of the filters $H(z)$ results in a pitch synchronous analysis, wherein a single period of the signal is analyzed by an array of harmonic detectors. As seen on the left-hand side, the results feed directly into an additive synthesis scheme.[1]

Figure 35.9 shows the result of a harmonic analysis of a clarinet note.

For the data inherent in this plot to be reduced, a piecewise linear fit is shown in Figure 35.10.

This line-segment approach, in which the segments are determined by human graphical interaction, appears to result in tones that resemble the original. An interesting application

---

[1] The analysis-based synthesis we have just described differs from the additive-synthesis algorithm of McAulay and Quatieri [3] in an important respect. The latter depends on a high-resolution DFT followed by a sophisticated peak-tracking algorithm that does not specifically track pitch harmonics, although many peaks found with this algorithm may correspond to harmonics. Their method works for music as well as speech.

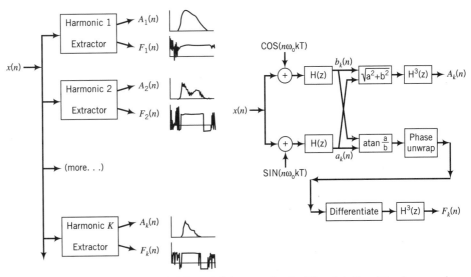

**FIGURE 35.8** Analysis model for additive synthesis. The details of the harmonic extractor are shown in the figure on the right. Note that the final smoothing is done using $H^3(z)$; that is, a cascade of three $H(z)$ section. From [4].

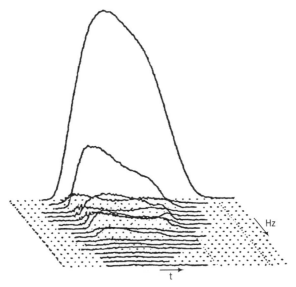

**FIGURE 35.9** Harmonic analysis of a clarinet note. From [4].

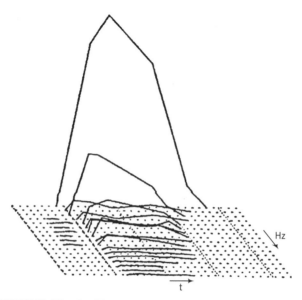

**FIGURE 35.10**   Piecewise linear fit to the clarinet note. From [4].

of this research is shown in Fig. 35.11, where the representation of the cello tone (1) undergoes successive transitions to culminate in a French horn tone (8).

Subtractive synthesis borrows directly from the work of speech research on vocoders. The classical speech-production model uses a wideband excitation to drive a vocal tract filter. The same model is used in subtractive synthesis for music. The excitation is wideband but the filters attempt to emulate the model of an actual musical instrument (possibly a novel one invented by the composer). As in additive synthesis, the major application of this method involves the intercession of the user to *modify* the results of the analysis to synthesize a variation of the input musical signal. As noted in Chapter 34, the time scale, pitch, and spectrum can be modified. In addition, the attack, steady state, and decay times of a note can also be changed.

## 35.5  OTHER TECHNIQUES FOR MUSIC SYNTHESIS

Musique concrete is synthesis based on connecting stored musical sounds in innovative ways. This method is somewhat analogous to voice answer-back machines that respond by concatenating stored words.

Real-time modeling of known and invented musical instruments is now a feasible approach to synthesis because hardware speeds make such modeling possible. This approach is analogous to that of articulatory models of the human speech-production mechanism (see Chapters 10–12). New DSP methods [8] enhance our ability to accurately model musical instruments.

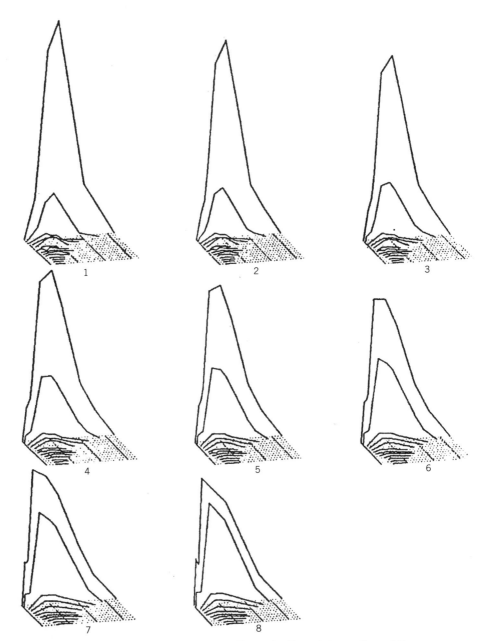

**FIGURE 35.11**   Transition from the cello to the French horn. From [4].

## 35.6  REVERBERATION

When we listen to music, the sound reaching our ears almost always includes reverberation; this is also true of speech when we are not listening to a telephone or headphones. In a large church, reverberation times of the order of seconds produce that special quality. In a concert hall, the architecture depends greatly on the desired acoustic response. Even in a small living room, listening to the radio or TV includes echoes. This has already been discussed in a very general way in Chapter 13, where we noted that essentially all sounds received at listener's ears are affected by the room acoustics.

When reverberation effects are simulated, there are two key aspects of the simulation from the standpoint of human hearing:   the temporal characteristics of the early (first 100 ms) echo response, and the overall energy envelope – the amplitude of indirect (reflected) energy, and how quickly it dies out. In the early work on the simulation of these effects, the first factor was generally ignored, and the focus was on simulating an energy decay that corresponded to a desired decay time.

Figure 35.12 shows some digital networks and the resultant pole-zero plots for some of the common early approaches. The left side of the figure shows simple feedback circuits: a first- and a second-order system. The correspondence to a physical reverberation is simple. After receiving an initial sound, one receives a set of reduced amplitude reflections every $m$ samples. Since these systems implement poles only, the resulting frequency response is quite bumpy, consisting of many peaks (commonly called a comb filter). Schroeder [7] proposed adding feedforward paths to these filters in order to add zeros, yielding an overall response that is all pass or with a flat frequency response. These are shown on the right side of the figure.

Two parameters control the quality of reverberation:   $m$ determines the echo time and $G$ controls the gain (and thus the closeness of the poles to the unit circle). These systems respond to an impulse by producing a set of impulses; as long as the gain is less than unity, successive impulses decrease in value. Figure 35.13 shows shows six impulse responses. Each reverberator is a cascade of three all-pass reverberators; the numbers in parentheses are the delays and gains of each all pass.

In practice, many listeners have found that there is no great difference in the artificial character of these reverberators, either with or without the feedforward paths. This is probably true because although the long-term frequency response for the all pass is flat, short-term spectra computed for the system output are anything but flat, and in fact do not greatly differ from the feedback-only case; note that when the feedforward impulse is out of the analysis window, the analyzed sequence is exactly the same for the left- and right-hand parts of the figure. Human hearing more closely corresponds to the short-term analysis.

Another key missing point in these early reverberation models was some representation of the early impulse response. Although a dense pattern of echos may not be a bad model for late reverberation, the first 100 ms or so of a reverberation pattern is better described as a series of discrete echos. They are generally too close in time to be heard as discrete echos by the listener, but still it can be demonstrated that much of the character of the apparent listening space (size of room, distance from source to receiver) can be inferred by the listener for modifications of the pattern of these early echos [5]. Digital reverberation

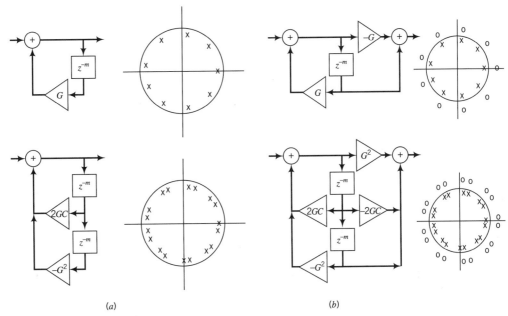

**FIGURE 35.12** Four variations of digital reverberators. Flow diagram and pole-plot for the two unit reverberators of (a). The upper figure is the standard all-pole comb filter. The root locus is shown for $m = 9$. The lower figure is an oscillatory filter. The impulse response of this unit will be a decaying sinusoid of frequency $\theta / m$. In both filters, $G$ controls the decay. The roots are all on a circle the radius of which is $G^{1/m}$. The term $C$ stands for $\cos(\theta)$. In an actual musical context, $m$ would be chosen to be a prime integer from about 50 to 2000. Flow diagram and pole-zero plot for the two all-pass unit reverberators of (b). The upper figure is the standard simple all-pass. It is equivalent to the form given by Schroeder. The root locus is shown for $m = 9$. The lower figure is an oscillatory filter. It is also an all-pass. The impulse response of this unit will be a decaying sinusoid of frequency $\theta / m$. In both filters, $G$ controls the decay. The poles are all on a circle the radius of which is $G^{1/m}$. The zeros lie on a circle of radius $G^{-1/m}$. The term $C$ stands for $\cos(\theta)$. In an actual musical context, $m$ would be chosen to be a prime integer from about 50 to about 2000. From [4].

devices that are commonly available today make use of both tailoring of early echos and of the decay pattern of the more dense later components.

## 35.7 SEVERAL EXAMPLES OF SYNTHESIS

The following examples are taken from Pierce [6]:

   **1.** Synthesis of a stringlike tone:   the frequency response in Fig. 35.14 is an approximation to the spectra we have encountered in Chapter 12.

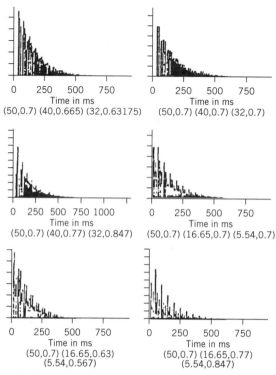

**FIGURE 35.13** Impulse responses of several reverberators. From [4].

**FIGURE 35.14** Spectrum for a synthetic stringlike tone. From [6].

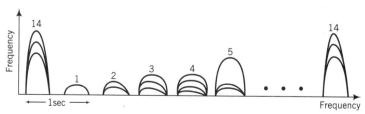

**FIGURE 35.15** Synthesis of brassy tones. From [6].

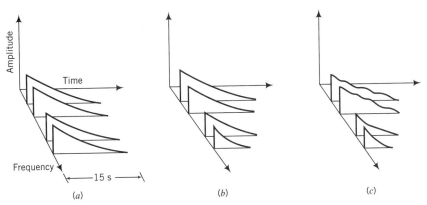

**FIGURE 35.16**  Bell-like tone experiment. From [6].

2. Synthesis of a set of brassy tones:  as illustrated in Fig. 35.15, the first item uses 15 harmonics; then a single harmonic is played with successive items including more and more harmonics; culminating in a repeat of the initial item.

3. Improvement of bell-like tone by trial and error:  Figure 35.16 shows frequency–time trajectories for three versions of a bell-like tone. In (a), all harmonics have equal trajectories (fast attack and slow decay). In (b), the higher harmonics have shorter decay times, and in (c), vibrato is added by splitting several harmonics into two closely spaced frequencies to produce beats.

## 35.8  EXERCISES

**35.1**  Devise a transformation scheme that would modify a trumpet note to sound like a cello note.

**35.2**  How would you modify the trumpet note to sound like a clarinet note?

**35.3**  What should $H(z)$ be for the system of Fig. 35.8 to be a truly pitch synchronous analysis?

**35.4**  Prove that some of the digital reverberators in the Fig. 35.12 networks are all-pass filters. Which ones?

**35.5**  From the basic FM equation, choose parameters to produce a harmonic series centered around the carrier frequency.

**35.6**  Now choose parameters to eliminate the odd harmonics of the series.

## 35.9  ACKNOWLEDGMENT

We thank Andy Moorer for permission to use the following figures:  35.1–35.4, and 35.7–35.13.

# BIBLIOGRAPHY

1. Chowning, J. M., "The synthesis of complex audio spectra by means of frequency modulation," *J. Audio Eng. Soc.* **21**: 26–534, 1973.
2. Mathews, M. V., *The Technology of Computer Music*, MIT Press, Cambridge, Mass., 1969.
3. McAulay, R. J., and Quatieri, T. F., "Speech analysis-synthesis based on a sinusoidal representation," *IEEE Trans. Acoust. Speech Signal Process.* **40**: 744–754, 1986.
4. Moorer, J. A., "Signal processing aspects of computer music:   a survey," *Proc. IEEE* **65**: 1108–1137, 1977.
5. Morgan, N., "Room acoustics simulation with discrete-time hardware," Ph.D. Dissertation, University of California at Berkeley, 1980.
6. Pierce, J., *The Science of Musical Sound,* Scientific American Books, New York, 1985.
7. Schroeder, M. R., "Improved quasi-stereophony and colorless artificial reverberation," *J. Acoust. Soc Am.* **33**: 1061, 1961.
8. Smith, J. O. III, "Physical modeling using digital wave guides," *Comput. Music J.* **16**: 74–91, 1992.
9. Vercoe, B., "Demonstration of human-computer performance," *private communication*, MIT Media Lab., Cambridge, 1992.

# CHAPTER 36

# SPEAKER VERIFICATION

## 36.1  INTRODUCTION

In Chapters 22–28, we introduced the basics of automatic speech-recognition systems. However, there are a number of related application areas that use many of the same tools and perspectives. One such class of applications is speaker recognition, for which speaker verification is a particularly important example. Here[1] we describe some of the basic principles of this application.

Speech contains many characteristics that are specific to each individual, many of which are independent of the linguistic message for an utterance. In Chapter 22 we discussed some of these characteristics from the perspective of speech recognition, for which they generally are a source of degradation. For instance, each utterance from an individual is produced by the same vocal tract, tends to have a typical pitch range (particularly for each gender), and has a characteristic articulator movement that is associated with dialect or gender. All of these factors have a strong effect on the speech that is highly correlated with the particular individual who is speaking. For this reason, listeners are often able to recognize the speaker identity fairly quickly, even over the telephone. Artificial systems recognizing speakers rather than speech have been the subject of much research over the past 20 years, and commercial systems are already in use.

Speaker recognition is a generic term for the classification of a speaker's identity from an acoustic signal. In the case of speaker identification, the speaker is classified as being one of a finite set of speakers. As in the case of speech recognition, this will require the comparison of a speech utterance with a set of references for each potential speaker. For the case of speaker verification, the speaker is classified as having the purported identity or not. That is, the goal is to automatically accept or reject an identity that is claimed by the speaker. In this case, the user will first identify herself or himself (e.g., by introducing or uttering a PIN code), and the distance between the associated reference and the pronounced utterance will be compared to a threshold that is determined during training. Speaker recognition can be based on text-dependent or text-independent utterances, depending on whether or not the recognition process is constrained to a predefined text or not.

Speaker recognition has many potential applications, including the secured use of access cards (e.g., calling and credit cards), access control to data bases (e.g., telephone and banking applications), access control to facilities, electronic commerce, information, and reservation services, remote access to computer networks, and so on.

---

[1]This chapter was largely written by Hervé Bourlard.

Speaker identification and verification each require the calculation of a score reflecting the distance between an utterance and set of references. One of the simplest approaches that was initially used was to represent each speaker by a single Gaussian (or a set of Gaussians) in the acoustic parameter space. The parameters were estimated on a training set that contained several sentences pronounced by each of the potential users. Assuming that all speakers were equiprobable and that the cost associated with an error was the same for every speaker, the decision rule consisted of assigning an utterance to the speaker with the closest density function. Variants of this approach are still used today and are briefly discussed in Section 36.5.

As for the case of speech recognition, speaker recognition has benefited from extensive applications of HMM technology in the 1990s. The resulting approaches for these two application areas are very similar. In speaker recognition, though, each speaker is represented by one or several specific HMMs. In the case of text-independent speaker recognition, these HMM models will often be ergodic (fully connected). For text-dependent speaker recognition, specific sentences can be used to model the lexical information in addition to the speaker characteristics. During speaker identification, these models will be used to compute matching scores associated with the input utterance, and the matching speaker will be the one associated with the closest reference. In the case of speaker verification, this matching score will be computed for the model associated with the claimed identity and some number of alternate models; the putative speaker model will then either be accepted or rejected based on some measure of its distance from the models of its rivals (e.g., from its closest rival). The decision also incorporates a threshold that is determined during the enrollment of each new speaker.

For a good introduction to speaker recognition, we refer the reader to [3] and [4].

## 36.2 ACOUSTIC PARAMETERS

In speech recognition, the main goal of the acoustic processing module is to extract features that are invariant to the speaker and channel characteristics and that are representative of the lexical content. In contrast, speaker recognition requires the extraction of speaker characteristic features, which may be independent of the particular words that were spoken. Such characteristics include the gross properties of the spectral envelope (such as the average formant positions over many vowels) or the average range of fundamental frequency. Unfortunately, since these features are often difficult to estimate reliably, particularly for a short enrollment period, current systems often use acoustic parameters that have been developed for use in speech recognition. However, LPC parameters (or LPC cepstra), which have fallen out of favor in ASR because of their strong dependence on individual speaker characteristics, tend to be preferred in speaker recognition for this very reason. In general, though, features that are based on some kind of short-term spectral estimate are used in speaker recognition much as they are in ASR. In addition, though, pitch information is sometimes used if it can be estimated reliably [1], [8].

Finally, the effects of transmission channel variability are usually reduced by the use of techniques initially proposed for speech recognition, such as cepstral mean subtraction or RASTA-PLP. However, these techniques also can filter out important speaker-specific characteristics.

## 36.3  SIMILARITY MEASURES

As we have seen for speech recognition, the problem of speaker recognition can be formulated in terms of statistical pattern classification, and the probability that a speaker $S_c$ (rather than some other speaker) has pronounced the sentence associated with the acoustic parameter sequence $X$ is given by

$$P(S_c \mid X) = \frac{P(X \mid S_c)P(S_c)}{\sum_{i=1}^{I} P(X \mid S_i)P(S_i)},\tag{36.1}$$

where $S_i$ represents the identity of the speaker, $S_c$ the identity being tested (or claimed, in the case of speaker verification), and $P(X \mid S_i)$ is the conditional probability of sequence $X$ given the speaker $S_i$. Ideally, the sum in the denominator should include all possible speakers. In general, this sum will be very large, and in the case of speaker verification, it should include all possible rival speakers, which is unfortunately impossible.

As for speech recognition, parameters for density estimation in speaker recognition are determined during a training phase. Once these parameters are determined, the denominator in Eq. 36.1 is independent of the class and can be neglected for speaker identification. Speaker verification, however, is a form of hypothesis test. In this case, we will verify the hypothesis that speaker $S_i$ is indeed the putative speaker $S_c$ if

$$P(S_c \mid X) > P(\overline{S_c} \mid X),\tag{36.2}$$

where the right-hand side is the probability of the speaker being anyone except $S_c$.

Typically, this is stated with some margin or threshold $\delta$, that is, speaker $S_i$ is taken to be the speaker $S_c$ if

$$\frac{P(S_c \mid X)}{P(\overline{S_c} \mid X)} > \delta,\tag{36.3}$$

Where $\delta$ is greater than one. Recall that

$$P(\overline{S_c} \mid X) = P(S_{i \neq c} \mid X) = \sum_{i \neq c} P(S_i \mid X),$$

if events $S_i$ are independent (which is the case) and collectively exhaustive (which will often be wrong). Using Eq. 36.1, and assuming uniform priors $P(S_i)$ over all speakers, we find that criterion 36.2 becomes (see Exercise 36.2)

$$S = S_c, \quad \text{if} \quad \frac{P(S_c \mid X)}{P(\overline{S_c} \mid X)} = \frac{P(X \mid S_c)}{\sum_{i \neq c} P(X \mid S_i)} > \delta,\tag{36.4}$$

defined as the likelihood ratio criterion, and where the sum over $i$ incorporates all the possible speakers.[2] From the logarithm of the likelihood ratio, we then have

$$S = S_c, \quad \text{if } \log P(X \mid S_c) - \log P(X \mid \overline{S_c}) > \Delta, \tag{36.5}$$

where $\Delta = \log \delta$.

Figure 36.1 represents the typical Gaussian approximation to the distributions of likelihoods $P(X \mid S_c)$ and $P(X \mid \overline{S_c})$ for a specific training and test set. The variability of these distributions shows the importance of using a similarity measure based on a likelihood ratio measure, as reported in [6] and [2].[3]

Two principal difficulties with this approach are the estimation of the optimal threshold (discussed in Section 36.7) and, similar to the case of the discriminant approaches described in Chapter 27, the estimation of the normalization factor $P(X \mid \overline{S_c})$. Several solutions have been proposed. In one approach, we assume that the set of reference speakers already enrolled in the data base is sufficiently representative of all possible speakers, and the normalization factor can then be estimated as

$$\log P(X \mid \overline{S_c}) \approx \log \sum_{S_i \in R, i \neq c} P(X \mid S_i), \tag{36.6}$$

where $R$ represents the set of speakers already enrolled in the system. One can also assume that the sum in approximation 36.6 is dominated by the closest rival speaker, yielding the approximation

$$\log P(X \mid \overline{S_c}) \approx \log \max_{S_i \in R, i \neq c} P(X \mid S_i). \tag{36.7}$$

These solutions are, however, not often practical, since:

1. In both cases, it will be necessary to estimate the conditional probabilities for all the reference speakers, which will often require too much computation.

2. In approximation 36.7, the value of the maximum conditional probability varies from speaker to speaker, depending on how close the nearest reference speaker is to the test speaker.

An alternative solution comprises considering a well-chosen subset of reference speakers, usually called a cohort, on which $P(X \mid \overline{S_c})$ will be estimated. The cohort is usually defined as the group of speakers whose models are determined to be close to or more competitive with the model of the target speaker $S_c$ [6], [14], [2]. A different cohort is thus assigned to every speaker and is automatically determined during the enrollment phase; it could also

---

[2]The hypothesized speaker could also be included in the sum. This sometimes yields better estimates and better performance.

[3]In addition to normalizing scores, the likelihood ratio will also reduce the effect of some parameters affecting the similarity measure, such as the variability caused by differences in the transmission channel, as well as changes in the speaker's voice over time.

eventually be updated during the enrollment of new users. For this approach, the following approximation is used:

$$\log P(X \mid \overline{S_c}) \approx \log \sum_{S_i \in R_c} P(X \mid S_i), \tag{36.8}$$

where $R_c$ represents the cohort associated with speaker $S_c$. Experimental results show that this kind of normalization improves speaker separability and reduces the sensitivity to the decision threshold. In the spirit of a better approximation to Eq. 36.1, it was recently shown in [9] that it can be advantageous to include the model of the hypothesized speaker in the cohort. This improves the behavior of the algorithm in the cases in which the acoustics for the actual speaker are rather different from the models for the claimed speaker identity (for instance, for the case of different gender), resulting in very small and unreliable likelihoods.

When HMMs are used, another solution consists of approximating $P(X \mid \overline{S_c}) \approx P(X \mid M)$, where $M$ is a speaker-independent model that is trained either on a large set of speakers or only on the set of reference speakers. Depending on whether the verification system is text dependent or text independent, $M$ will either be a fully connected (ergodic) Markov model or a model representing the sentence to be pronounced.

Following [4], we now briefly discuss some of the main speaker-verification approaches.

## 36.4 TEXT-DEPENDENT SPEAKER VERIFICATION

In text-dependent speaker verification, the system knows in advance the access password (or sentence) that will be used by the user. For each individual, there is a model that encodes both the speaker characteristics as well as the lexical content of the password. In this case, the techniques used for speaker verification are particularly similar to the methods used in speech recognition, namely the following.

**1.** DTW approach: in this case, the password of each user is simply represented as a small number of acoustic sequence templates corresponding to pronunciations of the password. During verification, the score associated with a new utterance of the password is computed by means of dynamic programming (and dynamic time warping) against the reference model(s). This approach is simple and requires relatively little computational resources during enrollment. It has been the basis of several commercial products.

**2.** HMM approach: in this case, the password associated with each user is represented by an HMM whose parameters are trained from several repetitions of the password. The amount of training required depends on the number of parameters, which can be a practical problem for larger models. Finally, the score associated with a new utterance is computed by either of the methods described in Chapter 25: the Viterbi algorithm, which finds the best state path, or the forward ($\alpha$) recurrence, taking all possible paths into account. As we have noted for speech recognition, HMM approaches have generally been found to be more accurate than simple DTW, but at the cost of higher computational requirements during training [15].

Given either of these approaches, a putative speaker identity can be verified by using the similarity measure defined in Section 36.3 and comparing it to a decision threshold.

## 36.5    TEXT-INDEPENDENT SPEAKER VERIFICATION

In the case of text-independent speaker verification, the lexical content of the utterance used for verification cannot be predicted. Since it is impossible to model all possible word sequences, different approaches have been proposed, including the following.

**1.** Methods based on long-term statistics such as the mean and variance calculated on a sufficiently long acoustic sequence. However, these statistics are a minimal representation of spectral characteristics, and they can also be sensitive to the variability of the transfer function of the transmission channel.

More recently, an alternative approach has been proposed in which the statistics of dynamic variables (e.g., in the cepstral domain) are used and modeled by a multidimensional autoregressive (AR) model [10]. In [5], different distance measures are compared for this AR approach, and it is shown that a performance similar to standard HMM approaches can be achieved. It is also shown that the optimal order of the AR process is approximately 2 or 3. Furthermore, correct normalization of the scores according to an *a posteriori* criterion seems essential to good performance.

**2.** Methods based on vector quantization:    as discussed earlier in this volume (for instance, in Chapter 26), vector quantization of spectral or cepstral vectors can be used to replace the original vector with an index to a codebook entry. In the case of speaker recognition, the spectral characteristics of each speaker can be modeled by one or more codebook entries that are representative of that speaker; see, for example, [17] for a typical reference. The score associated with an utterance is then defined as the sum of the distances between each acoustic vector in the sequence and its closest prototype vector from the codebook associated with the putative speaker (or codebooks associated with the cohort, for the normalization score). It is also possible to use a pitch detector and to define two sets of prototypes per speaker, one set each for voiced and unvoiced segments. For the voiced segments, pitch can then be added to the feature set to define the prototypes and compute the distance, requiring a choice of weights for the features.

Finally, an alternative to "memoryless" vector quantization (so called since each vector is quantized independently of its predecessors) was proposed in [7], in which source-coding algorithms were used.

**3.** Fully connected (ergodic) HMMs:    in this case, a fully connected HMM is trained during the enrollment of each user. The HMM states can then be defined in a completely arbitrary and unsupervised manner; in this approach, distances are stochastic and trained, but otherwise the approach is similar to the determination of codebook entries in vector quantization. Alternatively, states can be associated with specific classes, for example, phones or even coarse phonetic categories. Some temporal constraints will generally be included in the models, typically by introducing minimum duration constraints on each state. Finally, several solutions using different topologies, different probability density functions associated with each state, as well as different training criteria, have been proposed, including:

- HMMs trained according to a maximum likelihood criterion and having several (single or multiple) Gaussian states, or just a single multi-Gaussian state [9, 13]. Some discriminant training approaches typically used in speech recognition (e.g., MMI, as described in Chapter 27) have also been used in speaker verification to improve discrimination between users.

- Autoregressive HMMs: in this case, the probability distribution associated with each state is estimated by an AR process. Initially introduced by Poritz [12], this approach has been used with success by several laboratories [16]. Later on, this approach was also generalized to the class of HMMs using mixtures of AR processes [18].

**4.** Artificial neural networks: multilayer perceptrons have also been tested on speaker-verification problems [11]. A specific neural network that has one or two output units is associated with each speaker. The weights of each network are trained positively by using utterances from the corresponding speaker, and negatively on many utterances from rival speakers.

## 36.6  TEXT-PROMPTED SPEAKER VERIFICATION

Since verification is based on both the speaker characteristics and the lexical content of a secret password, text-dependent speaker verification systems are generally more robust than text-independent systems. However, both kinds of systems are susceptible to fraud, since for typical applications the voice of the speaker could be captured, recorded, and reproduced. In the case of a text-dependent system, even a password could be captured. To limit this risk, speaker-verification systems based on prompted text have been developed. In this case, for each access, a recorded or synthetic prompt will ask the user to pronounce a different random sentence [2], [9]. The underlying lexicon could either be very large, or even be limited to the 10 digits, which would then be used to generate random digit strings. The advantage of such an approach is that impostors cannot predict the prompted sentence. Consequently, prerecorded utterances from the customer will be of no use to the imposter. During each access, the system will prompt the user with a different sentence and a speech-recognition system will be used prior to verification to validate the utterance. Finally, even when the utterance is rejected, the user can still be prompted with an additional sentence. Since the new sentence will be different, the acoustic vector sequence will not be too correlated with the previous one, which will improve the quality of the estimators by accumulating uncorrelated evidence. This strategy would not be as useful in a text-dependent system, since the repeated sentence would have the same lexical content as the original one.

The speech recognition that is used before text-prompted verification is often based on phonetic HMMs, typically using Gaussian or multi-Gaussian distributions. These models are defined to cover the lexicon, and they are independently trained on each user. A key difficulty with this approach is that there is typically not much enrollment data available to train the HMMs. For this reason, single-Gaussian single-state phonetic models are often used. Given the enrollment data generally available in speaker-verification problems, such simple models have often performed as well as more complex models [2].

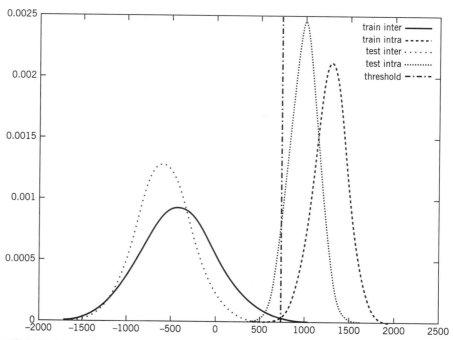

**FIGURE 36.1** Example of Gaussian approximations of the distributions of $P(X \mid \overline{S_c})$ (the left two Gaussians, respectively, for training and test set) and $P(X \mid S_c)$ (right Gaussians). The vertical line (750) represents the decision threshold corresponding to the EER as estimated on the training data. As shown here, the position of the Gaussian can vary from training to test data, depending on the variability of channel and speaker characteristics. This illustrates the importance of using normalized scores, as discussed in Section 36.3. Means and variances were computed on a set of real data corresponding to a specific speaker and a given set of impostors.

During verification, the system knows the prompted sentence and, using the phonetic transcription of the lexicon, can build the associated HMM model by simple concatenation of the constituent phones. The resulting model is then used to first validate the utterance (by computing the confidence level associated with the acoustic vector sequence) and then perform speaker verification. Given score normalization, a similar procedure can be used for the cohort speakers.

# 36.7 IDENTIFICATION, VERIFICATION, AND THE DECISION THRESHOLD

Each model discussed here can be used to compute a matching score (or a likelihood ratio) between some speaker model and a speech utterance. In the case of speaker identification, these scores are computed for all possible reference models and the identified speaker will be recognized as the one yielding the best score. In the case of speaker verification, scores are only computed for the putative speaker model and (when a normalization such as approximation 36.8 is used) the cohort models. As illustrated by Fig. 36.1, the resulting score will be compared to a threshold above which the speaker will be accepted. The estimation of the optimal threshold is often critical in good performance of the system. If the decision threshold is too high, too many customers will be rejected as imposters; such an error is referred to as a *false rejection*. Such a threshold will screen imposters very well, but at the cost of a high customer-rejection rate. If the threshold is too low, too many imposters will be accepted as customers; this kind of error is called a *false acceptance* or (in more general signal-detection parlance) a *false alarm*. Such a threshold will accept customers with little difficulty, but at the cost of a high imposter-acceptance rate. In any real task, the cost of these two kinds of errors must be assessed for the real application in order to evaluate the utility of the system. As a convenience for system comparisons, the performance of speaker-verification systems is often measured in terms of equal error rate (EER), corresponding to the decision threshold in which the false rejection rate is equal to the false acceptance rate.[4] Of course, in real systems, this measure will not be accessible (since in any one application the system operates with some particular scheme for setting the threshold), but EER is often approximated as half of the sum of the two error rates.

## 36.8 EXERCISES

**36.1**  Give some of the specific properties of mel cepstral or PLP analysis, as described in Chapter 22, that could be a poor match to the goals of speaker recognition.

**36.2**  Prove Eq. 36.4 given the earlier equations in Section 36.3.

**36.3**  Suppose that you have already trained a large-vocabulary speaker-independent recognizer on many speakers. Propose some ways that such a system could be used as the basis for a speaker-verification system.

## BIBLIOGRAPHY

1. Atal, B. S., "Automatic speaker recognition based on pitch contours," *J. Acoust. Soc. Am.* **52**: 1687–1697, 1972.
2. de Veth, J., and Bourlard, H., "Comparison of hidden Markov model techniques for automatic speaker verification in real-world conditions," *Speech Commun.* **17**: 81–90, 1995.
3. Doddington, G., "Speaker recognition-identifying people by their voices," *Proc. IEEE* **73**: 1651–1664, 1985.

[4]1998 speaker verification systems are reporting EER performance varying between 0.1% and 5%, depending on the conditions.

4. Furui, S., "An overview of speaker recognition technology," in C.-H. Lee, F. K. Soong, and K. K. Paliwal, eds., *Automatic Speech and Speaker Recognition*, Kluwer, Boston, Mass., pp. 31–56, 1996.

5. Griffin, C., Matsui, T., and Furui, S., "Distance measures for text-independent speaker recognition based on MAR model," in *Proc. IEEE Int. Conf. Acoust. Speech Signal Process.*, Adelaide, Australia, pp. I-309–312, 1994.

6. Higgins, A. L., Bahler, L., and Porter, J., "Speaker verification using randomized phrase prompting," *Digital Signal Process.* **1**: 89–106, 1991.

7. Juang, B.-H., and Soong, F. K., "Speaker recognition based on source coding approaches," in *Proc. IEEE Int. Conf. Acoust. Speech Signal Process.*, Albuquerque, N.M., pp. 613–616, 1990.

8. Matsui, T., and Furui, S., "Text-independent speaker recognition using vocal tract and pitch information," in *Proc. IEEE Int. Conf. Acoust. Speech Signal Process.*, Albuquerque, N.M., pp. 137–140, 1990.

9. Matsui, T., and Furui, S., "Concatenated phoneme models for text-variable speaker recognition," in *Proc. IEEE Int. Conf. Acoust. Speech Signal Process.*, Minneapolis, pp. II-391–394, 1993.

10. Montacie, C., Deleglise, P., Bimbot, F., and Caraty, M.-J., "Cinematic techniques for speech processing: temporal decomposition and multi-variate linear prediction," in *Proc. IEEE Int. Conf. Acoust. Speech Signal Process.*, San Francisco, pp. I-153–156, 1992.

11. Oglesby, J., and Mason, J. S., "Optimization of neural models for speaker identification," in *Proc. IEEE Int. Conf. Acoust. Speech Signal Process.*, Albuquerque, N.M., pp. 261–264, 1990.

12. Poritz, A. B., "Linear predictive hidden Markov models and the speech signal," in *Proc. IEEE Int. Conf. Acoust. Speech Signal Process.*, Paris, France, pp. 1291–1294, 1982.

13. Rose, R. C., and Reynolds, R. A., "Text independent speaker identification using automatic acoustic segmentation," in *Proc. IEEE Int. Conf. Acoust. Speech Signal Process.*, Albuquerque, N.M., pp. 293–296, 1990.

14. Rosenberg, A. E., DeLong, J., Lee, C.-H., Juang, B. H., and Soong, F. K., "The use of cohort normalized scores for speaker verification," in *Proc. Int. Conf. Spoken Lang. Process.*, Banff, Alberta, pp. 599–602, 1992.

15. Rosenberg, A. E., Lee, C.-H., and Gokcen, S., "Connected word talker verification using whole word hidden Markov models," in *Proc. IEEE Int. Conf. Acoust. Speech Signal Process.*, Toronto, pp. 381–384, 1991.

16. Savic, M., and Gupta, S. K., "Variable parameter speaker verification system based on hidden Markov modeling," in *Proc. IEEE Int. Conf. Acoust. Speech Signal Process.*, Albuquerque, N.M., pp. 281–284, 1990.

17. Soong, F. K., Rosenberg, A. E., Rabiner, L. R., and Juang, B.-H., "A vector quantization approach to speaker recognition," in *Proc. IEEE Int. Conf. Acoust. Speech Signal Process.*, Tampa, Fla., pp. 387–390, 1985.

18. Tishby, N. Z., "On the application of mixture AR hidden Markov models to text independent speaker recognition," *IEEE Trans. Acoust. Speech Signal Process.* **ASSP-30**: 563–570, 1991.

# INDEX